Vasily S. Beskin

MHD Flows in Compact Astrophysical Objects

Accretion, Winds and Jets

Translated from the Russian by N.A. Ivanova

Springer

AF774470

Prof. Vasily S. Beskin
P. N. Lebedev Physical Institute
I. E. Tamm Theoretical Dept.
Leninsky Prospect 53
Moscow
Russia 119991
beskin@lpi.ru

The work was first published in 2005 by МОСКВА ФИЗМАТДИТ with the following title: ОСЕСИММЕТРИЧНЫ Е СТ АЦИОНАРНЫ Е Т ЕЧЕХИЯ В АСТРОФИЗИКЕ.

Cover photo: The Inner Jet of the Radio Galaxy M87 located in the Virgo cluster at a distance of only 16 Megaparsec (52 million light years). The angular resolution of this false-color radio image made by the NRAO Very Long Baseline Array at 2 cm is fifty times better than the one of the Hubble Space Telescope. The linear resolution is about three light months. The image shows a limb brightened jet and a faint counter-jet. The central gap is consistent with the presence of a fast inner jet which is beamed away from the observer surrounded by a slower moving outer plasma seen by the VLBA. Image courtesy of NRAO/AUI and Y. Y. Kovalev, MPIfR and ASC Lebedev Institute.

ISSN 0941-7834
ISBN 978-3-642-26177-0 e-ISBN 978-3-642-01290-7
DOI 10.1007/978-3-642-01290-7
Springer Heidelberg Dordrecht London New York

© Springer-Verlag Berlin Heidelberg 2009
Softcover reprint of the hardcover 1st edition 2009
This work is subject to copyright. All rights are reserved, whether the whole or part of the material is concerned, specifically the rights of translation, reprinting, reuse of illustrations, recitation, broadcasting, reproduction on microfilm or in any other way, and storage in data banks. Duplication of this publication or parts thereof is permitted only under the provisions of the German Copyright Law of September 9, 1965, in its current version, and permission for use must always be obtained from Springer. Violations are liable to prosecution under the German Copyright Law.
The use of general descriptive names, registered names, trademarks, etc. in this publication does not imply, even in the absence of a specific statement, that such names are exempt from the relevant protective laws and regulations and therefore free for general use.

Cover design: eStudio Calamar S.L.

Printed on acid-free paper

Springer is part of Springer Science+Business Media (www.springer.com)

ASTRONOMY AND ASTROPHYSICS LIBRARY

Series Editors:
G. Börner, Garching, Germany
A. Burkert, München, Germany
W. B. Burton, Charlottesville, VA, USA and
Leiden, The Netherlands
M. A. Dopita, Canberra, Australia
A. Eckart, Köln, Germany
E. K. Grebel, Heidelberg, Germany
B. Leibundgut, Garching, Germany
A. Maeder, Sauverny, Switzerland
V. Trimble, College Park, MD, and Irvine, CA, USA

To the blessed memory of my teacher S.I. Syrovatskii

Preface

This monograph is based on the lectures I gave to the staff of the Theoretical Astrophysics Department at the National Astronomical Observatory (Mitaka, Japan) in 1998. Later I incorporated them as part of a 1-year course in magnetohydrodynamics at the Department of Physics and Astrophysics Problems of the Moscow Institute of Physics and Technology and at the Astronomy Department of the Moscow State University. The monograph deals with one of the analytical approaches in modern astrophysics that goes back to the equation first formulated by H. Grad and V.D. Shafranov for static magnetic configurations. In a rather simple language, this approach can describe axisymmetric stationary flows that occur in a variety of astrophysical objects.

A lot of people were fascinated by the elegance of the Grad–Shafranov method and thought it could be used as the basic instrument for building realistic models of astrophysical systems. The Grad–Shafranov method has indeed become a tool for describing the fundamental physics of many such systems; however, it turned out that other methods are often needed for constructing more detailed models. The present course should be regarded in this context. Its aim is to invite further investigation rather than sum up the results.

A few words should be said about the prerequisites for this lecture course. These include familiarity with the main notions used in General Relativity (a covariant derivative, tensor algebra). However, as we will see, the use of the $3 + 1$-splitting formalism admits the formulation of all laws in the language of three-dimensional vectors with a clear physical meaning and substantially simplifies the representation of even the most complex flows in the neighborhood of rotating black holes. As an introduction to the $3 + 1$-splitting formalism, I strongly recommend the book "Black Holes. The Membrane Paradigm" edited by K. Thorne, D. MacDonald, and R. Price. This monograph can, in a sense, be regarded as the continuation of the first four chapters of this remarkable book (however, as I will show, the membrane approach does not always provide the correct interpretation of the processes in the vicinity of the black hole horizon).

I would like to precede the book with my personal reminiscences. I was the last undergraduate student of Sergey I. Syrovatskii. He was my scientific advisor at the Department of Physics and Astrophysics Problems for 3 years. He died of a second heart attack in the autumn of 1979, which was the year when I graduated from the institute and started to study at the I.E. Tamm Theory Department of the P.N. Lebedev Physical Institute. So, I had no time to actually work with him. Nevertheless, bright memories of Sergey I. Syrovatskii are still in my heart. Moreover, after a number of years I realize that the influence my teacher had on me has become even stronger. Certainly, I think it is my duty to dedicate this book to the memory of Sergey I. Syrovatskii.

An article published in Physics Uspekhi says that the words "a life given to science" are not stereotypical when we speak about Sergey Syrovatskii. He belonged to the generation of the year 1925 and went to the front as many other 16-year-olds did (most of them gave their lives for the country). He was at the front throughout the war and was seriously wounded several times. Sergey I. Syrovatskii's heroic youth formed the major traits of his character that later helped him become one of the leading theoretical astrophysicists. Having suffered a most serious heart attack, he was torn away from his research for a few months but recovered his strength and continued to work as hard as before the illness. It is not accidental his portrait is next to the portraits of I.E. Tamm and A.D. Sakharov in the conference hall of the theory department.

Sergey I. Syrovatskii's scientific interests were broad and encompassed a variety of problems. He obtained the most important results in magnetohydrodynamics (classification of discontinuities and shock waves, the problem of their evolution, the stability analysis of tangential discontinuities), radio astronomy (the theory of synchrotron radiation that accounts for inhomogeneous distribution of electrons, their diffusion, and electron energy losses), cosmic ray astrophysics (the problems of the preferential acceleration of heavy nuclei and the universality of the spectrum), and solar physics. In 1964, he and V.L. Ginzburg wrote the fundamental monograph "The Origin of Cosmic Rays" that is still often cited, though great progress has been made in this area in the past 40 years.

I would like to stress the trait that, I think, truly characterizes Sergey Syrovatskii as a scientist. He liked exact solutions and spared no effort to study the two-dimensional flows in magnetohydrodynamics (complex variable methods can be used to efficiently obtain solutions in two dimensions, as opposed to three dimensions). His first significant work, which was on the evolutionarity of magnetohydrodynamic discontinuities, demonstrated the remarkable lucidity of his mind and the fundamentality of his scientific approach. Actually, all one needed was to accurately enumerate the number of equations and the unknowns (or, in the language of physics, the number of disturbances and waves that could transfer these disturbances) to obtain the result that was immediately included in the Landau–Lifshits course. When discussing scientific articles or student works in class, Sergey I. Syrovatskii would often stress the exactness (or, conversely, inexactness) of the formulation of physical problems and their boundary conditions. As we will see, it is the analysis of exact solutions that is the main theme of this book.

Sergey I. Syrovatskii believed that even the exact solutions of approximate equations are extremely important for forming our intuition that helps us qualitatively understand the basic properties of various physical processes. This shows, in particular, that he belonged to the I.E. Tamm school that maintained that any observational interpretation should be based on fundamental physics. Incidentally, as to this problem, he disagreed with Ya.B. Zeldovich who believed that, on the contrary, attention should principally be given to the analysis of approximate solutions of exact equations.

I stress that, in spite of a relatively small number of citations of Sergey I. Syrovatskii's journal articles, especially in recent years, he has been and remains a major authority in scientific circles. I think that, besides his high scientific potential, such traits of his character as the adherence to principles and kindness were crucial here. Besides, he had no envy of other scientists' advances in science. The issues of priority were of no interest to him at all. But he always stood up for the principles of scientific decency and respect for the work by others—principles he always strictly followed. Undoubtedly, the credit for forming the atmosphere of high scientific and moral standards, without which the truly golden age of Soviet astrophysics would be impossible, is given to S.I. Syrovatskii and S.B. Pikelner.

Apart from scientific research, Syrovatskii gave much of his time and effort to teaching. He established a scientific school united by the common aim—the development of a consistent theory of current sheets as applied to the flare processes on the Sun. S.V. Bulanov, V.A. Dogiel, A.G. Frank, B.V. Somov, and Yu. D. Zhugzhda are only a few of his disciples whose names speak for themselves. Certainly, his ability to unite and lead completely different people, see the positive potential in a heated argument, and settle the differences is just what was needed to establish a unique community of scientists able to challenge difficult scientific problems.

I remember how kind and polite he was with his students and disciples. Since I started to work at the department, I found myself in a peculiar atmosphere of creative work, its distinctive feature was friendly relations with the people around and complete equality before science. Needless to say that now working at the Department of Physics and Astrophysics Problems and giving the course of magnetohydrodynamics that was once read by S.I. Syrovatskii, I try to follow my teacher in many ways.

That is why, in recent years, when giving a lecture to a new audience, I often begin with the words: "I was Sergey Syrovatskii's last student..." And it is pleasant to see that the words inspire the audience, they act as a tuning fork helping me and the audience tune to the right state of mind.

Moscow, July 2009 *Vasily S. Beskin*

Acknowledgments

I would like to thank V.L. Ginzburg for his interest and assistance and A.V. Gurevich and Ya.N. Istomin who took part in our joint work on the physics of the radio pulsar magnetosphere. Their advice helped to achieve a deeper understanding of the basic properties of axisymmetric stationary flows. Over the 25-year period devoted to the study of the GS equation and the possibility to use it for concrete astrophysical objects, I was happy to participate in numerous discussions with the people who were the first to develop the approach. I should, first of all, mention V.D. Shafranov (who, strange as it may seem, did not know that his equation was widely used in astrophysics) and also R. Blandford, M. Camenzind, J. Heyvaerts, R. Lovelace, L. Mestel, I. Novikov, I. Okamoto, G. Pelletier, R. Pudritz, K. Thorne, and A. Tomimatsu. I express my thanks to them and their disciples and followers (H. Ardavan, I. Contopoulos, Ch. Fendt, J. Ferreira, K. Hirotani, O. Kaburaki, R. Khanna, A. Levinson, S. Nitta, M. van Putten, S. Shibata, M. Takahashi, K. Tsinganos, and N. Vlahakis). Also, I would like to thank G.S. Bisnovatyi-Kogan, S.V. Bogovalov, V.M. Chechenkin, V.I. Dokuchaev, B.V. Komberg, S.S. Komissarov, M.V. Konyukov, V.M. Lipunov, Yu.E. Lyubarsky, K.A. Postnov, A.N. Timokhin, and D. Usdensky, especially for stimulating discussions that were helpful in clarifying the details of the problem discussed. I would also like to thank my former students of the Problems of Physics and Astrophysics Department at the Moscow Institute of Physics and Technology, V.I. Pariev, Yu.M. Pidoprygora, L.M. Malyshkin, I.V. Kuznetsova, R.R. Rafikov, N.L. Zakamska, R.Yu. Kompaneetz, A.D. Tchekhovskoy, and E.E. Nokhrina, for the fruitful joint work that, we hope, also contributed to the development of the analytical methods in the GS equation theory.

Contents

Acronyms

AGN	active galactic nuclei
EMF	electromotive force
GJ	Goldreich–Julian
GR	General Relativity
GS	Grad–Shafranov
IC	inverse Compton
MHD	magnetohydrodynamics
QPO	quasiperiodic oscilations
YSO	young stellar objects
ZAMO	zero angular momentum observers

Introduction

Axisymmetric stationary flows considered within ideal magnetohydrodynamics (MHD) have been discussed in the context of many astrophysical sources. This class of flows includes accretion onto ordinary stars and black holes (Bondi and Hoyle, 1944; Bondi, 1952; Zel'dovich and Novikov, 1971; Shapiro and Teukolsky, 1983; Lipunov, 1992), axially symmetric stellar wind (Parker, 1958; Tassoul, 1978; Mihalas, 1978, Lammers and Cassinelli 1999; Bisnovatyi-Kogan 2001), jets from young stellar objects (Lada, 1985; Reipurth and Bally, 2001), and winds from the magnetospheres of radio pulsars (Michel, 1991; Beskin et al., 1993; Mestel, 1999). MHD models were actively developed in connection with the theory of magnetospheres of supermassive black holes which are assumed to be a "central engine" in active galactic nuclei and quasars (Blandford, 1976; Lovelace, 1976; Phinney, 1983; Begelman et al., 1984; Thorne et al., 1986).

The attractiveness of this model is due to its relative simplicity. The point is that in view of axisymmetry and the stationarity (and also a "frozen-in" condition) there are, in the general case, five "integrals of motion" that hold on axisymmetric magnetic surfaces. It is, first of all, the energy flux (the Bernoulli integral) and the z-component of the angular momentum and the electric potential, entropy, and the particle-to-magnetic flux ratio. This remarkable fact makes it possible to separate the problem of poloidal field structure (poloidal flow structure in hydrodynamics) from the problem of particle acceleration and electric current structure. The solution of the latter problem in the given poloidal field is expressed in terms of rather simple algebraic relations. It is important that this approach is generalized to flows in the vicinity of rotating black holes, because the Kerr metric is axisymmetric and stationary as well. As a result, it became possible to study quantitatively an extremely broad class of flows from the magnetized stellar (solar) wind (Weber and Davis, 1967; Mestel, 1968; Sakurai, 1990) and jet ejection from young stars (Blandford and Payne, 1982; Heyvaerts and Norman, 1989) to processes taking place in the magnetospheres of radio pulsars (Michel, 1969; Ardavan, 1976; Okamoto, 1978; Kennel et al., 1983) and supermassive black holes in active galactic nuclei (Camenzind, 1986; Takahashi et al., 1990; Chakrabarti, 1990). In particular, the possibility of energy extraction from a rotating black hole was demonstrated by Blandford and Znajek (1977) and Macdonald and Thorne (1982). In other words, some progress was made in this direction.

V.S. Beskin, *MHD Flows in Compact Astrophysical Objects,* Astronomy and Astrophysics Library, DOI 10.1007/978-3-642-01290-7_1,
© Springer-Verlag Berlin Heidelberg 2010

On the other hand, the problem of finding the poloidal field structure (hydrodynamic flow structure) encounters greater difficulties. First of all, it is the complex structure of the equation describing stationary axisymmetric flows. In the general case, it reduces to a nonlinear mixed-type equation that changes from a hyperbolic to an elliptic type on critical surfaces and, besides, contains integrals of motion in the form of free functions. Generally speaking, the analogous equations that date back to the classical Tricomi equation were discussed in the context of the problem of transonic hydrodynamic flows from the beginning of the 20th century (Guderley, 1957; von Mises, 1958). In particular, extremely fruitful for two-dimensional flows is a hodograph transformation technique (resulting in Chaplygin's linear equation) that substantially advanced the understanding of the studied processes (Frankl', 1945; Landau and Lifshits, 1987). In astrophysical literature, the axisymmetric stationary equilibrium equations were commonly called the Grad–Shafranov equation. A similar-type equation was formulated by them in connection with the problem of controlled thermonuclear fusion at the end of the 1950s (Shafranov, 1958; Grad, 1960). This equation, however, referred only to equilibrium static configurations and, generally speaking, its substantial transformation was needed when generalizing it to the case of transonic flows. The full version of this equation containing five integrals of motion was formulated by L.S. Soloviev in the fifth volume of the famous series of collections "Problems of Plasma Theory" and was well known to physicists. However, as is often the case, the Grad–Shafranov equation was little known in astrophysical literature so that it was repeatedly rediscovered anew. As to astrophysics, the Grad–Shafranov-type equations (in a force-free approximation and in the absence of gravity) was widely discussed in the 1970s in connection with the theory of the pulsar magnetospheres (Mestel, 1973; Scharlemann and Wagoner, 1973; Michel, 1973a; Okamoto, 1974; Mestel and Wang, 1979). The full nonrelativistic version was independently found in Okamoto (1975) and Heinemann and Olbert (1978). The relativistic generalization in flat space was found by Ardavan (1979) and then studied in tens and, may be, hundreds of papers dealing with various astrophysical objects (see e.g., Mestel, 1973; Lovelace et al., 1987; Bogovalov, 1990; Pelletier and Pudritz, 1992; Shu et al., 1994). Later in Lovelace et al. (1986) the case of the Schwarzschild metric was considered. Finally, the equilibrium equation was written in the most general Kerr metric as well (Nitta et al., 1991; Beskin and Pariev, 1993). Nevertheless, in spite of a great number of papers devoted to this subject, there was not much progress made in this direction.

The difficulty, as we will see, is that the very statement of the direct problem within the Grad–Shafranov approach turns out to be a nontrivial one. For example, in the hydrodynamic limit when there are three integrals of motion only, the problem requires four boundary conditions for the transonic flow. This implies that on some surface two thermodynamic functions and two velocity components are, for example, to be specified. However, to determine the Bernoulli integral, if we do not know which equilibrium equation cannot be naturally solved, we are to give all three velocity components, which is impossible because the third velocity component is to be found from the solution. This inconsistency is just one of the main difficulties of the Grad–Shafranov equation method. Moreover, it was quite clear that this approach cannot be generalized to the case of nonideal, nonaxisymmetric, and nonstationary flows and, therefore, it is impossible to take into account many processes of crucial importance for concrete astrophysical sources. These may include interaction between matter and self-radiation with accretion (Shvartsman, 1970; Bisnovatyi-Kogan and Blinnikov, 1980; Thorne et al., 1981; Nobili et al., 1991) and the formation of stellar (solar) wind (Tassoul, 1978; Mihalas, 1978), with account taken of viscous forces (Shakura, 1973; Shakura and Sunyaev, 1973; Novikov and Thorne, 1973) and radiation transport effects with disk accretion (Abramowicz et al., 1988; Narayan and Yi, 1995b) as well as the kinetic effects (Gurevich et al., 1993).

At the same time, in some cases, the ideal hydrodynamics approximation is quite adequate. Thus, for example, the radiation associated with the adiabatic heating of accreting matter appears smaller than the Eddington luminosity (Shapiro and Teukolsky, 1983); therefore, the matter entropy is considered to be constant. The ideal medium is provided by high plasma conductivity. These examples give one hope that the ideal (magnetic) hydrodynamics approximation can describe rather exactly the real astrophysical flows. Therefore, the ideal flows have been actively examined in the past 30 years. Of great importance here was Blandford and Payne's remarkable paper (Blandford and Payne, 1982), which considers a rather broad class of self-similar solutions to the Grad–Shafranov equation. As a result, the analysis of these self-similar solutions, which were obtained by reducing the Grad–Shafranov equation to an ordinary differential equation, resulted in one of the most popular lines of investigation (see e.g., Low and Tsinganos, 1986; Li et al., 1992; Sauty and Tsinganos, 1994; Sauty et al., 1999).

However, the difficulties associated with the application of the Grad–Shafranov equation method proved too serious. Because of the intrinsic inconsistency of this approach, which was mentioned above, it was impossible, in the general case, to solve direct problems, i.e., specify the flow structure in some region, given the physical parameters on its boundary. All this refers to self-similar solutions that require self-similarity of the boundary conditions. As a result, in the past 30 years no generally accepted model for particular astrophysical objects has been constructed, though the investigations in this area have been very intensive. Therefore, it is not surprising that the crushing majority of researchers who are, first of all, interested in astrophysical applications placed primary emphasis upon a quite different class of equations, viz., upon time-dependent problems the solution of which is possible by numerical methods only (Pneuman and Kopp, 1971; Hawley et al., 1984; Uchida and Shibata, 1984; Pudritz and Norman, 1986; Petrich et al., 1989; Ruffert and Arnett, 1994; Cao and Spruit, 1994; Ustyugova et al., 1995; Bogovalov and Tsinganos, 1999; Romanova et al., 2002).

Nevertheless, as will be shown, there is an approach that makes it possible to solve direct problems analytically even by the Grad–Shafranov method. This book is just devoted to a comprehensive introduction to the analytical methods of analysis of the Grad–Shafranov equation. It is necessary to make a reservation that the analytical methods discussed below allow us to obtain the solution in exceptional cases only. Therefore, our main goal is to clarify some key properties peculiar to MHD flows in the vicinity of the real cosmic sources rather than constructing self-consistent models of concrete compact objects. Simultaneously, using simple examples, we clarify the possibilities and set the limits of application of the Grad–Shafranov equation method. Thus, in a strict sense, the present book concerns purely physical aspects of the theory rather than astrophysical ones. However, all applications will be astrophysical ones. Moreover, the key physical results obtained by the Grad–Shafranov approach should be independent of computational methods.

Chapter 1
Hydrodynamical Limit—Classical Problems of Accretion and Ejection

Abstract There are several reasons why it is useful to start from the pure hydrodynamical case. First of all, the hydrodynamical version of the Grad–Shafranov equation is not as popular as the full MHD one. On the other hand, it has all the features of the full MHD version in the simplest form. In particular, within the hydrodynamical approach one can introduce the $3+1$-splitting language—the most convenient one for the description of the ideal flows in the vicinity of a rotating black hole. Starting from the well-known set of equations describing the nonrelativistic ideal flow, we will go step by step to more complicated cases up to the most general one corresponding to the axisymmetric stationary flows in the Kerr metric. Finally, several examples will be considered which demonstrate how the approach under study can be used to obtain the quantitative information of the real transonic flows in the vicinity of rotating black holes. The necessity of taking into account the effects of General Relativity is not so obvious for most compact sources. For instance, one cannot exclude that the black hole plays only a passive role in the jet formation process, and the effects of General Relativity in this case may be unimportant for flow description in the region of jet formation. At the same time, gravitational effects make, apparently, an appreciable contribution to the determination of physical conditions in compact objects. First, this is indicated by the hard spectra and the $e^{+}e^{-}$ annihilation line observed in galactic X-ray sources, which are believed to be solar mass black holes. Such characteristics are never observed in the X-ray sources which are firmly established to show accretion not onto a black hole but onto a neutron star. Another indication comes from superluminal motion in quasars which may be due to the relativistic plasma flow ejected along with a weakly relativistic jet. All these testify in favor of the existence of an additional mechanism for particle creation and acceleration for which the effects of General Relativity may be of principal importance. So, it is undoubtedly interesting to consider the flow structure in the most general conditions, i.e., in the presence of a rotating black hole.

V.S. Beskin, *MHD Flows in Compact Astrophysical Objects,* Astronomy and Astrophysics Library, DOI 10.1007/978-3-642-01290-7_2,

© Springer-Verlag Berlin Heidelberg 2010

1.1 Astrophysical Introduction—Accretion onto Compact Objects

1.1.1 Accretion Disks

Hydrodynamical accretion and ejection have been the focus of attention of astrophysics of compact objects since their advent. These phenomena are associated with the problem of the activity of galactic nuclei and quasars (Zel'dovich and Novikov, 1971), the mechanism of jet formation and its stability (Begelman et al., 1984), as well as the nature of galactic X-ray sources (Shakura and Sunyaev, 1973; Novikov and Thorne, 1973; Mirabel et al., 1992). The foundation of the consistent theory of these flows was laid even in the 1940s–1950s when the problem of the transonic accretion of an ideal gas onto a gravitational center (Bondi and Hoyle, 1944; Bondi, 1952) and the spherically symmetric transonic outflow from the star surface (Stanyukovich, 1955; Parker, 1958) were thoroughly studied.

The hydrodynamical accretion theory was rapidly developed after the discovery of galactic X-ray sources in the early 1970s (Giacconi et al., 1971) connected with accreting neutron stars and solar mass black holes, as well as quasars and other active galaxies in the center of which supermassive black holes are supposed to be located. Then it became clear that the nature of the activity of all these objects is connected with gravitational energy released by accretion. Thus, for accretion onto a neutron star the luminosity L is to be fully determined by the accretion rate $\dot{M}$:

$$L = \frac{GM\dot{M}}{R}. \tag{1.1}$$

The most important argument for the validity of the theory was the absence of X-ray sources with luminosity exceeding the Eddington limit:

$$L_{\text{Edd}} \approx 10^{38}\,\text{erg/s}\frac{M}{M_{\odot}}. \tag{1.2}$$

During accretion onto a black hole a major part of radiation, generally speaking, can be absorbed by a black hole. Nevertheless, to roughly estimate the energy released by an accretion disk one usually uses the formula

$$L \sim \eta \dot{M} c^2, \tag{1.3}$$

where the efficiency $\eta \approx 0.06$–0.4. Note that even for adiabatic flows of zero viscosity the exact solutions were obtained only for a number of special cases, for example, the spherically symmetric accretion. In this case, the flow structure is, in fact, specified—the motion is along the radius and the availability of the integrals of motion allows one to fully specify the flow characteristics. The above discussion deals with the problem of gas ejection from stars (mostly, of the early spectral types) which have no exact two-dimensional solutions either, though this problem was numerically studied in detail (Bjorkman and Cassinelli, 1993; Owocki et al., 1994). Recall that in the middle of the last century there were two viewpoints on the solar wind structure. They corresponded to two different asymptotic solutions of the hydrodynamical equations relating to supersonic ("wind") and subsonic ("breeze") outflow regimes. It was E. Parker who elucidated the separability of the transonic regime and showed that the gravitational field can act as a "nozzle" permitting the flow to pass the sonic surface. Clearly, the direct evidence for the supersonic nature of the solar wind was found only after the direct measurements made on artificial satellites.

Thus, the simplest adiabatic (including spherically symmetric) accretion and ejection models allowed one to substantially clarify most of the properties of the real astrophysical objects. Nevertheless, using these models, it was not, of course, always possible to achieve even a qualitative agreement with the observational data. In particular, it soon became obvious that, with adiabatic gas accretion onto a black hole, its radiation proves too low. And this result was already in contradiction with the luminosity of black hole candidates. Fortunately, the answer to this question was quickly found—a strong energy release can occur in accretion disks. These accretion disks naturally occur in binary systems in which the specific angular momentum of an accreting matter is rather large.

It is clear, however, that a gas flow in accretion disks cannot be described by the ideal hydrodynamical equations. Indeed, the accretion disk is a gas that approaches a compact object (a neutron star, a black hole) in a spiral due to viscous friction, its gravitational energy being released as heat and radiation. The disk is thin if the gas radiates a greater part of its energy and, hence, remains cold. The thin disk approximation holds good for external parts of the disk. The estimates show that this regime is realized for $r \gg r_{\rm g}$, where

$$r_{\rm g} = \frac{2GM}{c^2} \tag{1.4}$$

is a gravitational radius. Another extreme case is a thick disk in which the radiated energy is fully absorbed by an accreting matter. However, this case can occur at sufficiently large accretion rates only. The thick disk approximation can be valid only in the vicinity of a black hole ($r < 20\,r_{\rm g}$), where the main energy release occurs.

For active galactic nuclei (AGN), the X-ray spectrum associated with the central regions of an accretion disk ($r < 10\,r_{\rm g}$) can, generally, be represented as a sum of two components: a hard power-law component (up to 100 keV) and a soft component in the continuum and emission lines (largely, in the reflected iron line 6.4 keV). Therefore, it is more probable that both a hot ($T > 10^8$ K) gas and a relatively cold ($T \sim 10^6$ K) gas participate in the accretion onto the central object (Mushotzky et al., 1993). The hot and cold phases are most likely to interact with each other intensively. However, it does not seem possible to examine this interaction until the source geometry—the relative position of the hot and cold phases—is specified.

In the *disk + corona* model, the cold phase is a disk and the hot phase is an optically thin corona above this disk. In this model, the corona heating can be associated, for example, with the reconnection of the small-scale magnetic field removed from the disk region due to the turbulent diffusion. The assumption on the presence of the corona more consistently accounts for the properties of the disk' spectra (Bisnovatyi-Kogan and Blinnikov, 1977; Liang, 1977; Haardt and Marashi, 1991, 1993). Indeed, disk soft photons in this case undergo inverse Compton scattering on the hot corona matter, which results in the observed hard power-law spectrum. Part of the high-energy photons returns to the disk and heats it, which is observed as emission lines. The modeling shows that the energy distribution over the spectrum can be better explained if the corona is assumed to be inhomogeneous. Therefore, the energy is released in portions rather than uniformly in time as is the case in stellar coronas (Haardt et al., 1994). The corona, as we will see, can play an appreciable role in the formation of jets.

An alternative model accounts for some peculiarities of the spectrum under the assumption that the system is *cold clouds submerged in hot medium* (Guilbert and Rees, 1988; Celotti et al., 1992; Kuncic et al., 1997). In this case, in the vicinity of the central object the disk, as such, is absent. The main difference as compared to the previous model is that the iron line is not necessarily emitted in the vicinity of the central object. This model yields good results for the sources in our Galaxy, which are connected with the accretion disks. It can also account for some features of the galactic nuclei spectra.

The theory of the hydrodynamical disk accretion onto compact objects (neutron stars, black holes) has been developed since the end of the 1960s (Lynden-Bell, 1969; Shakura, 1973; Shakura and Sunyaev, 1973; Novikov and Thorne, 1973). However, most details are still to be cleared up. Therefore, one had to use simplified solutions such as the standard model (α-disk) (Shakura and Sunyaev, 1973) and also their various modifications (Abramowicz et al., 1988; Narayan and Yi, 1994).

Besides, one should remember that in the case of accretion onto a black hole the effects of General Relativity result in two new significant properties:

1. First, at small distances from a black hole, stable circular orbits are absent. For a Schwarzschild (nonrotating) black hole the radius r_0 of the marginally stable orbit is (Shapiro and Teukolsky, 1983)

$$r_0 = 3\, r_{\rm g}. \tag{1.5}$$

This implies that an accreting matter, which penetrated the region $r < r_0$ rather quickly, more exactly, in the dynamic time

$$\tau_{\rm d} \sim \left[\frac{v_r(r_0)}{c}\right]^{-1/3} \frac{r_{\rm g}}{c}, \tag{1.6}$$

approaches the horizon of a black hole. It is important that this motion occurs in the absence of viscosity as well. For the accretion onto a neutron star with a weak magnetic field, slow spiral motion occurs up to the surface itself. Note that, in most cases, the discussion was limited only by the use of Paczyński–Wiita model gravitational potential (Paczyński and Wiita, 1980)

$$\varphi_{\rm pw} = -\frac{GM}{r - r_{\rm g}}, \tag{1.7}$$

which, though it allows one, in some cases, to model the real situation (for example, to determine the location of the marginally stable orbit), does not fully correspond to the Schwarzschild metric.

2. Second, accretion onto a black hole must be of a transonic character. Indeed, as shown below, on the horizon of a black hole the flow must be supersonic. On the other hand, as we will see from the estimate (1.14), within the standard approach, the smallness condition of the radial velocity $v_r < c_{\rm s}$ remains valid up to the marginally stable orbit, at least, for not too large accretion rates. Therefore, the sonic surface on which, by definition, the poloidal velocity of matter becomes equal to the velocity of sound is located somewhere between the horizon and the marginally stable orbit. Thus, the problem of the structure of the accretion flow onto a black hole requires the consistent analysis of the transonic flow regime.

Problem 1.1 Show that for the Paczyński–Wiita potential (1.7)

- the radius of the marginally stable orbit, as in the Schwarzschild metric, is given by relation (1.5),
- the orbital velocity $v = v_\varphi$ on the marginally stable orbit differs from the relativistic value $v_{\rm rel} = c/2$:

$$v = \frac{\sqrt{3}}{2\sqrt{2}}\, c \tag{1.8}$$

(accordingly, the specific angular momentum $L = r v_\varphi$ is different too).

We can mention here one more extremely interesting question also associated with the use of the simplified equations. It concerns the problem of causality that arises in the case of the standard determination of viscous terms in the hydrodynamical equations (Narayan, 1992; Narayan et al., 1994). Using this standard approach, we had to give additional boundary conditions on the black hole horizon (Narayan et al., 1997; Chen et al., 1997), which cannot be causally connected with an accreting plasma.

One should note that the accretion disk theory has not become less interesting with time as the increasing sensitivity of the receivers provides new information about the accreting systems. Therefore, the number of publications devoted to this problem has increased as well (Igumenshchev et al., 2000; Artemova et al., 2001; de Villiers and Hawley, 2002; Krolik and Hawley, 2002). In particular, the accretion was studied by the Kerr metric (Riffert and Herold, 1995; Peitz and Appl, 1997, 1998; Gammie and Popham, 1998a,b; Beloborodov, 1998), which allowed one to include the effects of the black hole rotation. Much attention was given to the hydrodynamical equations with nonzero viscosity in which, however, there are no difficulties associated with causality (Kley and Papaloizou, 1997; Gammie and Popham, 1998a,b). Thus, it was possible to settle most of the questions connected with the problem of boundary conditions.

1.1.2 Standard Model

Let us recall two simplifying assumptions that form the basis of the standard model (see, e.g., Shapiro and Teukolsky, 1983):

1. The viscous stress tensor $t_{r\varphi}$, which results in a loss of the angular momentum of plasma and, hence, its accretion, can be represented in the model form

$$t_{r\varphi} = \alpha_{\rm SS} P, \tag{1.9}$$

   where P is the gas pressure and the phenomenological dimensionless parameter $\alpha_{\rm SS} < 1$ is considered to be a constant.
2. The assumption on the total reradiation in situ of the energy released by the viscous friction:

$$F^{+} = F^{-}. \tag{1.10}$$

The value F^{+} corresponds to the viscous plasma heating of a unit area of a disk

$$F^{+} \approx H t_{r\varphi} r \frac{\mathrm{d}\Omega}{\mathrm{d}r}, \tag{1.11}$$

and the thermal radiation of a unit area F^{-} proportional to the fourth power of temperature is generally estimated as

$$F^{-} \approx \frac{2}{3} \frac{a T^4 c}{\kappa \rho H}. \tag{1.12}$$

Here H is the accretion disk half-thickness, the angular velocity $\Omega = v_{\varphi}/r$, κ is a coefficient of opacity, and a is a radiation constant.

As is well known, the above assumptions turn out to be adequate for finding, in the case of the thin disk, all disk parameters by the simple analytical relations. In particular,

- the plasma rotates, practically, with the Keplerian velocity $v_{\varphi} \approx v_{\rm K} = (GM/r)^{1/2}$,
- the disk half-thickness H is determined by the vertical balance of the gravitational force and the pressure gradient

$$\frac{H}{r} \approx \frac{c_s}{v_K}, \tag{1.13}$$

- the radial velocity at distances larger than the radius of the sonic surface is small as compared to the toroidal one

$$\frac{v_r}{v_K} \approx \alpha_{SS} \frac{c_s^2}{v_K^2}, \tag{1.14}$$

- the sonic surface is located in the vicinity of the marginally stable orbit.

The values fully characterizing the accretion are natural physical parameters: the mass M of a compact object, the accretion rate $\dot{M}$, and the parameter α_{SS}. Because of the extreme simplicity of the standard model, it was regarded as a basic one over many years (Shapiro and Teukolsky, 1983; Lipunov, 1992).

However, the very nature of the key parameter—the value α_{SS}—still remains unclear. In any event, it cannot be connected with the ordinary molecular viscosity, because, in this case, α_{SS} would not be larger than 10^{-6} (Shapiro and Teukolsky, 1983). On the other hand, as follows from the observations, α_{SS} is to be much larger. It would be logical to connect it with (magnetic) turbulence (Balbus and Hawley, 1998). But only in recent years the first papers in which the parameter α_{SS} could be determined "from the first principles" have appeared (Brandenburg et al., 1995; Hawley et al., 1995; Stone et al., 1996; Arlt and Rüdiger, 2001).

Besides, in the past years, the problem of the so-called slow accretion arose—most of the nuclei of the galaxies were found to radiate much less than predicted by the standard model (up to $10^{-7}\dot{M}c^2$). It is exactly the case that is realized in our Galaxy (Narayan et al., 1998; Mahadevan, 1998) and in giant elliptic galaxies that are active but have low luminosity (Reynolds et al., 1996a; di Matteo and Fabian, 1997; Mahadevan, 1997). Relatively inefficient (in the sense of the observed radiation) is the supercritical accretion when the accretion rates are much larger than the Eddington limit and the equilibrium between the accreting matter pressure and the radiation pressure coming out from the inner layers is reached.

The low emissivity, in these cases, can be explained as follows. Black holes differ from other astrophysical objects in that they have no surface and absorb not only an accreting matter but also radiation. Therefore, if the gas is "forced" either not to radiate its gravitational or magnetic energy or to direct the radiated flux straight to a black hole, all energy is absorbed together with the matter. A distant observer will see that the gravitational energy is reprocessed into a receding radiation with low efficiency. The first case is realized if the accretion rates are very small. The second case can be realized in optically thick disks when the extracted heat energy fails to radiate and, finally, is captured by a black hole together with the matter.

An important observational test here would be to compare the radiation efficiency of weak sources associated with black holes with other accretors, for example, neutron stars. If it turned out that accreting black holes radiate much less efficiently, it would be a strong argument that the energy absorption by a black hole does occur.

However, there is no clarity in this question yet, because there are publications that both confirm (Garcia et al., 2001) and refute (Abramowicz et al., 2002) this assertion. This uncertainty is associated with the difficulties of taking into consideration the group of the weakest sources.

To explain the nature of weakly radiating objects one had to consider the radial energy transport. This was first done in Paczyński and Bisnovatyi-Kogan (1981). They showed that when averaging the two-dimensional (i.e., axisymmetric stationary) hydrodynamical equations in the direction perpendicular to the thin disk plane, they can be reduced to the system of ordinary differential equations in which all values depend on the radial coordinate r only. In the simplest nonrelativistic version, the corresponding system of equations looks like (Artemova et al., 2001)

- the continuity equation

$$\dot{M} = 4\pi r H \rho v, \tag{1.15}$$

- the r-component of the Euler equation

$$v\frac{\mathrm{d}v}{\mathrm{d}r} = -\frac{1}{\rho}\frac{\mathrm{d}P}{\mathrm{d}r} + \left(\Omega^2 - \Omega_{\mathrm{K}}^2\right) r, \tag{1.16}$$

- the φ-component of the Navier–Stokes equation which is integrated with respect to r

$$\frac{\dot{M}}{4\pi}(L_{\mathrm{n}} - L_{0}) + r^2 H t_{r\varphi} = 0, \tag{1.17}$$

- the θ-component of the Euler equation equivalent to (1.13)

$$H = \frac{c_{\mathrm{s}}}{\Omega_{\mathrm{K}}}, \tag{1.18}$$

- the energy equation

$$F^{+} - F^{-} = -\frac{\dot{M}}{2\pi r}\left[\frac{\mathrm{d}E}{\mathrm{d}r} + P\frac{\mathrm{d}}{\mathrm{d}r}\left(\frac{1}{\rho}\right)\right]. \tag{1.19}$$

Here

$$E = \frac{3}{2}\mathcal{R}T + \frac{aT^4}{\rho} \tag{1.20}$$

is the full energy per one particle ($\mathcal{R}$ is a gas constant), $L_{\mathrm{n}} = \Omega r^2$ is the specific angular momentum, and L_0 is an integration constant. The density ρ, the pressure P, the velocity $v = v_r$, and the energy E correspond to their values in the equatorial plane. In the 1990s, the analysis of Eqs. (1.15), (1.16), (1.17), (1.18), and (1.19) and their generalizations, which allow one to consider the effects of General Relativity, was one of the principal trends in the theory of the disk accretion onto compact objects (Riffert and Herold, 1995; Peitz and Appl, 1997; Gammie and Popham, 1998a,b; Beloborodov, 1998).

1.1.3 ADAF, ADIOS, etc.

As another solution of the problem of the low inefficiency by accretion, there were proposed flows with energy advection (the so-called ADAF—advection dominated accretion flows) (Ichimaru, 1977; Narayan and Yi, 1994). This flow must have the form of a thick quasispherical flow in which the radial velocity is close to the toroidal one: $v_r \sim v_\varphi$. This regime can take place for sufficiently slow ($\dot{M} \ll \dot{M}_{\rm Edd}$) accretion of an optically thin gas.

This model is based on the well-known property according to which the heating in a viscous flow results in an increase in the temperature of heavy particles (ions), whereas electrons, with which the radiation mechanism is associated, can remain cold (Braginsky, 1965). If we assume that ions and electrons exchange energy rather slowly, the viscous heating is not accompanied by the effective loss of energy. Therefore, the energy is transported together with an accreting matter and thus absorbed by a black hole. This model was thoroughly developed (Narayan and Yi, 1995a,b; Abramowicz et al., 1995; Igumenshchev et al., 1998; Medvedev and Narayan, 2000) and allowed one to explain most observational facts, for example, the luminosity and spectrum of the central source in our Galaxy (Narayan et al., 1998; Mahadevan, 1998). It was also discussed in connection with the active elliptic galaxies with low luminosity (Reynolds et al., 1996a; di Matteo and Fabian, 1997; Mahadevan, 1997), the galaxy NGC4258 (Lasota et al., 1996), and some other sources in our Galaxy. Sometimes, however, there were difficulties associated with the compatibility of this model with the observations. In particular, the predicted flow from the sources was not observed in the radiofrequency band (Herrnstein et al., 1998; di Matteo et al., 1999) (see also Celotti and Rees, 1999).

Besides, there are obvious gaps in the physical ground of the ADAF model as well.

1. It is assumed that the heating of ions is more effective than that of electrons and the energy exchange between two components is ineffective. Therefore, the disk matter must exist in the form of a two-temperature plasma so that in the vicinity of the inner edge of an accretion disk the temperatures of ions and electrons appreciably differ from one another ($T_{\rm i} \sim 10^{12}$ K, $T_{\rm e} \sim 10^9$ K) (Quataert, 1998; Gruzinov, 1998; Begelman and Chiueh, 1988; Blackman, 1999). However, these assumptions are not obvious. For example, the possibility of the electron heating due to the magnetic reconnection is not taken into account, though this process, undoubtedly, occurs (it is directly observed in solar flares) and is very efficient (Bisnovatyi-Kogan and Lovelace, 1997).

2. It is supposed that the total energy of an accreting gas is positive; hence, the disk as a whole is not gravitationally bounded. Therefore, in this system after the energy redistribution and after part of a gas is falling on a black hole, the outflowing jets, which are not taken into account within ADAF, can occur (Blandford and Begelman, 1999).
3. The solution under study is self-similar. Therefore, within its framework, the natural boundary conditions at infinity and in the vicinity of the gravitational center cannot be satisfied.
4. There are indications that this accretion regime is not stable to thermal perturbations, the matter concentration in clusters, etc.

There has recently been proposed a modified advection-dominated flow model that, in principle, solved the positive energy problem—the so-called ADIOS (advection-dominated inflow–outflow solutions) (Blandford and Begelman, 1999). Besides the total negative energy inflow, a more or less isotropic outflowing wind was introduced. Within this model, only a small part of the matter coming from large distances thus falls onto a black hole. In this way, the low radiation efficiency is explained and the problem of radiation excess in the radiofrequency band is solved (Beckert, 2000; Becker et al., 2001).

However, this model encounters some difficulties as well.

1. For matter ejection, it is necessary for its energy (the Bernoulli integral E) to be positive. However, with disk accretion, the binding energy of a gas rotating in the Keplerian orbits is necessarily negative. A comprehensive analysis showed that, for the sufficiently low viscosity parameter $\alpha_{SS} < 0.1$, this outflow regime appears impossible (Abramowicz et al., 2000). This is evident both from the regularity conditions on the critical surfaces, which uniquely show that the accreting matter energy must be negative, and from the direct numerical computations (see, e.g., Narayan et al., 1997; Chen et al., 1997; Abramowicz et al., 1988; Gammie and Popham, 1998a,b; Ogilvie, 1999).
2. The positive value of the binding energy is the necessary but not sufficient cause of the outflow (as is known, the Bernoulli integral in the spherically symmetric Bondi accretion is positive).

Thus, one can conclude that even in the simplest hydrodynamical approximation the theory of accretion onto compact objects is still very far from completion. In particular, within the purely hydrodynamical approach, one still failed to construct a sufficiently reliable model of the central engine in the active galactic nuclei, which would lead to the efficient matter outflow and, hence, give rise to jet ejections that carry away a considerable part of the extracted energy. The outflow could be connected with the strongly heated corona but, in this case, the X-ray luminosity of the active nuclei must have been much higher than is evident from the observations. Besides, in spite of various modifications, the advection-dominated flow models still fail to consistently account for the inefficiently radiating sources. However, undoubtedly, for the accretion to be fully described it is absolutely necessary to take into account the advection (Bisnovatyi-Kogan and Lovelace, 2001).

On the other hand, one cannot but emphasize that the detailed computations showed the sufficient stability of the standard model. In particular, for a broad class of accretion flows and in a wide range of distances from the gravitational center, the rotational velocity of an accreting matter slightly differs from the Keplerian velocity and the sonic surface is in the vicinity of the marginally stable orbit. A difference is observed only in the vicinity of the black hole horizon when the relativistic effects become substantial.

1.2 Main Properties of Transonic Hydrodynamical Flows

1.2.1 Basic Equations

Let us now proceed to the comprehensive description of the axisymmetric stationary hydrodynamical flows. Let us begin from the beginning. We write the ideal stationary ($\partial/\partial t = 0$) hydrodynamical equations in flat space (Landau and Lifshits, 1987):

- the continuity equation

$$\nabla \cdot (n\mathbf{v}) = 0, \tag{1.21}$$

- the Euler equation

$$(\mathbf{v} \cdot \nabla)\mathbf{v} = -\frac{\nabla P}{\rho} - \nabla \varphi_{\rm g}, \tag{1.22}$$

- the ideal condition

$$\mathbf{v} \cdot \nabla s = 0, \tag{1.23}$$

- the equation of state

$$P = P(n, s). \tag{1.24}$$

The latter relation can be rewritten as

$${\rm d}P = \rho {\rm d}w - nT{\rm d}s. \tag{1.25}$$

For the polytropic equation of state

$$P = k(s)n^{\Gamma}, \tag{1.26}$$

which we, for simplicity, use in the following, we have for $\Gamma = \text{const} \neq 1$

$$c_s^2 = \frac{1}{m_\mathrm{p}}\left(\frac{\partial P}{\partial n}\right)_s = \frac{1}{m_\mathrm{p}}\Gamma k(s) n^{\Gamma-1}, \tag{1.27}$$

$$w = \frac{c_s^2}{\Gamma - 1}, \tag{1.28}$$

$$T = \frac{m_\mathrm{p}}{\Gamma} c_s^2. \tag{1.29}$$

Here n (1/cm^3) is the concentration, m_p (g) is the mass of particles ($\rho = m_\mathrm{p} n$—the mass density), s is the entropy per one particle (dimensionless), w (cm^2/s^2) is the specific enthalpy, T (erg) is the temperature in energy units, and, finally, c_s (cm/s) is the velocity of sound.

A number of important comments are already necessary here.

- The Euler equation (1.22) together with relations (1.21), (1.23), and (1.25) can be rewritten as the conservation of the energy flux:

$$\nabla \cdot \left[n\mathbf{v}\left(\frac{v^2}{2} + w + \varphi_\mathrm{g}\right)\right] = 0. \tag{1.30}$$

Using now the continuity equation (1.21), we get

$$\mathbf{v} \cdot \nabla E_\mathrm{n} = 0, \tag{1.31}$$

where

$$E_\mathrm{n} = \frac{v^2}{2} + w + \varphi_\mathrm{g}. \tag{1.32}$$

It is the well-known nonrelativistic Bernoulli integral that, as we see, is to be constant on the streamlines.

- The energy equation (1.30) together with the Euler equation (1.22) can be rewritten as the four-dimensional energy–momentum conservation law

$$\nabla_\alpha T^{\alpha\beta} = 0, \tag{1.33}$$

where for $\varphi_\mathrm{g} = 0$

$$T^{\alpha\beta} = \begin{pmatrix} \dfrac{\rho v^2}{2} + \varepsilon & \rho\mathbf{v}\left(\dfrac{v^2}{2} + w\right) \\ \rho v^i & P\delta^{ik} + \rho v^i v^k \end{pmatrix}. \tag{1.34}$$

Here ε is the internal energy density. In the following, the Greek indices α, β, γ stand for the four-dimensional values, whereas the Latin indices i, j, k stand for three-dimensional ones.

- The hydrodynamical equations are the system of five nonlinear equations per five unknowns: two thermodynamic functions and three velocity components $\mathbf{v}$.

Problem 1.2 Using the thermodynamic identity (1.25) and explicit expressions (1.26), (1.27), (1.28), and (1.29), show that for $\Gamma = \text{const} \neq 1$ the function $k(s)$ must have the fully definite form

$$k(s) = k_0 e^{(\Gamma-1)s}. \tag{1.35}$$

Problem 1.3 How can $c_s \neq 0$ for $s = 0$ be explained?

Problem 1.4 Check relations (1.30), (1.31), (1.32), and (1.34).

1.2.2 Spherically Symmetric Flow

As the simplest but very important example, we consider the spherically symmetric flow. Since, as we saw, the ideal hydrodynamical equations can be written as conservation laws, we have for the purely radial flow $v = v_r$

- the continuity equation

$$\Phi = 4\pi r^2 n(r) v(r) = \text{const}, \tag{1.36}$$

- the ideal condition

$$s = \text{const}, \tag{1.37}$$

- the energy equation

$$E_{\mathrm{n}} = \frac{v^2(r)}{2} + w(r) + \varphi_{\mathrm{g}}(r) = \text{const}. \tag{1.38}$$

As a result, given three parameters Φ, s, and E_{n}, we can determine all physical characteristics of the flow. Indeed, having rewritten the Bernoulli integral (1.38) as

$$E_{\mathrm{n}} = \frac{\Phi^2}{32\pi^2 n^2 r^4} + w(n, s) + \varphi_{\mathrm{g}}(r), \tag{1.39}$$

we see that this equation contains only one unknown value—the concentration n. Hence, this algebraic equation in implicit form specifies the concentration n as a function of three invariants and the radius r:

$$n = n(E_{\rm n}, s, \Phi; r). \tag{1.40}$$

Along with the entropy s, this relation allows us to define all other thermodynamic functions and according to relation (1.36) the flow velocity v as well.

One should stress that Eq. (1.39) has a singularity on the sonic surface. To show this we determine the derivative $\mathrm{d}n/\mathrm{d}r$. Differentiating Eq. (1.39) with respect to r, we have for the gravitational potential $\varphi_{\rm g} = -GM/r$

$$\frac{\mathrm{d}n}{\mathrm{d}r}\left[\left(\frac{\partial w}{\partial n}\right)_s - \frac{\Phi^2}{16\pi^2 n^3 r^4}\right] - \frac{\Phi^2}{8\pi^2 n^2 r^5} + \frac{GM}{r^2} = 0. \tag{1.41}$$

As a result, using the thermodynamic relation (1.25), we obtain for the logarithmic derivative η_1

$$\eta_1 = \frac{r}{n}\frac{\mathrm{d}n}{\mathrm{d}r} = \frac{2v^2 - \dfrac{GM}{r}}{c_s^2 - v^2} = \frac{2 - \dfrac{GM}{rv^2}}{-1 + \dfrac{c_s^2}{v^2}} = \frac{N}{D}. \tag{1.42}$$

We see that the derivative (1.42) has a singularity when the matter velocity is equal to the velocity of sound: $v = c_{\rm s} = c_*$ ($D = 0$). This implies that for smooth transition through the sonic surface $r = r_*$, the additional condition is to be satisfied:

$$N(r_*) = 2 - \frac{GM}{r_* c_*^2} = 0. \tag{1.43}$$

In other words, the transonic flows are two-parameter ones. As shown in Fig. 1.1, the sonic surface is the X-point on the (distance r)–(velocity v) plane.

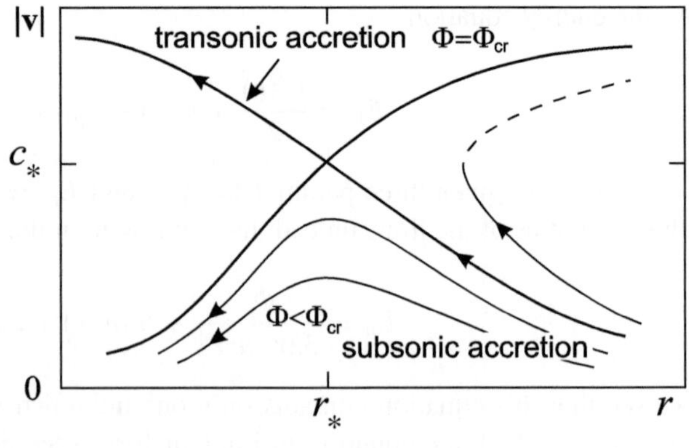

Fig. 1.1 Spherically symmetric accretion structure for the given values n_∞ and c_∞ and the different values Φ. The transonic flow (*bold curves*) corresponds to the critical accretion rate $\Phi = \Phi_{\rm cr}$ (1.56). The *fine curves* below the X-point correspond to the subsonic accretion with $\Phi < \Phi_{\rm cr}$

Problem 1.5 For the case of the spherically symmetric transonic accretion (the so-called Bondi accretion), when an accreting matter has a zero velocity for $r \to \infty$, the Bernoulli integral $E_{\rm n}$ can be expressed in terms of the velocity of sound at infinity:

$$E_{\rm n} = w_{\infty} = \frac{c_{\infty}^2}{\Gamma - 1}.$$

Using now relations (1.36), (1.37), (1.38), and (1.43), find the well-known expressions for the velocity of sound c_* and the concentrations n_* on the sonic radius r_* (Bondi, 1952):

$$c_*^2 = \left(\frac{2}{5 - 3\Gamma}\right) c_{\infty}^2, \tag{1.44}$$

$$n_* = \left(\frac{2}{5 - 3\Gamma}\right)^{1/(\Gamma - 1)} n_{\infty}, \tag{1.45}$$

$$r_* = \left(\frac{5 - 3\Gamma}{4}\right) \frac{GM}{c_{\infty}^2}. \tag{1.46}$$

Problem 1.6 Show that

$$\eta_1(r_*) = \frac{-4 \pm \sqrt{10 - 6\Gamma}}{\Gamma + 1}, \tag{1.47}$$

where plus stands for the accretion and minus for the ejection.

Problem 1.7 Show that for the spherically symmetric accretion with $\Gamma < 5/3$

- for $r \gg r_*$ (the subsonic regime) the flow can be considered to be incompressible:

$$n(r) \approx \text{const}, \tag{1.48}$$

$$v(r) \propto r^{-2}. \tag{1.49}$$

- for $r \ll r_*$ (the supersonic flow) the particle motion is close to free fall:

$$n(r) \propto r^{-3/2}, \tag{1.50}$$

$$v(r) \approx \left(\frac{2GM}{r}\right)^{1/2}. \tag{1.51}$$

Problem 1.8 Show that for the spherically symmetric transonic outflow [the Parker outflow (Parker, 1958)]

- the physical parameters on the sonic surface $r = r_*$, where

$$r_* = \frac{GM}{2c_*^2}, \tag{1.52}$$

  are connected with the corresponding values on the star surface $r = R$ as

$$c_*^2 = \left(\frac{2}{5-3\Gamma}\right)c_R^2 + \left(\frac{\Gamma-1}{5-3\Gamma}\right)\left(v_R^2 - \frac{2GM}{R}\right), \tag{1.53}$$

$$n_* = n_R\left(\frac{c_*^2}{c_R^2}\right)^{1/(\Gamma-1)}. \tag{1.54}$$

- the radial velocity on the star surface is to be

$$v_R = c_*\left(\frac{c_*^2}{c_R^2}\right)^{1/(\Gamma-1)}\left(\frac{r_*}{R}\right)^2. \tag{1.55}$$

Being an extremely simplified model, the radial one-dimensional flow, nevertheless, allows one to formulate several important properties, while most of them, as we will see, remain valid for the Grad-Shafranov (GS) equation as well.

- The flow can pass the sonic surface smoothly in the gravitational field only. Indeed, the numerator N in (1.42) can be zero only if the gravitational term GM/rv^2 is available.
- The solutions (1.44), (1.45), (1.46), and (1.53) have a singularity for $\Gamma = 5/3$. This implies that for $\Gamma = 5/3$ an increase/decrease in the velocity of sound due to the adiabatic heating/cooling exactly coincides with the change in the velocity of matter. As a result, in the nonrelativistic case for $\Gamma \geq 5/3$, the transonic flow cannot be realized.
- The transonic flow is a two-parameter one. This implies that for the full definition of the transonic flow one can specify two boundary conditions, for example, the density $\rho_\infty = m_p n_\infty$ and the sound velocity c_∞ at infinity. Then all other parameters can be expressed in terms of these values. For example, we have for the full accretion rate $\Phi_{tot} = \Phi_{cr}$, where

$$\Phi_{\rm cr} = 4\pi r_*^2 c_* n_* = \pi \left(\frac{2}{5-3\Gamma}\right)^{(5-3\Gamma)/2(\Gamma-1)} \frac{(GM)^2}{c_\infty^3} n_\infty. \tag{1.56}$$

On the other hand, for the given values of n_∞ and c_∞, there is the infinite number of subsonic flows with $\Phi < \Phi_{\rm cr}$ (sec Fig. 1.1).

- For the given flow structure, the number of integrals of motion is enough to determine all flow parameters from the algebraic relations.

The latter property is, in fact, the key property of the approach studied. Indeed, algebraic relations (1.36), (1.37), and (1.38) together with the equation of state make it possible to determine all physical parameters of the flow (the velocity $v(r)$, the temperature $T(r)$, etc.) in terms of the invariants $E_{\rm n}$ and s, as well as the stream function Φ. This property remains valid for arbitrary two-dimensional flows. On the other hand, it is clear that, in the general case, the structure of the flow itself (i.e., the function $\Phi(r, \theta)$) is not known beforehand. To determine it one has to use all five hydrodynamical equations.

1.2.3 Potential Plane Flow

As the simplest example of a flow whose structure is not known beforehand, we consider a potential plane flow without gravitation. Then the velocity $\mathbf{v}$ located in the xy-plane can be determined from the condition

$$\mathbf{v} = \nabla\phi(x, y), \tag{1.57}$$

where $\phi(x, y)$ is a scalar potential. Besides, we assume, for simplicity, that the integrals $E_{\rm n}$ and s are constant in the whole space:

$$E_{\rm n} = \text{const}, \qquad s = \text{const}. \tag{1.58}$$

Then the continuity equation $\nabla \cdot (n\mathbf{v}) = 0$ can be rewritten as

$$\nabla^2\phi + \frac{\nabla n \cdot \nabla\phi}{n} = 0. \tag{1.59}$$

Finally, using the Euler equation for determining $\nabla n \cdot \nabla\phi$

$$\mathbf{v} \cdot \nabla\left(\frac{v^2}{2}\right) + c_{\rm s}^2 \frac{\nabla n \cdot \nabla\phi}{n} = 0,$$

we find

$$\phi_{xx} + \phi_{yy} + \frac{(\phi_y)^2\phi_{xx} - 2\phi_x\phi_y\phi_{xy} + (\phi_x)^2\phi_{yy}}{(\nabla\phi)^2 D} = 0. \tag{1.60}$$

Here again

$$D = -1 + \frac{c_s^2}{v^2}, \tag{1.61}$$

and the subscripts indicate the partial derivatives with respect to the corresponding coordinates.

The second-order partial differential equation (1.60) is well known and can be found in many textbooks (see, e.g., Landau and Lifshits, 1987). It has properties most of which, as we see below, remain valid for the GS equation as well.

- For determining c_s^2, Eq. (1.60) must be supplemented with Bernoulli's equation (1.38). For the polytropic equation of state, the velocity of sound c_s can be explicitly expressed in terms of $E_{\rm n}$ and ϕ:

$$c_s^2 = (\Gamma - 1)E_{\rm n} - \frac{\Gamma - 1}{2}(\nabla\phi)^2. \tag{1.62}$$

- Together with Bernoulli's equation, Eq. (1.60) contains only the potential $\phi(x, y)$ and the invariant $E_{\rm n}$ (it does not contain the entropy s, but s is necessary in order to specify the concentration n).
- For $n = \text{const}$ ($c_s^2 \to \infty$) the equation becomes linear.
- It is nonlinear in the general case, however, always linear in the higher derivatives.
- Equation (1.60) is of an elliptic type for the subsonic flow $D > 0$.
- Equation (1.60) is of an hyperbolic type for the supersonic flow $D < 0$.
- When the flow structure is known (i.e., for known $\phi(x, y)$, $E_{\rm n}$, and s), all the physical parameters are determined from the algebraic relations.
- Equation (1.60) does not comprise the coordinates x and y explicitly.

The latter property is known to allow one to perform the so-called hodograph transformation, i.e., the change of the variables from the physical plane (x, y) to the velocity plane (v_x, v_y), where $v_x = v\cos\theta$, $v_y = v\sin\theta$. The other potential $\phi_v(v, \theta)$ is thus introduced so that $\mathbf{r} = \nabla_{\mathbf{v}}\phi_v$. As a result, Eq. (1.60) can be rewritten as

$$\frac{\partial^2\phi_v}{\partial\theta^2} + \frac{v^2}{1 - \dfrac{v^2}{c_s^2}}\frac{\partial^2\phi_v}{\partial v^2} + v\frac{\partial\phi_v}{\partial v} = 0. \tag{1.63}$$

This linear equation was first obtained in 1902 by S.A. Chaplygin and bears his name.

The hodograph transformation method was the main direction when analyzing the potential plane flows over the 20th century (Frankl', 1945; von Mises, 1958). Here we formulate two important results obtained in this area, which are necessary in the following:

1. It is impossible, in the general case, to solve the direct problem for the transonic flow (i.e., determine the flow structure by the known form of the boundary, for example, giving the form of a nozzle or a wing).
2. On the other hand, it is possible to solve the inverse problem.

This fact is based on the fundamental theorem:

Theorem 1.1 *The transonic flow is analytical at the critical point (the only point where the sonic surface is orthogonal to the streamline, see Fig. 1.2)* (Frankl', 1945; Landau and Lifshits, 1987).

Let us comment on these two assertions. The most obvious example that clarifies the unavailability of the regular (i.e., noniterative) procedure for solving Eq. (1.60) for the transonic flow is as follows. As is known, the number of boundary conditions b for an arbitrary (not necessarily purely hydrodynamical) transonic flow can be specified as (Beskin, 1997; Bogovalov, 1997a)

$$b = 2 + i - s'. \tag{1.64}$$

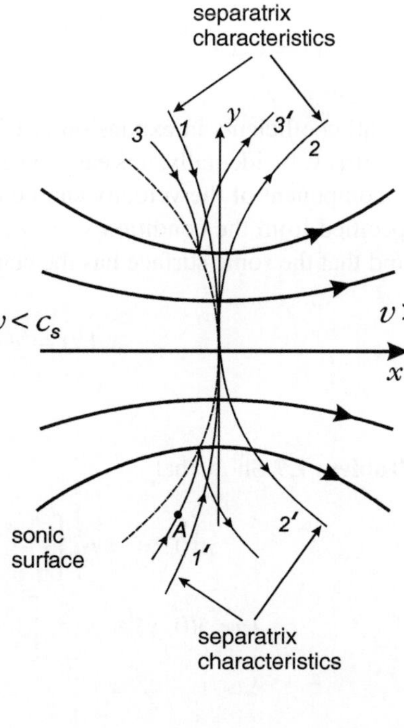

Fig. 1.2 "Analytical nozzle" structure in the vicinity of the singular point $x = y = 0$—the only point at which the sonic surface $v = c_s$ (1.67) (*dashed line*) is orthogonal to the streamline (*bold*). The characteristic surfaces (*fine lines*), which intersect the sonic surface not at the singular point, have a cusp on it. The separatrix characteristics (incoming 1, 1′ and outgoing 2, 2′) are tangent to the sonic surface at the singular point. The flow disturbance from the supersonic point A located within the separatrix characteristic affects the subsonic region

Here i is the number of invariants and s' is the number of critical surfaces. In hydrodynamics, the only singularity is the sonic surface. Therefore, for the transonic flow $s' = 1$. Further, for the two-dimensional flow we have two invariants, $E_{\rm n}$ and s, so that $i = 2$. Therefore, for the transonic flow structure to be specified it is necessary to give three boundary conditions on some surface. These can be two thermodynamic functions and also the tangential velocity component prescribing the potential ϕ on this surface. The second (and the last one for the two-dimensional flow) velocity component is to be determined from the solution. But for the solution of Eq. (1.60) one must know the Bernoulli integral $E_{\rm n} = v^2/2 + w$, i.e., both the velocity components on this surface. Consequently, in the general case, even the equation itself describing the flow structure cannot be formulated. For the subsonic and supersonic flows ($s' = 0$, so that $b = 4$) this difficulty is absent.

On the other hand, the transonic flow structure can be found by expanding the solution in the vicinity of the singular point (at which we put $x = y = 0$). Indeed, in addition to the invariants $E_{\rm n}$ and s (the latter is needed to specify the concentration n), we can give the x-component of the velocity $v_x(x, 0)$ along the x-axis. In the first approximation, it is enough to know only the first two terms of the expansion

$$v_x(x, 0) = c_* + kx + \cdots . \tag{1.65}$$

Here $c_*^2 = 2E_{\rm n}(\Gamma - 1)/(\Gamma + 1)$ (this relation follows from Bernoulli's equation (1.62) for $v = c_{\rm s} = c_*$) so that this value is also determined directly from the boundary conditions. Thus, as can be verified by direct substitution, the first terms in the expansion of the potential $\phi(x, y)$ look like (Landau and Lifshits, 1987)

$$\phi(x, y) = c_* x + \frac{kx^2}{2} + \frac{k^2(\Gamma + 1)}{2c_*} xy^2 + \frac{k^3(\Gamma + 1)^2}{24c_*^2} y^4 + \cdots . \tag{1.66}$$

Given all coefficients in expansion (1.65), we can reconstruct the potential ϕ with any accuracy. Incidentally, it is easy to verify that in the vicinity of the singular point, the y-component of the velocity can be neglected. Therefore, the sonic surface can be specified from the condition $v_x = c_*$. Using now the explicit expression (1.66), we find that the sonic surface has the standard parabolic form:

$$x_*(y) = -\frac{k(\Gamma + 1)}{2c_*} y^2. \tag{1.67}$$

Problem 1.9 Show that

$$n(0, y) \approx n_* \left[1 - \frac{k^2(\Gamma + 1)}{2c_*^2} y^2 + \cdots \right], \tag{1.68}$$

$$(nv_x)(0, y) \approx n_* c_* \left[1 - \frac{k^4(\Gamma + 1)^3}{8c_*^4} y^4 + \cdots \right], \tag{1.69}$$

so that the typical scale of the variation of the values in a perpendicular direction $\delta y \approx c_*/k$ turns out to be the same as in a longitudinal direction.

Problem 1.10 Show that, besides the symmetry axis, the geometric locus at which the velocity vector is parallel to the x-axis is also a parabola:

$$x_0(y) = -\frac{k(\Gamma + 1)}{6c_*} y^2. \quad (1.70)$$

Consequently, the sonic surface is in the confluence region ($|x_*| > |x_0|$).

As to the properties of the singular points, one cannot but mention the structure of the characteristic surfaces that occur in the hyperbolic (i.e., supersonic) domain of Eq. (1.60). As is well known, for the second-order partial differential equation written in the canonical form $\mathcal{A}\phi_{xx} + 2\mathcal{B}\phi_{xy} + \mathcal{C}\phi_{yy} + \cdots = 0$, the differential equation for the characteristic surfaces looks like (Korn and Korn, 1968)

$$\frac{\mathrm{d}x}{\mathrm{d}y} = \frac{\mathcal{B} \pm \sqrt{\mathcal{B}^2 - \mathcal{A}\mathcal{C}}}{\mathcal{C}}. \quad (1.71)$$

Using the explicit form of the coefficients $\mathcal{A}$–$\mathcal{C}$, we obtain the equation for specifying the characteristic surfaces:

$$\frac{\mathrm{d}x}{\mathrm{d}y} = \frac{-\phi_x\phi_y \pm \left(\phi_x^2 + \phi_y^2\right)\sqrt{-D(D+1)}}{\phi_x^2 + D\left(\phi_x^2 + \phi_y^2\right)}. \quad (1.72)$$

Since, in the general case, on the sonic surface the partial derivative ϕ_y is not zero, according to (1.72), the derivative $\mathrm{d}(x - x_*)/\mathrm{d}y$ is also different from zero here. This implies that in this region the sonic surface is not orthogonal to the streamlines. Accordingly, only one characteristic having a cusp here passes through each of such points on the sonic surface (unlike any point in the hyperbolic domain) (Landau and Lifshits, 1987). In Fig. 1.2, it is associated with the curves 3 and 3′. This structure is due to the degeneration of the Mach cone on the sonic surface into the plane and to the existence of the characteristics only in the hyperbolic domain of Eq. (1.60). However, at the singular points at which the partial derivative $\phi_y = 0$ (i.e., in the vicinity of the point $x = y = 0$), a comprehensive analysis is necessary to specify the behavior of the characteristic surfaces. It can again be done by expanding the solution in terms of powers of small displacements x and y. Introducing the new variable

$$R = \frac{x - x_*}{D_1}, \tag{1.73}$$

where $D_1 = -\,\partial D/\partial x|_{x=y=0}$ (so that $D_1 > 0$ here), we can rewrite Eq. (1.72) as

$$\frac{\mathrm{d}R}{\mathrm{d}y} = ay \pm \sqrt{R}, \tag{1.74}$$

where

$$a = -\frac{\partial^2 D/\partial y^2|_{x=y=0}}{D_1^2} - \frac{\phi_{yy}|_{x=y=0}}{\phi_x|_{x=y=0}\, D_1}. \tag{1.75}$$

The exact solution of Eq. (1.74) can be found by substitution:

$$R(y) = w^2(y)\, y^2. \tag{1.76}$$

Substituting expression (1.76) in Eq. (1.74), we obtain in the implicit form

$$w(y) = w_1 + C\,[w_2 - w(y)]^{w_2/w_1}\, y^{(w_2 - w_1)/w_1}. \tag{1.77}$$

Here C is an integration constant and the values w_1 and w_2 correspond to two separated solutions $R_{1,2}(y) = w_{1,2}^2 y^2$, in which the coefficients $w_{1,2}$ are independent of the y-coordinate. They can be obtained by the direct substitution of the definition (1.76) in Eq. (1.74). As a result, we have

$$2w^2 \pm w - a = 0, \tag{1.78}$$

and, hence,

$$w_{1,2}^2 = \frac{1 + 4a \pm \sqrt{1 + 8a}}{8}. \tag{1.79}$$

One should specially stress here that, for the potential plane flow, the condition $a > 0$ always turns out to be satisfied. This condition corresponding to the standard singular point shows that the streamlines pass the sonic surface in the confluence region ($|x_*| > |x_0|$). In this case, two characteristics (two incoming 1, 1′ and two outgoing 2, 2′ branches) corresponding to two roots (1.79) pass through the singular point. Indeed, for $w(y) \approx w_1$ Eq. (1.77) yields

$$w(y) \approx w_1 + C(w_2 - w_1)^{w_2/w_1}\, y^m, \tag{1.80}$$

where the exponent is

$$m = -\frac{1 + 8a + \sqrt{1 + 8a}}{4a}. \tag{1.81}$$

Therefore, for $a > 0$, when $m < 0$, the second term in (1.80) diverges for $y \to 0$. Hence, for $w \approx w_1$ only the characteristics corresponding to the constant $C = 0$ pass through the origin.

It is easy to find for the case considered here

$$D_1|_{x=y=0} = (\Gamma + 1)\,\frac{k}{c_*}, \tag{1.82}$$

$$\left.\frac{\partial^2 D}{\partial y^2}\right|_{x=y=0} = -(\Gamma + 1)^2\,\frac{k^2}{c_*^2}, \tag{1.83}$$

$$\phi_{yy}\big|_{x=y=0} = 0, \tag{1.84}$$

so that

$$a = 1. \tag{1.85}$$

Hence,

$$w_1^2 = \frac{1}{4}, \tag{1.86}$$

$$w_2^2 = 1. \tag{1.87}$$

Thus, the two characteristics tangent to the sonic surface at the singular point actually pass through the singular point. Their incoming and outgoing branches also have a parabolic form (Landau and Lifshits, 1987)

$$x^{(1,1')} = -\frac{k(\Gamma + 1)}{4c_*}\,y^2, \tag{1.88}$$

$$x^{(2,2')} = \frac{k(\Gamma + 1)}{2c_*}\,y^2, \tag{1.89}$$

while the incoming branches are associated with $R(y) = w_1^2 y^2$ and the outgoing ones with the solution $R(y) = w_2^2 y^2$. For $a < 0$ (which can take place in the presence of the gravitational field only), the situation appears much more complex. This nonstandard singular point is considered in detail in Sect. 1.3.3.3.

Problem 1.11 Find expressions (1.84), (1.85), (1.86), (1.87), (1.88), and (1.89).

In conclusion, we emphasize one more extremely important circumstance. As seen from Fig. 1.2, the perturbation from the A-point, which is located in the supersonic region, reaches along the characteristic sonic surface and can, hence, affect the flow structure in the whole subsonic region. This implies that it is the separatrix characteristic rather than the sonic surface that divides two causally unconnected regions.

To summarize, we emphasize once again that, in the general case, there is no direct procedure for solving Eq. (1.60). Note that this property is common to the similar class of equations. In particular, it is valid for the GS equation. Moreover, within the potential plane flow

- it is impossible to consider the case $E \neq \mathrm{const}$, $s \neq \mathrm{const}$;
- it is impossible to consider the nonpotential flows with $\nabla \times \mathbf{v} \neq 0$;
- it is impossible to include the gravitation (which, in the general case, is not two-dimensional).

1.3 Nonrelativistic Axisymmetric Stationary Flows

1.3.1 Basic Equations

We will show now how this approach can be applied to the axisymmetric stationary flows. This implies that we, as before, assume that all values depend on two variables r and θ only. But now all three velocity components can differ from zero. Therefore, the axisymmetric stationary flows are much richer than the two-dimensional ones.

In the axisymmetric stationary case, we can introduce the stream function $\Phi(r, \theta)$ connected with the poloidal velocity $\mathbf{v}_\mathrm{p}$ as

$$n\mathbf{v}_\mathrm{p} = \frac{\nabla \Phi \times \mathbf{e}_\varphi}{2\pi r \sin\theta}. \tag{1.90}$$

This definition results in the following properties (see Fig. 1.3):

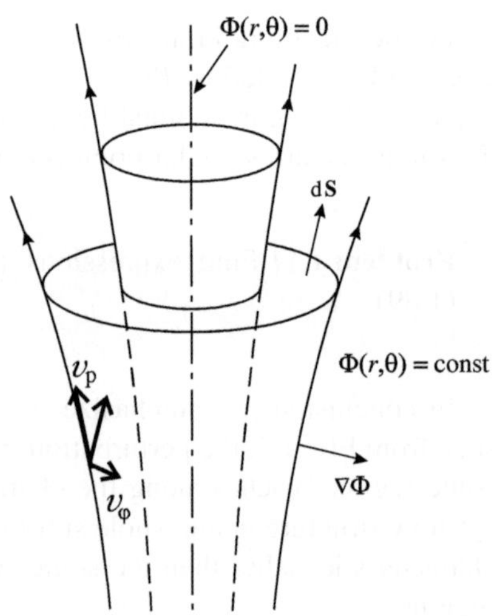

Fig. 1.3 The surfaces of the constant flow $\Phi(r, \theta) =$ const. The velocity vectors $\mathbf{v}$ are always on these surfaces; therefore, the total particle flux is conserved inside each tube

- The continuity equation is satisfied automatically: $\nabla \cdot (n\mathbf{v}) = 0$.
- It is easy to verify that $\mathrm{d}\Phi = n\mathbf{v} \cdot \mathrm{d}\mathbf{S}$, where $\mathrm{d}\mathbf{S}$ is an area element. As seen, $\Phi(r, \theta)$ is a particle flux through the circle $r, \theta, 0 < \varphi < 2\pi$. In particular, the total flux through the surface of the sphere of radius r is

$$\Phi_{\mathrm{tot}} = \Phi(r, \pi). \tag{1.91}$$

- As $\mathbf{v} \cdot \nabla\Phi = 0$, the velocity vectors $\mathbf{v}$ are located on the surfaces $\Phi(r, \theta) = \mathrm{const}$.

We now formulate the conservation laws which must be valid for the axisymmetric stationary flows. As before, the component $\beta = t$ and the projection onto the direction $\mathbf{v}_{\mathrm{p}}$ of the energy–momentum conservation law $\nabla_\alpha T^{\alpha\beta} = 0$ yield

$$E_{\mathrm{n}} = E_{\mathrm{n}}(\Phi) = \frac{v^2}{2} + w + \varphi_{\mathrm{g}}, \tag{1.92}$$

$$s = s(\Phi). \tag{1.93}$$

However, as we see, it seems much simpler to describe the case in which the integrals themselves are different on different streamlines, because this property is formulated by the explicit dependence of the integrals of motion of the function Φ.

There is new information from the $\beta = \varphi$-component of the energy–momentum conservation law (or, what is the same, from the φ-component of the Euler equation)

$$\nabla_\varphi \left(\frac{v^2}{2}\right) - [\mathbf{v} \times (\nabla \times \mathbf{v})]_\varphi + \frac{\nabla_\varphi P}{\rho} + \nabla_\varphi \varphi_{\mathrm{g}} = 0. \tag{1.94}$$

Indeed, for the axisymmetric flow considered, all the gradients ∇_φ are zero. The term $[\mathbf{v} \times (\nabla \times \mathbf{v})]_\varphi$, as is readily checked, can be rewritten as

$$\mathbf{v} \cdot \nabla(v_\varphi r \sin\theta) = 0. \tag{1.95}$$

Consequently, in the axisymmetric case, the z-component of the angular momentum

$$L_{\mathrm{n}}(\Phi) = v_\varphi r \sin\theta \tag{1.96}$$

is the third integral of motion.

Problem 1.12 Show that the total energy W_{tot} and angular momentum K_{tot} losses can be defined as

$$W_{\mathrm{tot}} = m_{\mathrm{p}} \int E_{\mathrm{n}}(\Phi)\mathrm{d}\Phi, \qquad K_{\mathrm{tot}} = m_{\mathrm{p}} \int L_{\mathrm{n}}(\Phi)\mathrm{d}\Phi. \tag{1.97}$$

1.3.2 Mathematical Intermezzo—the Covariant Approach

Since, in the following, we are to generalize our approach to the case of strong gravitational fields, it seems advisable already now to rewrite all relations in covariant form. Recall that in flat space the metric tensor g_{ik} ($\mathrm{d}l^2 = g_{ik}\mathrm{d}x^i\mathrm{d}x^k$) in the spherical coordinates $x^1 = r, x^2 = \theta, x^3 = \varphi$ has the form

$$g_{rr} = 1, \qquad g_{\theta\theta} = r^2, \qquad g_{\varphi\varphi} = r^2 \sin^2\theta, \tag{1.98}$$

and all the other components are zero. Using expression (1.34)

$$T_i^k = P\delta_i^k + \rho v^k v_i, \tag{1.99}$$

we obtain for the φ-component of the energy–momentum conservation law

$$\begin{aligned} \nabla_k T_\varphi^k &= \nabla_k\left(\delta_\varphi^k P\right) + \nabla_k\left(\rho v^k v_\varphi\right) \\ &= \frac{\partial P}{\partial\varphi} + \frac{\partial}{\partial x^k}\left(\rho v^k v_\varphi\right) + \Gamma_{ik}^k\,\rho v^i v_\varphi - \Gamma_{\varphi i}^k\,\rho v^i v_k = 0. \end{aligned} \tag{1.100}$$

Here Γ_{jk}^i are Christoffel symbols.

It is easy to verify that the last term in (1.100) is zero: $\Gamma_{\varphi i}^k\,\rho v^i v_k = \Gamma_{k,\varphi i}\,\rho v^i v^k = 0$. Analogously, the first term is also zero because of the axisymmetry of the problem (all the values are independent of the angle φ). Using the continuity equation

$$\nabla_k(\rho v^k) = \frac{1}{\sqrt{g}}\frac{\partial}{\partial x^k}\left(\sqrt{g}\,\rho v^k\right) = \frac{\partial}{\partial x^k}\left(\rho v^k\right) + \Gamma_{ik}^i\,\rho v^k = 0, \tag{1.101}$$

where $g = \det g_{ik} = g_{rr}g_{\theta\theta}g_{\varphi\varphi}$, we see that the condition (1.100) can again be rewritten as the conservation law: $\nabla_k T_\varphi^k = \rho\mathbf{v}\cdot\nabla v_\varphi = 0$. Consequently, the third invariant has the form

$$L_\mathrm{n}(\Phi) = v_\varphi. \tag{1.102}$$

Problem 1.13 Check relations (1.100), (1.101), and (1.102).

Problem 1.14 How can the contradiction between (1.96) and (1.102) be explained?

To understand the difference between expressions (1.96) and (1.102), we must return to the basic relations of the covariant approach. We have so far dealt with the physical components of the vectors only. Below in the relativistic relations these components are indicated by caps over the corresponding indices so that $v_{\hat{\varphi}} = v^{\hat{\varphi}}$ is the physical component of the toroidal velocity and its dimension is cm/s. However, in covariant relations (1.100), (1.101), and (1.102) we, in fact, deal with the other objects—contravariant components v^i and the covariant components v_k. Using now the vector length definition $\mathbf{v}^2 = g_{ik}v^i v^k = g^{ik}v_i v_k$, we obtain for the diagonal metric (1.98)

$$(v_{\hat{\varphi}})^2 = g_{\varphi\varphi}(v^{\varphi})^2 = g^{\varphi\varphi}(v_{\varphi})^2, \tag{1.103}$$

and the same for the other components. Therefore, the contravariant v^{φ} and covariant v_{φ} velocity components are expressed in terms of the physical component $v_{\hat{\varphi}}$ according to the relations

$$v^{\varphi} = \frac{1}{\sqrt{g_{\varphi\varphi}}} v_{\hat{\varphi}}, \qquad v_{\varphi} = \sqrt{g_{\varphi\varphi}}\, v_{\hat{\varphi}}. \tag{1.104}$$

In particular, this implies that the dimension of the covariant and contravariant components can differ from that of the physical quantity itself. Comparing relations (1.96) and (1.102), their difference becomes obvious: the physical component of the toroidal velocity is available in (1.96), whereas in (1.102) its covariant component is available.

1.3.3 Two-Dimensional Flow Structure

1.3.3.1 Basic Equations

To obtain an equation for the stream function $\Phi(r,\theta)$, one must return to the poloidal component of the Euler equation. It turns out that along with the definition of the invariants $E_{\mathrm{n}}(\Phi)$, $L_{\mathrm{n}}(\Phi)$, and $s(\Phi)$, this vector equation can be written as a product of the scalar factor [GS] and the vector $\nabla\Phi$:

$$[\mathrm{Euler}]_{\mathrm{p}} = [\mathrm{GS}]\, \nabla\Phi. \tag{1.105}$$

In view of this, in most papers concerned with magnetohydrodynamical flows the GS equation [GS] $= 0$ was obtained as a projection of the poloidal equation onto the direction parallel to $\nabla\Phi$. For the purely hydrodynamical flows the corresponding projection has the form

$$\frac{1}{(\nabla\Phi)^2}\nabla\Phi \cdot \left[(\mathbf{v}\nabla)\mathbf{v} + \frac{\nabla P}{\rho} + \nabla\varphi_{\mathrm{g}}\right] = 0. \tag{1.106}$$

Using definitions (1.90), (1.92), and (1.96), we find

$$-\varpi^2\nabla_k\left(\frac{1}{\varpi^2}\nabla^k\Phi\right)+\frac{1}{n}\nabla_k n\cdot\nabla^k\Phi-4\pi^2 L_{\rm n}\frac{{\rm d}L_{\rm n}}{{\rm d}\Phi}$$
$$+4\pi^2\varpi^2 n^2\frac{{\rm d}E_{\rm n}}{{\rm d}\Phi}-4\pi^2\varpi^2 n^2\frac{T}{m_{\rm p}}\frac{{\rm d}s}{{\rm d}\Phi}=0. \qquad (1.107)$$

Here and to the end of the book

$$\varpi=\sqrt{g_{\varphi\varphi}}, \qquad (1.108)$$

so that for flat metric $\varpi = r\sin\theta$.

As in the case of the plane parallel flow, in order for the system to be closed, i.e., for the product $(\nabla n \cdot \nabla\Phi)$ to be determined Eq. (1.107) is to be supplemented with Bernoulli's equation (1.92) which can now be rewritten as (cf. (1.39))

$$E_{\rm n}=\frac{(\nabla\Phi)^2}{8\pi^2\varpi^2 n^2}+\frac{1}{2}\frac{L_{\rm n}^2}{\varpi^2}+w(n,s)+\varphi_{\rm g}. \qquad (1.109)$$

We see that Bernoulli's equation (1.109), along with n, again comprises only the invariants $E_{\rm n}$, $L_{\rm n}$, and s and also the stream function Φ. Consequently, it, as before, specifies the concentration n in implicit form by the stream function Φ and the integrals of motion:

$$n=n(\nabla\Phi;E_{\rm n},L_{\rm n},s;r,\theta). \qquad (1.110)$$

On the other hand, the implicit algebraic equation (1.109) can be written in the explicit differential form

$$\nabla_k n=n\frac{N_k}{D}, \qquad (1.111)$$

where now

$$D=-1+\frac{c_{\rm s}^2}{v_{\rm p}^2}, \qquad (1.112)$$

and

$$N_k=-\frac{\nabla^i\Phi\cdot\nabla_i\nabla_k\Phi}{(\nabla\Phi)^2}+\frac{1}{2}\frac{\nabla_k\varpi^2}{\varpi^2}-4\pi^2\varpi^2 n^2\frac{\nabla_k\varphi_{\rm g}}{(\nabla\Phi)^2}$$
$$+4\pi^2\varpi^2 n^2\frac{{\rm d}E_{\rm n}}{{\rm d}\Phi}\frac{\nabla_k\Phi}{(\nabla\Phi)^2}-4\pi^2 n^2 L_{\rm n}\frac{{\rm d}L_{\rm n}}{{\rm d}\Phi}\frac{\nabla_k\Phi}{(\nabla\Phi)^2}$$
$$+\frac{2\pi^2 n^2 L_{\rm n}^2}{\varpi^2}\frac{\nabla_k\varpi^2}{(\nabla\Phi)^2}-4\pi^2\varpi^2 n^2\left[\frac{T}{m_{\rm p}}+\frac{1}{\rho}\left(\frac{\partial P}{\partial s}\right)_n\right]\frac{{\rm d}s}{{\rm d}\Phi}\frac{\nabla_k\Phi}{(\nabla\Phi)^2}. \qquad (1.113)$$

Problem 1.15 Show that

$$\nabla^i \Phi \cdot \nabla_i \nabla_k \Phi = \frac{1}{2}\,\nabla_k (\nabla \Phi)^2. \tag{1.114}$$

Problem 1.16 Show that in the spherically symmetric case the value N_r corresponds to N (1.43) and $N_\theta = 0$.

As a result, the stream equation can be written as (Beskin and Pariev, 1993)

$$\begin{aligned}
-\varpi^2 \nabla_k \left(\frac{1}{\varpi^2}\nabla^k \Phi\right) &- \frac{\nabla^i \Phi \cdot \nabla^k \Phi \cdot \nabla_i \nabla_k \Phi}{D(\nabla\Phi)^2} + \frac{\nabla \varpi^2 \cdot \nabla \Phi}{2D\varpi^2} \\
&-4\pi^2\varpi^2 n^2 \frac{\nabla \varphi_{\rm g} \cdot \nabla \Phi}{D(\nabla\Phi)^2} - 4\pi^2 n^2 \frac{D+1}{D} L_{\rm n} \frac{{\rm d}L_{\rm n}}{{\rm d}\Phi} \\
&+2\pi^2 n^2 \frac{\nabla \varpi^2 \cdot \nabla \Phi}{D\varpi^2 (\nabla\Phi)^2} L_{\rm n}^2 + 4\pi^2 \varpi^2 n^2 \frac{D+1}{D}\frac{{\rm d}E_{\rm n}}{{\rm d}\Phi} \\
&-4\pi^2 \varpi^2 n^2 \left[\frac{D+1}{D}\frac{T}{m_{\rm p}} + \frac{1}{D\rho}\left(\frac{\partial P}{\partial s}\right)_n\right]\frac{{\rm d}s}{{\rm d}\Phi} = 0,
\end{aligned} \tag{1.115}$$

or, in compact form, as (cf. Heyvaerts, 1996)

$$-\varpi^2 \nabla_k \left(\frac{1}{\varpi^2 n}\nabla^k \Phi\right) - 4\pi^2 n L_{\rm n}\frac{{\rm d}L_{\rm n}}{{\rm d}\Phi} + 4\pi^2\varpi^2 n \frac{{\rm d}E_{\rm n}}{{\rm d}\Phi} - 4\pi^2\varpi^2 n \frac{T}{m_{\rm p}}\frac{{\rm d}s}{{\rm d}\Phi} = 0. \tag{1.116}$$

At first sight, the stream equation (1.115) is much more complex than Eq. (1.60) for the plane parallel flow. Nevertheless, they have much in common. As in Eq. (1.60), the stream equation (1.115) begins with a linear elliptic term and a nonlinear term of the same form. The third term, of course, is not available in (1.60)—it results from writing Eq. (1.115) in arbitrary coordinates. However, all the other terms are not to be regarded as a complication. They allow one to include into consideration not only the gravitation but also a much wider class of flows in which the invariants are different on different flow surfaces.

In other respects, the stream equation is quite analogous to Eq. (1.60).

- Equation (1.115) must be supplemented with Bernoulli's equation (1.109).
- Along with Bernoulli's equation, Eq. (1.115) contains only the stream function $\Phi(r,\theta)$ and the invariants $E_{\rm n}(\Phi)$, $L_{\rm n}(\Phi)$, and $s(\Phi)$ (i.e., it has the form of the GS equation).
- For n = const ($c_s^2 \to \infty$), $E_{\rm n}$ = const, s = const, and $L_{\rm n} = 0$ the equation becomes linear.

- In the general case, it is nonlinear; however, it remains linear in the higher derivatives.
- Equation (1.115) is of an elliptic type for the subsonic flow $D > 0$.
- Equation (1.115) is of a hyperbolic type for the supersonic flow $D < 0$.
- For the given flow structure (i.e., for the given stream function Φ) and the invariants $E_{\rm n}(\Phi)$, $L_{\rm n}(\Phi)$, and $s(\Phi)$, all physical parameters are determined from the algebraic relations.

We should stress the following circumstance. The denominator $D = -1 + c_{\rm s}^2/v_{\rm p}^2$ (1.112) contains the poloidal velocity $v_{\rm p}$ rather than the total one. This implies that the sonic surface corresponds to the position where the poloidal velocity rather than the total one coincides with the velocity of sound. This property is a direct consequence of our main assumption on the flow axisymmetry. As a result, all perturbations (waves) must also be axisymmetric, i.e., they can propagate in the poloidal direction only. Therefore, the flow singularity occurs at the position where the poloidal velocity coincides with the perturbation velocity.

Problem 1.17 Introducing for the plane parallel flow the stream function $\psi(x, y)$ such as $n\mathbf{v} = \nabla\psi \times \mathbf{e}_z$, show that the first terms of its expansion in the vicinity of the singular point $x = y = 0$ corresponding to the solution (1.66) are

$$\psi(x, y) = n_* c_* \left[y - \frac{k^2(\Gamma+1)}{2c_*^2} x^2 y - \frac{k^3(\Gamma+1)^2}{6c_*^3} x y^3 - \frac{k^4(\Gamma+1)^3}{40c_*^4} y^5 + \cdots \right]. \tag{1.117}$$

Problem 1.18 Show that the first terms of the expansion for the potential $\phi(\varpi, z)$ ($\mathbf{v} = \nabla\phi$) in the cylindrical coordinates ϖ, z in the vicinity of the singular point for an ordinary axisymmetric nozzle (i.e., in the absence of the gravitation and for $L = 0$) have the form

$$\phi(\varpi, z) = c_* z + \frac{kz^2}{2} + \frac{k^2(\Gamma+1)}{4c_*} z\varpi^2 + \frac{k^3(\Gamma+1)^2}{64c_*^2} \varpi^4 + \cdots . \tag{1.118}$$

1.3.3.2 Eigenfunctions

In the following, we repeatedly deal with the linear operator

$$\hat{\mathcal{L}} = \varpi^2 \nabla_k \left(\frac{1}{\varpi^2} \nabla^k \right) = \frac{\partial^2}{\partial r^2} + \frac{\sin\theta}{r^2} \frac{\partial}{\partial\theta} \left(\frac{1}{\sin\theta} \frac{\partial}{\partial\theta} \right). \tag{1.119}$$

As seen from Eq. (1.107), the function Φ, which satisfies the condition $\hat{\mathcal{L}}\Phi = 0$, describes an incompressible flow. Therefore, it is advisable to discuss at once the properties of this operator in more detail. We first consider the angular operator

$$\hat{\mathcal{L}}_\theta = \sin\theta \frac{\partial}{\partial\theta}\left(\frac{1}{\sin\theta}\frac{\partial}{\partial\theta}\right). \tag{1.120}$$

It has the eigenfunctions

$$Q_0 = 1 - \cos\theta, \tag{1.121}$$
$$Q_1 = \sin^2\theta, \tag{1.122}$$
$$Q_2 = \sin^2\theta\cos\theta, \tag{1.123}$$
$$\dots$$
$$Q_m = \frac{2^m m!(m-1)!}{(2m)!}\sin^2\theta\,\mathcal{P}'_m(\cos\theta), \tag{1.124}$$

and the eigenvalues

$$q_m = -m(m+1). \tag{1.125}$$

Here $\mathcal{P}_m(x)$ are the Legendre polynomials and the dash indicates their derivatives. Thus, neglecting their dimension, the eigenfunctions of the full operator $\hat{\mathcal{L}}$ have the form

1. $m = 1$

- $\Phi_1^{(1)} = r^2\sin^2\theta$—a homogeneous flow (Fig. 1.4a),

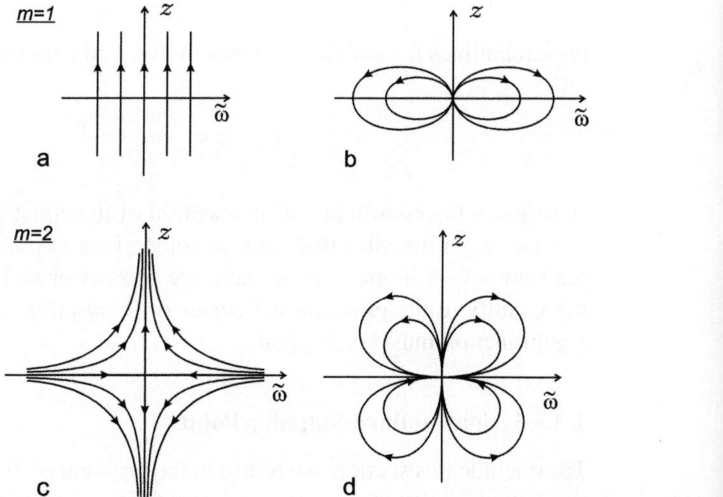

Fig. 1.4 Eigenfunctions of the operator $\hat{\mathcal{L}}$ for $m = 1$ and $m = 2$. (**a**) Homogeneous flow. (**b**) Dipole flow. (**c**) Flow in the vicinity of the zero point. (**d**) Quadrupole flow

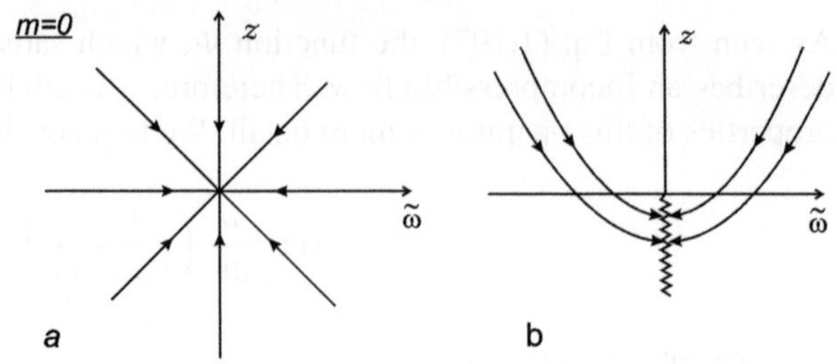

Fig. 1.5 Eigenfunctions of the operator $\hat{\mathcal{L}}$ for $m = 0$. (**a**) Spherically symmetric flow. (**b**) Parabolic flow which can occur only in the presence of the volume sources or sinks of matter, for example, on the axis $\theta = \pi$

- $\Phi_1^{(2)} = \sin^2\theta/r$—a dipole flow (Fig. 1.4b),

2. $m = 2$

- $\Phi_2^{(1)} = r^3 \sin^2\theta \cos\theta$—a zero point (Fig. 1.4c),
- $\Phi_2^{(2)} = \sin^2\theta \cos\theta/r^2$—a quadrupole flow (Fig. 1.4d),

3. ...

At first sight, the problem is quite clear and no difficulties can be encountered here. Nevertheless, this is not the case. Indeed, let us consider the eigenfunctions corresponding to $m = 0$. The first eigenfunction is obvious: the function

$$\Phi_0^{(1)} = (1 - \cos\theta) \tag{1.126}$$

describes the spherically symmetric accretion or the ejection (see Fig. 1.5a). Incidentally, only this harmonic determines the accretion or the ejection rate because for all the other eigenfunctions with $m > 0$ we have $\Phi_m(r, \pi) = 0$. The uncertainty results from the second eigenfunction

$$\Phi_0^{(2)} = r(1 - \cos\theta), \tag{1.127}$$

the streamlines for which, as shown in Fig. 1.5b, are parabolas

$$z = \frac{\varpi^2 - \varpi_0^2}{2\varpi_0}. \tag{1.128}$$

Here ϖ_0 is the coordinate of intersection of the equatorial plane by the streamline. For this eigenfunction $\Phi_0(r, \pi) \neq \text{const}$. This implies that this harmonic can be realized only if in the volume there are sources or sinks of matter (i.e., not only in the vicinity of the gravitational center or at infinity). In all other cases, the second eigenfunction must be dropped.

1.3.3.3 Nonstandard Singular Point

To conclude this section, we return to the problem of the behavior of the characteristics in the vicinity of the singular points for the axisymmetric stationary flows. Since

Eq. (1.115) is still quasilinear (i.e., linear in the higher derivatives), the equation for the characteristics can again be written in the standard form:

$$\frac{\mathrm{d}r}{\mathrm{d}\theta} = \frac{\mathcal{B} \pm \sqrt{\mathcal{B}^2 - \mathcal{A}\mathcal{C}}}{\mathcal{C}}. \tag{1.129}$$

Using the explicit form of the coefficients $\mathcal{A}$–$\mathcal{C}$, we have

$$\frac{\mathrm{d}r}{\mathrm{d}\theta} = \frac{r^2(\partial\Phi/\partial r)(\partial\Phi/\partial\theta) \pm r\left[r^2(\partial\Phi/\partial r)^2 + (\partial\Phi/\partial\theta)^2\right]\sqrt{-D(D+1)}}{(\partial\Phi/\partial\theta)^2 + D\left[r^2(\partial\Phi/\partial r)^2 + (\partial\Phi/\partial\theta)^2\right]}. \tag{1.130}$$

Introducing now dimensionless variable

$$R = \frac{r_* - r}{r_* D_1}, \tag{1.131}$$

where $D_1 = r_*\, \partial D/\partial r|_{r=r_*}$, Eq. (1.130) can be rewritten as

$$\frac{\mathrm{d}R}{\mathrm{d}\vartheta} = a\vartheta \pm \sqrt{R}. \tag{1.132}$$

Here now

$$a = -\frac{\partial^2 D/\partial\theta^2|_{r=r_*,\vartheta=0}}{D_1^2} - \frac{r_*\, \partial^2\Phi/\partial r\partial\theta|_{r=r_*,\vartheta=0}}{\partial\Phi/\partial\theta|_{r=r_*,\vartheta=0}\, D_1}, \tag{1.133}$$

the angle $\vartheta = \theta - \theta_*$, and the partial derivatives are taken at the singular point $r = r_*$, $\theta = \theta_*$. We point out that the variable R remains positive in the hyperbolic domain for the case of both the ejection ($r > r_*$, $D_1 < 0$) and the accretion ($r < r_*$, $D_1 > 0$).

The exact solution of Eq. (1.132) can again be found by the substitution

$$R(\vartheta) = w^2(\vartheta)\vartheta^2, \tag{1.134}$$

where the implicit solution for the function $w(\vartheta)$ has the form

$$w(\vartheta) = w_1 + C\,[w_2 - w(\vartheta)]^{w_2/w_1}\, \vartheta^{(w_2-w_1)/w_1}. \tag{1.135}$$

The values w_1 and w_2, which correspond to two directrix parabolas shown in Fig. 1.2, are, as before, given by the relation

$$w_{1,2}^2 = \frac{1 + 4a \pm \sqrt{1+8a}}{8}. \tag{1.136}$$

However, we should stress here two important distinctions from the above case of the plane parallel flows. First, for the spherically symmetric flows all streamlines intersect the sonic surface at a right angle. Thus, everywhere here $a = 0$, so that all points on the sonic surface turn out to be singular ones and the solution to Eq. (1.132) yields only the outgoing characteristics

$$R = \pm\frac{1}{4}\,\vartheta^2. \tag{1.137}$$

Second, for the nonspherical flows on the sonic surface not only the standard singular point with $a > 0$ but also the completely new nonstandard singular point with $a < 0$ must inevitably occur. Indeed, a indicates the slope of the streamlines with respect to the sonic surface. The positive value a corresponds to the convergent streamlines. This case shown in Fig. 1.2 was considered in the previous section. However, for the axisymmetric case in which the flow is not bounded by the walls and must fill all the space, at least one singular point in the vicinity of which the streamlines diverge must inevitably occur on the sonic surface (for $\theta_* \neq 0, \pi$, this "singular point" is associated with a circle).

The most important property of the nonstandard singular points is the bifurcation of the characteristics (Bogovalov, 1994; Beskin and Kuznetsova, 1998). When satisfying the condition $-1/8 < a < 0$ corresponding to the real roots of Eq. (1.136) the exponent m (1.81) in the solution $w(\vartheta) \approx w_1 + C_1\vartheta^m$ appears positive. Hence, the infinite number of characteristics passes through the singular point because now $w(\vartheta) \rightarrow w_1$ as $\vartheta \rightarrow 0$ for all values of constant C_1. On the other hand, for $a < -1/8$, when the roots (1.136) become complex, the structure of the characteristic surfaces abruptly changes, with the result that not a single characteristic passes through the singular point.

The behavior of the characteristic surfaces for the nonstandard case is shown on the left in Fig. 1.6. On the right, there are shown standard singular points with $a > 0$, two characteristics corresponding to two branches of solutions (1.136) pass through these points. Thus, we see that if the flow rather strongly differs from the spherically symmetric one, the entire structure of the characteristic surfaces, including the location of the separatrix characteristic (which, as was already mentioned, divides two causally unconnected domains), abruptly changes. In particular, if, for $a > -1/8$, the separatrix characteristic connects both the singular points, this is not the case for $a < -1/8$.

As a result, if the parameter a changes slowly, the whole domain located directly over the nonstandard singular point for $a = -1/8$ suddenly begins to affect the elliptic domain of the stream equation because, as seen from Fig. 1.6b, a perturbation from this domain along the characteristics now reaches the sonic surface. Note also that if in the vicinity of the standard singular point the location of the separatrix characteristic is exactly defined by the solution $R(\vartheta) = w_1^2\vartheta^2$, for the nonstandard singular point the form of the separatrix characteristic (the constant C in relation (1.135), with $a > -1/8$) cannot be defined locally. To define it we must integrate the full equation (1.130) up to the standard singular point.

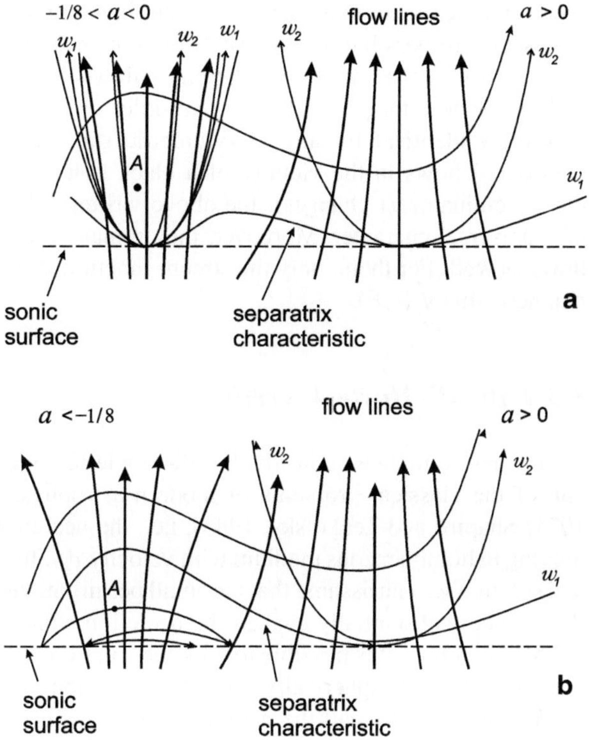

Fig. 1.6 Flow structure (*bold lines*) in the vicinity of singular points. (**a**) The behavior of the characteristic surfaces when the parameter a for the nonstandard singular point (*left*) satisfies the condition $-1/8 < a < 0$. The *heavier lines* indicate the separatrix characteristic and the solutions corresponding to $w = w_1$ and $w = w_2$. The standard singular point $a > 0$ is shown on the *right*. The point A does not affect the subsonic region. (**b**) The same for $a < -1/8$. The perturbation from the point A along the characteristic reaches the sonic surface

One should also note that at the moment of the bifurcation of the characteristics (i.e., with the abrupt change in the separatrix characteristic form) no changes in the procedure of the solution construction in the vicinity of the singular point occur. As in the case of the plane parallel flow, given the integrals of motion, it is necessary to specify for the solution construction only one more function, for example, the velocity along the streamline passing through the singular point (Beskin and Kuznetsova, 1998). Otherwise, for both $a > -1/8$ and $a < -1/8$, the GS equation requires the equal number of boundary conditions. And only for a much stronger distortion of the flow when $a \to -\infty$ (i.e., $D_1 \to 0$), the flow structure significantly changes. The point is that D_1 on the sonic surface must be determined by resolving the singularity of the type 0/0 in the expression for the concentration gradient $\nabla_k n = nN_k/D$ (this procedure was, in fact, used when determining the logarithmic derivative $\eta_1 = (r_*/n_*)\mathrm{d}n/\mathrm{d}r$ (1.47)). As a result, D_1 is found as the solution of the quadratic equation of the form $D_1^2 = F$, which has no real roots for $F < 0$ (see Beskin and Kuznetsova (1998) for more details). Recall that for the spherically symmetric flow $D_1^2 = 10 - 6\Gamma$.

The unavailability of the real roots for D_1 implies that our initial assumption on the possibility to expand the solution in integer powers of deviation from the singular point is not valid. Otherwise, the solution at the singular point ceases to be an analytical one. Following Landau and Lifshits (1987) this implies that weak discontinuities occur in the vicinity of the singular point so that the smooth transonic flow seems impossible. Below we discuss the behavior of the separatrix characteristics for various axisymmetric stationary flows.

Finally, note that since the second-order operator in the GS equation remains absolutely identical for any axisymmetric stationary flows, including the below considered flows in the vicinity of a black hole (only the explicit expression for the denominator D changes), the above written relations (1.132) and (1.133) are of a universal character. Moreover, they remain valid for the strongly magnetized flows as well. For them, only the stream function $\Phi(r, \theta)$ is to be substituted by the magnetic flux $\Psi(r, \theta)$.

1.3.4 Bondi–Hoyle Accretion

As the first example we consider Bondi–Hoyle accretion (Bondi and Hoyle, 1944)—one of the classical problems of modern astrophysics (Zel'dovich and Novikov, 1971; Shapiro and Teukolsky, 1983), i.e., the accretion onto a gravitational center moving in homogeneous medium with velocity $v_{\rm gc}$. In order to construct a nonspherical solution we can assume that the small perturbations of the spherically symmetric flow cannot substantially change the accretion structure (Beskin and Pidoprygora, 1995). Therefore, it is possible to seek the solution of the stream equation as a small perturbation of the spherically symmetric solution.

We should first recall the main qualitative results of the theory. It is convenient to carry out computations in the frame of reference moving together with the gravitational center. Then the homogeneous plasma moves with velocity $v_\infty = v_{\rm gc}$. Comparing now the Bondi accretion rate $4\pi r_*^2 n_* c_* \sim (GM)^2 n_\infty / c_\infty^3$ (1.56) with the flux $\Phi \sim \pi R_{\rm cap}^2 n_\infty v_\infty$ captured within the capture radius $R_{\rm cap}$, we can estimate $R_{\rm cap}$ as

$$R_{\rm cap} \sim \varepsilon_1^{-1/2} r_*, \tag{1.138}$$

where

$$\varepsilon_1 = \frac{v_\infty}{c_\infty}. \tag{1.139}$$

Hence, for $\varepsilon_1 \ll 1$ the capture radius $R_{\rm cap}$ is much larger than the sonic surface radius r_* so that for $r \ll R_{\rm cap}$ the flow can be assumed to be close to a spherically symmetric one. Therefore, we can seek the solution of Eq. (1.115) in the form

$$\Phi(r, \theta) = \Phi_0[1 - \cos\theta + \varepsilon_1 f(r, \theta)]. \tag{1.140}$$

For the gravitational center at rest, i.e., for $\varepsilon_1 = 0$, we return to the spherically symmetric flow.

Since Eq. (1.115) contains all $i = 3$ invariants so that $b = 2 + 3 - 1 = 4$, it is necessary to give four boundary conditions, for example, two velocity components at infinity and two thermodynamic functions. Turning the z-axis opposite to the velocity of an incoming flow, we have for the homogeneous environment

1. $v_z = -v_\infty = \text{const}$;
2. $v_\varphi = 0$ (hence, $L_\mathrm{n} = 0$);
3. $s_\infty = \text{const}$;
4. $E_\infty = c_\infty^2/(\Gamma - 1) = \text{const}$.

In the latter relation we dropped the terms $\sim\varepsilon_1^2$, because, with the nonzero velocity v_∞, the Bernoulli integral $E_\mathrm{n} = w_\infty + v_\infty^2/2$ differs from that in the case of the gravitational center at rest $E_\mathrm{n}^{(0)} = w_\infty$ by the value of order ε_1^2:

$$E_\mathrm{n} = E_\mathrm{n}^{(0)}\left(1 + \frac{\Gamma - 1}{2}\varepsilon_1^2\right). \tag{1.141}$$

As a result, Eq. (1.115) can be linearized:

$$-\varepsilon_1 D\frac{\partial^2 f}{\partial r^2} - \frac{\varepsilon_1}{r^2}(D+1)\sin\theta\frac{\partial}{\partial\theta}\left(\frac{1}{\sin\theta}\frac{\partial f}{\partial\theta}\right) + \varepsilon_1\left(\frac{2}{r} - \frac{GM}{v^2 r^2}\right)\frac{\partial f}{\partial r} = 0. \tag{1.142}$$

This equation has the following properties:

- It is linear.
- The angular operator coincides with the operator $\hat{\mathcal{L}}_\theta$ (1.120).
- Since all terms of the equation comprise the small parameter ε_1, the functions D, c_s, n, etc., can be taken from the zero approximation.
- Since for the spherically symmetric flow the functions D, c_s, n, etc., are independent of θ, the solution of Eq. (1.142) can be expanded in terms of the eigenfunctions of the operator $\hat{\mathcal{L}}_\theta$.

Therefore, the solution of Eq. (1.142) can be represented as

$$f(r,\theta) = \sum_{m=0}^{\infty} g_m(r) Q_m(\theta), \tag{1.143}$$

and the equations for the radial functions $g_m(r)$ are written as

$$-r^2 D\frac{\mathrm{d}^2 g_m}{\mathrm{d}r^2} + \left(2r - \frac{GM}{v^2}\right)\frac{\mathrm{d}g_m}{\mathrm{d}r} + m(m+1)(D+1)g_m = 0. \tag{1.144}$$

As for the boundary conditions, they can be given as follows:

1. The absence of a singularity on the sonic surface, where by definition (1.42) $D = -1 + c_\mathrm{s}^2/v^2 = 0$, $r^2 N_r = 2r - GM/v^2 = 0$, which yields for $m \neq 0$

$$g_m|_{r=r_*} = 0. \tag{1.145}$$

2. The flow homogeneity condition at infinity $\Phi = \pi n_\infty v_\infty r^2 \sin^2\theta$, which yields

$$g_1 \to \frac{1}{2}\frac{n_\infty c_\infty}{n_* c_*}\frac{r^2}{r_*^2}, \qquad g_2,\ g_3,\ \cdots = 0. \tag{1.146}$$

We now determine g_0 that fixes the change in the accretion rate. We write the exact values of the sonic surface radius $r_*(\theta)$ and the thermodynamic functions $n_*(\theta) = n(r_*, \theta)$, $w_*(\theta) = w(r_*, \theta)$, $c_*(\theta) = c_s(r_*, \theta)$ as

$$\begin{aligned} r_*(\theta) &= r_*^{(0)}[1 + \varepsilon_1 d(\theta)], \\ c_*(\theta) &= c_*^{(0)}[1 + \varepsilon_1 b(\theta)], \\ n_*(\theta) &= n_*^{(0)}[1 + \varepsilon_1 q(\theta)], \\ w_*(\theta) &= w_*^{(0)}[1 + \varepsilon_1 p(\theta)], \end{aligned}$$

where the indices "0" stand for the unperturbed values. The first two equations connecting the dimensionless functions $b(\theta)$, $p(\theta)$, and $q(\theta)$

$$p - q = 0, \tag{1.147}$$
$$2b - (\Gamma - 1)p = 0, \tag{1.148}$$

follow from thermodynamic relations (1.27) and (1.28). Equations $D(r_*) = 0$ and $N_r(r_*) = 0$, in which the expansion up to the values of order ε_1 is to be done, are now written as

$$b + 2d + p = 0, \tag{1.149}$$
$$4b + 2d = \frac{r_*^{(0)}}{\sin\theta}\left.\frac{\partial^2 f}{\partial r \partial\theta}\right|_{r=r_*}. \tag{1.150}$$

Finally, we can write the fifth relation that follows from Bernoulli's equation (1.109), in which the expansion up to the values of order ε_1 is also to be done:

$$b + 2d + q = \frac{1}{\sin\theta}\left.\frac{\partial f}{\partial\theta}\right|_{r=r_*}. \tag{1.151}$$

Here we used relation (1.141) according to which the Bernoulli integral up to the terms $\sim\varepsilon_1$ remains the same as in the case of the spherically symmetric accretion. Substituting on the right-hand side of Eq. (1.151) the expression

$$\frac{1}{\sin\theta}\left.\frac{\partial f}{\partial\theta}\right|_{r=r_*} = g_0 + 2g_1(r_*)\cos\theta, \tag{1.152}$$

we obtain from the compatibility condition of five equations (1.147), (1.148), (1.149), (1.150), and (1.151) that

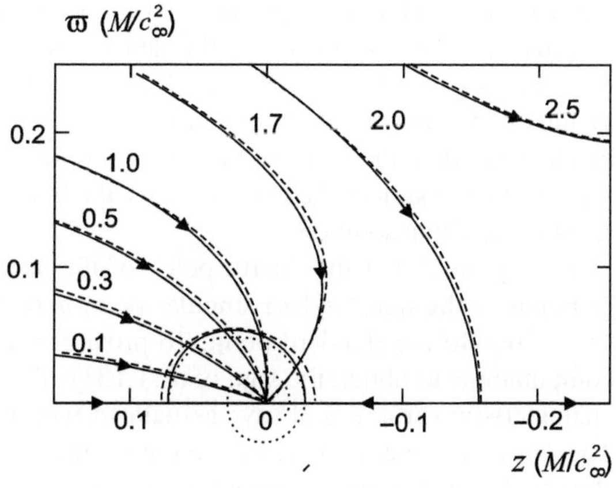

Fig. 1.7 Flow structure and the form of the sonic surface by accretion onto the moving gravitational center for $\Gamma = 4/3$, $\varepsilon_1 = 0.6$ (Beskin and Pidoprygora, 1995). The numbers on the curves indicate the ratio Φ/Φ_0, and the *dashed curves* indicate the form of the streamlines and the sonic surface, which were obtained numerically in Hunt (1979). *Dotted curve* indicates one of the separatrix characteristics

$$g_0 = 0. \tag{1.153}$$

Thus, in the first order of ε_1 the accretion rate onto the moving gravitational center does not change.

As a result, all the radial functions g_m, except for g_1, turn out to be zero so that the complete solution can be represented as

$$\Phi(r,\theta) = \Phi_0[1 - \cos\theta + \varepsilon_1 g_1(r)\sin^2\theta]. \tag{1.154}$$

The radial function $g_1(r)$ is the solution of the ordinary differential equation (1.144) for $m = 1$ with the boundary conditions (1.145) and (1.146).

At the present level of PC technology, this implies that we were able to construct the analytical solution of the posed problem, which allows us to obtain the exhaustive information on the flow structure. The sonic surface, for example, has the nonspherical form now:

$$r_*(\theta) = r_*^{(0)}\left[1 + 2\varepsilon_1\left(\frac{\Gamma+1}{D_1^2}\right)k_1\cos\theta\right], \tag{1.155}$$

where again $D_1^2 = 10 - 6\Gamma$, and the numerical coefficient $k_1 = r_* g_1'(r_*) > 0$ can be obtained directly from the solution of the ordinary differential equation (1.144) (see Beskin and Pidoprygora (1995) for more details and also Table 1.2). As shown in Fig. 1.7, the analytical solution is in good agreement with the results of the numerical computations (Hunt, 1979) though the parameter $\varepsilon_1 = 0.6$ is rather large here.

In view of the above-obtained solution, one additional comment is necessary. As is readily seen, beyond the capture radius our main assumption—the small perturbation of the spherically symmetric flow—is not true. Nevertheless, the construction of the solution remains valid. This remarkable property is associated with the already mentioned fact that for the constant concentration n the stream equation becomes linear. But as the analysis of the spherically symmetric Bondi accretion shows (see

(1.48)), far from the sonic surface $r \gg r_*$ the density of an accreting matter remains roughly constant. Accordingly, the density is constant for the homogeneous flow as well. As a result, provided that $R_{\rm cap} \gg r_*$, which is satisfied for $\varepsilon_1 \ll 1$, in the vicinity of and beyond the capture radius (where the perturbation $\sim\varepsilon_1 g_1(r)$ becomes much larger than the value of the zero approximation), Eq. (1.115) becomes linear. As a result, the sum of the two solutions, the homogeneous and spherically symmetric ones, is also a solution.

Finally, note that the "north pole" of the sonic surface ($\theta = 0$, $\Phi = 0$) corresponds to the nonstandard singular point and its opposite "south pole" ($\theta = \pi$, $\Phi = 2\Phi_0$) to the standard point. To prove it we are not to carry out complicated computations to obtain the sign of a (1.133). The point is that the motion along the characteristic surface is always associated with the motion together with the flow. Therefore, in the case of accretion the standard singular point must be located at a shorter distance from a compact object. As a result, as shown in Fig. 1.7, the separatrix characteristics coming out from the nonstandard singular point and moving practically along the sonic surface are again tangent to it at the standard singular point and only later start a spiral motion to the gravitational center. As to a, at the nonstandard singular point it can be written as (Beskin and Kuznetsova, 1998)

$$a = -\frac{2b_1 + b_3(\Gamma + 1)}{D_1^2}, \tag{1.156}$$

where

$$b_1 = 2k_1\varepsilon_1, \tag{1.157}$$

$$b_3 = \frac{4k_1}{\sqrt{10 - 6\Gamma}}\left(\frac{4 - \sqrt{10 - 6\Gamma}}{\Gamma + 1} - 1\right)\varepsilon_1, \tag{1.158}$$

$$D_1^2 = 4(2 - b_1)^2 - (\Gamma + 1)(6 - 6b_1 + b_1^2 + 2b_3). \tag{1.159}$$

Therefore, for the subsonic motion $\varepsilon_1 \ll 1$ under study we have $a > -1/8$, so that at the nonstandard singular point there is no bifurcation of the characteristics.

1.3.5 *Outflow from a Slowly Rotating Star*

Another interesting example of the nonrelativistic flow is a transonic ejection from a slowly rotating star (Tassoul, 1978; Lammers and Cassinelli, 1999). We should point out at once that this example is of an illustrative character only, because the radiation pressure, which cannot be successively included in the consideration within the approach studied, actually, plays a crucial role in stars. Nevertheless, the analysis of this case helps us clarify most problems associated with the GS equation method (Beskin and Pidoprygora, 1998).

It is logical to consider, as a zero approximation, the well-known Parker solution for the spherically symmetric transonic outflow (1.52), (1.53), (1.54), and (1.55).

This implies that all the parameters of the spherically symmetric flow (including the sonic radius r_*, the velocity of sound on the sonic surface c_*, the radial velocity v_R on the star surface $r = R$) are assumed to be known beforehand. For the polytropic equation of state (1.26), which is considered in the following, they are given by relations (1.52), (1.53), and (1.54):

$$c_*^2 = \frac{2}{5-3\Gamma} c_R^2 + \frac{\Gamma - 1}{5 - 3\Gamma}\left(v_R^2 - \frac{2GM}{R}\right), \tag{1.160}$$

$$r_* = \frac{GM}{2c_*^2}, \tag{1.161}$$

$$n_* = n_R \left(\frac{c_*^2}{c_R^2}\right)^{1/(\Gamma-1)}, \tag{1.162}$$

where the values with the index "R" stand for the star surface. Further, it is clear that the ejection rate Φ_0 in the expression $\Phi = \Phi_0(1 - \cos\theta)$ can be written as

$$\Phi_0 = 2\pi r_*^2 c_* n_*. \tag{1.163}$$

Finally, since the gas ejection velocity v_R on the star surface $r = R$, under the continuity condition, is defined as (1.55)

$$v_R = c_* \left(\frac{c_*^2}{c_R^2}\right)^{1/(\Gamma-1)} \left(\frac{r_*}{R}\right)^2,$$

this relation together with (1.160) implicitly defines the velocity of sound c_* as the functions n_R and c_R^2:

$$\frac{c_*^2}{2}\left(\frac{c_*^2}{c_R^2}\right)^{2/(\Gamma-1)} \left(\frac{GM}{2c_*^2 R}\right)^4 + \frac{c_R^2}{\Gamma - 1} - \frac{GM}{R} = \frac{5 - 3\Gamma}{2(\Gamma - 1)} c_*^2. \tag{1.164}$$

At large distances from a star $r \gg r_*$, where a free plasma outflow occurs, we have

$$v_r = (2E_{\mathrm{n}})^{1/2} = v_\infty, \tag{1.165}$$

$$n(r) = n_* \frac{c_*}{v_\infty}\left(\frac{r_*}{r}\right)^2, \tag{1.166}$$

where

$$E_{\mathrm{n}} = \frac{v_R^2}{2} + \frac{c_R^2}{\Gamma - 1} - \frac{GM}{R}. \tag{1.167}$$

As we see, all the flow parameters are completely defined by two thermodynamic functions n_R and c_R^2 given on the star surface.

Problem 1.19 Show that in the considered approximation for the transonic outflow regime to exist sufficiently hard conditions are to be satisfied (see, e.g., Leer and Axford, 1972):

$$(\Gamma - 1)\frac{GM}{R} < c_R^2 < \frac{GM}{2R}. \tag{1.168}$$

The right-hand side inequality corresponds to the existence condition of the transonic flow. Otherwise, the flow, starting from the very star surface, would be a supersonic one. The violation of the left-hand side inequality would imply (for $v_R \ll c_R$) the absence of the outflow from the star surface (the Bernoulli integral $E_{\rm n} < 0$).

As before, we seek the solution of the nonrelativistic GS equation (1.115) in the form

$$\Phi(r, \theta) = \Phi_0[1 - \cos\theta + \varepsilon_2^2 f(r, \theta)], \tag{1.169}$$

where the small parameter now is

$$\varepsilon_2^2 = \frac{\Omega^2 R^3}{GM}. \tag{1.170}$$

Here Ω is the angular velocity of a star. In the problem studied, all $i = 3$ invariants are to be determined. Hence, $b = 2 + 3 - 1 = 4$ so that four boundary conditions must be given on the star surface $r = r_R(\theta)$, which now differs from a sphere

$$r_R(\theta) = R[1 + \varepsilon_2^2 \rho(\theta)]. \tag{1.171}$$

In (1.171), the dimensionless parameter $\rho(\theta) \sim 1$ was introduced.

We emphasize that, at first sight, there is an obvious contradiction here. Indeed, in the example considered, we added one degree of freedom (the toroidal velocity $v_\varphi \neq 0$), whereas the problem required two additional functions as compared to the axisymmetric case. We will try to answer this question below.

It is important that for the small values of the parameter ε_2 all three integrals of motion can be determined by the real physical parameters on the star surface, i.e., by two thermodynamic functions (for example, T and n) and two velocity components (for example, v_r and v_φ). It is convenient to express them in terms of four dimensionless functions $\tau(\theta)$, $\eta(\theta)$, $\omega(\theta)$, and $h(\theta)$:

$$T(r_R,\theta) = T_R[1 + \varepsilon_2^2\tau(\theta)], \tag{1.172}$$

$$n(r_R,\theta) = n_R[1 + \varepsilon_2^2\eta(\theta)], \tag{1.173}$$

$$v_\varphi(r_R,\theta) = \varepsilon_2\left(\frac{GM}{R}\right)^{1/2}\omega(\theta)\sin\theta, \tag{1.174}$$

$$v_r(r_R,\theta) = v_R[1 + \varepsilon_2^2 h(\theta)]. \tag{1.175}$$

Here the parameter $\omega(\theta)$ defined as

$$\Omega(r_R,\theta) = \Omega\omega(\theta) \tag{1.176}$$

describes the differential rotation of the star.

Using the thermodynamic relation

$$\mathrm{d}s = \frac{1}{\Gamma-1}\frac{\mathrm{d}T}{T} - \frac{\mathrm{d}n}{n}, \tag{1.177}$$

we can obtain for the first invariant $s(\theta)$ (the constant term is dropped)

$$\delta s(\theta) = \varepsilon_2^2\left[\frac{1}{\Gamma-1}\tau(\theta) - \eta(\theta)\right]. \tag{1.178}$$

Accordingly, two other invariants can also be defined by the boundary conditions:

$$\delta E_{\mathrm{n}}(\theta) = \varepsilon_2^2 v_R^2 h(\theta) + \varepsilon_2^2\frac{GM}{2R}\omega^2(\theta)\sin^2\theta + \varepsilon_2^2\frac{\Gamma}{\Gamma-1}\frac{T}{m_\mathrm{p}}\tau(\theta) + \delta\varphi_\mathrm{g}, \tag{1.179}$$

$$L_\mathrm{n}^2(\theta) = \varepsilon_2^2 R^2\frac{GM}{R}\omega^2(\theta)\sin^4\theta. \tag{1.180}$$

When, for example, the main star mass is concentrated in its center, we can use the expression for the gravitational potential perturbation on the star surface

$$\delta\varphi_\mathrm{g}(r_R,\theta) = \varepsilon_2^2\frac{GM}{R}\rho(\theta). \tag{1.181}$$

It is very important that the possibility to perform a successive step, i.e., write the GS equation itself, is associated with the simplicity of the zero approximation. Indeed, since in the zero approximation $\Phi = \Phi_0(1-\cos\theta)$, i.e., the stream function depends only on the angle θ (and, besides, all the derivatives $\mathrm{d}E_\mathrm{n}/\mathrm{d}\Phi$, $\mathrm{d}L_\mathrm{n}/\mathrm{d}\Phi$, and $\mathrm{d}s/\mathrm{d}\Phi$, as well as L_n itself, are zero for a nonrotating star), we can use the relation

$$\mathrm{d}\Phi = \Phi_0\sin\theta\mathrm{d}\theta. \tag{1.182}$$

It allows us to determine, with adequate accuracy, the derivatives $\mathrm{d}E_\mathrm{n}/\mathrm{d}\Phi$, $\mathrm{d}L_\mathrm{n}/\mathrm{d}\Phi$, and $\mathrm{d}s/\mathrm{d}\Phi$.

As a result, the stream equation can again be linearized, while the equation for the perturbation function $\varepsilon_2^2 f(r,\theta)$ is written as

$$-\varepsilon_2^2\Phi_0^2 D\frac{\partial^2 f}{\partial r^2} - \frac{\varepsilon_2^2}{r^2}\Phi_0^2(D+1)\sin\theta\frac{\partial}{\partial\theta}\left(\frac{1}{\sin\theta}\frac{\partial f}{\partial\theta}\right) + \varepsilon_2^2\Phi_0^2 N_r\frac{\partial f}{\partial r} =$$
$$-4\pi^2 n^2 r^2\sin\theta(D+1)\frac{\mathrm{d}E_\mathrm{n}}{\mathrm{d}\theta} + 4\pi^2 n^2(D+1)\frac{L_\mathrm{n}}{\sin\theta}\frac{\mathrm{d}L_\mathrm{n}}{\mathrm{d}\theta}$$
$$-4\pi^2 n^2\frac{\cos\theta}{\sin^2\theta}L_\mathrm{n}^2 + 4\pi^2 n^2 r^2\sin\theta\left[(D+1)\frac{T}{m_\mathrm{p}} + \frac{\Gamma-1}{\Gamma}c_s^2\right]\frac{\mathrm{d}s}{\mathrm{d}\theta}. \quad (1.183)$$

Here again $D = -1 + c_\mathrm{s}^2/v^2$, $N_r = 2/r - 4\pi^2 n^2 r^2 GM/\Phi_0^2$, and we used the polytropic equation of state $P = k(s)n^\Gamma$ (1.26) and the explicit form of the function $k(s)$ (1.35) to determine the partial derivative $(\partial P/\partial s)_n$.

The properties of Eq. (1.183) are analogous to those of Eq. (1.142).

- It is linear.
- The angular operator coincides with the operator $\hat{\mathcal{L}}_\theta$ (1.120).
- Since all terms of the equation comprise the small parameter ε_2^2, the functions D, c_s, n, etc., can be taken from the zero approximation.
- Since for the spherically symmetric flow the functions D, c_s, n, etc. are independent of θ, the solution of Eq. (1.183) can be expanded in terms of the eigenfunctions of the operator $\hat{\mathcal{L}}_\theta$.

Hence, we can again seek the solution in the form

$$f(r,\theta) = \sum_{m=0}^{\infty} g_m(r)Q_m(\theta). \quad (1.184)$$

Introducing now the dimensionless variables

$$x = \frac{r}{r_*}, \quad u = \frac{n}{n_*}, \quad l = \frac{c_s^2}{c_*^2}, \quad (1.185)$$

we can write the ordinary differential equations describing the radial functions $g_m(r)$:

$$(1 - x^4 l u^2)\frac{\mathrm{d}^2 g_m}{\mathrm{d}x^2} + 2\left(\frac{1}{x} - x^2u^2\right)\frac{\mathrm{d}g_m}{\mathrm{d}x} + m(m+1)x^2 l u^2 g_m =$$
$$\kappa_m\frac{R^2}{r_*^2}x^4 l u^4 - \lambda_m\frac{R^2}{r_*^2}u^2 - \sigma_m x^6 l u^4 + \frac{1}{\Gamma}\nu_m x^6 l^2 u^4 + \frac{\Gamma-1}{\Gamma}\nu_m x^2 l u^2, \quad (1.186)$$

where κ_m, λ_m, σ_m, and ν_m are defined as the expansion coefficients:

$$\sin\theta\frac{\mathrm{d}E_{\mathrm{n}}}{\mathrm{d}\theta} = \varepsilon_2^2 c_*^2 \sum_{m=0}^{\infty} \sigma_m Q_m(\theta), \tag{1.187}$$

$$\frac{\cos\theta}{\sin^2\theta} L_{\mathrm{n}}^2 = \varepsilon_2^2 c_*^2 r_*^2 \sum_{m=0}^{\infty} \lambda_m Q_m(\theta), \tag{1.188}$$

$$\frac{L_{\mathrm{n}}}{\sin\theta}\frac{\mathrm{d}L_{\mathrm{n}}}{\mathrm{d}\theta} = \varepsilon_2^2 c_*^2 r_*^2 \sum_{m=0}^{\infty} \kappa_m Q_m(\theta), \tag{1.189}$$

$$\sin\theta\frac{\mathrm{d}s}{\mathrm{d}\theta} = \varepsilon_2^2 \sum_{m=0}^{\infty} \nu_m Q_m(\theta). \tag{1.190}$$

Finally, the functions $l(x)$ and $u(x)$ corresponding to the spherically symmetric flow for the polytropic equation of state (1.26) are connected by the relation $l = u^{\Gamma-1}$. The function $u(x)$, because of (1.111), (1.112), and (1.113) can be found from the ordinary differential equation

$$\frac{\mathrm{d}u}{\mathrm{d}x} = -2\frac{u}{x}\frac{1 - x^3u^2}{1 - x^4lu^2} \tag{1.191}$$

with the boundary conditions (cf. (1.47))

$$u(x)|_{x=1} = 1, \qquad \left.\frac{\mathrm{d}u}{\mathrm{d}x}\right|_{x=1} = -\frac{4 + \sqrt{10 - 6\Gamma}}{\Gamma + 1}. \tag{1.192}$$

As for the boundary conditions for the system of equations (1.186), they are quite analogous to the case of the Bondi–Hoyle accretion (Beskin and Pidoprygora, 1998).

1. The condition on the star surface. Since

$$\mathrm{d}\Phi = 2\pi r^2 n v_r \sin\theta \mathrm{d}\theta = 2\pi R^2 n_R v_R [1 + \varepsilon_2^2(\eta + h + 2\rho)] \sin\theta \mathrm{d}\theta, \tag{1.193}$$

we have

$$g_m(R/r_*) = \frac{(2m)!}{2^m(m+1)!m!}(\eta_m + h_m + 2\rho_m). \tag{1.194}$$

Here η_m, h_m, and ρ_m are the coefficients of the expansion in terms of the Legendre polynomials, for example, $\eta(\theta) = \sum_m \eta_m P_m(\cos\theta)$. If we specify on the star surface the meridional velocity component v_θ, using definitions (1.90) and (1.143), we have

$$n v_\theta = -\frac{\partial\Phi/\partial r}{2\pi R\sin\theta} = \varepsilon_*^2 \frac{\Phi_0}{2\pi R r_* \sin\theta} \sum_{m=0}^{\infty} g_m'|_{r=R}\, Q_m(\theta). \tag{1.195}$$

As a result, $v_\theta(R,\theta)$ defines, in fact, the derivative $g'_m = \mathrm{d}g_m/\mathrm{d}x$ on the star surface

$$r_* \frac{v_\theta(R,\theta)}{v_R} = \varepsilon_*^2 \frac{R}{\sin\theta} \sum_{m=0}^{\infty} g'_m\big|_{r=R}\, Q_m(\theta), \tag{1.196}$$

and, hence,

$$g'_m\big|_{r=R} = \frac{r_*}{\varepsilon_*^2 R v_R} (v_\theta)_m \,. \tag{1.197}$$

Here $(v_\theta)_m$ are to be determined from the condition

$$\sum_{m=0}^{\infty} (v_\theta)_m \; Q_m(\theta) = v_\theta(R,\theta)\sin\theta. \tag{1.198}$$

In particular, if the meridional convection is absent, so that $v_\theta(R,\theta) = 0$, we just have

$$g'_m\big|_{r=R} = 0. \tag{1.199}$$

As we see, in both cases, to specify the boundary condition for the radial function $g_m(R/r_*)$ we must specify on the star surface two more functions as compared to the spherically symmetric flow.

2. The absence of the singularity on the sonic surface $N_\theta = 0$. This condition yields

$$\varepsilon_2^2 g_m(1) = \frac{(2m)!}{2^m(m+1)!m!} \left[\frac{(\delta E_\mathrm{n})_m}{c_*^2} - (\delta s)_m - \frac{(L_\mathrm{n}^2/\sin^2\theta)_m}{2c_*^2 r_*^2} \right], \tag{1.200}$$

where again $(\ldots)_m$ stands for the expansion in terms of the Legendre polynomials, which can be found from relations (1.178), (1.179), and (1.180).

As a result, Eqs. (1.186) together with the boundary conditions (1.194) or (1.197) and (1.200) make it possible to solve the direct problem, i.e., determine the flow structure from the physical boundary conditions on the star surface.

It is necessary to emphasize two important circumstances here.

- We were able to formulate the regularity condition on the sonic surface $N_\theta = 0$ (and, thus solve the direct problem) again only due to the simple geometry of the zero approximation. In particular, within the approximation considered, the sonic surface location itself could be taken from the spherically symmetric solution. In the general case, the location of the singular surface is not known so that the condition $N_\theta = 0$ cannot be expressed in terms of the known functions $(\delta E_\mathrm{n})_m$, $(\delta s)_m$, etc., on the star surface.

- The occurrence of the "additional" boundary condition becomes clear now. The point is that, as was already noted, for $m = 0$ we must choose only one particular solution, viz., $g_0 = \text{const}$. The other particular solution is a nonphysical one. Hence, for $m = 0$ we have the additional relation

$$g_0(R) = g_0(r_*). \tag{1.201}$$

This condition defines h_0, which, therefore, is no longer a free parameter. In other words, we are not quite free in choosing the function $h(\theta)$ that specifies the radial velocity: its zero harmonic must be determined from the condition (1.201). However, as was already mentioned, it is the zero harmonic g_0 connected with h_0 that defines the ejection rate. Therefore, the ejection rate is a function of only three parameters, viz., two zero harmonics of the thermodynamic functions η_0 and τ_0 and the toroidal velocity v_φ. For the spherically symmetric flow $v_\varphi = 0$, and we return to two functions specifying the ejection rate. As to the higher harmonics with $m > 0$, they are completely free and we must know four functions on the star surface to specify them. Thus, the spherically symmetric flow is degenerated and extreme care must be taken in extending its properties to the two-dimensional flow.

To conclude this section, we consider a simple example of the outflow. We assume

1. the total mass of a star is concentrated in its center so that $\varphi_g = -GM/r$,
2. the absence of the differential rotation: $\omega(\theta) = 1$,
3. the von Zeipel law is valid for the temperature on the star surface: $T(R,\theta) \propto g_{\text{eff}}^{1/4}$, where $\mathbf{g}_{\text{eff}} = -\nabla\varphi_{\text{eff}}$ and $\varphi_{\text{eff}} = \varphi_g + L_n^2/\varpi^2$.
4. the absence of the meridional convection on the star surface: $v_\theta(r_R,\theta) = 0$ (this implies that we give the meridional velocity $v_\theta(r_R,\theta)$ rather than the radial velocity $v_r(r_R,\theta)$ here, so that the coefficients h_0, h_1, etc., must be found from the solution).

Problem 1.20 Show that the perturbations of the star surface $\rho(\theta)$ in (1.171) and the temperature $\tau(\theta)$ in (1.172) have the form

$$\rho(\theta) = \frac{1}{2}\sin^2\theta, \qquad \tau(\theta) = -\frac{1}{2}\sin^2\theta. \tag{1.202}$$

Problem 1.21 Show that in expansion (1.184) only the harmonics with $m = 0$ and $m = 2$ are available, while the expansion coefficients in (1.187), (1.188), (1.189), and (1.190) are defined as

$$\sigma_2 = 2\frac{r_*}{R} - \frac{5-3\Gamma}{2(\Gamma-1)} + \frac{1}{2}\frac{v_R^2}{c_*^2} - 3\frac{v_R^2}{c_*^2}h_2, \tag{1.203}$$

$$\lambda_2 = 2\frac{R}{r_*}, \quad \kappa_2 = 4\frac{R}{r_*}, \quad \nu_2 = -\frac{\Gamma}{\Gamma-1}, \tag{1.204}$$

and $\sigma_0, \ldots, \nu_0 = 0$. Recall that $h(\theta) = h_0 + h_1 \cos\theta + h_2 \mathcal{P}_2(\cos\theta) + \cdots$.

Problem 1.22 Find the form of the sonic surface and, qualitatively, the behavior of the separatrix characteristic.

As a result, having solved Eq. (1.186) for $m = 2$ and using the condition (1.201), we find the following:

1. The ejection velocity can be represented as

$$\Phi_{\rm tot} = 2\Phi_0\left[1 + \frac{\Omega^2 R^3}{GM}(1+h_0)\right]. \tag{1.205}$$

   Here h_0 can be found from relation (1.201) (see Table 1.1)

$$h_0 = -\frac{1}{6} + \frac{2}{3}\frac{\dfrac{r_*}{R} - \dfrac{R}{r_*}}{1 - \dfrac{v_R^2}{c_*^2}}. \tag{1.206}$$

   As was expected, the rotation increases the ejection rate.
2. Far from the sonic surface $r \gg r_*$ the stream function has the form

$$\lim_{r\to\infty}\frac{\Phi(r,\theta)}{\Phi_0} = (1-\cos\theta) + \frac{\Omega^2 R^3}{GM}(1+h_0)(1-\cos\theta) + \frac{\Omega^2 R^3}{GM}q_2 \sin^2\theta\cos\theta, \tag{1.207}$$

Table 1.1 Parameters of transonic outflows for different models

The model	$1+h_0$	h_2	q_2	b_0	b_2
$r_*/R = 1.1,\ \Gamma = 4/3$	2.9	−0.8	−0.40	2.2	−0.41
$r_*/R = 2.0,\ \Gamma = 4/3$	3.2	−3.3	−0.71	2.5	−0.47
$r_*/R = 10,\ \Gamma = 4/3$	8.0	−56.0	−2.18	5.7	−0.92
$r_*/R = 1.1,\ \Gamma = 1.1$	1.7	−0.8	−0.15	1.8	−0.37
$r_*/R = 2.0,\ \Gamma = 1.1$	2.1	−2.3	−0.18	2.2	−0.40
$r_*/R = 10,\ \Gamma = 1.1$	7.4	−26.0	−0.40	7.2	−0.58

where the coefficients q_2 are also given in Table 1.1. Since in expression (1.207) there is no dependence on r, we can conclude that the flow at large distances becomes radial.

3. Accordingly, the asymptotic expression for the concentration n has the form

$$\lim_{r\to\infty}\frac{n(r,\theta)}{n_*} = \frac{c_*}{v_\infty}\frac{r_*^2}{r^2}\left[1+\frac{\Omega^2R^3}{GM}b_0+\frac{1}{2}\frac{\Omega^2R^3}{GM}b_2(3\cos^2\theta-1)\right], \quad (1.208)$$

where $v_\infty^2 = v_R^2 - 2GM/R$. As seen from Table 1.1, the condition $b_2 < 0$ is satisfied for all examples considered. This implies that the rotation leads to the occurrence of a dense disk in the equatorial plane; this result is well known (Lammers and Cassinelli, 1999), but it was earlier obtained on the basis of the numerical computations only.

4. Finally, the negative values of q_2 in (1.207) show that for $\varepsilon_2 \geq 1$ the larger part of the matter flux is also concentrated in the vicinity of the equatorial plane.

Certainly, the visible difference from the spherically symmetric outflow takes place for large enough $\varepsilon_2 > 1$, when the very approach under consideration is not valid.

1.4 Axisymmetric Stationary Flows in the Vicinity of a Black Hole

1.4.1 Physical Intermezzo—(3 + 1)-Splitting in the Kerr Metric

Let us see now how the GS equation method can be used for the axisymmetric stationary flows in the vicinity of a rotating black hole. Recall that one of the main difficulties of General Relativity is the necessity to work with four-dimensional objects. As a result, we cannot often use our three-dimensional intuition when analyzing the relativistic processes.

Nevertheless, there was found a simple language—the so-called (3 + 1)-splitting—which makes it possible to work with three-dimensional values even in the framework of General Relativity (Thorne and Macdonald, 1982). A comprehensive introduction can also be found in the book "Black Holes. The Membrane Paradigm" edited by K. Thorne, D. MacDonald, and R. Price (1986). The main idea of this approach is that for the stationary metrics the proper time τ is uniquely connected with the "time at infinity" t. This fact allows one to separate the time t from the space coordinates x^i ($i = 1, 2, 3$). As a result, all equations can be rewritten in simple three-dimensional form whose physical meaning remains absolutely transparent. We give the main results of this approach below.

Let us first recall the basic relations for the Kerr metric—the metric of a rotating black hole. In the Boyer–Lindquist coordinates t, r, θ, and φ it has the form

$$ds^2 = -\alpha^2 dt^2 + g_{ik}(dx^i + \beta^i dt)(dx^k + \beta^k dt), \quad (1.209)$$

where the value

$$\alpha = \frac{\rho_K}{\Sigma}\sqrt{\Delta} \tag{1.210}$$

is a lapse function (gravitational red shift) and the vector $\boldsymbol{\beta}$ is toroidal:

$$\beta^r = \beta^\theta = 0, \qquad \beta^\varphi = -\omega. \tag{1.211}$$

Here

$$\omega = \frac{2aMr}{\Sigma^2} \tag{1.212}$$

—the so-called Lense–Thirring angular velocity (recall that β^φ is the contravariant component of the vector $\boldsymbol{\beta}$). Finally, M and a are the mass and the specific angular momentum of a black hole ($a = J/M$). Besides, we introduced the standard notation:

$$\begin{aligned} \Delta &= r^2 + a^2 - 2Mr, \qquad & \rho_K^2 &= r^2 + a^2\cos^2\theta, \\ \Sigma^2 &= (r^2 + a^2)^2 - a^2\Delta\sin^2\theta, \qquad & \varpi &= \frac{\Sigma}{\rho_K}\sin\theta. \end{aligned} \tag{1.213}$$

Further, everywhere in this section we use the units in which $c = G = 1$. Finally, it is important that the three-dimensional metric g_{ik} in (1.209) is a diagonal one

$$g_{rr} = \frac{\rho_K^2}{\Delta}, \qquad g_{\theta\theta} = \rho_K^2, \qquad g_{\varphi\varphi} = \varpi^2. \tag{1.214}$$

The Kerr metric has the following properties:

- It is axisymmetric and stationary. It is just what is needed to use the GS approach.
- The Kerr metric is a two-parameter one, i.e., it depends on two parameters: the mass M and the specific angular momentum a.
- The Kerr metric becomes the Schwarzschild metric for a nonrotating black hole: $g_{rr} = \alpha^{-2}$, $g_{\theta\theta} = r^2$, $g_{\varphi\varphi} = r^2\sin^2\theta$. Here $\alpha^2 = 1 - 2M/r$.
- At large distances $r \gg 2M$ the Boyer–Lindquist coordinates coincide with the spherical coordinates: $g_{rr} = 1$, $g_{\theta\theta} = r^2$, $g_{\varphi\varphi} = r^2\sin^2\theta$.

As we see, one of the main parameters of the Kerr metric is the lapse function α. This quantity allows us to determine the main space–time characteristics in the vicinity of a rotating black hole.

- For nonrotating black hole the lapse function α describes the delay between the proper time τ and the time at infinity t: $\mathrm{d}\tau = \alpha \mathrm{d}t$.
- The condition $\alpha = 0$ specifies the location of the event horizon, i.e., the black hole radius

$$r_{\rm g} = M + \sqrt{M^2 - a^2}. \tag{1.215}$$

- The Boyer–Lindquist coordinates do not describe the space–time inside the horizon; for $r = r_{\rm g}$ the Kerr metric has a coordinate singularity.

Let us also itemize the main properties of the Lense–Thirring angular velocity ω which is the second key parameter in the Kerr metric.

- The Lense–Thirring angular velocity ω corresponds to the proper space motion around a black hole.
- By definition, $\Omega_{\rm H} = \omega(r_{\rm g})$ is the angular velocity of the black hole rotation (it is independent of the angle θ).
- For any rotational velocities

$$\Omega_{\rm H} = \frac{a}{2Mr_{\rm g}}. \tag{1.216}$$

- For small angular velocities $\omega \propto a$.
- As seen from relation (1.215), the parameter a is bounded above: $a \leq M$.

Finally, it is convenient to introduce the special coordinate system—ZAMO (zero angular momentum observers) (Thorne and Macdonald, 1982)—which has the following properties:

- ZAMO observers are located at the constant radius r = const, θ = const, but they rotate with the Lense–Thirring angular velocity $\mathrm{d}\varphi/\mathrm{d}t = \omega$.
- For ZAMO, the four-dimensional metric $g_{\alpha\beta}$ is a diagonal one, and its three-dimensional part g_{ik} coincides with (1.214).
- In the ZAMO local experiment, there is no gyroscopic precession.
- For ZAMO, $\mathrm{d}\tau = \alpha \mathrm{d}t$ for arbitrary $\Omega_{\rm H}$.

Problem 1.23 Show that the ergosphere surface of the black hole

$$r_{\rm erg} = M + \sqrt{M^2 - a^2 \cos^2\theta} \tag{1.217}$$

(within which there is no body at rest) is given by the simple condition

$$\alpha^2 = \omega^2 \varpi^2. \tag{1.218}$$

Check it by direct substitution.

To clarify the physical meaning of α and ω we consider the motion of a particle in the gravitational field of a rotating black hole. It turns out that the four-dimensional equation of motion

$$\frac{d^2x^\alpha}{ds^2} + \Gamma^\alpha_{\beta\gamma}\frac{dx^\beta}{ds}\frac{dx^\gamma}{ds} = 0 \tag{1.219}$$

can be rewritten in the simple three-dimensional form:

$$\frac{dp_i}{d\tau} = \frac{m_p}{\sqrt{1-v^2}}g_i + H_{ik}\frac{m_p v^k}{\sqrt{1-v^2}}, \tag{1.220}$$

where

$$\mathbf{g} = -\frac{1}{\alpha}\nabla\alpha, \tag{1.221}$$

$$H_{ik} = \frac{1}{\alpha}\nabla_i\beta_k. \tag{1.222}$$

Recall that

- the Greek indices α, β, and γ stand for the four-dimensional values, whereas the Latin indices i, j, and k stand for the three-dimensional ones;
- τ is the proper time, and all the three-dimensional physical quantities are measured by ZAMO;
- ∇_i is a covariant derivative in the three-dimensional metric (1.214).

It turns out that in the weak gravitational field, i.e., far from a black hole there is a remarkable analogy between the gravitational and electromagnetic equations. Indeed, the equation of motion (1.220) can be rewritten as

$$m_p\frac{d^2\mathbf{r}}{d\tau^2} = m_p\left(\mathbf{g} + \frac{d\mathbf{r}}{d\tau}\times\mathbf{H}\right), \tag{1.223}$$

where

$$\mathbf{g} = -\nabla\alpha, \qquad \mathbf{H} = \nabla\times\boldsymbol{\beta}, \tag{1.224}$$

and, as we see, α and $\boldsymbol{\beta}$ act as scalar and vector potentials, respectively. Moreover, time-independent Einstein's equations in the weak gravitational field are also fully equivalent to Maxwell's equations (recall that $G = c = 1$, and $v \ll c$)

$$\nabla\cdot\mathbf{g} = -4\pi\rho, \tag{1.225}$$

$$\nabla\times\mathbf{g} = 0, \tag{1.226}$$

$$\nabla\cdot\mathbf{H} = 0, \tag{1.227}$$

$$\nabla\times\mathbf{H} = -16\pi\rho\mathbf{v}. \tag{1.228}$$

The difference is only in the signs of the first and last equations (the like charges in the gravitation are attracted). In other words, the gravitational field $\mathbf{g}$ is analogous to the electric field, and the new (so-called gravitomagnetic) field $\mathbf{H}$ to the magnetic field which is proportional to the angular velocity of a black hole. The source of the gravitoelectric field **g** is masses and the source of the gravitomagnetic field **H** is mass fluxes. The occurrence of an extra gravitational force is an important consequence of the motion of bodies in General Relativity.

A rotating sphere with mass M and angular momentum **J**, for example, produces the fields (Thorne et al., 1986)

$$\mathbf{g} = -\frac{M}{r^2}\mathbf{e}_{\hat{r}}, \tag{1.229}$$

$$\mathbf{H} = 2\,\frac{\mathbf{J} - 3\mathbf{e}_{\hat{r}}(\mathbf{J}\mathbf{e}_{\hat{r}})}{r^3}, \tag{1.230}$$

i.e., the ordinary radial gravitational field **g** and the dipole gravitomagnetic field **H**. Another example—an infinitely long massive cylinder at rest (which, for simplicity, is oriented along the vertical axis) produces a gravitational field

$$\mathbf{g} = -\frac{2}{\varpi}\mu_{\mathrm{g}}\mathbf{e}_{\hat{\varpi}}, \tag{1.231}$$

where μ_{g}(g/cm) is the mass per unit length. If the cylinder is now moving along its axis with the velocity $V \ll c$, in addition to the gravitoelectric field **g** the toroidal gravitomagnetic field **H** occurs, its value according to (1.228) can be determined exactly by the Biot–Savart law:

$$\mathbf{II} = -\frac{8V}{\varpi}\mu_{\mathrm{g}}\mathbf{e}_{\hat{\varphi}}. \tag{1.232}$$

The main difference from the electrodynamics is that in the gravitation theory there are particles of the same sign of charge only. Therefore, the gravitomagnetic force always is vV/c^2 times smaller than the gravitoelectric one. In electrodynamics, it may happen that a current carrying wire has no full charge, i.e., there is only a magnetic field and no electric field at all.

1.4.2 Basic Equations

Thus, the (3 + 1)-splitting allows us to describe the physical processes in a simple three-dimensional language. If, besides, we use ZAMO as basic observers, in this case, we are able to write the equations of motion in more compact form. The point is that ZAMO is the analogue of the inertial reference frame, in any event, relative to the toroidal motion. Thus, within the (3 + 1)-splitting

- all three-dimensional vectors must be defined by the local ZAMO measurements,
- all computations must be carried out in the three-dimensional diagonal metric (1.214), for example (Korn and Korn, 1968) ($g = g_{rr}g_{\theta\theta}g_{\varphi\varphi}$),

$$\nabla \cdot \mathbf{A} = \frac{1}{\sqrt{g}} \frac{\partial}{\partial x^i} \left(\sqrt{g} A^i \right), \tag{1.233}$$

$$\nabla \times \mathbf{A} = \frac{1}{\sqrt{g}} \begin{pmatrix} \sqrt{g_{rr}} \mathbf{e}_{\hat{r}} & \sqrt{g_{\theta\theta}} \mathbf{e}_{\hat{\theta}} & \sqrt{g_{\varphi\varphi}} \mathbf{e}_{\hat{\varphi}} \\ \partial/\partial r & \partial/\partial \theta & \partial/\partial \varphi \\ \sqrt{g_{rr}} A_{\hat{r}} & \sqrt{g_{\theta\theta}} A_{\hat{\theta}} & \sqrt{g_{\varphi\varphi}} A_{\hat{\varphi}} \end{pmatrix}, \tag{1.234}$$

- all vector relations remain the same as in flat space, for example, $\nabla \times (\nabla a) = 0$, $\nabla \cdot (\nabla \times \mathbf{A}) = 0$.

On the other hand, all thermodynamic functions within the $(3+1)$-splitting must be specified in the comoving coordinate system. In fact, there is only one difficulty here: in the relativistic case, we have to deal with the relativistic enthalpy μ involving the rest mass

$$\mu = \frac{\varepsilon_{\rm m} + P}{n} \approx m_{\rm p} c^2 + m_{\rm p} w + \cdots . \tag{1.235}$$

Here $\varepsilon_{\rm m}$ is the relativistic internal energy density including rest mass of particles. For the polytropic equation of state $P = k(s) n^{\Gamma}$ (1.26), we have for $c = 1$ (Shapiro and Teukolsky, 1983)

$$\mu = m_{\rm p} + \frac{\Gamma}{\Gamma - 1} k(s) n^{\Gamma - 1}, \tag{1.236}$$

$$c_{\rm s}^2 = \frac{1}{\mu} \left(\frac{\partial P}{\partial n} \right)_s = \frac{\Gamma}{\mu} k(s) n^{\Gamma - 1}. \tag{1.237}$$

Finally, the relativistic energy–momentum tensor has the same symmetric form as in flat space:

$$T^{\alpha\beta} = \begin{pmatrix} \varepsilon & \mathbf{S} \\ \mathbf{S} & T^{ik} \end{pmatrix} = \begin{pmatrix} (\varepsilon_{\rm m} + P v^2)\gamma^2 & (\varepsilon_{\rm m} + P)\gamma \mathbf{u} \\ (\varepsilon_{\rm m} + P)\gamma \mathbf{u} & (\varepsilon_{\rm m} + P) u^i u^k + P g^{ik} \end{pmatrix}. \tag{1.238}$$

Recall that γ is the Lorentz factor of the medium measured by ZAMO. Using the relativistic expression for the energy–momentum conservation law $\nabla_\alpha T^{\alpha\beta} = 0$, we get (Thorne and Macdonald, 1982)

$$-\frac{1}{\alpha^2} \nabla \cdot (\alpha^2 \mathbf{S}) + H_{ik} T^{ik} = 0, \tag{1.239}$$

$$\nabla_k T_i^k + \frac{1}{\alpha} S_\varphi \frac{\partial \omega}{\partial x^i} + (\varepsilon \delta_i^k + T_i^k) \frac{1}{\alpha} \frac{\partial \alpha}{\partial x^k} = 0. \tag{1.240}$$

The additional terms in the energy (1.239) and momentum (1.240) equations are due to the action of the gravitomagnetic force. In particular, the poloidal component of Eq. (1.240) (i.e., simply the poloidal component of the relativistic Euler equation) is written as (Frolov and Novikov, 1998)

$$nu^b \nabla_b(\mu u_a) + \nabla_a P - \mu n (u_{\hat{\varphi}})^2 \frac{1}{\varpi} \nabla_a \varpi + \frac{1}{\alpha} \mu n \gamma \varpi u_{\hat{\varphi}} \nabla_a \omega + \frac{1}{\alpha} \mu n \gamma^2 \nabla_a \alpha = 0. \quad (1.241)$$

Here the indices a and b only run from r to θ.

Problem 1.24 Show that for a Schwarzschild black hole the velocity along a circular orbit is

$$v_{\hat{\varphi}}^2 = \frac{M}{r\alpha^2}. \quad (1.242)$$

How can we explain that for $r < 3r_{\rm g}/2$ the velocity $v_{\hat{\varphi}} > 1$?

Problem 1.25 Show that for a Kerr black hole the angular velocity $\Omega = {\rm d}\varphi/{\rm d}t$ measured by a distant observer for a circular orbit in the equatorial plane has the form (Shapiro and Teukolsky, 1983)

$$\Omega = \frac{M^{1/2}}{r^{3/2} \pm a M^{1/2}}. \quad (1.243)$$

We proceed to the study of the axisymmetric stationary flows. We introduce, as in the flat space, the stream function $\Phi(r, \theta)$ through the poloidal four-velocity component $\mathbf{u}_{\rm p}$ as

$$\alpha n \mathbf{u}_{\rm p} = \frac{\nabla \Phi \times \mathbf{e}_{\hat{\varphi}}}{2\pi \varpi}. \quad (1.244)$$

Since $|\mathbf{e}_{\hat{\varphi}}| = 1$, this vector definition involves the following relations for the physical components:

$$\alpha n u_{\hat{r}} = \frac{1}{2\pi\varpi} (\nabla \Phi)_{\hat{\theta}}, \qquad \alpha n u_{\hat{\theta}} = -\frac{1}{2\pi\varpi} (\nabla \Phi)_{\hat{r}}. \quad (1.245)$$

The physical components of the gradient $\nabla \Phi$ are defined as

$$(\nabla \Phi)_{\hat{r}} = \frac{1}{\sqrt{g_{rr}}} \frac{\partial \Phi}{\partial r}, \qquad (\nabla \Phi)_{\hat{\theta}} = \frac{1}{\sqrt{g_{\theta\theta}}} \frac{\partial \Phi}{\partial \theta}, \quad (1.246)$$

because the derivatives $\partial \Phi / \partial r$ and $\partial \Phi / \partial \theta$ are the covariant components of the gradient $\nabla \Phi$.

From the definition (1.244) it also follows that the continuity equation

$$\nabla \cdot (\alpha n \mathbf{u}) = 0 \quad (1.247)$$

holds automatically. Let us clear up the occurrence of the additional factor α in Eq. (1.247). It is due to the circumstance that the three-dimensional continuity equation (1.247) is actually a consequence of the four-dimensional equation for the flux $N^{\beta} = nu^{\beta}$

$$\nabla_{\beta} N^{\beta} = \frac{1}{\sqrt{-g^{(4)}}} \frac{\partial}{\partial x^{\beta}} \left(\sqrt{-g^{(4)}} N^{\beta} \right) = 0, \tag{1.248}$$

where $g^{(4)}$ is the determinant of the four-dimensional metric $g_{\alpha\beta}$. It is easy to verify that for the metric (1.209) we have $-g^{(4)} = \alpha^2 g$, where $g = g_{rr} g_{\theta\theta} g_{\varphi\varphi}$, which results in relation (1.247).

Using now definitions (1.238) and (1.244), we can rewrite the energy equation (1.239) and the φ-component of Eq. (1.240) as

$$\mathbf{u} \cdot \nabla(\alpha\mu\gamma) + \mu u_{\varphi} \mathbf{u} \cdot \nabla\omega = 0, \tag{1.249}$$

$$\mathbf{u} \cdot \nabla(\mu u_{\varphi}) = 0. \tag{1.250}$$

Hence, two integrals of motion can be represented as

$$E(\Phi) = \alpha\mu\gamma + \mu\omega\varpi u_{\hat{\varphi}}, \tag{1.251}$$

$$L(\Phi) = \mu\varpi u_{\hat{\varphi}}. \tag{1.252}$$

Problem 1.26 Find expressions (1.249), (1.250), (1.251), and (1.252).

Problem 1.27 Show that inside the ergosphere the relativistic energy E (1.251) can be negative.

Expressions (1.251) and (1.252) are the generalization of nonrelativistic relations (1.92) and (1.96) to the case of a rotating black hole. Indeed, for $\omega = 0$ we have, for example, for the Bernoulli integral (with dimension taken into account)

$$E = \gamma\mu\alpha \approx \left(1 + \frac{1}{2}\frac{v^2}{c^2} + \cdots\right)\left(m_{\rm p}c^2 + m_{\rm p}w + \cdots\right)\left(1 - \frac{GM}{c^2 r} + \cdots\right)$$

$$\approx m_{\rm p}c^2 + m_{\rm p}\left(\frac{v^2}{2} + w + \varphi_{\rm g}\right) + \cdots . \tag{1.253}$$

As we see, the dimension of the relativistic Bernoulli integral E, as well as the invariant L, differs from the nonrelativistic values $E_{\rm n}$ and $L_{\rm n}$ by the factor $m_{\rm p}$. As to the third invariant, it is still the entropy

$$s = s(\Phi). \tag{1.254}$$

As a result, the relativistic Bernoulli's equation ($\gamma^2 = 1 + u_{\hat{\varphi}}^2 + u_{\rm p}^2$) can now be written as

$$(E - \omega L)^2 = \alpha^2 \mu^2 + \frac{\alpha^2}{\varpi^2} L^2 + \frac{\hat{\mathcal{M}}^4}{64\pi^4 \varpi^2} (\nabla \Phi)^2, \tag{1.255}$$

where we introduced the thermodynamic function

$$\hat{\mathcal{M}}^2 = \frac{4\pi \mu}{n}. \tag{1.256}$$

By analogy with (1.111) Bernoulli's equation can be rewritten in differential form as

$$\nabla_k \hat{\mathcal{M}}^2 = -\hat{\mathcal{M}}^2 \frac{N_k}{D}. \tag{1.257}$$

Here

$$N_k = -\frac{\nabla^i \Phi \cdot \nabla_i \nabla_k \Phi}{(\nabla \Phi)^2} + \frac{\nabla_k' F}{2(\nabla \Phi)^2}, \tag{1.258}$$

$$F = \frac{64\pi^4}{\hat{\mathcal{M}}^4} \left[\varpi^2 (E - \omega L)^2 - \alpha^2 L^2 - \varpi^2 \alpha^2 \mu^2 \right], \tag{1.259}$$

and the operator ∇_k' acts on all variables except for $\hat{\mathcal{M}}^2$. Recall that in (1.255) and (1.258) the relativistic enthalpy μ must be regarded as the function $\hat{\mathcal{M}}^2$ and s: $\mu = \mu(\hat{\mathcal{M}}^2, s)$. In the general case, the corresponding differential relation has the form (Beskin and Pariev, 1993)

$$\mathrm{d}\mu = -\frac{c_{\rm s}^2}{1 - c_{\rm s}^2} \mu \frac{\mathrm{d}\hat{\mathcal{M}}^2}{\hat{\mathcal{M}}^2} + \frac{1}{1 - c_{\rm s}^2} \left[\frac{1}{n} \left(\frac{\partial P}{\partial s} \right)_n + T \right] \mathrm{d}s. \tag{1.260}$$

Therefore, as in the nonrelativistic case, Bernoulli's equation (1.255) implicitly defines $\hat{\mathcal{M}}^2$ in terms of the stream function Φ and three integrals of motion: $\hat{\mathcal{M}}^2 = \hat{\mathcal{M}}^2(\nabla \Phi; E, L, s)$.

Using now the invariants E, L, and s, we can again write the poloidal component of the relativistic Euler equation (1.241) as $[\mathrm{Euler}]_{\rm p} = [\mathrm{GS}] \nabla \Phi$, where the stream equation $[\mathrm{GS}] = 0$ has the form

$$-\hat{\mathcal{M}}^2\left[\alpha\varpi^2\nabla_k\left(\frac{1}{\alpha\varpi^2}\nabla^k\Phi\right)+\frac{\nabla^i\Phi\cdot\nabla^k\Phi\cdot\nabla_i\nabla_k\Phi}{(\nabla\Phi)^2D}\right]+\frac{\hat{\mathcal{M}}^2\nabla_k' F\cdot\nabla^k\Phi}{2(\nabla\Phi)^2D}$$
$$+\frac{32\pi^4}{\hat{\mathcal{M}}^2}\frac{\partial}{\partial\Phi}\left[\varpi^2(E-\omega L)^2-\alpha^2L^2\right]-16\pi^3\alpha^2\varpi^2 nT\frac{\mathrm{d}s}{\mathrm{d}\Phi}=0. \qquad (1.261)$$

Here the operator $\partial/\partial\Phi$ acts only on the invariants $E(\Phi)$, $L(\Phi)$, and $s(\Phi)$. Finally, now the denominator D looks like

$$D=-1+\frac{1}{u_\mathrm{p}^2}\frac{c_\mathrm{s}^2}{1-c_\mathrm{s}^2}, \qquad (1.262)$$

and the physical component of the poloidal four-velocity u_p can be obtained from the definition (1.244). In compact form, the relativistic equation is

$$-\alpha\varpi^2\nabla_k\left(\frac{\hat{\mathcal{M}}^2}{\alpha\varpi^2}\nabla^k\Phi\right)+\frac{32\pi^4}{\hat{\mathcal{M}}^2}\frac{\partial}{\partial\Phi}\left[\varpi^2(E-\omega L)^2-\alpha^2L^2\right]$$
$$-16\pi^3\alpha^2\varpi^2 nT\frac{\mathrm{d}s}{\mathrm{d}\Phi}=0. \qquad (1.263)$$

The hydrodynamical version of the GS equation in the Kerr metric was first obtained in Anderson (1989) within the standard four-dimensional formalism and also in Beskin and Pariev (1993) within the (3 + 1)-splitting.

Finally, it is convenient to use the other form for the poloidal physical four-velocity u_p:

$$u_\mathrm{p}^2=\frac{(E-\omega L)^2-\alpha^2L^2/\varpi^2-\alpha^2\mu^2}{\alpha^2\mu^2}. \qquad (1.264)$$

Hence, $u_\mathrm{p}\to\infty$ as α^{-1} when approaching the horizon. As was already stressed, this behavior results from the choice of the coordinate system that has a coordinate singularity for $r=r_\mathrm{g}$. Note, finally, that relation (1.264) results in an important conclusion—in the vicinity of the black hole horizon the flow must be supersonic ($v_\mathrm{p}>c_\mathrm{s}$).

Using the definitions (1.244), (1.245), and (1.246), we easily show that when approaching the black hole horizon only the radial component of the poloidal four-velocity diverges, whereas the θ-component remains finite:

$$|u_{\hat{r}}|=O(\alpha^{-1}), \qquad |u_{\hat{\theta}}|=O(1). \qquad (1.265)$$

Accordingly, as L is finite, the toroidal four-velocity component remains finite as well:

$$|u_{\hat{\varphi}}|=O(1). \qquad (1.266)$$

The latter relation shows that in the vicinity of the horizon $|v_{\hat{\varphi}}| = O(\alpha)$ and, therefore, for a distant observer all bodies in the vicinity of the horizon rotate with the Lense–Thirring angular velocity ω. As we will see, this important property remains valid for magnetized flows as well.

Problem 1.28 Show that when approaching the horizon $v_{\hat{r}} \to -1$, $v_{\hat{\theta}} \to 0$, so that the motion from the ZAMO viewpoint becomes purely radial.

1.4.3 Exact Solutions

1.4.3.1 Spherically Symmetric Accretion

If the gas velocity at infinity is zero ($\gamma_\infty = 1$) and the thermodynamic functions are isotropic, then $E = \mu_\infty = \text{const}$ and $s = s_\infty = \text{const}$. Consequently, as in the nonrelativistic case, two thermodynamic functions specify two integrals of motion E and s. Finally, for the spherically symmetric flow we can put $L = 0$. Under these conditions the stream equation (1.261) has the trivial solution $\Phi = \Phi_0(1 - \cos\theta)$, where the accretion rate $2\Phi_0$ must be determined from the critical conditions on the sonic surface $r = r_*$.

As a result, we can obtain the expression for the radius of the sonic surface

$$r_* = \frac{M}{2}\left(\frac{1}{c_*^2} + 3\right), \tag{1.267}$$

so that for $c_*^2 \ll 1$ we return to the nonrelativistic expression (1.52). As to the value c_*^2 itself, it can be expressed in terms of the velocity of sound at infinity c_∞ from the implicit relation (Michel, 1972)

$$\left(1 - \frac{c_\infty^2}{\Gamma - 1}\right)^2 = (1 + 3c_*^2)\left(1 - \frac{c_*^2}{\Gamma - 1}\right)^2. \tag{1.268}$$

In the limit $c_\infty \ll 1$, it becomes the well-known expression (1.44). Further, $\hat{\mathcal{M}}_*^2$ and μ_* on the sonic surface are defined as

$$\hat{\mathcal{M}}_*^2 = \hat{\mathcal{M}}_\infty^2 \left(\frac{c_\infty^2}{c_*^2}\right)^{1/(\Gamma-1)} \left(\frac{\Gamma - 1 - c_*^2}{\Gamma - 1 - c_\infty^2}\right)^{(2-\Gamma)/(\Gamma-1)}, \tag{1.269}$$

$$\mu_* = \mu_\infty \frac{\Gamma - 1 - c_\infty^2}{\Gamma - 1 - c_*^2}. \tag{1.270}$$

Thus, the accretion rate can be written as $\dot{M} = 2m_{\rm p}|\Phi_{\rm cr}|$, where in the relativistic case we have

$$\Phi_{\rm cr} = -\frac{8\pi^2 r_*^2 E c_*}{\hat{\mathcal{M}}_*^2}. \tag{1.271}$$

Problem 1.29 Find expressions (1.267), (1.268), and (1.271) using the explicit form of the numerator and the denominator in (1.257) for $E = \text{const}$, $L = 0$, $s = \text{const}$, and $\omega = 0$.

Problem 1.30 Show that, with account taken of the effects of General Relativity, the transonic accretion occurs for $\Gamma = 5/3$ as well.

Finally, in the supersonic region $r \ll r_*$ we have for $\Gamma \neq 1$ and $c_* \ll 1$

$$\frac{\hat{\mathcal{M}}^2}{\hat{\mathcal{M}}_*^2} \approx 2\left(\frac{r}{r_*}\right)^{3/2}, \tag{1.272}$$

and

$$\frac{c_s^2}{c_*^2} \approx \frac{1}{2^{\Gamma-1}}\left(\frac{r}{r_*}\right)^{-3(\Gamma-1)/2}. \tag{1.273}$$

In particular, on the horizon

$$c_{\rm s}^2(r_{\rm g}) = \frac{1}{16^{\Gamma-1}} c_*^{5-3\Gamma}. \tag{1.274}$$

Hence, for $c_*^2 \approx c_\infty^2 \ll 1$ and $\Gamma < 5/3$ the sound velocity remains small ($c_{\rm s} \ll 1$) up to the horizon of a black hole.

We should point out that accretion onto black holes considerably differs from the nonrelativistic flows. The point is that in the relativistic case, all subsonic solutions existing by accretion onto ordinary stars have the singularity $v(r \to r_{\rm g}) = 0$, $n(r \to r_{\rm g}) = \infty$ on the horizon (see Fig. 1.8). Otherwise, for the subsonic stationary accretion regime to be maintained the infinite gravitational force in the vicinity of the horizon should be balanced by the infinite pressure gradient. Clearly, this regime of accretion cannot be realized. Thus, we can formulate a theorem:

Theorem 1.2 *The only physically acceptable accretion regime onto the black hole is a transonic accretion (Michel, 1972).*

As will be demonstrated below, this conclusion remains true for magnetohydrodynamic accretion as well.

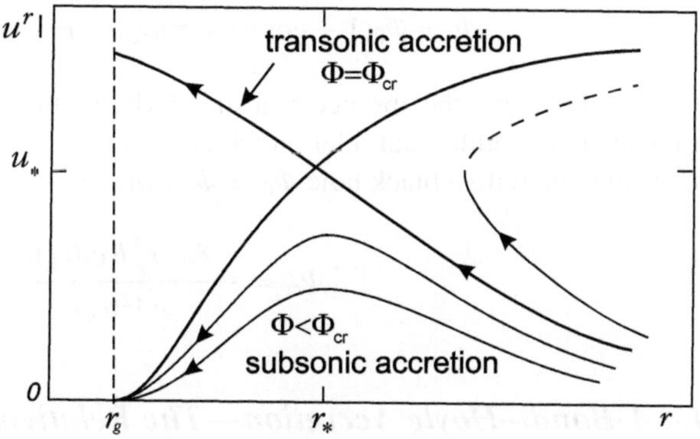

Fig. 1.8 The structure of spherically symmetric accretion onto a black hole. The transonic flow (*bold curve*) again corresponds to the critical accretion rate $\Phi = \Phi_{cr}$ (1.271). All *fine curves* below the X-point corresponding to the subsonic accretion with $\Phi < \Phi_{cr}$ have a singularity on the horizon $v_{\hat{r}} \to 0$ (i.e., $n \to \infty$)

1.4.3.2 Dust Accretion ($P = 0$)

.

For the accretion of matter with zero pressure (dust accretion) the streamlines must coincide with the trajectories of particles freely moving in the gravitational field of a rotating black hole. For the case $L = 0$ ($u_\varphi = 0$) and the zero kinetic energy at infinity $\gamma_\infty = 1$, these trajectories are "straight lines" $\theta =$ const for the arbitrary rotation parameter a (Frolov and Novikov, 1998). Moreover, for the zero pressure $P = 0$ the streamlines density can be arbitrary. Otherwise, the arbitrary function

$$\Phi = \Phi(\theta) \tag{1.275}$$

must be the solution of the GS equation. In particular, this implies that the accretion rate can be arbitrary. It is not surprising because the flow is supersonic in the whole space.

Problem 1.31 Using Bernoulli's equation (1.255) and the compact form of the relativistic stream equation (1.263), check that for $E = \mu =$ const, $L = 0$ and $s = 0$, the arbitrary function $\Phi(\theta)$ is really the solution.

1.4.3.3 Gas Accretion with $c_s = 1$

It follows from relation (1.262) that for $c_s = 1$ we have $D^{-1} = 0$. Hence, for $E =$ const, $L = 0$, and $s =$ const, the stream equation becomes linear:

$$\frac{\Delta}{\rho_K^2}\frac{\partial^2 \Phi}{\partial r^2} + \frac{\sin\theta}{\rho_K^2}\frac{\partial}{\partial\theta}\left(\frac{1}{\sin\theta}\frac{\partial\Phi}{\partial\theta}\right) = 0. \tag{1.276}$$

As a result, its solution can again be expanded in terms of the eigenfunctions of the operator $\hat{\mathcal{L}}_\theta$ (Petrich et al., 1988). Thus, for a moving black hole we find

$$\Phi = \Phi_0(1 - \cos\theta) + \pi n_\infty v_\infty (r^2 - r_g r)\sin^2\theta. \quad (1.277)$$

On the other hand, the absence of a singularity on the horizon for the concentration n results in the additional relation which, as in the Bondi accretion, again fixes the accretion rate onto a black hole $\Phi_0 = \Phi_{cr}$, where

$$\Phi_{cr} = -\frac{8\pi^2 r_g^2 E c_s(r_g)}{\hat{\mathcal{M}}^2(r_g)}. \quad (1.278)$$

1.4.4 Bondi–Hoyle Accretion—The Relativistic Regime

We first consider the accretion onto a moving black hole, i.e., the relativistic version of the Bondi–Hoyle accretion. The small parameter of the problem is again $\varepsilon_1 = v_\infty / c_\infty$. In the relativistic case, the linearized stream equation for the flux function $\Phi = \Phi_0[1 - \cos\theta + \varepsilon_1 f(r,\theta)]$ can be written as (Beskin and Pidoprygora, 1995)

$$-\varepsilon_1\alpha^2 D\frac{\partial^2 f}{\partial r^2} - \frac{\varepsilon_1}{r^2}(D+1)\sin\theta\frac{\partial}{\partial\theta}\left(\frac{1}{\sin\theta}\frac{\partial f}{\partial\theta}\right) + \varepsilon_1\alpha^2 N_r\frac{\partial f}{\partial r} = 0, \quad (1.279)$$

where now

$$N_r = \frac{2}{r} - \frac{\mu^2}{E^2 - \alpha^2\mu^2}\frac{M}{r^2}. \quad (1.280)$$

We see that this equation has the same properties as the nonrelativistic equation (1.142), viz.,

- Eq. (1.279) is linear;
- the angular operator coincides with the operator $\hat{\mathcal{L}}_\theta$ (1.120);
- because all terms of the equation contain the small parameter ε_1, the functions D, c_s, etc., can be taken from the zero approximation;
- because for the spherically symmetric flow the functions D, c_s, etc., are independent of θ, the solution of Eq. (1.279) can be expanded in terms of the eigenfunctions of the operator $\hat{\mathcal{L}}_\theta$.

However, Eq. (1.279) has another remarkable property. According to (1.262) and (1.264)

$$D + 1 = \frac{\alpha^2\mu^2}{E^2 - \alpha^2\mu^2}\frac{c_s^2}{1 - c_s^2}, \quad (1.281)$$

so that the factor α^2 is present in all terms of Eq. (1.279). Therefore, Eq. (1.279) has no singularity on the horizon. In particular, this implies that no additional boundary condition is to be specified on the horizon. It is not surprising because the horizon is associated with the supersonic region that cannot affect the flow structure at large distances from a black hole.

As a result, the solution of Eq. (1.279) can again be sought in the form

$$\Phi(r,\theta) = \Phi_0[1 - \cos\theta + \varepsilon_1 g_1(r)\sin^2\theta], \tag{1.282}$$

and the equation for the radial function $g_1(r)$ now looks like

$$-D\frac{\mathrm{d}^2 g_1}{\mathrm{d}r^2} + N_r\frac{\mathrm{d}g_1}{\mathrm{d}r} + 2\frac{\mu^2}{E^2 - \alpha^2\mu^2}\frac{c_\mathrm{s}^2}{1 - c_\mathrm{s}^2}\frac{g_1}{r^2} = 0. \tag{1.283}$$

As in the nonrelativistic case, the accretion rate does not change in the first order of ε_1.

Introducing now dimensionless variables

$$x = \frac{r}{r_*}, \quad u = \frac{\hat{\mathcal{M}}_*^2}{\hat{\mathcal{M}}^2} \approx \frac{n}{n_*}, \quad l = \frac{c_\mathrm{s}^2}{c_*^2}, \tag{1.284}$$

one can rewrite Eq. (1.283) for $c_\mathrm{s} \ll 1$ in the form

$$\left(1 - x^4 l u^2\right)\frac{\mathrm{d}^2 g_1}{\mathrm{d}x^2} + 2\left(\frac{1}{x} - x^2 u^2\right)\frac{\mathrm{d}g_1}{\mathrm{d}x} + 2x^2 l u^2\, g_1 = 0. \tag{1.285}$$

Here, according to polytropic equation of state (1.26) $l = u^{\Gamma - 1}$ and dimensionless value $u(x)$, due to (1.257) and (1.258), again satisfies Eq. (1.191)

$$\frac{\mathrm{d}u}{\mathrm{d}x} = -2\frac{u}{x}\frac{1 - x^3u^2}{1 - x^4 l u^2}, \tag{1.286}$$

but now with boundary conditions

$$u(x)|_{x=1} = 1, \qquad \left.\frac{\mathrm{d}u}{\mathrm{d}x}\right|_{x=1} = \frac{-4 + \sqrt{10 - 6\Gamma}}{\Gamma + 1}. \tag{1.287}$$

As to the boundary conditions for the equation for the radial function $g_1(r)$, they are formulated as follows:

1. The regularity condition on the sonic surface. It yields $g_1(r_*) = 0$.
2. At infinity

$$g_1(r) \to K(\Gamma)\frac{r^2}{r_*^2}, \tag{1.288}$$

where

$$K(\Gamma) = \frac{1}{2}\frac{\hat{\mathcal{M}}_*^2}{\hat{\mathcal{M}}_\infty^2}\frac{c_\infty}{c_*}. \tag{1.289}$$

As a result, we have (see Table 1.2)

$$K(\Gamma) = \frac{1}{2}\left(\frac{5-3\Gamma}{2}\right)^{(\Gamma+1)/2(\Gamma-1)}, \quad \Gamma \neq 5/3, \tag{1.290}$$

$$K(\Gamma) = \frac{9}{8}c_\infty^2, \quad \Gamma = 5/3. \tag{1.291}$$

More exactly, Table 1.2 gives the values for $c_\infty \ll 1$; in the general case, these values depend on c_∞ as well.

Equation (1.283) with boundary conditions 1 and 2 fully determines the structure of the Bondi–Hoyle accretion onto a nonrotating black hole. In particular, the sonic surface radius still looks like

$$r_*(\theta) = r_*^{(0)}\left[1 + 2\varepsilon_1\left(\frac{\Gamma+1}{D_1^2}\right)k_1(\Gamma)\cos\theta\right], \tag{1.292}$$

where now

$$\begin{aligned} D_1^2 &= 10 - 6\Gamma + 18c_*^2, \qquad \Gamma \neq 5/3, \\ D_1^2 &= 12c_\infty, \qquad \Gamma = 5/3, \end{aligned} \tag{1.293}$$

and again $k_1(\Gamma) = g_1'(r_*)r_*$. The values of $k_1(\Gamma)$ are given in Table 1.2.

Table 1.2 Parameters of the Bondi–Hoyle accretion for different polytropic indices Γ

Γ	1.01	1.1	1.2	1.333	1.5	1.6
$K(\Gamma)$	0.11	0.09	0.07	0.044	0.016	0.003
$k_1(\Gamma)$	0.66	0.56	0.46	0.31	0.12	0.026
$K_{in}(\Gamma)$	−0.16	−0.090	−0.026	0.025	0.008	0.0002

The flow structure as a whole slightly differs from the nonrelativistic case shown in Fig. 1.7. We can stress only one new property. The point is that for $r \ll r_*$ the radial function $g_1(r)$ has the asymptotic behavior

$$g_1(r) \approx K_{\rm in}(\Gamma)\left(\frac{r}{r_*}\right)^{-1/2} \tag{1.294}$$

(the values of $K_{\rm in}$ can also be found in Table 1.2). Consequently, for $\varepsilon_1 > (M/r_*)^{1/2}$ in the vicinity of a black hole (i.e., for $r < \varepsilon_1^2 K_{\rm in}^2 r_*$) the linear approximation is not valid so that in this domain it is necessary to analyze already the full nonlinear equation (1.261). Since the sign of the coefficient $K_{\rm in}(\Gamma)$ depends on the polytropic index Γ, the confluence region of the streamlines occurs either in the frontal part of a black hole for $\Gamma > 1.27$ or in its rear part for $\Gamma < 1.27$. This concentration of the streamlines corresponds, however, to the supersonic flow region that does not affect the above-constructed solution. On the other hand, the intersection of ballistic trajectories can result in additional heating which, in principle, can be detected (Shcherbakov, 2005).

1.4.5 Accretion onto a Slowly Rotating Black Hole

We now consider the accretion of a gas with zero intrinsic angular momentum $L = 0$ (i.e., $i = 2$ and $b = 2 + 2 - 1 = 3$) onto a slowly rotating black hole. In this case, the small parameter is

$$\varepsilon_3 = \frac{a}{M} \ll 1. \tag{1.295}$$

The components of the metric g_{ik} (1.214) differ from the metric of a nonrotating black hole by the values of order ε_3^2. On the other hand, we assume that the thermodynamic functions at infinity s_∞ and μ_∞ remain the same as for the spherically symmetric accretion. Therefore, we again can seek the solution of Eq. (1.261) in the form

$$\Phi(r,\theta) = \Phi_0[1 - \cos\theta + \varepsilon_3^2 f(r,\theta)], \tag{1.296}$$

where the flux Φ_0 corresponds to the unperturbed case $a = 0$. Substituting expression (1.296) in (1.261), we obtain in the first order of ε_3^2

$$-\varepsilon_3^2\alpha_0^2 D\frac{\partial^2 f}{\partial r^2} - \frac{\varepsilon_3^2}{r^2}(D+1)\sin\theta\frac{\partial}{\partial\theta}\left(\frac{1}{\sin\theta}\frac{\partial f}{\partial\theta}\right) + \varepsilon_3^2\alpha_0^2 N_r\frac{\partial f}{\partial r}$$
$$= \frac{a^2}{r^4}\left(1 - \frac{2M}{r}\right)\left(1 - 2\frac{\mu^2}{E^2 - \alpha_0^2\mu^2}\frac{M}{r}\right)\sin^2\theta\cos\theta, \tag{1.297}$$

where N_r is, as before, defined by relation (1.280), and $\alpha_0^2 = 1 - 2M/r$.

Clearly, the properties of Eq. (1.297) are quite analogous to those of (1.279). In particular, Eq. (1.297) has no singularity on the horizon. As a result, as for the ejection from a slowly rotating star, the flux $\Phi(r,\theta)$ contains only two harmonics $m=0$ and $m=2$. Therefore, the stream function $\Phi(r,\theta)$ can be represented as (Beskin and Pidoprygora, 1995)

$$\Phi(r,\theta) = \Phi_0[(1-\cos\theta) + \varepsilon_3^2 g_0(1-\cos\theta) + \varepsilon_3^2 g_2(r)\sin^2\theta\cos\theta]. \qquad (1.298)$$

The equation for the radial function $g_2(r)$ now has the form

$$-D\frac{\mathrm{d}^2 g_2}{\mathrm{d}r^2} + N_r\frac{\mathrm{d}g_2}{\mathrm{d}r} + 6\frac{\mu^2}{E^2-\alpha_0^2\mu^2}\frac{c_\mathrm{s}^2}{1-c_\mathrm{s}^2}\frac{g_2}{r^2} = \frac{M^2}{r^4}\left(1-2\frac{\mu^2}{E^2-\alpha_0^2\mu^2}\frac{M}{r}\right). \qquad (1.299)$$

Accordingly, the regularity condition on the sonic surface $N_\theta(r_*)=0$ is

$$g_2(r_*) = -\frac{1}{2}\frac{M^2}{r_*^2}\alpha_0^2(r_*). \qquad (1.300)$$

On the other hand, the condition at infinity (which is the third boundary condition besides s_∞ and μ_∞) yields $g_2(r\to\infty)=0$.

Introducing again dimensionless variables (1.284) and

$$y(x) = \frac{r_*^2}{M^2}\, g_2(x r_*), \qquad (1.301)$$

we can rewrite Eq. (1.299) for $c_\mathrm{s} \ll 1$ in the form

$$\left(1-x^4 l u^2\right)\frac{\mathrm{d}^2 y}{\mathrm{d}x^2} + 2\left(\frac{1}{x} - x^2u^2\right)\frac{\mathrm{d}y}{\mathrm{d}x} + 6x^2 l u^2 y = \frac{1}{x^4} - \frac{4u^2}{x}, \qquad (1.302)$$

with the boundary conditions $y(0)=-1/2$, $y(\infty)=0$. Analyzing now the system of equations (1.302) and (1.286), one can find that the radial function $g_2(r)$ for $r \ll r_*$ can be written as

$$g_2(r) = -G(\Gamma)\frac{M^2}{r_*^2}\left(\frac{r}{r_*}\right)^{(1-3\Gamma)/2}, \qquad (1.303)$$

where $G(\Gamma) \sim 1$ (see Table 1.3). Hence, on the horizon $g_2(r_\mathrm{g}) \sim (M/r_*)^{(5-3\Gamma)/2}$. Therefore, the perturbation of the spherically symmetric flow remains small up to the horizon of a black hole ($\varepsilon_3^2 g_2(r) \ll 1$). On the other hand, since $g_2(r) < 0$, the rotation of a black hole leads to the concentration of the streamlines in the equatorial plane. Finally, the complementary consideration similar to that for the nonrelativistic Bondi–Hoyle accretion shows that (Beskin and Pidoprygora, 1995)

Table 1.3 Parameters of the accretion onto a rotating black hole for different polytropic indices Γ

Γ	1.01	1.1	1.2	1.333	1.5	1.6
$G(\Gamma)$	9.31	4.14	1.89	0.79	0.35	0.27
$k_2(\Gamma)$	0.889	0.890	0.891	0.896	0.910	0.935
$k_3(\Gamma)$	0.22	0.23	0.24	0.24	0.22	0.17
$k_4(\Gamma)$	4.61	4.09	3.52	2.73	1.67	0.91

$$g_0 = -2\frac{M^3}{r_*^3}. \tag{1.304}$$

Hence, the black hole rotation decreases the accretion rate. However, under the normal conditions, generally, $c_\infty^2 \ll 1$ (and, therefore, $M/r_* \ll 1$), so that the effects of the black hole rotation prove extremely small.

Finally, the expression for the sonic surface radius has the form

$$r_*(\theta) = r_*^{(0)}\left\{1 + \varepsilon_3^2\frac{M^2}{D_1^2 r_*^2}\left[k_3(\Gamma) - k_4(\Gamma)\cos^2\theta\right]\right\}. \tag{1.305}$$

Here D_1^2 is again defined from relations (1.293), and the coefficients

$$k_3(\Gamma) = [1 - k_2(\Gamma)](\Gamma + 1), \tag{1.306}$$

$$k_4(\Gamma) = 10 - 6\Gamma + 3[1 - k_2(\Gamma)](\Gamma + 1) \tag{1.307}$$

depend on the derivative $k_2(\Gamma) = r_*^3 g_2'(r_*)/M^2$ (see Table 1.3). As we see, $k_2(\Gamma)$ changing in the range 0.89–0.93 is practically independent of the polytropic index Γ. The availability of the additional small parameter M^2/r_*^2 shows that the form of the sonic surface slightly differs from the spherical one even for the parameter $\varepsilon_3 \approx 1$. Nevertheless, we see that the sonic surface turns out to be flattened at the poles and extended at the equator $[r_*(0) < r_*(\pi/2)]$. This implies that the standard singular point is located at the poles and the nonstandard one in the equatorial plane. As a result, the separatrix characteristics come out from the points located at the equator, are again tangent to the sonic surface in the polar regions, and then bent in the direction of a black hole.

1.4.6 Accretion of Matter with Small Angular Momentum

We now consider the problem of matter accretion with small angular momentum onto a nonrotating black hole (Anderson, 1989; Beskin and Malyshkin, 1996). To obtain the analytical solution we, following the computing method developed earlier, believe that the angular momentum of an accreting matter L is sufficiently small so that the radial velocity $v_{\hat{r}}$ is always larger than the toroidal velocity $v_{\hat{\varphi}}$. In this case, it is logical to assume that the flow structure slightly differs from the spherically symmetric accretion.

Let us point to one important difference of this problem from the case of matter accretion with zero angular momentum. The point is that the angular momentum conservation law $L \propto r v_{\hat{\varphi}}$ shows that at large distances $v_{\hat{\varphi}} \propto r^{-1}$. On the other hand, for the spherically symmetric accretion for $r \gg r_*$, we have $v_{\hat{r}} \propto r^{-2}$. Thus, for any arbitrary small angular momentum L at large distances from the gravitational center the flow structure already greatly differs from the case of the spherically symmetric accretion. Therefore, we restrict our consideration only to the domain within some outer radius R, inside which the radial velocity $v_{\hat{r}}$ is always much larger than the toroidal velocity $v_{\hat{\varphi}}$. Incidentally, this statement of the problem is more realistic because it is logical to consider only the domain in which the velocity of turbulent pulsations in the external medium $v_{\rm turb}$ is everywhere smaller than the hydrodynamical velocity v. As we will see, the conditions on the outer boundary $r = R$ do not essentially affect the accretion structure in the supersonic region.

Thus, we consider the axisymmetric stationary accretion of an ideal gas onto the nonrotating (Schwarzschild) black hole. Following (1.64), this problem also requires four boundary conditions. These boundary conditions are, for example, two thermodynamic functions s and $c_{\rm s}$, the angular momentum L, and also one more function on the outer boundary $r = R$. Here, for simplicity, the entropy s is assumed to be the same for all the streamlines. We suppose also that far from a black hole $c_{\rm s} \ll 1$, so that $\mu \approx m_{\rm p}$. Finally, to specify the angular momentum L on the outer boundary $r = R$ we assume that the gas rotates homogeneously, i.e., $v_{\hat{\varphi}}(R,\theta) \propto \sin\theta$, $v_{\hat{\theta}}(R,\theta) = 0$. Hence,

$$L(R,\theta) = m_{\rm p} R v_{\hat{\varphi}} \sin\theta = L_0 \sin^2\theta, \tag{1.308}$$

$$E(R,\theta) = E_0 + \frac{m_{\rm p} v_{\hat{\varphi}}^2}{2} = E_0 + \frac{L_0^2}{2R^2E_0}\sin^2\theta, \tag{1.309}$$

$$v_{\hat{\theta}}\,(R,\theta) = 0. \tag{1.310}$$

As we see, the small parameter of our problem is ε_L^2, where

$$\varepsilon_L = \frac{L_0}{E_0 r_{\rm g}}. \tag{1.311}$$

We seek, as before, the solution of our problem in the form of

$$\Phi(r,\theta) = \Phi_0\left[1 - \cos\theta + \varepsilon_L^2 f(r,\theta)\right], \tag{1.312}$$

where the latter term is a perturbation of the spherically symmetric flow with the small parameter ε_L^2. If we linearize Eq. (1.261) with respect to small values ε_L^2, L^2, and $\mathrm{d}E/\mathrm{d}\theta$ and again go from the derivatives $\mathrm{d}/\mathrm{d}\Phi$ to the derivatives $\mathrm{d}/\mathrm{d}\theta$ according to expression $\mathrm{d}\Phi = \Phi_0 \sin\theta \mathrm{d}\theta$, we obtain (cf. (1.183))

$$
-\varepsilon_L^2 D\frac{\partial^2 f}{\partial r^2} - \varepsilon_L^2\frac{D+1}{\alpha^2 r^2}\sin\theta\frac{\partial}{\partial\theta}\left(\frac{1}{\sin\theta}\frac{\partial f}{\partial\theta}\right) + \varepsilon_L^2 N_r\frac{\partial f}{\partial r} =
$$

$$
-\frac{D+1}{\alpha^2}\frac{E}{E^2-\alpha^2\mu^2}\frac{\sin\theta}{r^2}\frac{\mathrm{d}E}{\mathrm{d}\theta} - \frac{1}{E^2-\alpha^2\mu^2}\frac{1}{r^4}\left(\frac{\cos\theta}{\sin^2\theta}L^2 - \frac{D+1}{\sin\theta}L\frac{\mathrm{d}L}{\mathrm{d}\theta}\right) \tag{1.313}
$$

It is clear that in this equation all values (except for L, $\mathrm{d}L/\mathrm{d}\theta$, and $\mathrm{d}E/\mathrm{d}\theta$) are to be taken from the spherically symmetric solution. In particular,

$$
D = -1 + \frac{1}{u_\mathrm{p}^2}\frac{c_\mathrm{s}^2}{1-c_\mathrm{s}^2} = -1 + \frac{\alpha_0^2\mu^2}{E_0^2-\alpha_0^2\mu^2}\frac{c_\mathrm{s}^2}{1-c_\mathrm{s}^2}, \tag{1.314}
$$

$$
N_r = \frac{2}{r} - \frac{\mu^2}{E_0^2-\alpha_0^2\mu^2}\frac{M}{r^2}, \tag{1.315}
$$

where $\alpha_0^2 = 1 - 2M/r$. We should again emphasize the fact that since $D+1 \propto \alpha_0^2$, as in the above examples, Eq. (1.313) has no singularity on the event horizon.

As a result, Eq. (1.313) can be rewritten as

$$
-D\frac{\partial^2 f}{\partial r^2} - \frac{D+1}{\alpha_0^2 r^2}\sin\theta\frac{\partial}{\partial\theta}\left(\frac{1}{\sin\theta}\frac{\partial f}{\partial\theta}\right) + N_r\frac{\partial f}{\partial r} =
$$

$$
\frac{4E_0^2}{E_0^2-\alpha_0^2\mu^2}\frac{M^2}{r^2}\left(\frac{2D+1}{r^2} - \frac{D+1}{\alpha_0^2 R^2}\right)\sin^2\theta\cos\theta. \tag{1.316}
$$

If we expand the function $f(r,\theta)$ in terms of the eigenfunctions of the operator $\hat{\mathcal{L}}_\theta$ by analogy with what was done earlier and substitute this series in Eq. (1.316), we obtain the system of ordinary differential equations for the functions $g_m(r)$:

$$
-D\frac{\mathrm{d}^2 g_m}{\mathrm{d}r^2} + N_r\frac{\mathrm{d}g_m}{\mathrm{d}r} - q_m\frac{\mu^2}{E_0^2-\alpha_0^2\mu^2}\frac{c_\mathrm{s}^2}{1-c_\mathrm{s}^2}\frac{g_m}{r^2} = 0, \qquad m \neq 2, \tag{1.317}
$$

$$
-D\frac{\mathrm{d}^2 g_2}{\mathrm{d}r^2} + N_r\frac{\mathrm{d}g_2}{\mathrm{d}r} + 6\frac{\mu^2}{E_0^2-\alpha_0^2\mu^2}\frac{c_\mathrm{s}^2}{1-c_\mathrm{s}^2}\frac{g_2}{r^2} =
$$

$$
\frac{4E_0^2}{E_0^2-\alpha_0^2\mu^2}\frac{M^2}{r^2}\left(\frac{2D+1}{r^2} - \frac{D+1}{\alpha_0^2 R^2}\right), \qquad m = 2. \tag{1.318}
$$

Here $q_m = -m(m+1)$ are the eigenvalues of the operator $\hat{\mathcal{L}}_\theta$.

As a result, as in the above examples, only two radial functions $g_0 = \mathrm{const}$ and $g_2(r)$ appear different from zero. For $R \gg r_*$, $c_* \ll 1$, we get

$$g_0 = -\frac{16}{3}c_*^2, \tag{1.319}$$

$$g_2(r_*) = \frac{8}{3}c_*^2. \tag{1.320}$$

Together with the condition on the outer boundary $u_{\hat{\theta}}(R) = 0$, which is equivalent to

$$\left.\frac{dg_2}{dr}\right|_{r=R} = 0, \tag{1.321}$$

Eqs. (1.318), (1.319), and (1.320) completely solve the problem posed. In particular, the negative value g_0 shows that the matter rotation slows down the accretion rate. Further, the stream function can be written as

$$\Phi(r,\theta) = \Phi_0\left[(1+\varepsilon_L^2 g_0)(1-\cos\theta) + \varepsilon_L^2 g_2(r)\sin^2\theta\cos\theta\right]. \tag{1.322}$$

Finally, the sonic surface radius $r_*(\theta)$ can be represented as

$$r_*(\theta) = r_*^{(0)}\left\{1 + \varepsilon_L^2 c_*^2\left[k_6(\Gamma)\cos^2\theta - k_7(\Gamma)\right]\right\}, \tag{1.323}$$

where the values

$$k_6(\Gamma) = \frac{3(\Gamma+1)}{10-6\Gamma}k_5(\Gamma) + \frac{8(3-\Gamma)}{5-3\Gamma}, \tag{1.324}$$

$$k_7(\Gamma) = \frac{\Gamma+1}{10-6\Gamma}k_5(\Gamma) + \frac{8(3-\Gamma)}{5-3\Gamma}, \tag{1.325}$$

as well as $k_5(\Gamma) = r_* g_2'(r_*)/c_*^2$, are given in Table 1.4. Clearly, in the studied case $r_* \gg r_g$ ($c_* \ll 1$), the sonic surface perturbation is extremely small even for $\varepsilon_L^2 \sim 1$. We see that the sonic surface is extended along the rotation axis ($k_6 > 0$). Therefore, in this case, the nonstandard singular points are located at the poles and the standard ones in the equatorial plane.

For further investigation, it is again convenient to introduce the dimensionless variables

$$y(x) = \frac{1}{c_*^2}g_2(xr_*), \quad x = \frac{r}{r_*}, \quad u = \frac{\hat{\mathcal{M}}_*^2}{\hat{\mathcal{M}}^2} \approx \frac{n}{n_*}, \quad l = \frac{c_s^2}{c_*^2}. \tag{1.326}$$

Table 1.4 Parameters of the accretion of matter with angular momentum for different polytropic indices Γ

Γ	1.1	1.2	1.3	1.4	1.5	1.6	1.65
$k_5(\Gamma)$	3.8	3.2	2.6	1.8	0.75	−0.80	−2.4
$k_6(\Gamma)$	16.0	17.8	20.5	24.1	26.6	40.4	57.2
$k_7(\Gamma)$	11.3	12.8	15.1	18.7	25.9	50.8	376.4

Using these variables, we can rewrite Eq. (1.318) as (cf. (1.186))

$$\left(1 - x^4 l u^2\right) \frac{\mathrm{d}^2 y}{\mathrm{d}x^2} + 2\left(\frac{1}{x} - x^2 u^2\right) \frac{\mathrm{d}y}{\mathrm{d}x} + 6x^2 l u^2 y$$
$$= -16u^2 \left[1 - 2x^4 l u^2 \left(1 - \frac{x^2}{2} \frac{r_*^2}{R^2}\right)\right], \tag{1.327}$$

with the boundary conditions

$$y(x)|_{x=1} = \frac{8}{3}; \qquad \left.\frac{\mathrm{d}y}{\mathrm{d}x}\right|_{x=R/r_*} = 0. \tag{1.328}$$

Here again $l = u^{\Gamma - 1}$, and we, for simplicity, assume that $c_* \ll 1$, so that $r_* \gg r_{\mathrm{g}}$ and $\alpha^2(r_*) = 1$.

As an illustration, let us consider in more detail the asymptotic behavior of Eq. (1.327) in the supersonic region $r < r_*$, i.e., for $x \ll 1$. According to Bernoulli's equation (1.255), we have

$$u = \frac{1}{2x^{3/2}}. \tag{1.329}$$

Then from (1.327) it follows that the function $y(x)$ must satisfy the equation

$$\frac{\mathrm{d}^2 y}{\mathrm{d}x^2} + \frac{3}{2x} \frac{\mathrm{d}y}{\mathrm{d}x} + \frac{3l}{2x} y = -\frac{4}{x^3}. \tag{1.330}$$

As a result, for small x the solution of this equation is universal (i.e., it is independent of the boundary conditions) and given by the equality

$$y(x) = -\frac{8}{x}. \tag{1.331}$$

This property is due to the circumstance that both particular solutions of the homogeneous equation increase slower than the solution of the nonhomogeneous equation. Accordingly, the stream function in the vicinity of a black hole has the form

$$\Phi(r, \theta) = \Phi_0 \left[\left(1 - \frac{16}{3} \varepsilon_L^2 c_*^2\right)(1 - \cos\theta) - 2\varepsilon_L^2 \frac{r_{\mathrm{g}}}{r} \sin^2\theta \cos\theta\right]$$
$$\approx \Phi_0 \left(1 - \cos\theta - 2\varepsilon_L^2 \frac{r_{\mathrm{g}}}{r} \sin^2\theta \cos\theta\right), \tag{1.332}$$

i.e., it is dependent on the small parameter ε_L only and independent of the conditions on the outer boundary $r = R$ and the polytropic index Γ.

Finally, the gas concentration n in the vicinity of the black hole horizon (for obtaining which we should use relation (1.264)) for the nonrelativistic temperatures $c_{\mathrm{s}} \ll 1$ can be written as

$$n = \frac{|\Phi_0|}{2\pi\sqrt{r_g r^3}}\left[1 - \frac{\varepsilon_L^2}{2}\frac{r_g}{r}\left(13\cos^2\theta - 5 + \frac{r_g}{r}\sin^2\theta\right)\right]. \tag{1.333}$$

As we see, for nonzero angular momentum of an accreting matter, the gas density in the vicinity of the equator becomes larger than the density in the vicinity of the rotation axis. Analogously, for the four-velocity components we obtain

$$u_{\hat{r}} = -\frac{1}{\sqrt{1-r_g/r}}\left(\frac{r_g}{r}\right)^{1/2}\left[1 - \frac{\varepsilon_L^2}{2}\frac{r_g}{r}\left(1 - \frac{r_g}{r}\right)\sin^2\theta\right], \tag{1.334}$$

$$u_{\hat{\theta}} = 2\varepsilon_L^2\left(\frac{r_g}{r}\right)^{3/2}\sin\theta\cos\theta, \tag{1.335}$$

$$u_{\hat{\varphi}} = \varepsilon_L\frac{r_g}{r}\sin\theta. \tag{1.336}$$

Clearly, relations (1.334), (1.335), and (1.336) can be used only when $\varepsilon_L^2 r_g/r \ll 1$.

In Fig. 1.9, the streamlines $\Phi(r,\theta) = \text{const}$ are drawn in the interval of the angles $0° < \theta < 90°$ for the distances from a black hole $r_g < r < 3r_g$ and the values of ε_L^2 equal to 0.1 and 0.3. The points indicate the streamlines for the spherically symmetric flow. As to the domain of parameters $\varepsilon_L > 1$, in this case at the distances $r \approx r_g\varepsilon_L^2$, within ideal hydrodynamics, the gas stop point $v_{\hat{r}} = 0$ must inevitably appear so that the considered approximation $v_{\hat{r}} \gg v_{\hat{\varphi}}$ is not valid.

Thus, the GS equation method allows us to find the exact solution for the case of matter accretion with small angular momentum. We are able to obtain the analytical asymptotic solution (1.331) in the vicinity of a black hole, or, more exactly, in the whole supersonic region $r \ll r_*$. The solution obtained confirms the well-known

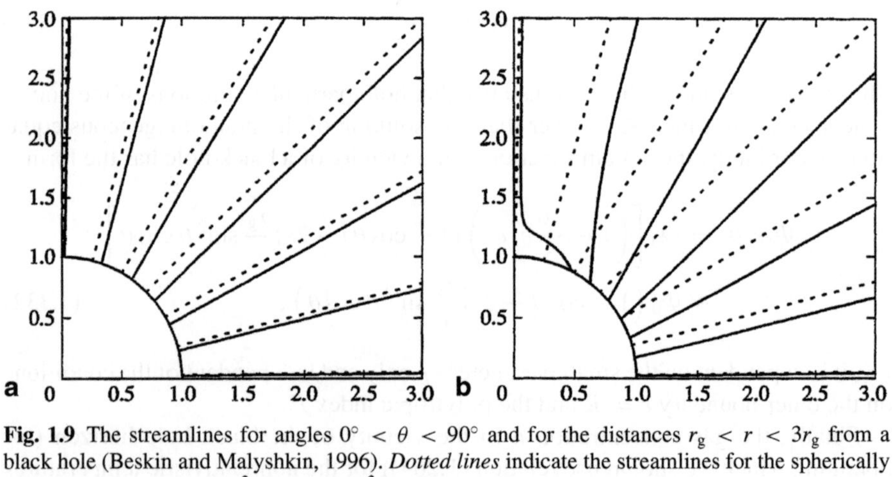

Fig. 1.9 The streamlines for angles $0° < \theta < 90°$ and for the distances $r_g < r < 3r_g$ from a black hole (Beskin and Malyshkin, 1996). *Dotted lines* indicate the streamlines for the spherically symmetric accretion. **(a)** $\varepsilon_L^2 = 0.1$. **(b)** $\varepsilon_L^2 = 0.3$

fact that, already within ideal hydrodynamics, the presence of the angular momentum in an accreting matter leads to the formation of a disk.

Certainly, for $\varepsilon_L \ll 1$, for which the solution was found in the whole space, we can only speak about the trend of its formation. On the other hand, in the domain $r > r_g \varepsilon_L^2$ ($\varepsilon_L > 1$), where the viscosity and the thermal conductivity do not play a substantial role, solutions (1.322) and (1.331) still remain valid; therefore, they can be used in the external regions of the real accretion disks in which the condition $\varepsilon_L \gg 1$ is, generally, satisfied. Note, finally, that for the nonrelativistic temperatures $c_* \ll 1$ (hence, for $r_* \gg r_g$), the obtained solution can be directly applied to the problem of accretion onto the Newtonian gravitational center, i.e., to the external regions of most other sources in which the transonic accretion regime occurs. They can include, in particular, young stellar objects and, possibly, some X-ray sources in which there is accretion onto a neutron star with a weak magnetic field.

To conclude this section, note that a similar approach was also used to construct the solution for the accretion of a gas with the nonrelativistic temperature ($c_\infty \ll 1$) and the zero intrinsic angular momentum ($L = 0$) onto a black hole rotating at an arbitrary angular velocity (Pariev, 1996). The construction of this solution becomes possible because

- for $c_\infty \ll 1$ the radius of the sonic surface is much larger than that of a black hole; therefore, the effects of the nonradial gravitational field appear small here;
- at low temperature the matter accretion onto an arbitrarily rotating black hole, as we saw, is along the radial coordinate ($\theta = \text{const}$).

As a result, the solution of the GS equation can be sought as a small perturbation to the spherically symmetric solution.

To sum up, we can say that the simple zero approximation—the spherically symmetric flow—allows us to find the analytical solution for a number of important astrophysical cases. Unfortunately, the above flows are so far the only examples of how it was possible to obtain the analytical solution to the two-dimensional flows in the vicinity of a black hole.

1.4.7 Thin Transonic Disk

We consider, as the last example, the inner two-dimensional structure of a thin accretion disk. This accretion regime is realized if the angular momentum of an accreting gas is rather large ($\varepsilon_L \gg 1$) (Shapiro and Teukolsky, 1983). Here, for simplicity, we consider only a nonrotating (Schwarzschild) black hole (Beskin and Tchekhovskoy, 2005).

According to the standard model (Shakura, 1973; Shakura and Sunyaev, 1973; Novikov and Thorne, 1973) for $\varepsilon_L \gg 1$ an accreting matter forms a disk rotating around the gravitational center with the Keplerian velocity $v_K(r) = (GM/r)^{1/2}$. The disk is thin provided that its temperature is sufficiently low ($c_s \ll v_K$), because, as is evident from the vertical balance of the gravitational force and the pressure gradient,

$H \approx r c_s / v_K$. According to the estimate $v_r / v_K \approx \alpha_{SS} c_s^2 / v_K^2$ (1.14), for $c_s \ll v_K$ the radial velocity v_r remains much smaller than both the Keplerian velocity v_K and the velocity of sound c_s.

The effects of General Relativity lead to two important properties:

- the absence of stable circular orbits for $r < r_0 = 3r_g$;
- the transonic accretion regime.

It is important that the fast gas accretion inside the last stable orbit occurs in the absence of viscosity. Note also that for the small radial four-velocity on the last stable orbit $u_0 \ll 1$, a free particle, according to (1.6), makes a great number of rotations $\sim u_0^{-1/3}$ before it reaches the black hole horizon.

Clearly, the GS equation method is not applicable in the stable orbit region, where the dissipative viscous forces play a leading role. However, as we saw, on passing the last stable orbit the viscosity effect must no longer be a crucial one. Therefore, we can suppose that the ideal hydrodynamical approximation is quite good when describing the flows in the inner regions of an accretion disk.

As was noted, in the great number of theoretical papers dealing with accretion disks the vertical averaging procedure was used, the velocity $v_{\hat{\theta}}$ in the vertical balance of forces (1.18) assumed to be zero. Therefore, the vertical component of the dynamic force $nu^b \nabla_b(\mu u_\theta)$ in (1.241) was also assumed to be small up to the horizon of a black hole. Thus, it was concluded that the disk thickness must be defined by the pressure gradient in the supersonic region (Abramowicz et al., 1997). It is shown here that the assumption $u_{\hat{\theta}} = 0$ in the vicinity of the sonic surface is not valid. As a result, as in the case of the Bondi accretion, the dynamic force can become substantial in the vicinity of the sonic surface.

We first consider the subsonic flow region in the vicinity of the last stable orbit $r_0 = 3r_g$, where the poloidal velocity u_p is even much smaller than the velocity of sound. In this case, Eq. (1.261) can be substantially simplified. Indeed, for $u_p \ll c_s$, we can drop the terms proportional to $D^{-1} \sim u_p^2/c_s^2$. As a result, we have

$$-\mathcal{M}^2 \frac{1}{\alpha} \nabla_k \left(\frac{1}{\alpha \varpi^2} \nabla^k \Phi \right) + \frac{64\pi^4}{\alpha^2 \varpi^2 \mathcal{M}^2} \left(\varpi^2 E \frac{\mathrm{d}E}{\mathrm{d}\Phi} - \alpha^2 L \frac{\mathrm{d}L}{\mathrm{d}\Phi} \right) - 16\pi^3 n T \frac{\mathrm{d}s}{\mathrm{d}\Phi} = 0. \tag{1.337}$$

This equation describing the subsonic flow is of an elliptic type.

To specify the two-dimensional subsonic flow structure ($s' = 0$, and, hence, $b = 2 + 3 - 0 = 5$) we must give five values (three velocity components and two thermodynamic functions) on some surface $r = r_0(\theta)$. It is logical to choose, as such a surface, the surface of the last stable orbit $r_0 = 3r_g$, where $\alpha(r_0) = \sqrt{2/3}$, $u_{\hat{\varphi}}(r_0) = 1/\sqrt{3}$, and $\gamma(r_0) = \sqrt{4/3}$ (Shapiro and Teukolsky, 1983). We consider, for simplicity, the case in which the radial velocity is constant on the surface $r = r_0$, and the toroidal velocity is exactly equal to $u_{\hat{\varphi}}(r_0)$:

$$u_{\hat{r}}(r_0, \Theta) = -u_0, \tag{1.338}$$

$$u_{\hat{\Theta}}(r_0, \Theta) = \Theta u_0, \tag{1.339}$$

$$u_{\hat{\varphi}}(r_0, \Theta) = 1/\sqrt{3}. \tag{1.340}$$

Here the condition for $u_{\hat{\Theta}} \ll |u_{\hat{r}}|$ corresponds to the plane parallel flow on the last stable orbit. We introduced the new angular variable $\Theta = \pi/2 - \theta$ ($\Theta_{\text{disk}} \sim c_0$) counted from the equator in a vertical direction.

Further, we assume that the velocity of sound is also constant on the whole surface $r = r_0$

$$c_s(r_0, \Theta) = c_0. \tag{1.341}$$

For the polytropic equation of state $P = kn^{\Gamma}$ (1.26), this implies that both the temperature $T_0 = T(r_0)$ and the relativistic enthalpy $\mu_0 = \mu(r_0)$ are also constant on this surface. As a result, according to (1.14), for the nonrelativistic temperatures $c_s \ll 1$ the small parameter of the problem can be defined as

$$\varepsilon_5 = \frac{u_0}{c_0} \sim \alpha_{\text{SS}} c_0 \ll 1. \tag{1.342}$$

Finally, as the last fifth boundary condition, it is convenient to choose the entropy $s(\Phi)$.

We emphasize that the availability of the small parameter $\varepsilon_5 \ll 1$ is fully based on relation (1.14) for the radial velocity of a gas flow in an accretion disk. In the vicinity of the last stable orbit this estimate is not, evidently, realistic (Igumenshchev et al., 2000; Artemova et al., 2001). Nevertheless, below, we consider the parameter ε_5 to be small because

- the availability of the small parameter allows us to analytically study the flow in more detail;
- when the small parameter is available, the discussed effect turns out to be more pronounced.

To conclude this section, we formulate the general properties that remain valid for the arbitrary radial velocity of the matter on the last stable orbit.

Introducing the values $e_0 = \alpha(r_0)\gamma(r_0) = \sqrt{8/9}$ and $l_0 = u_{\hat{\varphi}}(r_0)r_0 = \sqrt{3}r_g$, we can write the invariants $E(\Phi)$ and $L(\Phi)$ as

$$E(\Phi) = \mu_0 e_0 = \text{const}, \tag{1.343}$$
$$L(\Phi) = \mu_0 l_0 \cos\Theta_m. \tag{1.344}$$

Here $\Theta_m = \Theta_m(\Phi)$ is a latitude for which $\Phi(r_0, \Theta_m) = \Phi(r, \Theta)$. Otherwise, the function $\Theta_m(r, \Theta)$ has the meaning of the Lagrange coordinate Θ on the marginally stable orbit connected with the given point (r, Θ) by the streamline $\Phi(r, \Theta) =$ const. In particular, $\Theta_m(r_0, \Theta) = \Theta$.

We first see that the condition $E =$ const (1.343) makes it possible to rewrite Eq. (1.337) even in the simpler form

$$\frac{\partial^2 \Phi}{\partial r^2} + \frac{\cos\Theta}{\alpha^2 r^2}\frac{\partial}{\partial\Theta}\left(\frac{1}{\cos\Theta}\frac{\partial\Phi}{\partial\Theta}\right) = -4\pi^2 n^2 \frac{L}{\mu^2}\frac{\mathrm{d}L}{\mathrm{d}\Phi} - 4\pi^2 n^2 r^2 \cos^2\Theta \frac{T}{\mu}\frac{\mathrm{d}s}{\mathrm{d}\Phi}. \tag{1.345}$$

Further, as shown in Appendix A, for $r = r_0$ the right-hand side of Eq. (1.345) describes the transverse balance of forces of the pressure gradient and the gravitational potential, whereas the left-hand side corresponds to the dynamic force $(\mathbf{u}\cdot\nabla)\mathbf{u}$. On the marginally stable orbit it is of a u_0^2/c_0^2-order infinitesimal and, hence, can be dropped. Therefore, it is logical to choose the entropy $s(\Phi)$ from the condition of the transverse balance of forces on the surface $r = r_0$

$$r_0^2 \cos^2\Theta_m \frac{\mathrm{d}s}{\mathrm{d}\Theta_m} = -\frac{\Gamma}{c_0^2}\frac{L}{\mu_0^2}\frac{\mathrm{d}L}{\mathrm{d}\Theta_m}, \tag{1.346}$$

where the invariant $L(\Theta_m)$ is defined from the boundary condition (1.344). Thus, we have

$$s(\Theta_m) = s(0) - \frac{\Gamma}{3c_0^2}\ln(\cos\Theta_m). \tag{1.347}$$

In view of (1.237), this implies that for $c_\mathrm{s} = c_0 = \mathrm{const}$ relation (1.347) corresponds to the standard density profile

$$n(r_0, \Theta) \approx n_0 \exp\left(-\frac{\Gamma}{6c_0^2}\Theta^2\right). \tag{1.348}$$

Problem 1.32 Using relations (1.25) and (1.26), show that the exact expression for $n(r_0, \Theta)$ has the form $n(r_0, \Theta) = n_0(\cos\Theta)^{\Gamma/3c_0^2}$.

Finally, note that the definition (1.244) leads to the relation between the functions Φ and Θ_m

$$\mathrm{d}\Phi = 2\pi\alpha(r_0) r_0^2 n(r_0, \Theta_m) u_0 \cos\Theta_m \mathrm{d}\Theta_m. \tag{1.349}$$

Hence, because of (1.344), (1.348), and (1.349), the invariant $L(\Phi)$ can also be directly expressed in terms of the boundary conditions.

Equation (1.345) together with the boundary conditions (1.344), (1.347), (1.348), and (1.349) and also with relation (1.339) fixing the derivative $\partial\Phi/\partial r$ specifies the structure of the ideal subsonic flow immediately after passing the marginally stable orbit. For example, for the nonrelativistic temperatures $c_\mathrm{s} \ll 1$ we have

$$u_\mathrm{p}^2 = u_0^2 + w^2 + \frac{1}{3}\left(\Theta_m^2 - \Theta^2\right) + \frac{2}{\Gamma - 1}(c_0^2 - c_\mathrm{s}^2) + \cdots. \tag{1.350}$$

Here $w(r)$ defined as

$$w^2(r) = \frac{e_0^2 - \alpha^2 l_0^2/r^2 - \alpha^2}{\alpha^2} \approx \frac{1}{6}\left(\frac{r_0 - r}{r_0}\right)^3 \qquad (1.351)$$

and dependent only on the coordinate r is the poloidal four-velocity of a free particle that has the zero poloidal velocity for $r = r_0$. As we see, w^2 increases very slowly when receding from the marginally stable orbit. Therefore, for $u_0 \ll c_0$ its contribution can, generally, be disregarded.

Problem 1.33 Show that the exact expression for $w(r)$ is

$$w^2(r) = \frac{1}{9}\frac{(r_0 - r)^3}{r^2(r - r_g)}. \qquad (1.352)$$

An extremely important conclusion can be drawn directly from the analysis of relation (1.350) in which in the equatorial plane we can put $\Theta_m = \Theta = 0$. Assuming that $u_p = c_s = c_*$ and disregarding w^2, we find that the velocity of sound c_* on the sonic surface $r = r_*$, $\Theta = 0$ is to be of the same order as that on the marginally stable orbit

$$c_* \approx \sqrt{\frac{2}{\Gamma + 1}}\, c_0. \qquad (1.353)$$

It is important that this conclusion remains true for the other streamlines, because the characteristic value of the angles Θ is not larger than c_0 either. Since the entropy s remains rigorously constant along the streamlines, the gas concentration must also remain roughly constant on the surfaces $\Phi = \text{const}$ $[n(r_*, \Theta) \sim n(r_0, \Theta_m)]$. In other words, as was already shown for the Bondi accretion for $\Gamma \neq 5/3$, the subsonic flow can be regarded as an incompressible one.

On the other hand, since the density remains roughly constant and the radial velocity changes from u_0 to $c_* \sim c_0$, i.e., for $\varepsilon_5 \ll 1$ changes by several orders, because of the continuity equation the disk thickness H must change in the same proportion (see Fig. 1.10)

$$H(r_*) \approx \frac{u_0}{c_0} H(3r_g). \qquad (1.354)$$

As a result, the fast change of the disk thickness must be accompanied by the appearance of a vertical velocity component that must be taken into account in Euler equation (1.241). Recall once again that in this domain both the radial and vertical velocity components remain much smaller than the toroidal one.

Indeed, analyzing the asymptotic solution of Eq. (1.345), we find that when approaching the sonic surface located at

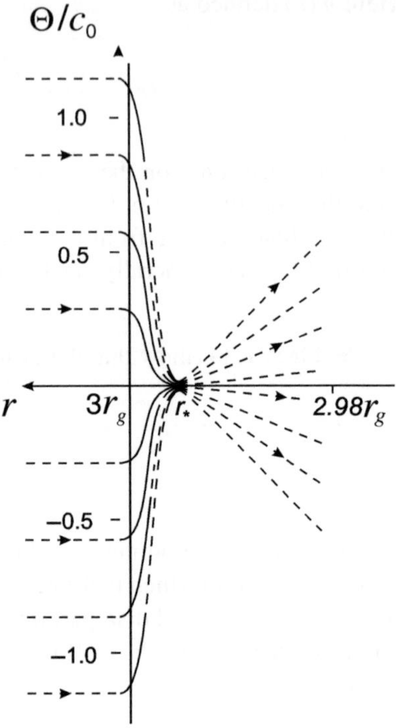

Fig. 1.10 Thin accretion disk structure (real scale) after passing the marginally stable orbit $r = 3r_g$, which results from the numerical solution of Eq. (1.345) for $c_0 = 10^{-2}$, $u_0 = 10^{-5}$. The *solid lines* correspond to the domain of parameters $u_p^2/c_0^2 < 0.2$, for which the solution does not differ from the solution of the full equation (1.261). The *dashed lines* show the extrapolation of the solution to the sonic surface region. In the vicinity of the sonic surface the flow has the form of a flat nozzle (Beskin et al., 2002)

$$r_* = r_0 - \Lambda u_0^{2/3} r_0, \tag{1.355}$$

where the logarithmic factor $\Lambda \approx (3/2)^{2/3}[\ln(c_0/u_0)]^{2/3} \approx 5$–7, the four-velocity components of the matter $\mathbf{u}$ and the pressure gradient $\nabla_{\hat{\theta}} P$ in order of magnitude tend to the values (Beskin et al., 2002)

$$u_{\hat{\Theta}} \to -\frac{c_0}{u_0}\Theta, \tag{1.356}$$

$$u_{\hat{r}} \to -c_*, \tag{1.357}$$

$$-\frac{\nabla_{\hat{\theta}} P}{\mu} \to \frac{c_0^2}{u_0^2}\frac{\Theta}{r}. \tag{1.358}$$

In the vicinity of the sonic surface the longitudinal scale δr defining the radial derivatives is of the order of the transversal disk dimension: $\delta r \approx H(r_*) \approx u_0 r_0$, so that the logarithmic derivative of the concentration is

$$\eta_1 = \frac{r}{n}\frac{\partial n}{\partial r} \approx u_0^{-1}. \tag{1.359}$$

As a result, in the vicinity of the sonic surface both the dynamic force components

$$\frac{u_{\hat{\Theta}}}{r}\frac{\partial u_{\hat{\Theta}}}{\partial \Theta} \to \frac{c_0^2}{u_0^2}\frac{\Theta}{r}, \tag{1.360}$$

$$-u_{\hat{r}}\frac{\partial u_{\hat{\Theta}}}{\partial r} \to \frac{c_0^2}{u_0^2}\frac{\Theta}{r}, \tag{1.361}$$

in order of magnitude become of the same order as the pressure gradient (1.358).

To verify the validity of our conclusion we consider the flow structure in the vicinity of the sonic surface in more detail. Using again Theorem 1.1 (the transonic flow is analytical at the critical point), we can write (cf. (1.65) and (1.66))

$$n = n_*\left(1 + \eta_1 h + \frac{1}{2}\eta_3\Theta^2 + \cdots\right), \tag{1.362}$$

$$\Theta_m = a_0\left(\Theta + a_1 h\Theta + \frac{1}{2}a_2 h^2\Theta + \frac{1}{6}b_0\Theta^3 + \cdots\right), \tag{1.363}$$

where $h = (r - r_*)/r_*$. We suppose here that three invariants E, L, and s are known beforehand, i.e., $i = 0$ and $b = 2 + 0 - 1 = 1$. Hence, as for the plane parallel flow, our problem needs one more boundary condition. If we compare the corresponding coefficients in Bernoulli's equation (1.255) and in the full GS equation (1.263), we find by disregarding the terms $\sim u_0^2/c_0^2$ and the difference r_* from r_0

$$a_0 = \left(\frac{2}{\Gamma + 1}\right)^{(\Gamma+1)/2(\Gamma-1)}\frac{c_0}{u_0}, \tag{1.364}$$

$$a_1 = 2 + \frac{1 - \alpha_*^2}{2\alpha_*^2} \approx 2.25, \tag{1.365}$$

$$a_2 = -(\Gamma + 1)\eta_1^2, \tag{1.366}$$

$$b_0 = \left(\frac{\Gamma + 1}{6}\right)\frac{a_0^2}{c_0^2}, \tag{1.367}$$

$$\eta_3 = -\frac{2}{3}(\Gamma + 1)\eta_1^2 - \left(\frac{\Gamma - 1}{3}\right)\frac{a_0^2}{c_0^2}, \tag{1.368}$$

where $\alpha_*^2 = \alpha^2(r_*) \approx 2/3$. Unlike the potential plane flow, all coefficients are expressed in terms of the radial logarithmic derivative η_1 (1.359) that acts as the last boundary condition.

The coefficients (1.364), (1.365), (1.366), (1.367), and (1.368) have a clear physical meaning. Thus, a_0 defines the degree of compression of the streamlines: $a_0 = H(r_0)/H(r_*)$. According to the estimate (1.354) we have $a_0 \approx c_0/u_0$. Further, a_1 defines the slope of the streamlines relative to the equatorial plane. Since $a_1 > 0$, in the vicinity of the sonic surface there is already an increase rather than a decrease in the angular thickness of the accretion disk. On the other hand, since $a_1 \ll u_0^{-1}$, for $r = r_0$ the divergence of the streamlines is still very weak. Consequently, for $u_0 \ll c_0$ the flow has the form of the ordinary flat nozzle (see Fig. 1.2). Finally, since $a_2 \sim \eta_3 \sim b_0 \sim u_0^{-2}$, we can conclude that the transversal scale of the inhomogeneity $\sim H(r_*)$ is really of the same order as the longitudinal scale. The last conclusion results in an extremely important consequence. As we see, the flow in the vicinity of the sonic surface is essentially a two-dimensional one so that it cannot be analyzed within the standard one-dimensional approach.

One should stress that the logarithmic derivative $\eta_1 = \eta_1(r_*)$ itself cannot be explicitly expressed in terms of the physical boundary conditions on the surface of the marginally stable orbit $r = r_0$ (it is necessary to know all expansion coefficients in (1.362) and (1.363)). In particular, it is impossible to formulate a relationship between five boundary conditions (1.338), (1.339), (1.340), (1.341), and (1.347) resulting from the critical condition on the sonic surface. Nevertheless, the estimate (1.359) allows us to determine the parameter η_1 sufficiently reliably. According to (1.364), (1.365), (1.366), (1.367), and (1.368), the relation between all the other expansion coefficients can be defined exactly.

Using now expansions (1.362) and (1.363), it is easy to define all the main characteristics of the transonic flow. In particular, we have

$$u_{\rm p}^2 = c_*^2\left[1 - 2\eta_1 h + \frac{1}{6}(\Gamma - 1)\frac{a_0^2}{c_0^2}\Theta^2 + \frac{2}{3}(\Gamma + 1)\eta_1^2\Theta^2 + \cdots\right],$$

$$c_{\rm s}^2 = c_*^2\left[1 + (\Gamma - 1)\,\eta_1 h + \frac{1}{6}(\Gamma - 1)\frac{a_0^2}{c_0^2}\Theta^2 - \frac{1}{3}(\Gamma - 1)(\Gamma + 1)\eta_1^2\Theta^2 + \cdots\right].$$

As a result, the sonic surface $u_{\rm p} = c_{\rm s}$ has the standard parabolic form

$$h = \frac{\Gamma + 1}{3}\eta_1\Theta^2. \tag{1.369}$$

Clearly, in this region the motion is mostly of an azimuth type ($u_\varphi \gg u_{\rm p}$).

In conclusion, we would like to say a few words about the supersonic region $r < r_*$. Here, as we saw, the role of the pressure gradient in the common balance of forces becomes unessential. Therefore, the θ-component in the Euler equation (1.241) can be rewritten as (cf. Abramowicz et al., 1997)

$$\alpha u_{\hat{r}}\frac{\partial(r u_{\hat{\Theta}})}{\partial r} + \frac{(r u_{\hat{\Theta}})}{r^2}\frac{\partial(r u_{\hat{\Theta}})}{\partial\Theta} + (u_{\hat{\varphi}})^2\tan\Theta = 0. \tag{1.370}$$

Using the explicit expression (1.252) for the invariant $L(\Psi)$, we can rewrite $u_{\hat{\varphi}}$ as $u_{\hat{\varphi}} = \sqrt{3}/x$, where $x = r/r_{\rm g}$. On the other hand, it is clear that in the vicinity of the equator the velocity $u_{\hat{\Theta}}$ must be the odd function Θ. Therefore, it is convenient to introduce the dimensionless functions $f(x)$ and $g(x)$

$$\Theta f(x) = x u_{\hat{\Theta}}, \tag{1.371}$$

$$g(x) = -\alpha u_{\hat{r}} > 0. \tag{1.372}$$

Thus, Eq. (1.370) can be written as a simple ordinary differential equation for the function $f(x)$

$$\frac{\mathrm{d}f}{\mathrm{d}x} = \frac{f^2+3}{x^2 g(x)}. \tag{1.373}$$

Integrating Eq. (1.373), we get

$$f(x) = \sqrt{3}\tan\left[\sqrt{3}\int_{x_*}^{x}\frac{\mathrm{d}\xi}{\xi^2 g(\xi)} + \frac{\pi}{2}\right]. \tag{1.374}$$

According to (1.350), in the vicinity of a black hole $u_{\mathrm{p}}^2 \to w^2$. On the other hand, $u_{\mathrm{p}} \approx c_* \approx c_0$ for $r \sim r_*$. Therefore, in the whole supersonic domain $r_{\mathrm{g}} < r < r_*$, with good accuracy, we can put

$$g(x) \approx \sqrt{(\alpha w)^2 + (\alpha c_*)^2}, \tag{1.375}$$

and according to (1.352),

$$(\alpha w)^2 = \frac{(3-x)^3}{9x^3}. \tag{1.376}$$

As a result, as shown in Fig. 1.11, the accretion disk in the supersonic flow region must have the quasiperiodic structure. After the expansion stage, the flow caused by the effect of the vertical component of the gravitational field begins to compress again. By definition (1.371), the locations of the nodes correspond to the condition $f(x_n) = \pm\infty$. Therefore, the node coordinates can be found from the obvious condition

$$\sqrt{3}\int_{x_n}^{x_*}\frac{\mathrm{d}\xi}{\xi^2 g(\xi)} = n\pi, \tag{1.377}$$

where the node $n = 0$ corresponds to the sonic surface. For the case $c_0 \ll 1$, we can also analytically estimate the distance $(\Delta r)_1 = r_* - r_1$ between the sonic surface

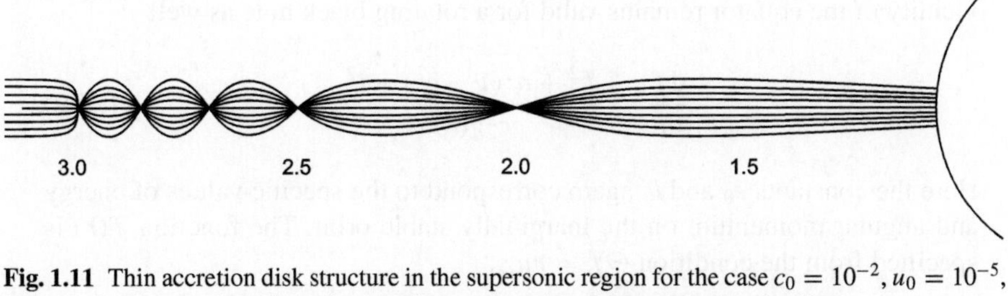

Fig. 1.11 Thin accretion disk structure in the supersonic region for the case $c_0 = 10^{-2}$, $u_0 = 10^{-5}$, and $a = 0$. The numbers show the coordinate r in units of the gravitational radius r_{g}

and the first node. Assuming that $g(x)$ remains constant and equal to its value on the sonic surface $\xi = x_*$, we find

$$(\Delta r)_1 = \frac{\pi}{\sqrt{3}} x_*^2 g(x_*) r_g \approx 3\sqrt{3}\pi \alpha_* c_* r_g \approx \frac{6\pi}{\sqrt{\Gamma + 1}} c_0 r_g. \tag{1.378}$$

We emphasize that expression (1.378) does not comprise u_0. Therefore, in the studied approximation $u_0 \ll c_0$, the locations of the nodes shown in Fig. 1.12 are defined by the parameter c_0 only. As we see, the visible vertical oscillations of the disk thickness take place for a very thin disk only: $c_0 < 10^{-2}$.

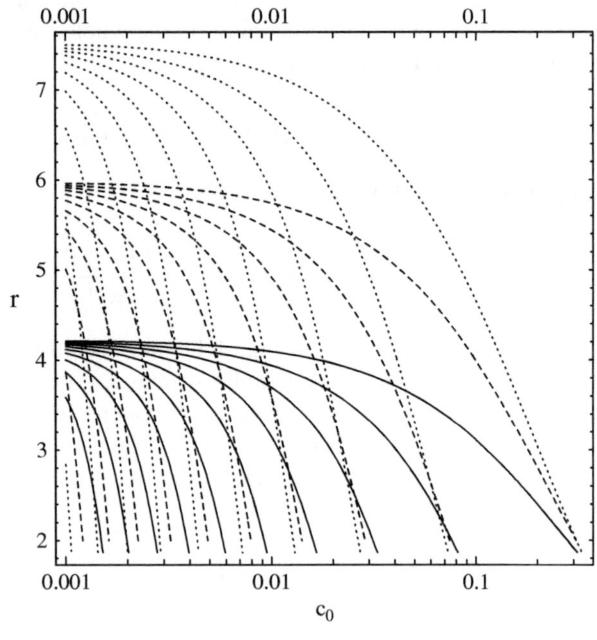

Fig. 1.12 The location of the nodes for the different values of c_0 for $a = 0$ (*dashed lines*), $a = -M/2$ (*dotted lines*), and $a = M/2$ (*solid lines*) for $\Gamma = 4/3$ (Beskin and Tchekhovskoy, 2005)

Problem 1.34 Show that the simple form of the Euler equation (1.373) in the vicinity of the equator remains valid for a rotating black hole as well

$$\frac{\mathrm{d}f}{\mathrm{d}r} = \frac{f^2 + a^2(1 - e_0^2) + l_0^2}{r^2 g(r)}. \tag{1.379}$$

Here the constants e_0 and l_0 again correspond to the specific values of energy and angular momentum on the marginally stable orbit. The function $f(r)$ is specified from the condition $\Theta f = r u_{\Theta}$.

Problem 1.35 Show that the characteristic thickness of a disk in the supersonic region is close to its thickness in the stable orbit region: $H \sim r c_s / v_K$.

Thus, the hydrodynamical GS equation (1.261) makes it possible to specify the internal structure of a thin transonic disk. As was shown, for $\varepsilon_5 = u_0/c_0 \ll 1$, an abrupt decrease in the thickness of a disk when approaching the sonic surface must inevitably give rise to the vertical velocity component of an accreting matter. As a result, we can no longer disregard the contribution of the dynamic term $(\mathbf{v} \cdot \nabla)\mathbf{v}$ to the vertical balance of forces. In this sense, the situation is quite similar to the Bondi accretion where the contribution of the dynamic term becomes substantial in the vicinity of the sonic surface and crucial in the supersonic flow region. It is clear that this property remains valid for the arbitrary radial flow velocity, i.e., even when the transversal disk compression is not very pronounced.

On the other hand, in the Bondi accretion case, the dynamic term $(\mathbf{v} \cdot \nabla)\mathbf{v}$ has only the single component $v_r \partial v_r / \partial r$ that in the vicinity of the sonic surface is of the same order as both the pressure gradient and the gravitational force. For a thin accretion disk we already have two components of the dynamic force $[(\mathbf{v} \cdot \nabla)\mathbf{v}]_\theta$, (1.360), and (1.361), and both of them on the sonic surface are, in order of magnitude, compared with the pressure gradient. However, the gravitational force is still defined as $\nabla_\theta \varphi_g \sim \Theta / r$, i.e., now it is c_0^2/u_0^2 times less than the leading terms.

As a result, for the small parameter $\varepsilon_5 \ll 1$, the transonic disk in the vicinity of the sonic surface is quite analogous to the ordinary flat nozzle in which the gravitation does not play a crucial role. For $\varepsilon_5 \sim 1$, all terms turn out to be of the same order so that the vertical oscillations become less pronounced. However, in any case, taking account of the dynamic forces gives rise to two additional degrees of freedom connected with the higher derivatives in the GS equation. In view of this, one more general conclusion independent of the value of the parameter ε_5 can be made. In a thin accretion disk, the critical condition on the sonic surface no longer fixes the accretion rate but specifies the bending of the streamlines in the vicinity of the sonic surface. Evidently, given the accretion rate, the critical condition thus gives the vertical velocity v_θ on the marginally stable orbit $r = r_0$, which slightly affects the transonic flow characteristics.

Further, regardless of ε_5, if the vertical velocity is taken into account, in the sonic surface region the small transversal scale $\delta r_\parallel \approx H_*$ inevitably occurs, which for a thin disk (i.e., under the condition $c_0 \ll 1$) proves much smaller than the distance to a black hole. Only in this case, the dynamic contribution $\mu v_r \partial v_\theta / \partial r$ may be of the same order as the pressure gradient and, hence, the flow can pass the sonic surface. In the standard one-dimensional approach, this small parameter does not occur.

Finally, it is obvious that the time it takes for the supersonic matter flux to pass among the nodes must be exactly equal to half the time of the orbital motion of a free particle on the corresponding radius, because both the spiral motion for $r < r_*$ and oscillations in the vertical plane are due to the effect of the gravitational field of a black hole. Therefore, for $c_0 \ll 1$ the flight time between the nodes must not, generally, depend on the properties of an accreting gas.

In conclusion, note that in the studied approximation of the harmonic oscillations all streamlines must intersect at the same point. Clearly, Eq. (1.373) is thus inapplicable in the node domain, where again the pressure gradient cannot be disregarded. In particular, the nodes can become domains of additional energy release, which, in turn, must result in a decrease in the oscillation amplitude in the vertical plane. It is clear, however, that the thorough analysis of all these problems is beyond the scope of our discussion (see Beskin and Tchekhovskoy 2005 for details).

1.5 Conclusion

Thus, in this chapter we formulated the basic equations describing the axisymmetric stationary hydrodynamical flows in the vicinity of a rotating black hole. Having begun with the equations for the nonrelativistic ideal flows, we gradually turned to the general case in which the effects of General Relativity were self-consistently taken into account. It was possible to construct the analytical solution for a number of important, from the astrophysical viewpoint, examples, in particular, for the transonic accretion onto rotating black holes.

There were several reasons why the purely hydrodynamical flows were studied in detail. Above all, the hydrodynamical version of the GS equation is not as popular as its full MHD version. On the other hand, it essentially contains all information concerning the full GS equation structure. Finally, already within the hydrodynamical approach, one can introduce the $(3+1)$-splitting language that is most convenient when considering the flows in the vicinity of a rotating black hole.

As a result, analysis of the hydrodynamical version made it possible to demonstrate the potentialities and the bounds of applicability of the GS equation method. As was shown, in some simplest cases this method allows us to construct the exact analytical solution of the problem. In particular, this approach is very convenient when studying the analytical properties of the transonic flows and specifying the number of the necessary boundary conditions. On the other hand, in the general case, within the GS equation method the consistent procedure for constructing the solution does not exist. The point is that the location of the critical surfaces on which the critical conditions are to be formulated is not known beforehand and is to be defined from the solution of the problem.

Chapter 2
Force-Free Approximation—The Magnetosphere of Radio Pulsars

Abstract The general view of the radio pulsar activity seems to have been established over many years. On the other hand, some fundamental problems are still to be solved. It is, first of all, the problem of the physical nature of the coherent radio emission of pulsars. In particular, as in the 1970s, there is no common view of the problem of the coherent radio emission mechanism of a maser or an antenna type. Moreover, there is no common view of the pulsar magnetosphere structure. The point is that the initial hypothesis for the magnetodipole energy loss mechanism is, undoubtedly, unrealistic. Therefore, the problem of the slowing-down mechanism can be solved only if the magnetosphere structure of neutron stars is established. However, a consistent theory of radio pulsar magnetospheres has not yet been developed. Thus, the structure of longitudinal currents circulating in the magnetosphere has not been specified and, hence, the problems of neutron star braking, particle acceleration, and energy transport beyond the light cylinder have not been solved either. The theory of the inner structure of neutron stars is also far from completion. Naturally, it is impossible to dwell on all these problems here and, therefore, we discuss in detail only the problems directly associated with the main theme of this book, viz., the theory of radio pulsar magnetospheres. The first two sections consider the basic physical processes in neutron star magnetospheres and the secondary plasma generation mechanism. Then we formulate a pulsar equation, i.e., the force-free Grad–Shafranov equation in flat space providing the correct determination of the energy losses of radio pulsars. Further, the exact analytical solutions obtained for radio pulsar magnetospheres are also discussed in detail. It is demonstrated that, within the force-free approximation, a self-consistent theory cannot be formulated. Finally, the current pulsar magnetosphere models are analyzed.

2.1 Astrophysical Introduction

It would be no exaggeration to say that the discovery of radio pulsars at the end of the 1960s—sources of cosmic pulse radio emission with characteristic period $P \sim 1$ s (Hewish et al., 1968)—can be called one of the most important events in astrophysics in the 20th century. Indeed, the new class of space sources connected with neutron stars was first discovered, the existence of which was even predicted

V.S. Beskin, *MHD Flows in Compact Astrophysical Objects*, Astronomy and Astrophysics Library, DOI 10.1007/978-3-642-01290-7_3,

© Springer-Verlag Berlin Heidelberg 2010

in the 1930s (Baade and Zwicky, 1934; Landau, 1932). Most of the other compact objects discovered later [X-ray pulsars, X-ray novae (Giacconi et al., 1971)] showed that neutron stars, even if they are not the richest ones, are really one of the most active populations in Galaxy. It is not surprising, therefore, that A. Hewish was awarded the Noble Prize for this discovery in 1974.

Neutron stars (mass M of the order of solar mass $M_{\odot} = 2 \times 10^{33}$ g with the radius R of only 10–15 km) are to evolve from the catastrophic compression (collapse) of ordinary massive stars at the later stage of their evolution or, for example, from white dwarves that exceeded, due to the accretion, the Chandrasekhar limit of mass $1.4\,M_{\odot}$. The simplest interpretation of both the small rotation periods P (the smallest known period $P = 1.39$ ms) and the superstrong magnetic fields $B_0 \sim 10^{12}$ G is based on exactly this generation mechanism (Kardashev, 1964; Pacini, 1967). Indeed, if the neutron star is supposed to evolve from a normal star (radius $R_{\rm s} \sim 10^{11}$ cm, the rotation period $P_{\rm s} \sim$ 10–100 years) with the magnetic field $B_{\rm s} \sim 1$ G, from the laws of angular momentum and magnetic flux conservation

$$M R_{\rm s}^2 \Omega_{\rm s} = M R^2 \Omega, \tag{2.1}$$

$$R_{\rm s}^2 B_{\rm s} = R^2 B_0, \tag{2.2}$$

it follows that, when compressed to the sizes R, the rotation period P and the magnetic field B_0 of the neutron star are of order

$$P \sim \left(\frac{R}{R_{\rm s}}\right)^2 P_{\rm s} \sim (0.01 - 1)\,{\rm s} \tag{2.3}$$

and

$$B_0 \sim \left(\frac{R_{\rm s}}{R}\right)^2 B_{\rm s} \sim 10^{12}\,{\rm G}. \tag{2.4}$$

It is interesting to note that the basic physical processes specifying the observed radio pulsar activity were actually identified immediately after their discovery. Thus, it was clear that the extremely regular pulsations of the observed radio emission are connected with the neutron star rotation (Gold, 1968). In some pulsars, the frequency stability on the scale of a few years is even larger than that of the atomic standards; therefore, work is underway on the development of a new pulsar timescale (Ilyasov et al., 1998). Further, the energy source of radio pulsars is due to the rotational energy, and the energy release mechanism is connected with their superstrong magnetic field $B_0 \sim 10^{12}$ G. Indeed, when estimated by the simple magnetodipole formula (Pacini, 1967), the energy losses

$$W_{\rm tot} = -I_{\rm r}\Omega\dot{\Omega} \approx \frac{1}{6}\frac{B_0^2\Omega^4 R^6}{c^3}\sin^2\chi, \tag{2.5}$$

where $I_r \sim MR^2$ is the moment of inertia of the star, χ is the inclination angle of the magnetic dipole axis to the rotation axis, and $\Omega = 2\pi/P$ is the angular velocity, amount to 10^{31}–10^{34} erg/s for most pulsars.

This energy release is just responsible for the observed slowdown $\dot{P} \sim 10^{-15}$, which corresponds to the dynamical age $\tau_D = P/2\dot{P} \sim 1$–10 mln years. The radio pulsars are thus the only space objects whose evolution is fully specified by the electrodynamic forces. Recall that the intrinsic radio emission is only 10^{-4}–10^{-6} of the total energy losses. For most pulsars, this corresponds to 10^{26}–10^{28} erg/s, which is 5–7 orders less than the luminosity of the Sun. Moreover, the extremely high brightness temperature $T_{br} \sim 10^{25}$–10^{28} K uniquely shows that the radio emission of pulsars is generated by a coherent mechanism (Ginzburg et al., 1969; Ginzburg, 1971).

As was noted, the possibility for existence of these objects has already been the subject for study since the 1930s. Moreover, since the early 1960s, the possibility of superfluidity and superconductivity in the interior regions of neutron stars has been actively discussed (see, e.g., Ginzburg and Kirzhniz 1968). Nevertheless, it was believed that because of their small size, neutron stars were actually impossible to detect. Accordingly, in spite of a number of papers (Kardashev, 1964; Pacini, 1967), before the discovery of radio pulsars it was not understood that neutron stars must rotate so fast that the main source of radiated energy is their kinetic rotational energy. As a result, no attempts were actually made to detect the pulsating radiation of the known objects. This was in spite of the fact that by that time an unusual optical star coinciding with an unusual radio source had already been detected in the Crab Nebula. The activity of this star was exactly responsible for the energy release $W_{tot} \approx 5 \times 10^{38}$ erg/s needed to supply the Crab Nebula with relativistic electrons (Rees and Gunn, 1974). Otherwise, the Crab Nebula would have ceased to glow long ago.

Only when it was clear that this unusual source is really connected with a rotating neutron star, the analysis of variability of its optical flux was made (Wampler et al., 1969). It turned out that the optical radiation also reaches us in the form of separate pulses, the period of which ($P \approx 0.033$ s) exactly coincides with the period specified by the data in the radio band. The truth was found after the rotational slowdown $\dot{P}$ of the pulsar in the Crab Nebula was measured, and it was clear that

1. the rate of the energy loss of the rotating neutron star, which was determined by the slowdown of the angular rotational velocity $W = -I_r\Omega\dot{\Omega}$, coincides with $W_{tot} \approx 5 \times 10^{38}$ erg/s;
2. the dynamical age of the radio pulsar $\tau_D = \Omega/2|\dot{\Omega}| \approx 1000$ years coincides with that of the Crab Nebula that came into existence, as is known, during the explosion of the historical supernova AD 1054.

Most radio pulsars are single neutron stars. Of over 1800 pulsars discovered by mid-2008, only about 100 of them belong to binary systems. However, in all these cases, it is known with certainty that in these binary systems there is not any substantial flux of matter from a star-companion onto the neutron star. Since, as we noted, the radio luminosity of pulsars is not high, the present-day receivers' accuracy allows one to observe pulsars only up to distances of order 3–5 kpc, which is less than the distance to the center of Galaxy. Therefore, we have the possibility to observe only a small part of all "working" radio pulsars. The total number of neutron stars in our Galaxy is 10^8–10^9. This large number of extinct neutron stars is naturally connected with their short lifetime mentioned above.

The discovery of neutron stars was, undoubtedly, an upheaval in astrophysics. Besides the emergence of new purely theoretical problems [magnetosphere structure and the coherent radio emission mechanism (Michel, 1991; Beskin et al., 1993; Lyubarskii, 1995; Mestel, 1999), the theory of accreting sources in close binary systems (Shapiro and Teukolsky, 1983; Lipunov, 1992), the theory of the inner structure and the surface layers of neutron stars (Baym and Pethick, 1979; Sedrakyan and Shakhabasyan, 1991; Liberman and Johansson, 1995; Kirzhnits and Yudin, 1995)], which gave impetus to theoretical research, the radio pulsars are used for concrete astrophysical measurements. This was possible due to the unique properties of the impulse emission of radio pulsars that make it possible, in particular, to control not only the frequency but also the signal phase. Here we can mention, for example,

- the determination of the electron density in the interstellar medium by the time delay of the arrival of pulses at different frequencies (Lyne and Graham-Smith, 1998; Johnston et al., 1999);
- the determination of the galactic magnetic field by the polarization plane rotation at different frequencies (Lyne and Graham-Smith, 1998; Brown and Taylor, 2001);
- the refined diagnostics of the GR effects in close binary systems (Taylor and Weisberg, 1989);
- the search for relic gravitational waves (Sazhin, 1978).

Thus, the general pattern of the radio pulsar activity seems to have been established over many years. On the other hand, some fundamental problems are still to be solved. It is, first of all, the problem of the physical nature of the coherent radio emission of pulsars. In particular, as in the 1970s, there is no common view of the problem of the coherent radio emission mechanism of a maser or an antenna type (Blandford, 1975; Melrose, 1978; Beskin et al., 1988; Lyubarskii, 1995; Usov and Melrose, 1996; Lyutikov et al., 1999). Besides, there is no common viewpoint on the structure of the pulsar magnetosphere (Michel, 1991; Beskin et al., 1993; Lyubarskii, 1995; Mestel, 1999). The point is that the initial hypothesis for the magnetodipole energy loss mechanism (2.5) is, undoubtedly, unrealistic. Strictly speaking, this chapter primarily deals with the proof of this assertion. We only stress here that low-frequency waves with frequency $\nu = 1/P$ cannot propagate in the interstellar medium for which the plasma frequency is, on average, several kilohertz (Lipunov, 1992). Therefore, the problem of the slowing-down mechanism can be solved only by determining the magnetosphere structure of the neutron star. However, the consistent theory of the radio pulsar magnetosphere has not been constructed yet. Thus, the structure of the longitudinal currents circulating in the magnetosphere is not specified and, hence, the problem of the neutron star braking, particle acceleration, and energy transport beyond the light cylinder still remains unsolved. The theory of the inner structure of neutron stars is also far from completion. Naturally, it seems impossible to discuss all these problems here. Therefore, we discuss in detail only the problems directly connected with the main theme of this book, viz., the theory of the pulsar magnetosphere. The main problems to be discussed are the following:

1. the magnetosphere structure of a rotating neutron star;
2. the determination of the energy loss mechanism of radio pulsars;
3. the energy transport from the rotating neutron star within the magnetosphere; and
4. the determination of the particle acceleration mechanism in the pulsar wind.

2.2 Basic Physical Processes

2.2.1 Vacuum Approximation

Before proceeding to the discussion of the consistent theory of radio pulsars, we consider the basic physical processes taking place in the magnetosphere. We should make a reservation that in this chapter we do not actually discuss the GR effects, the exception is one of the particle generation mechanisms. Though the GR effects on the neutron star surface can amount to 20% (Kim et al., 2005), they are not, generally, taken into account in the development of the pulsar magnetosphere theory. The point is that the electromagnetic force $F_{\rm em} \sim eE$ acting on a charged particle near the neutron star surface turns out to be many orders greater than the gravitational force $F_{\rm g} = GMm/R^2$. This condition allows us to disregard the electromagnetic field distortion connected with the space curvature in the vicinity of the neutron star.

We first discuss the simplest vacuum model which, even if very far from reality, gives an insight into the key properties of the real magnetosphere of the neutron star. Thus, we consider a homogeneous magnetized star rotating in vacuum. The basic parameters defining the properties of the magnetosphere are the magnetic field B_0, the star radius R, and the angular rotational velocity Ω. For a well-conducting star, we find that in its interior

$$\mathbf{E}_{\rm in} + \frac{\boldsymbol{\Omega} \times \mathbf{r}}{c} \times \mathbf{B}_{\rm in} = 0. \tag{2.6}$$

In this chapter, we, as usual, restore the dimension. The condition (2.6) simply implies that the electric field in the coordinate system rotating with the star is zero: $\mathbf{E}' = 0$.

Suppose now that the star rotation axis is parallel to the magnetization axis. Then the problem is stationary and, therefore, the electric field is fully defined by the potential $\Phi_{\rm e}$ ($\mathbf{E} = -\nabla\Phi_{\rm e}$), which inside the star can be written as

$$\Phi_e(r < R, \theta) = \frac{1}{2}\frac{\Omega B_0}{c} r^2 \sin^2\theta. \tag{2.7}$$

Hence, on the star surface

$$\Phi_e(R, \theta) = \Phi_0(\theta) = -\frac{1}{3}\frac{\Omega B_0}{c} R^2 \mathcal{P}_2(\cos\theta) + \text{const}, \tag{2.8}$$

where $\mathcal{P}_2(x) = (3x^2 - 1)/2$ is the Legendre polynomial. The electric potential beyond the star can be found from the solution of the Laplace equation $\nabla^2\Phi_e = 0$ with the boundary conditions

1. $\Phi_e(R, \theta) = \Phi_0(\theta)$;
2. $\Phi_e(r, \theta) \to 0$ for $r \to \infty$.

The solution corresponding to the zero total electric charge of the star has the form

$$\Phi_e(r > R, \theta) = -\frac{1}{3}\frac{\Omega B_0}{c}\frac{R^5}{r^3}\mathcal{P}_2(\cos\theta). \tag{2.9}$$

As shown in Fig. 2.1, the rotation of homogeneously magnetized star gives rise to a quadrupole electric field beyond it. As to the magnetic field, for an axisymmetric rotator, it is exactly equal to the dipole magnetic field

$$\mathbf{B}(r > R) = \frac{3(\mathbf{mn})\mathbf{n} - \mathbf{m}}{r^3}, \tag{2.10}$$

where $\mathbf{n} = \mathbf{r}/r$, and $|\mathbf{m}| = B_0 R^3/2$ is the star magnetic moment.

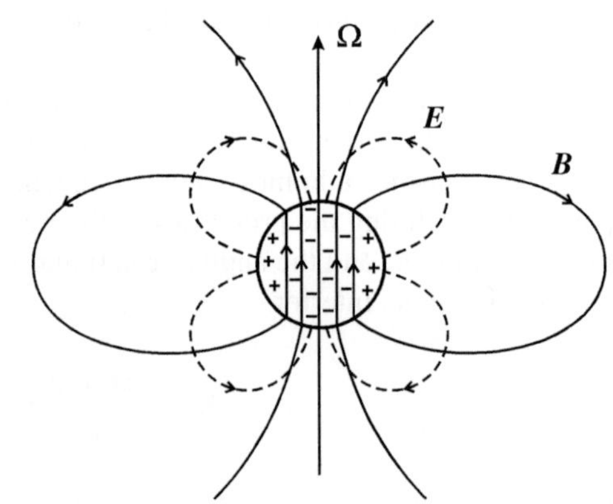

Fig. 2.1 The structure of the axisymmetric vacuum magnetosphere of the neutron star. The rotating homogeneously magnetized star generates the dipole magnetic field (*solid lines*) and the quadrupole electric field **E** (*dashed lines*)

Problem 2.1 Show that the surface charge density σ_s defined by the jump of the normal electric field component $4\pi\sigma_s = \{E_n\}$ has the form (Mestel, 1971)

$$\sigma_s(\theta) = \frac{1}{8\pi}\frac{\Omega R}{c}B_0(3 - 5\cos^2\theta). \tag{2.11}$$

Explain why the total surfacc charge is different from zero

$$Q_* = \int \sigma_s(\theta)\mathrm{d}o = \frac{2}{3}\frac{\Omega B_0}{c}R^3 \neq 0. \tag{2.12}$$

Using the simplest vacuum model, we can make a number of general conclusions.

- The longitudinal electric field $E_{\parallel} = (\mathbf{E}\cdot\mathbf{B})/B$ in the vicinity of the star surface can be estimated as

$$E_{\parallel} \sim \frac{\Omega R}{c}B_0. \tag{2.13}$$

- In the axisymmetric case (and for the zero total electric charge), the sign of the product $(\mathbf{E}\cdot\mathbf{B})(\mathbf{B}\cdot\mathbf{n})$ remains the same over the neutron star surface.

The latter conclusion is very important. The particles in the strong magnetic field can move along the magnetic field only (see below). This implies that for the axisymmetric rotator, particles of the same sign are ejected from both magnetic poles of the neutron star. As we will see, this important property retains in the case of the plasma-filled magnetosphere.

For an arbitrary inclination angle χ, the problem was solved by Deutsch (1955) long before the discovery of pulsars. In this case, the electromagnetic fields are a sum of the fields of the rotating magnetic dipole and the electric quadrupole, and the quadrupole moment can be represented as

$$Q_{ik} = \frac{R^2}{c}\left[m_i\Omega_k + m_k\Omega_i - \frac{2}{3}(\mathbf{m}\cdot\boldsymbol{\Omega})\delta_{ik}\right]. \tag{2.14}$$

The electromagnetic fields for the arbitrary distance r in the limit $R \to 0$ for $\chi = 90°$ are described by the known expressions (Landau and Lifshits, 1989)

$$B_r = \frac{|\mathbf{m}|}{r^3}\sin\theta\,\mathrm{Re}\left(2 - 2\mathrm{i}\frac{\Omega r}{c}\right)\exp\left(\mathrm{i}\frac{\Omega r}{c} + \mathrm{i}\varphi - \mathrm{i}\Omega t\right), \tag{2.15}$$

$$B_\theta = \frac{|\mathbf{m}|}{r^3}\cos\theta\,\mathrm{Re}\left(-1 + \mathrm{i}\frac{\Omega r}{c} + \frac{\Omega^2 r^2}{c^2}\right)\exp\left(\mathrm{i}\frac{\Omega r}{c} + \mathrm{i}\varphi - \mathrm{i}\Omega t\right), \tag{2.16}$$

$$B_\varphi = \frac{|\mathbf{m}|}{r^3}\mathrm{Re}\left(-\mathrm{i} - \frac{\Omega r}{c} + \mathrm{i}\frac{\Omega^2 r^2}{c^2}\right)\exp\left(\mathrm{i}\frac{\Omega r}{c} + \mathrm{i}\varphi - \mathrm{i}\Omega t\right), \tag{2.17}$$

$$E_r = E_r^Q, \tag{2.18}$$

$$E_\theta = \frac{|\mathbf{m}|\Omega}{r^2 c}\,\mathrm{Re}\left(-1 + \mathrm{i}\frac{\Omega r}{c}\right)\exp\left(\mathrm{i}\frac{\Omega r}{c} + \mathrm{i}\varphi - \mathrm{i}\Omega t\right) + E_\theta^Q, \tag{2.19}$$

$$E_\varphi = \frac{|\mathbf{m}|\Omega}{r^2 c}\cos\theta\,\mathrm{Re}\left(-\mathrm{i} - \frac{\Omega r}{c}\right)\exp\left(\mathrm{i}\frac{\Omega r}{c} + \mathrm{i}\varphi - \mathrm{i}\Omega t\right) + E_\varphi^Q. \tag{2.20}$$

Here $\mathbf{E}^Q$ is the quadrupole static electric field

$$\mathbf{E}^Q = -\nabla\Phi_{\mathrm{e}}^Q, \quad \Phi_{\mathrm{e}}^Q = \frac{Q_{ik} n_i n_k}{2r^3}. \tag{2.21}$$

At distances much smaller than the wavelength $r \ll c/\Omega$, the electromagnetic fields are close to the sum of the fields of the magnetic dipole and the electric quadrupole at rest, and at large distances $r \gg c/\Omega$, they correspond to a spherical wave. Since, according to (2.13), the quadrupole electric field on the star surface is much smaller than the magnetic field and, on the other hand, the quadrupole electric field decreases with distance faster than the dipole magnetic field, the electric quadrupole does not make a real contribution to the energy loss of the rotating star. Consequently, the energy losses are determined, with adequate accuracy, by the standard expression (2.5). Therefore, we restrict ourselves in (2.18), (2.19), and (2.20) to the static part of the electric quadrupole field only.

One should stress here that the magnetodipole radiation turned out to result in the change of not only the rotation period $P = 2\pi/\Omega$ but also the evolution of the inclination angle χ, since, for the magnetodipole losses the invariant $\mathcal{I}_{\mathrm{md}}$ remains constant (Davis and Goldstein, 1970)

$$\mathcal{I}_{\mathrm{md}} = \Omega\cos\chi. \tag{2.22}$$

Hence, for the magnetodipole losses, the inclination angle of the rotating magnetized star must decrease with the characteristic time τ_χ coinciding with the dynamical lifetime $\tau_{\mathrm{D}} = P/2\dot{P}$. As a result, a decrease in the energy release is due not only to an increase in the rotation period but also to a decrease in the inclination angle χ.

Unfortunately, the only direct observational channel permitting us to judge the radio pulsar energy release mechanism is the so-called braking index

$$n_{\mathrm{br}} = \frac{\ddot{\Omega}\Omega}{\dot{\Omega}^2} = 2 - \frac{\ddot{P}P}{\dot{P}^2}, \tag{2.23}$$

which, as is easily checked, coincides with the exponent in the slowing-down dependence on the angular velocity, viz., $\dot{\Omega} \propto \Omega^{n_{\mathrm{br}}}$. As we see, to determine the braking index, we must know the second derivative of the period $\ddot{P}$. However, for most radio pulsars, we fail to identify the second derivative of the noise background associated with faster (than the slowing-down time) variations of the rotation period of the neutron star (Johnston and Galloway, 1999). Therefore, it is possible to determine the breaking index only for the fastest radio pulsars. As seen from Table 2.1, in all

Table 2.1 Braking index n_{br} for fast radio pulsars

PSR	P (s)	$\dot{P}(10^{-15})$	n_{br}
$B0531+21$	0.033	421	2.51 ± 0.01
$B0540-693$	0.050	479	2.14 ± 0.01
$J1119-6127$	0.408	4022	2.91 ± 0.05
$B1509-58$	0.150	1490	2.84 ± 0.01
$J1846-0258$	0.324	7083	2.65 ± 0.01

cases, the braking index is less than 3, whereas the dipole slowing-down law (2.5) yields $n_{br} = 3$.

Problem 2.2 Show that in a more realistic model taking into account the evolution of the inclination angle χ (2.22), the braking index is even larger than 3 (Davis and Goldstein, 1970)

$$n_{br} = 3 + 2\cot^2\chi . \tag{2.24}$$

Problem 2.3 Integrate the evolution equation (2.5), with account taken of the integral of motion (2.22), and show that the period of the pulsar $P(t)$ exponentially fast (with characteristic time $\tau_D = P_0/2\dot{P}_0$) approaches the maximum value of $P_{max} = P_0/\cos\chi_0$ and the angle χ approaches 0°.

Thus, we can conclude from the analysis of the braking index that the simple magnetodipole mechanism cannot, evidently, be responsible for the observed slowing down of the radio pulsar rotation. Therefore, there were numerous attempts to correct relation (2.24) for example, by the magnetic field evolution (Blandford and Romani, 1988; Chen et al., 1998) or the interaction of the superfluid component in the neutron star nucleus with its hard crust (Allen and Horvath, 1997; Baykal et al., 1999) (see also Melatos, 1997; Xu and Qiao, 2001). It turned out, however, that most of the similar effects can lead to insignificant corrections only and cannot change the value appreciably (2.24). In any event, the determination of the braking index of other neutron stars and also the second-order braking index $n_{br}^{(2)} = \Omega^2\dddot{\Omega}/\dot{\Omega}^3$ [this parameter is now known only for Crab pulsar (Lyne and Graham-Smith, 1998)] would make it possible to greatly clarify the nature of the radio pulsar slowing down. On the other hand, almost immediately after the discovery of the radio pulsars, it was obvious that the vacuum model is not a good zero approximation to describe the neutron star magnetosphere. And the reason, strange as it may seem, is that a superstrong magnetic field exists.

2.2.2 Particle Generation in the Strong Magnetic Field

The superstrong magnetic field $B \sim 10^{12}$ G leads to a number of important consequences.

- The synchrotron lifetime (Landau and Lifshits, 1989)

$$\tau_{\rm s} \approx \frac{1}{\omega_B}\left(\frac{c}{\omega_B r_{\rm e}}\right) \sim 10^{-15}\ {\rm s} \tag{2.25}$$

  ($\omega_B = eB/m_{\rm e}c$—electron cyclotron frequency, $r_{\rm e} = e^2/m_{\rm e}c^2$—the classical electron radius) appears much smaller than the time it takes for a particle to escape the magnetosphere. Consequently, the charged particle motion in the neutron star magnetosphere includes the motion along the magnetic field lines and the electric drift in a transverse direction.
- Since the dipole magnetic field lines are curved, the relativistic particle motion along a curved trajectory gives rise to the emission of hard γ-quanta due to the so-called curvature radiation (Zheleznyakov, 1996). This process is quite analogous to the ordinary synchrotron radiation, because the nature of the accelerated motion is unessential and for relativistic particles the formation length $\delta r \sim R_{\rm c}\gamma^{-1}$ is much smaller than the curvature radius $R_{\rm c}$. Therefore, all formulae for the synchrotron radiation can be used to describe the curvature radiation with the only change, viz., the Larmor radius $r_B = m_{\rm e}c^2\gamma/eB$ is to be replaced by the radius of curvature of the magnetic field line $R_{\rm c}$. In particular, the frequency corresponding to the maximum radiation now looks like

$$\omega_{\rm cur} = 0.44\frac{c}{R_{\rm c}}\gamma^3. \tag{2.26}$$

  The extra degree γ as compared to the synchrotron radiation case $\omega_{\rm syn} = 0.44\omega_B\gamma^2$ is associated here with the fact that for the synchrotron losses the Larmor radius r_B itself is proportional to the particle energy.
- Finally, the importance of the one-photon generation of electron–positron pairs in the superstrong magnetic field $\gamma + B \rightarrow e^+ + e^- + B$ was understood, which occurs when photons in their motion cross the magnetic field lines (Sturrock, 1971). Indeed, the probability (per unit length) of the conversion of a photon with energy $\varepsilon_{\rm ph}$ propagating at an angle of θ to the magnetic field $\mathbf{B}$ far from the threshold (i.e., for $\varepsilon_{\rm ph} \gg 2m_{\rm e}c^2$) is (Berestetsky et al., 1982)

$$w = \frac{3\sqrt{3}}{16\sqrt{2}}\frac{e^3 B\sin\theta}{\hbar m_{\rm e}c^3}\exp\left(-\frac{8}{3}\frac{B_\hbar}{B\sin\theta}\frac{m_{\rm e}c^2}{\varepsilon_{\rm ph}}\right). \tag{2.27}$$

  Here the value

$$B_\hbar = \frac{m_{\rm e}^2c^3}{e\hbar} \approx 4.4\times 10^{13}\ {\rm G} \tag{2.28}$$

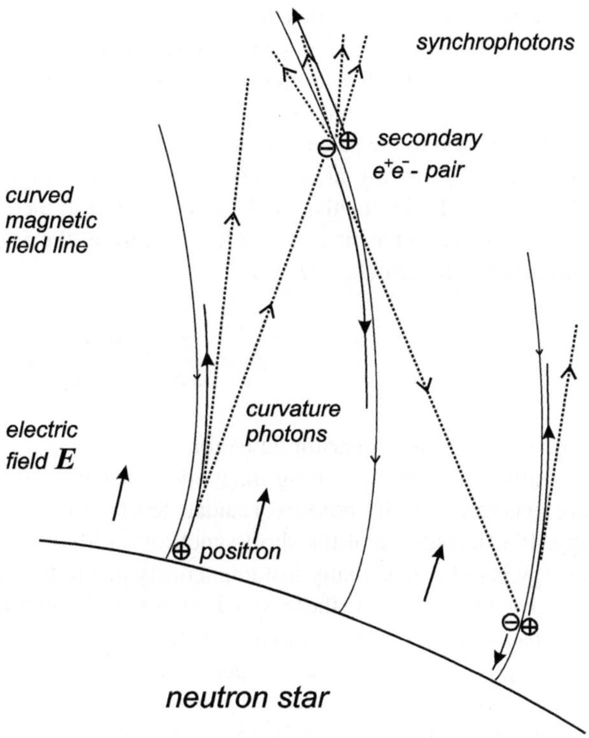

Fig. 2.2 Structure of the acceleration region and particle generation in the vicinity of the neutron star surface. The primary particles that penetrated the nonzcro longitudinal electric field region are accelerated along the curved magnetic field lines and begin to radiate hard γ-quanta. These curvature photons (*dotted lines*) propagating in the curved magnetic field reach the particle generation threshold and create electron–positron pairs. Secondary particles radiate synchrophotons and, after acceleration, start to radiate new generation of curvature γ-quanta

corresponds to the critical magnetic field for which the energy gap between two Landau levels reaches the rest energy of an electron, viz., $\hbar\omega_B = m_e c^2$. Recall that, unlike the electric field, the magnetic field itself cannot generate particles. However, it can act as a catalyst that ensures the fulfillment of the laws of energy and momentum conservation for the process studied.

As we see, the characteristic magnetic fields of neutron stars are not much smaller than the critical magnetic field $B_\hbar$. Therefore, the neutron star magnetosphere appears nontransparent even to low-energy photons with energy $\varepsilon_{ph} \sim$ 2–3 MeV, i.e., in the vicinity of the particle generation threshold. We thus have the chain of processes (see Fig. 2.2).

1. The primary particle acceleration by the longitudinal electric field existing, as was shown, in the vacuum approximation.
2. The emission of curvature photons with characteristic frequencies $\omega \leq \omega_{cur}$ (2.26).
3. The photons propagation in the curved magnetic field up to the generation of the secondary electron–positron pairs.
4. The acceleration of secondary particles, the emission of curvature photons, which, in turn, give rise to the generation of new secondary particles.
5. The screening of the longitudinal electric field by the secondary plasma.

Thus, we can conclude that the vacuum magnetosphere of the neutron star with magnetic field $B_0 \sim 10^{12}$ G proves unstable to the charged particle generation.

Some comments for correcting the above-formulated pattern are necessary. Note first that though the curvature photons are actually emitted parallel to the magnetic field lines, due to the same curvature of the magnetic lines a γ-quantum in its propagation starts moving at an increasingly greater angle of θ to the magnetic field. On the other hand, for the small, as compared to the curvature radius, photon free path l_γ, we can take $\sin\theta \approx l_\gamma/R_c$. Therefore, the γ-quantum free path l_γ can be estimated as (Sturrock, 1971)

$$l_\gamma = \frac{8}{3\Lambda} R_c \frac{B_\hbar}{B} \frac{m_e c^2}{\varepsilon_{ph}}, \tag{2.29}$$

where $\Lambda \approx 20$ is a logarithmic factor.

Further, for not too strong magnetic fields $B < 10^{13}$ G, the secondary particles are generated on the nonzero Landau levels (Beskin, 1982; Daugherty and Harding, 1983). Because of the short synchrotron lifetime τ_s (2.25), all the "transverse" energy is radiated actually instantaneously due to the synchrotron emission. It turns out that the energy of these synchrophotons is high enough for these photons to be absorbed by the strong magnetic field and generate secondary particles. As to primary particles, they can be generated by the cosmic background radiation. A comprehensive analysis showed (Shukre and Radhakrishnan, 1982) that the cosmic γ-ray background leads to the generation of 10^5 primary particles per second. This is quite enough for the neutron star magnetosphere to be effectively filled with an electron–positron plasma.

Problem 2.4 Having determined the free path length l_γ as $\int_0^{l_\gamma} w(l)\mathrm{d}l = 1$, show that

$$\Lambda \approx \ln\left[\frac{e^2}{\hbar c}\frac{\omega_B R_c}{c}\left(\frac{B_\hbar}{B}\right)^2\left(\frac{m_e c^2}{\varepsilon_{ph}}\right)^2\right]. \tag{2.30}$$

Problem 2.5 Show that if a photon of energy $\varepsilon_{ph} \gg m_e c^2$ generates a pair moving at an angle of θ to the magnetic field, after the secondary particles descend to the lower Landau level, their Lorentz factors are

$$\gamma \approx \frac{1}{\theta} \approx \frac{R_c}{l_\gamma}. \tag{2.31}$$

Problem 2.6 Using the law of motion of a relativistic particle

$$\frac{d\varepsilon_e}{dl} = eE_\parallel - \frac{2}{3}\frac{e^2}{R_c^2}\gamma^4, \tag{2.32}$$

where the first term on the right-hand side corresponds to the acceleration in the electric field and the second one to the radiation reaction, show that for the standard radio pulsar ($B_0 = 10^{12}$ G, $P = 1$ s) the primary electron energy ε_e (and the positron one) can amount to 10^8 MeV, and the energy of curvature photons to 10^7 MeV.

2.2.3 Magnetosphere Structure

Thus, the important conclusion is that the plasma-filled magnetosphere model rather than the vacuum model is a more natural zero approximation. This implies that in the zero approximation the longitudinal electric field can be considered to be zero

$$E_\parallel = 0. \tag{2.33}$$

Physically, this condition implies that light electrons and positrons can always be redistributed so as to screen the longitudinal electric field. The occurrence of the longitudinal field in some magnetosphere region immediately leads to an abrupt plasma acceleration and to the explosive generation of secondary particles.

As a result, we can determine the main features defining the pulsar magnetosphere.

Corotation. Due to the presence of plasma in the pulsar magnetosphere, the frozen-in condition (2.6)

$$\mathbf{E} + \frac{\boldsymbol{\Omega} \times \mathbf{r}}{c} \times \mathbf{B} = 0 \tag{2.34}$$

is, with adequate accuracy, satisfied not only in the interior of the neutron star but also in the whole magnetosphere. As a result, the drift velocity

$$\mathbf{U}_{\mathrm{dr}} = c\frac{\mathbf{E} \times \mathbf{B}}{B^2} = \boldsymbol{\Omega} \times \mathbf{r} + j_\parallel \mathbf{B} \tag{2.35}$$

($j_\parallel$—a scalar function) consists of the motion along the magnetic field and the rigid corotation with the neutron star. This corotation is present in the magnetosphere of the Earth and large planets.

Light cylinder. It is clear that the rigid corotation becomes impossible at large distances from the rotation axis $\varpi > R_L$, where the light cylinder radius R_L is defined as

$$R_L = \frac{c}{\Omega}. \tag{2.36}$$

Actually, this scale defines the magnetosphere boundary. For the ordinary pulsars $R_L \sim 10^9$–10^{10} cm, i.e., the light cylinder is at distances several thousand times larger than the neutron star radius.

Light surface. As we see in the following, of great importance in the radio pulsar magnetosphere structure is the so-called light surface—the surface on which the electric field becomes equal to the magnetic one, viz., $|\mathbf{E}| = |\mathbf{B}|$. In the presence of longitudinal currents, this surface does not coincide with the light cylinder but is at larger distances and extends to infinity for rather high longitudinal currents. The light surface defines the magnetosphere boundary more correctly, because the drift approximation (2.34) and (2.35) becomes inapplicable beyond its boundaries and so does the MHD approximation.

Polar cap. Since in the polar coordinates r, θ the dipole magnetic field lines are described by the relation $r = r_{max} \sin^2 \theta_m$, where r_{max} is the maximum distance of the given field line from the star center, we can estimate the polar cap radius at the pulsar magnetic pole $R_0 = R \sin \theta_0$ from which the magnetic field lines extend beyond the light cylinder. Substituting for r_{max} the light cylinder radius R_L, we get

$$R_0 = R \left(\frac{\Omega R}{c}\right)^{1/2}, \tag{2.37}$$

where the factor

$$\varepsilon_A = \left(\frac{\Omega R}{c}\right)^{1/2} \sim 10^{-2} \tag{2.38}$$

is, as we will see, the main small parameter in the theory of the pulsar magnetosphere. Thus, for ordinary radio pulsars the polar cap size is only several hundreds of meters. And on this extremely small, on a cosmic scale, area comparable with the stadium size, the basic processes responsible for the observed activity of radio pulsars occur.

Open and closed field lines. As shown in Fig. 2.3, the magnetic field lines going beyond the light cylinder can diverge and extend to infinity. Since, as was noted, the particle motion is possible only along the magnetic field, two groups of magnetic field lines stand out in the magnetosphere. One group passing through the polar cap intersects the light cylinder and extends to infinity. The other group located far from the magnetic axis is closed within the light cylinder. The plasma located on the closed magnetic lines turns out to be captured, whereas the plasma filling the open magnetic lines can escape the neutron star magnetosphere.

Critical charge density. Finally, it is very important that the charge density in the magnetosphere of the rotating neutron star must be different from zero. Indeed, using relation (2.34), we find $\rho_e \approx \rho_{GJ}$ where

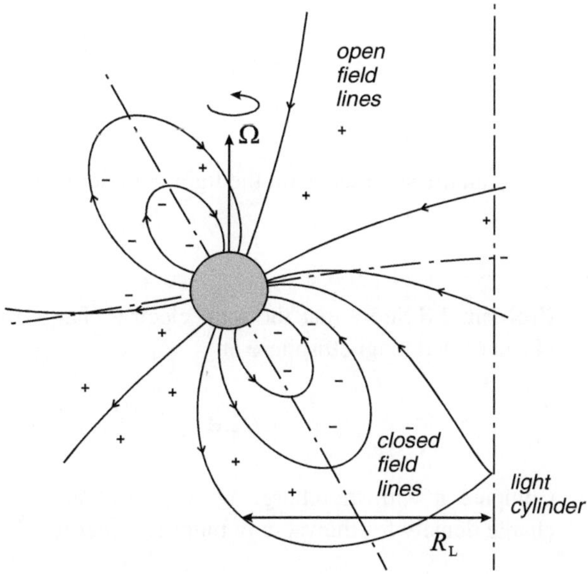

Fig. 2.3 The magnetosphere structure of radio pulsars. The open field lines coming out from the magnetic poles cross the light cylinder (*dashed and dotted line*). The charge density ϱ_{GJ} (2.39) changes the sign on the surface on which the magnetic lines are orthogonal to the angular velocity vector $\mathbf{\Omega}$

$$\rho_{GJ} = \frac{1}{4\pi}\mathrm{div}\mathbf{E} \approx -\frac{\mathbf{\Omega}\cdot\mathbf{B}}{2\pi c}. \tag{2.39}$$

This expression was first obtained in P. Goldreich and P. Julian's pioneer paper (Goldreich and Julian, 1969). Therefore, the critical charge density (2.39) is, generally, called the Goldreich–Julian (GJ) charge density. For ordinary pulsars, the appropriate concentration $n_{GJ} = |\rho_{GJ}|/e$ near the star surface is 10^{10}–10^{12} $1/\mathrm{cm}^3$. Accordingly, the characteristic value of the current density can be written as $j_{GJ} = \rho_{GJ}c$. Finally, the characteristic value of the total electric current in the magnetosphere can be estimated as a product of the polar cap area, the GJ charge density, and the velocity of light:

$$I_{GJ} = \pi R_0^2 \rho_{GJ} c. \tag{2.40}$$

The physical meaning of the GJ charge density is simple—it is the charge density needed to screen the longitudinal electric field. The perpendicular electric field occurs and its value, as we saw, turns out to be exactly the value of the electric drift in the crossed fields to generate the plasma corotation.

Problem 2.7 Show that for the case of the total corotation (i.e., when the poloidal currents are absent in the neutron star magnetosphere and, therefore, the total current $\mathbf{j}$ can be written as $\mathbf{j} = \rho_e \mathbf{\Omega} \times \mathbf{r}$), the exact expression for the GJ charge density has the form

$$\rho_{\rm GJ} = -\frac{\mathbf{\Omega}\cdot\mathbf{B}}{2\pi c\left(1-\dfrac{\Omega^2\varpi^2}{c^2}\right)}. \tag{2.41}$$

How can the singularity on the light cylinder be explained?

Problem 2.8 Show that the total electric charge of the neutron star for the plasma-filled magnetosphere is

$$Q_{*} = \frac{1}{3}\frac{\Omega B_0}{c}R^3 \neq 0. \tag{2.42}$$

Compare it with the charge Q_{*} (2.12) obtained by integrating the surface charge density for the vacuum magnetosphere.

Some explanation is also necessary here. First of all, as is evident from relation (2.35), the light cylinder is the real boundary of the magnetosphere only for the zero toroidal magnetic field, i.e., for the zero longitudinal electric current. As we will see, for the sufficiently large longitudinal current (and, hence, for the large enough toroidal magnetic field), the drift motion can occur at distances much larger than the light cylinder radius $R_{\rm L}$. However, as shown in Fig. 2.4, in this case, there is almost the full compensation of the corotational velocity $\mathbf{\Omega}\times\mathbf{r}$ and the toroidal slip velocity along the magnetic field $j_{\parallel}B_{\varphi}$, so that the drift velocity $\mathbf{U}_{\rm dr}$ is directed radially from the star. Therefore, beyond the light cylinder, in spite of the validity of the drift approximation, the particle motion is actually perpendicular to the magnetic field lines.

Further, relation (2.37) for the polar cap radius is only an estimate in order of magnitude. The point is that the electric currents connected with electric charges filling the pulsar magnetosphere in the vicinity of the light cylinder begin to disturb the dipole magnetic field. Therefore, the exact form of the polar cap can be found together with the solution of the complete problem of the neutron star magnetosphere. On the other hand, expression (2.37) allows us to estimate the maximum value of the voltage drop in the vicinity of the magnetic poles $\psi_{\rm max} = E(R_0)R_0$:

$$\psi_{\rm max} = \left(\frac{\Omega R}{c}\right)^2 R B_0. \tag{2.43}$$

For ordinary pulsars, it yields $\psi_{\rm max} \sim 10^7$–10^8 MeV.

Finally, important consequences follow from expression (2.39) for the GJ charge density. As shown in Fig. 2.3, in the vicinity of the neutron star, the charge density $\rho_{\rm GJ}$ changes sign on the surface, where $\mathbf{\Omega}\cdot\mathbf{B} = 0$. Therefore, except for the

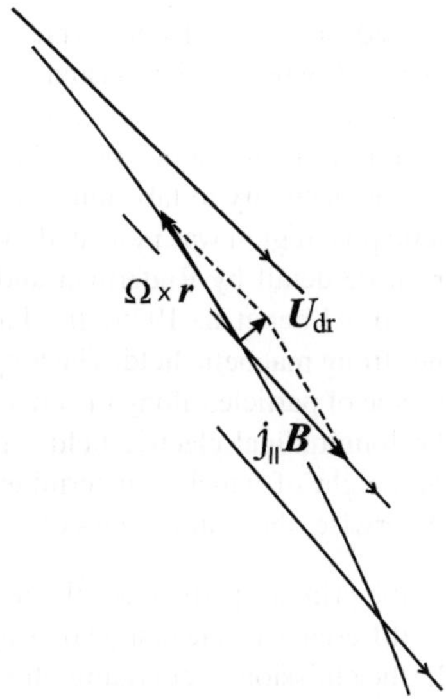

Fig. 2.4 The drift motion of a charged particle beyond the light cylinder in the presence of the strong toroidal field $B_\varphi \gg B_p$ is nearly in a radial direction. The velocity $\mathbf{U}_{dr}$ (which is, naturally, smaller than the velocity of light) can be formally resolved into the corotation velocity $\boldsymbol{\Omega} \times \mathbf{r}$ and the slip velocity along the magnetic field $j_\parallel \mathbf{B}$, each of them can be much larger than the velocity of light. The rotation axis is perpendicular to the figure plane

orthogonal rotator $\chi = 90°$, the GJ charge density has the same sign in the vicinity of both magnetic poles (in fact, this property is directly associated with the already mentioned property of the vacuum magnetosphere—the radial electric field in the region of the magnetic poles is identical). This implies that an inverse current flowing in the vicinity of the boundary of the closed and open magnetic field lines is sure to occur—only, in this case, the total charge of the neutron star does not change. We should call attention to this property since it is the key property in the development of the theory of the neutron star magnetosphere.

Problem 2.9 Show that the light cylinder (where the corotation velocity approaches the velocity of light) is just the scale on which

- the electric field is compared in magnitude with the poloidal magnetic field;
- the toroidal electric currents flowing in the magnetosphere begin to disturb the poloidal magnetic field of the neutron star;
- the toroidal magnetic field connected with the longitudinal GJ current is compared in magnitude with the poloidal magnetic field.

2.3 Secondary Plasma Generation

2.3.1 "Inner Gap"

Thus, in the radio pulsar magnetosphere, two substantially different regions must develop, viz., the regions of open and closed magnetic field lines. The particles located on the field lines which do not intersect the light cylinder turn out to be captured, whereas the plasma on the field lines intersecting the light cylinder can extend to infinity. Consequently, the plasma must be continuously generated in the region of the magnetic poles of a neutron star.

The necessity to take into account the secondary plasma generation in the magnetic pole region was indicated by Sturrock (1971) and then this process was studied in more detail by Ruderman and Sutherland (1975), and also by V.Ya. Eidman's group (Al'ber et al., 1975). It is based on the above one-photon particle generation in the strong magnetic field. The longitudinal electric field is generated by a continuous escape of particles along the open field lines beyond the magnetosphere. As a result, the longitudinal electric field region forms in the vicinity of the magnetic poles, the height of which is determined by the secondary plasma generation condition. Otherwise, the chain of processes is (see again Fig. 2.2)

1. the primary particle acceleration by the longitudinal electric field induced by the difference of the charge density ρ_{e} from the GJ charge density ρ_{GJ};
2. the emission of curvature photons with characteristic frequency $\omega \leq \omega_{\mathrm{cur}}$ (2.26);
3. the photons propagation in the curved magnetic field up to the secondary electron–positron pair generation;
4. the secondary particles acceleration, the emission of curvature photons, which, in turn, give rise to the new generation of secondary particles.

It is important that a greater part of secondary particles is generated already over the acceleration region, where the longitudinal electric field is rather small, so that the secondary plasma can escape the neutron star magnetosphere.

To estimate the longitudinal electric field we consider, for simplicity, only the one-dimensional equation

$$\frac{\mathrm{d}E_{\parallel}}{\mathrm{d}h} = 4\pi(\rho_{\mathrm{e}} - \rho_{\mathrm{GJ}}), \tag{2.44}$$

which can be used if the gap height H is much smaller than the size of the polar cap R_0 (2.37). Unfortunately, this approximation is valid for the fastest pulsars only. Nevertheless, it contains all information concerning the inner gap structure. In spite of its outward simplicity, Eq. (2.44) comprises a number of substantial uncertainties. The main uncertainty is, undoubtedly, in the expression for the charge density ρ_{e}, which depends on the particle generation mechanism, which, in turn, is defined by the value of the longitudinal electric field.

We discuss the basic properties of Eq. (2.44). Thus, for the models with the non-free particle escape from the neutron star surface, which are, generally, called the Ruderman–Sutherland model (see the next section), we can take $|\rho_{\mathrm{e}}| \ll |\rho_{\mathrm{GJ}}|$ in the zero approximation, and the electric field on the star surface can be different from zero. As a result, we have (Ruderman and Sutherland, 1975)

$$E_{\parallel} \approx E_{\mathrm{RS}} \frac{H - h}{H}, \tag{2.45}$$

where

$$E_{\rm RS} = 4\pi\rho_{\rm GJ}H, \tag{2.46}$$

and H is the height of the longitudinal electric field region. Its value should just be determined from the condition for the onset of the secondary plasma generation. Indeed, for $H < H_{\rm cr}$ the longitudinal electric field is not strong enough to effectively generate particles, whereas for $H > H_{\rm cr}$, the secondary plasma results in the fast screening of the acceleration region. Besides, for the solid star surface, this event can occur for the antiparallel directions of the magnetic and rotation axes, when near the polar caps $\rho_{\rm GJ} > 0$, and positively charged particles are to be ejected from the surface. Within this model, the longitudinal current I, generally speaking, can be arbitrary, but, certainly, not larger than the GJ current $I_{\rm GJ}$.

Problem 2.10 Using expression (2.46) connecting the longitudinal electric field with the gap height H and relations (2.26) and (2.29) for the characteristic energy and the free path of curvature photons, find the expressions for gap height H and potential drop $\psi = E_{\parallel}H$ (Ruderman and Sutherland, 1975)

$$H_{\rm RS} \sim \lambda_{\rm C}^{2/7} R_{\rm c}^{2/7} R_{\rm L}^{3/7} \left(\frac{B}{B_{\hbar}}\right)^{-4/7}, \tag{2.47}$$

$$\psi_{\rm RS} \sim \frac{m_{\rm e}c^2}{e} \lambda_{\rm C}^{-3/7} R_{\rm c}^{4/7} R_{\rm L}^{-1/7} \left(\frac{B}{B_{\hbar}}\right)^{-1/7}. \tag{2.48}$$

Here $\lambda_{\rm C} = \hbar/m_{\rm e}c$ is the Compton wavelength.
(Hint: the gap height H can be estimated as a sum of primary particle acceleration length $l_{\rm acc}$ and free path of emitted curvature photon l_{γ}. For small acceleration lengths $l_{\rm acc}$, the primary particle energy $\varepsilon_{\rm e} = eE_{\parallel}l_{\rm acc}$ and, therefore, the emitted photon energy $\varepsilon_{\rm ph}$ are low, and the free path of such low-energy photons appears significant. The short free paths can be realized only for the sufficiently high energy of photons, for the emission of which a primary particle is to pass a large distance. Therefore, the minimum value of the sum $l_{\rm acc} + l_{\gamma}$ is the scale on which the secondary plasma generation starts, which can screen the longitudinal electric field. This value is taken as an estimate of the gap height H.)

On the other hand, if particles can freely escape from the neutron star surface, it is logical to take here

$$E_{\parallel}(h = 0) = 0, \tag{2.49}$$

and the charge density $\rho_{\rm e}$ is close to $\rho_{\rm GJ}$. The longitudinal electric field must also be zero on the upper boundary of the acceleration region

$$E_{\parallel}(h = H) = 0. \tag{2.50}$$

Otherwise, the secondary particles of one of the signs would fail to extend to infinity. As we see, in this model the longitudinal electric current I is to be very close to GJ current $I_{\rm GJ}$. As a result, in the free particle escape model, the longitudinal electric field is specified only by a small difference between the charge density $\rho_{\rm e}$ and the critical density $\rho_{\rm GJ}$. Indeed, the GJ charge density can be written as

$$\rho_{\rm GJ} = -\frac{\Omega B \cos\theta_b}{2\pi c}, \tag{2.51}$$

where θ_b is an angle between the magnetic field and the rotation axis. On the other hand, for the relativistic plasma moving with velocity $v \approx c$, we have within the same accuracy

$$\rho_{\rm e} = C(\Psi)B, \tag{2.52}$$

where $C(\Psi)$ is constant along the magnetic field lines. As we see, the charge densities (2.51) and (2.52) change differently along the magnetic field line. Thus, the GJ charge density (2.51), besides the factor B, also contains the geometric factor $\cos\theta_b$. As a result, the charge-separated relativistic plasma in its motion fails to satisfy the condition $\rho_{\rm e} = \rho_{\rm GJ}$, which gives rise to the particle acceleration in the longitudinal electric field. The longitudinal electric field gives rise to particle acceleration, to hard photon emission, and, hence, to secondary electron–positron plasma generation. Therefore, beyond the acceleration region, the field must already be close to zero.

Note that the conditions (2.49) and (2.50) can be satisfied simultaneously only if the electric charge density on the acceleration region boundaries does not coincide with the GJ density, i.e., when the derivative $\mathrm{d}E_{\parallel}/\mathrm{d}h$ is different from zero here (see Fig. 2.5). As a result, Eq. (2.44) can be rewritten as

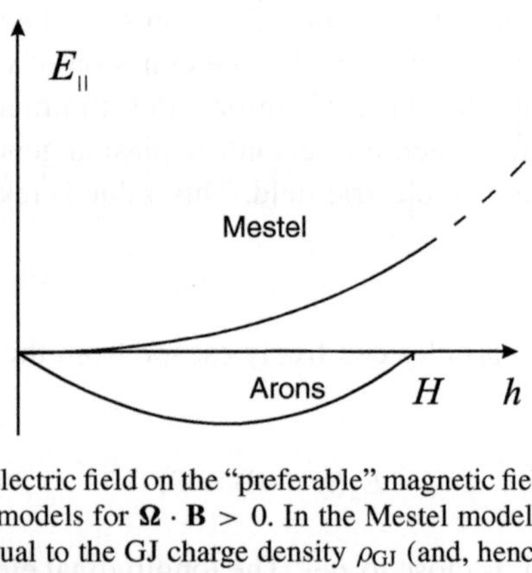

Fig. 2.5 The longitudinal electric field on the "preferable" magnetic field lines $A_a > 0$ in the Arons (1981) and Mestel (1999) models for $\mathbf{\Omega} \cdot \mathbf{B} > 0$. In the Mestel model, the plasma charge density $\rho_{\rm e}$ on the star surface is equal to the GJ charge density $\rho_{\rm GJ}$ (and, hence, $\mathrm{d}E/\mathrm{d}h = 0$), whereas in the Arons model, the charge density for $h = 0$, due to the presence of a particle backflow, differs from $\rho_{\rm GJ}$. As a result, though in both cases the electric field is zero on the star surface, the electric field direction and, hence, the particle acceleration appear different

$$\frac{\mathrm{d}E_{\parallel}}{\mathrm{d}h} = A_a \left(h - \frac{H}{2}\right), \tag{2.53}$$

where

$$A_a = 4\pi \left. \frac{\mathrm{d}(\rho_\mathrm{e} - \rho_\mathrm{GJ})}{\mathrm{d}h} \right|_{h=H/2}. \tag{2.54}$$

Finally, we have for $\chi > \varepsilon_\mathrm{A}$

$$A_a = \frac{3}{2} \frac{\Omega B_0}{cR} \theta_\mathrm{m} \cos \varphi_\mathrm{m} \sin \chi. \tag{2.55}$$

Here $\theta_\mathrm{m} \sim \varepsilon_\mathrm{A}$ is the polar angle and φ_m is an azimuthal angle relative to the magnetic dipole axis. The solution to Eq. (2.53) has the form

$$E_{\parallel} = -E_\mathrm{A} \frac{h(H-h)}{H^2}, \tag{2.56}$$

where

$$E_\mathrm{A} \approx \frac{3\pi}{2} \,|\rho_\mathrm{GJ}|\, \frac{H^2}{R} \theta_\mathrm{m} \cos \varphi_\mathrm{m} \tan \chi \sim \varepsilon_\mathrm{A} \frac{H}{R} E_\mathrm{RS}, \tag{2.57}$$

so that $|E_\mathrm{A}| \ll |E_\mathrm{RS}|$. Therefore, for this solution to exist, a particle backflow is needed; the value of which can be determined from Eq. (2.44):

$$\frac{j_\mathrm{back}}{j_\mathrm{GJ}} \approx \varepsilon_\mathrm{A} \frac{H}{R} \sim 10^{-4}. \tag{2.58}$$

This model was first studied by J. Arons' group (Fawley et al., 1977; Scharlemann et al., 1978; Arons and Scharlemann, 1979).

Note that the acceleration regime (when the generated longitudinal electric field accelerates particles from the star surface) can occur only on the northern half of the polar cap $-\pi/2 < \varphi_\mathrm{m} < \pi/2$ ($A_a > 0$), for which the magnetic field lines bend in the direction of the rotation axis and, hence, $\cos\theta_b$ increases with distance from the star surface. In this case, the generated longitudinal electric field accelerates particles from the star surface. These field lines were called the "preferable" lines. In the domain $\pi/2 < \varphi_\mathrm{m} < 3\pi/2$ ($A_a < 0$), where the magnetic field lines, on the contrary, tend to be perpendicular to the rotation axis, the generated longitudinal electric field would lead to the deceleration of particles rather than to their acceleration. As a result, within this model, the acceleration and the generation of the secondary particles occur only in one-half of the region of the open field lines and, accordingly, the radiation directivity pattern should also have the form of a semicircle (Arons and Scharlemann, 1979). However, this conclusion contradicts the observational data (Lyne and Graham-Smith, 1998).

If the bulk particle backflow is absent, Eq. (2.44) yields the completely different solution

$$E_{\parallel} \approx \frac{3\pi}{2}\,|\rho_{\rm GJ}|\,\theta_{\rm m}\cos\varphi_{\rm m}\tan\chi\,\frac{h^2}{R} \sim E_{\rm A}\frac{h^2}{H^2}, \tag{2.59}$$

in which the longitudinal electric field turns out to be in the opposite direction. Clearly, relation (2.59) can be used only up to distances $h \ll R_0$; at larger distances the longitudinal electric field tends to zero. Consequently, the particle acceleration is possible only on the "unpreferable" magnetic field lines. Exactly this model, in which the particle backflow must naturally be rather small, had been developed for many years by L. Mestel (Mestel and Wang, 1979; Fitzpatrick and Mestel, 1988; Mestel and Shibata, 1994; Mestel, 1999). Thus, only the consistent kinetic model can choose between these two realizations [the thorough investigation of this problem can be found in Shibata (1997) and Shibata et al. (1998)].

2.3.2 Neutron Star Surface

The problem of the neutron star surface structure, which is of interest by itself, is directly associated with the theory of the radio pulsar magnetosphere. Indeed, as was mentioned, the inner gap structure greatly depends on the work function $\varphi_{\rm w}$ for electrons (the cohesive energy for nuclei) on the neutron star surface. Recall that in the 1970s, the nonfree particle escape model was mainly developed. It was based on a series of theoretical papers on the matter structure in the superstrong magnetic field, in which the work function had a rather large value $\varphi_{\rm w} \sim 1$–5 keV (Kadomtsev and Kudryavtsev, 1971; Ginzburg and Usov, 1972; Chen et al., 1974; Hillebrandt and Müller, 1976; Flowers et al., 1977). However, from the early 1980s, when due to the more accurate computations the work function reduced to $\varphi_{\rm w} \sim 0.1$ keV, the free particle escape models grew in popularity (Müller, 1984; Jones, 1980; Neuhauser et al., 1986).

We stress that the problem remains unsolved. The point is that the accuracy of determination of work function and cohesive energy is not high enough yet (Usov and Melrose, 1996). It turned out that even the chemical composition of the neutron star surface layers is not known—possibly, they do not consist of iron atoms, as was supposed in most papers. The point is that the chemical composition of the surface layers on the polar caps can greatly change because of their bombardment by energetic particles accelerated by the longitudinal electric field in the gap. Besides, and it is the subject of wide speculation now, iron atoms (which, being the most stable nuclei, are, undoubtedly, copiously produced) could have been "sunk" by the action of the gravitational field within the first few years after the formation of the neutron star when its surface was not solid yet (Salpeter and Lai, 1997). It is not improbable, therefore, that, in reality, the neutron star surface layers consist of much lighter atoms rather than iron atoms—hydrogen and helium ones. Since the melting temperature roughly estimated by the formula (Shapiro and Teukolsky, 1983)

$$T_{\rm m} \approx 3.4\times 10^7\,{\rm K}\left(\frac{Z}{26}\right)^{5/3}\left(\frac{\rho}{10^6{\rm g/cm^3}}\right) \tag{2.60}$$

depends on the atomic number Z, the neutron star surface at temperature $T \sim 10^6$ K characteristic of ordinary radio pulsars should be liquid and, in any event, must not prevent the free particle escape. The radio pulsar thermal radiation models are just based on this pattern (Zavlin and Pavlov, 2002; Haensel et al., 2007).

2.3.3 Propagation of γ-Quanta in the Superstrong Magnetic Field

We now proceed with a brief discussion of the effects of the propagation of high-energy photons in the superstrong magnetic field in the vicinity of the neutron star surface. This problem is directly associated with the particle generation mechanism in the polar regions of radio pulsars. The quantum effects in the magnetic field, the value of which is close to the critical value $B_\hbar = 4.4 \times 10^{13}$ G (2.28), were known long ago (Berestetsky et al., 1982), but only after the discovery of radio pulsars there was hope of their direct observation. These may include, for example, the photon splitting process $\gamma + B \rightarrow \gamma + \gamma + B$ (Bialynicka-Birula and Bialynicka-Birula, 1970; Adler, 1971), the change in the cross-section of the two-photon pair generation $\gamma + \gamma \rightarrow e^+ + e^-$, especially near the generation threshold (Kozlenkov and Mitrofanov, 1986), the quantum synchrotron cooling connected with the fast particle transition to the lower Landau level (Mitrofanov and Pozanenko, 1987), as well as the propagation effects due to both the vacuum refraction (Bialynicka-Birula and Bialynicka-Birula, 1970) and the peculiarities of the photon trajectories in the vicinity of the generation threshold of secondary electron–positron pairs (Shabad and Usov, 1984, 1985, 1986). As a result, in the 1970s, the possibility of the direct detection of the effects connected with a quantizing magnetic field (2.28) seemed absolutely real (Mészáros, 1992). Nevertheless, these effects for most radio pulsars appeared rather weak. The point is that, for example, the expression for the refraction index in the strong magnetic field (the formula corresponds to one of the linear polarizations)

$$n = 1 + \frac{7\alpha_{\rm fin}}{90\pi}\left(\frac{B}{B_\hbar}\right)^2 \tag{2.61}$$

comprises the fine structure constant $\alpha_{\rm fin} = e^2/\hbar c \approx 1/137$; therefore, we can expect the occurrence of considerable quantum effects only in the fields $B > 10^{14}$ G. For most neutron stars observed as radio pulsars, we can, with adequate accuracy, suppose that γ-quanta propagate rectilinearly.

However, in the context of the discovery of magnetars (pulsating X-ray sources, the periods of which amount to a few seconds and the magnetic field estimated by formula (2.5) reaches 10^{14}–10^{15} G (Thompson and Duncan, 1993; Kouveliotou et al., 1998)), this problem has recently become an urgent one. Therefore, the new thorough computations of both the secondary particle generation process (Weise and Melrose, 2002) and the photon splitting (Baring and Harding, 1997; Chistyakov et al., 1998), and the determination of the trajectories of hard γ-quanta near the particle generation threshold (Shaviv et al., 1999) were carried out. In particular, it was shown that for sufficiently large magnetic fields $B \sim 10^{14}$–10^{15} G, the process of the γ-quanta conversion due to the photon splitting can be considerably suppressed (Baring and Harding, 1998). Consequently, the secondary plasma generation process can be considerably suppressed as well. It is not surprising, therefore, that most magnetars are not manifested as radio pulsars.

On the other hand, it was shown (Usov, 2002) that the splitting of $\|$-polarized photons (i.e., those with the electric vector located in the plane containing the external magnetic field and the wave vector) below the pair production threshold is strictly forbidden in arbitrary magnetic fields. Solving the system of kinetic equations for splitting photons and taking into account their polarization, it was shown that the photon splitting, which was earlier considered as a suppression factor for the secondary electron–positron plasma generation, is not suppressed at all (Istomin and Sobyanin, 2007). Moreover, the plasma density in the magnetar magnetosphere can be even higher than that in the magnetosphere of a pulsar with a weak magnetic field. Thus, some light can be shed on the recent discovery of the pulsed radio emission from several magnetars (Malofeev et al., 2007).

But, in general, the new qualitative phenomena that could be helpful in the observation of the quantum effects in the superstrong magnetic field were not found, and the earlier obtained results were only refined in the computations.

2.3.4 General Relativity Effects

We consider the GR effects which, unlike the quantizing magnetic field effects, can, undoubtedly, greatly affect the particle generation process in the vicinity of radio pulsars. It turned out that in the model of free particle escape from the neutron star surface, the GR effects must be of vital importance. Recall that the gravitational potential $\varphi_{\rm g}$ on the pulsar surface is rather large

$$\varepsilon_{\rm g} = \frac{2|\varphi_{\rm g}|}{c^2} \approx \frac{2GM}{Rc^2} \sim 0.2, \tag{2.62}$$

and any computations whose accuracy is better than 20% must be carried out, with account taken of the relativistic effects. However, in the nonfree particle escape models, taking account of these effects does not ensure substantial corrections, because the qualitative structure of the electrodynamic equations does not change. On the other hand, in the free particle escape model in Eq. (2.44), besides the small geometric factor $\varepsilon_{\rm A}$ (2.38), the purely relativistic factor $\varepsilon_{\rm g}$ appears, which is associated with the frame-dragging (Lense–Thirring) effect (Thorne et al., 1986). For most radio pulsars with $P \sim 1$ s, the relativistic correction $\varepsilon_{\rm g}$ turns out to be, at least in order of magnitude, larger than $\varepsilon_{\rm A}$ so that the GR effects are to be taken into consideration.

Indeed, as was already mentioned, in the Arons model, the occurrence of longitudinal electric field in the gap region is due to the difference in the plasma charge density ρ_e from the GJ charge density ρ_{GJ} (2.39). In the general relativistic case, Eq. (2.44) is to be rewritten as (Thorne et al., 1986)

$$\frac{d}{dh}\left(\frac{1}{\alpha}E_\parallel\right) = 4\pi(\rho_e - \rho_{GJ}), \tag{2.63}$$

and the GJ density has the form (see Sect. 3.2.5 for details)

$$\rho_{GJ} = -\frac{1}{8\pi^2}\nabla_k\left(\frac{\Omega - \omega}{\alpha c}\nabla^k\Psi\right). \tag{2.64}$$

Here again α is the lapse function, ω is the Lense–Thirring angular velocity, and Ψ is a magnetic flux. Within the necessary accuracy, they can be written as

$$\alpha^2 = 1 - \frac{r_g}{r}, \tag{2.65}$$

$$\omega = \Omega\frac{r_g I_r}{Mr^3}, \tag{2.66}$$

$$\Psi = \frac{1}{2}B_0 R^3\frac{\sin^2\theta_m}{r}, \tag{2.67}$$

where B_0 is the magnetic field at the neutron star pole and I_r is its moment of inertia. In the linear order with respect to the small values ε_A and ε_g, we now have

$$\rho_{GJ} = -\frac{(\Omega - \omega)B\cos\theta_b}{2\pi c\alpha}, \tag{2.68}$$

where θ_b is again an angle between the magnetic field line and the rotation axis. On the other hand, the expression for the charge density of the relativistic plasma has the form

$$\rho_e = C(\Psi)\frac{B}{\alpha}, \tag{2.69}$$

where, as before, $C(\Psi)$ is constant along the magnetic field lines. As a result, the GJ charge density (2.68), besides the factor B/α identical to the density ρ_e (2.69), as well as the geometric factor $\cos\theta_b$, also contains the factor $(\Omega - \omega)$, which changes by the dependence of $\omega(r)$ on r. As a result, for $\sin\chi > \varepsilon_A$ and $\cos\chi > \varepsilon_A$, the constant A_a in Eq. (2.53) has the form (Muslimov and Tsygan, 1990; Beskin, 1990; Muslimov and Tsygan, 1992)

$$A_a = \frac{3}{2}\frac{\Omega B_0}{cR}\left[4\frac{\omega}{\Omega}\cos\chi + \theta_m\cos\varphi_m\sin\chi + O(\varepsilon_g^2) + \cdots\right]. \tag{2.70}$$

As we see, taking account of the GR effects leads to the additional term, proportional to $\omega/\Omega \sim \varepsilon_g$. According to (2.70), for $4\,\omega/\Omega > \varepsilon_A\tan\chi$, the major contribution to A_a is made by the gravitational term. For the homogeneous density of the star when on its surface

$$\frac{\omega}{\Omega} = \frac{2}{5}\varepsilon_{\rm g}, \tag{2.71}$$

this condition can be rewritten as

$$P > 10^{-3}\,{\rm s}\left(\frac{R}{10^6\,{\rm cm}}\right)^2\left(\frac{M}{M_{\odot}}\right)^{-2}. \tag{2.72}$$

Hence, the GR effects are of vital importance for all observed pulsars. The most important consequence of expression (2.70) is that all open field lines prove "preferable" (Beskin, 1990), because the first term in (2.70) proves positive. Thus, allowance for the GR effects qualitatively changes the conclusions of the first version of the Arons model. The stationary generation becomes possible over the entire polar cap surface.

2.3.5 Particle Generation in the Magnetosphere

We discuss how all the above physical processes affect the particle generation in the vicinity of the neutron star surface. We first consider the effects of the superstrong magnetic field $B > 10^{14}$ G characteristic of magnetars. As was noted, only for these magnetic fields, the pronounced effects of the quantizing magnetic field should be expected (Baring and Harding, 1997; Shaviv et al., 1999). First of all, it was obvious long ago that the strong magnetic field must suppress the secondary plasma generation process. First, with the fields larger than 10^{13} G, a secondary electron–positron pair is to be produced at the lower Landau level, which results in the suppression of the synchrotron radiation (Beskin, 1982; Daugherty and Harding, 1983). Second, the nontrivial vacuum permeability in the vicinity of the generation threshold at the zero Landau level with the transverse photon momentum close to $2m_{\rm e}c$ can give rise to the deflection of the γ-quanta along the magnetic field. As a result, instead of two free particles, their bound state is generated, viz., positronium (Shabad and Usov, 1985, 1986). Third, as was mentioned, the photon splitting process $\gamma \to \gamma + \gamma$ becomes significant, which results in a decrease in their energy and the suppression (though incomplete) of the secondary particle generation (Baring and Harding, 1998). However, most radio pulsars have insufficiently large magnetic fields for these effects to be detected.

On the other hand, for ordinary radio pulsars, the interaction process of primary particles accelerated in the gap, with X-ray photons radiated by the heated neutron star surface, may appear substantial; Kardashev et al. (1984) first pointed to the importance of inverse Compton (IC) scattering in the particle generation region. As it turned out, the hard γ-quanta generated by this interaction have enough energy to produce electron–positron pairs and, hence, affect the inner gap structure (Cheng et al., 1986; Hirotani and Shibata, 2001). Finally, as was already noted, the value of the work function φ_w also substantially affects the electric field structure.

Nevertheless, in this part of the theory, new important results have recently been obtained. In particular, one should mention A. Harding and A. Muslimov (1998, 2002) who studied both the GR effects and the process of (the nonresonance and resonance) IC scattering of X-ray photons emitted by the neutron star surface. It is interesting to note that in this model, the acceleration region may not be adjacent to the neutron star surface, but it is as if suspended over the magnetic poles of the pulsar. However, as was noted, for a comprehensive analysis, it is necessary to take into account the kinetic effects, as it was first done by Gurevich and Istomin (1985), for the acceleration region in vicinity of the neutron star surface within the nonfree particle escape model (see also Hirotani and Shibata, 2001). Recall that analysis of the kinetic effects is needed, in particular, for the determination of particle backflow, which, in turn, is directly associated with the problem of constructing the plasma generation region.

In conclusion, we emphasize that the general properties of the secondary electron–positron plasma outflowing from the magnetosphere appeared, as a whole, to be low-sensitive to the details of the acceleration region structure. For most models (Ruderman and Sutherland, 1975; Daugherty and Harding, 1982; Gurevich and Istomin, 1985), both the density and the energy spectra of the outflowing plasma appear rather universal. Therefore, it is safe to say that the plasma flowing along the open field lines in the pulsar magnetosphere consists of a beam of primary particles with energy $\varepsilon \approx 10^7$ MeV and density close to the GJ density $n_{\rm GJ}$ and also of the secondary electron–positron component. Its energy spectrum, within adequate accuracy, has the power form

$$N(\varepsilon_{\rm e}) \propto \varepsilon_{\rm e}^{-2}, \tag{2.73}$$

and the energies are enclosed in the range from $\varepsilon_{\rm min} \sim 100$ MeV to $\varepsilon_{\rm max} \sim 10^4$–$10^5$ MeV (true, if we suppose the presence of a strong nondipole component near the magnetic poles, the minimum energies can be reduced to 10 MeV and even 3 MeV). Note that the minimum energy $\varepsilon_{\rm min}$ directly follows from the estimate (2.31), where for most low-energy particles we should take $l_\gamma = R$, because for longer free paths the decrease in the magnetic field with distance from the neutron star surface is substantial. The total secondary plasma density, as the numerous calculations show, is to be 10^3–10^4 times greater than the GJ density:

$$\lambda = \frac{n_{\rm e}}{n_{\rm GJ}} \sim 10^3 - 10^4. \tag{2.74}$$

Exactly this model was studied in a great number of papers devoted to the pulsar radio emission theory. It is important that the electron and positron distribution functions must be shifted from one another [this was already shown in Ruderman and Sutherland (1975)]. Only in this case, the outflowing plasma charge density coincides with the GJ charge density.

2.3.6 "Hollow Cone" Model

As was noted, there is no common viewpoint on the nature of the pulsar coherent radio emission now. Nevertheless, it turned out that the basic observed properties of the radio emission can be interpreted by the above particle generation pattern. It is the so-called hollow cone model (Radhakrishnan and Cooke, 1969), which was proposed already at the end of the 1960s and perfectly accounted for the basic geometric properties of the radio emission. Indeed, as was shown, the secondary particle generation is impossible in the rectilinear magnetic field when, first, the intensity of the curvature radiation is low and, second, the photons emitted by relativistic particles propagate at small angles to the magnetic field. Therefore, as shown in Fig. 2.6, in the central regions of the open magnetic field lines, a decrease in secondary plasma density should be expected.

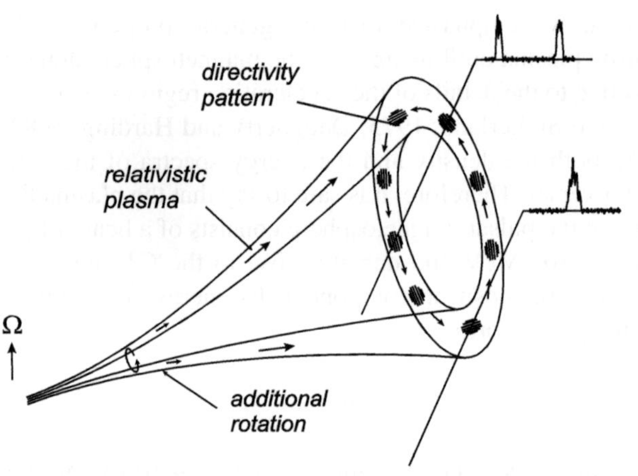

Fig. 2.6 The hollow cone model. If the intensity of the radio emission is directly connected with the outflowing plasma density, in the center of the directivity pattern there must be a decrease in the radio emission. Therefore, we should expect a single mean profile in pulsars whose line of sight intersects the directivity pattern far from its center and the double profile for the central passage. The plasma rotation around the magnetic axis leads to the observed subpulse drift

If we make a rather reasonable assumption that the radio emission must be directly connected with the outflowing plasma density, there must be a decrease in the radio emission intensity in the center of the directivity pattern. Therefore, without going into details (actually, the mean profiles have a rather complex structure (Rankin, 1983, 1990; Lyne and Graham-Smith, 1998)), we should expect a single (one-hump) mean profile in pulsars in which the line of sight intersects the directivity pattern far from its center and the double (two-hump) profile for the central passage. It is exactly what is observed in reality (Beskin et al., 1993; Lyne and Graham-Smith, 1998).

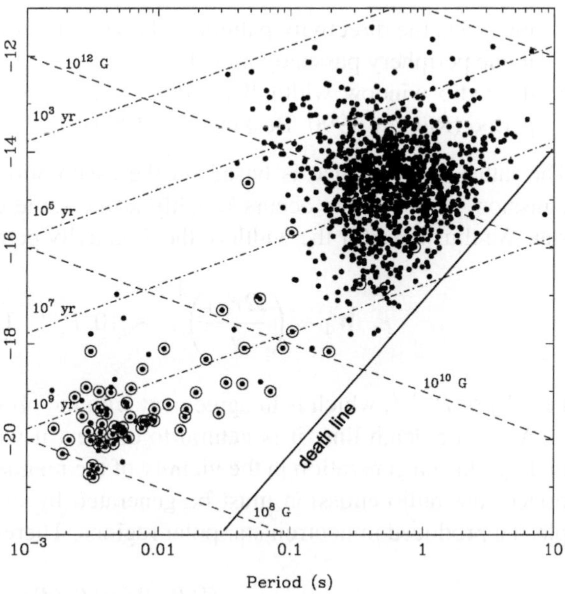

Fig. 2.7 Pulsar distribution in the P–$\dot{P}$ diagram. *Encircled dots* indicate radio pulsars in binary systems. *Dashed lines* indicate magnetic field B_0 evaluated by magnetodipole formula (2.5), *dashed and dotted lines* indicate dynamical age τ_D (Seiradakis and Wielebinski, 2004). The *death line* corresponds to the relation $H = R_0$

As a result, it was possible to explain all the basic properties of the pulsar radio emission such as

- the death line in the P–$\dot{P}$ diagram (see Fig. 2.7);
- the statistical distribution of pulsars with single and double mean profiles (double profiles are mainly observed in pulsars in the vicinity of the death line when particles can be generated only in a thin ring in the vicinity of the polar cap boundary) (Beskin et al., 1993);
- the characteristic S-shaped change in the position angle of the linear polarization along the mean profile (Radhakrishnan and Cooke, 1969) (as shown in Fig. 2.8, the complete change in the position angle is close to 180° if the line of sight

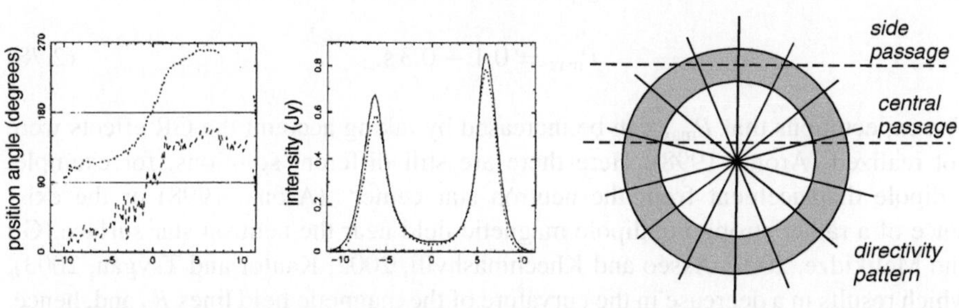

Fig. 2.8 The change in the position angle (*left panel*) of two linear polarizations along the double mean profile, which is naturally connected with the change in the magnetic field orientation (*right panel, radial lines*) in the picture plane. With the central passage of the directivity pattern, the change in the position angle is close to 180° (with side passage, it is much less)

intersects the directivity pattern in the vicinity of its center and the small change in the periphery passage); and also

- the radio window width W_d and even its statistical dependence on the pulsar period (Rankin, 1990; Beskin et al., 1993).

The latter circumstance is based on the assumption that the generation of radio emission in all pulsars occurs roughly at the same distance r_rad from the neutron star. We thus have for the width of the directivity pattern W_d

$$W_\mathrm{d} \approx \left(\frac{\Omega r_\mathrm{rad}}{c}\right)^{1/2} \approx 10^\circ P^{-1/2} \left(\frac{r_\mathrm{rad}}{10R}\right)^{1/2}, \tag{2.75}$$

i.e., $W_\mathrm{d} \propto P^{-1/2}$, which is in agreement with the observations.

As to the death line, it is natural to connect it with the termination of the secondary plasma generation in the vicinity of the magnetic poles. Indeed, as was mentioned, the radio emission must be generated by the secondary electron–positron plasma produced in neutron star polar regions. Therefore, the condition

$$H(P, B) = R_0(P) \tag{2.76}$$

(i.e., $\psi = \psi_\mathrm{max}$) can be regarded as an “ignition condition” dividing the active and passive ranges of parameters when the neutron star does not manifest itself as a radio pulsar. In the nonfree particle escape model, relation (2.76) can be rewritten as (Ruderman and Sutherland, 1975; Beskin et al., 1984)

$$P_\mathrm{max} \approx 1\,\mathrm{s} \left(\frac{B_0}{10^{12}\,\mathrm{G}}\right)^{8/15} \approx 1\text{–}3\,\mathrm{s}. \tag{2.77}$$

This condition is usually represented as a “death line” in the P–$\dot{P}$ diagram. This satisfactory agreement can, unconditionally, be regarded as the confirmation of the pattern discussed here. For the free particle escape model, because of the much smaller values of the accelerating potential, the limit period must be smaller:

$$P_\mathrm{max} = 0.1 - 0.3\,\mathrm{s}. \tag{2.78}$$

The expectations that P_max can be increased by taking account the GR effects were not realized (Arons, 1998). Here there are still different solutions, for example, a dipole displacement from the neutron star center (Arons, 1998) or the existence of a rather strong nondipole magnetic field near the neutron star surface (Gil and Melikidze, 2002; Asséo and Khechinashvili, 2002; Kantor and Tsygan, 2003), which results in a decrease in the curvature of the magnetic field lines R_c and, hence, in an increase in the particle generation efficiency. Nevertheless, as we see, the free particle escape models encounter certain difficulties.

Note also that for the nonfree particle escape models, it is convenient to introduce the dimensionless parameter Q

$$Q = 2\left(\frac{P}{1\,s}\right)^{11/10}\left(\frac{\dot{P}}{10^{-15}}\right)^{-4/10}, \tag{2.79}$$

determined, as we see, directly from the observations. It turns out to be an extremely convenient parameter characterizing the main characteristics of radio pulsars (Beskin et al., 1984; Taylor and Stinebring, 1986; Rankin, 1990). For example, the ratios of the inner radius of the hollow cone near the star surface $r_{\rm in}$ and the inner gap height H to the polar cap radius R_0 are written as

$$\frac{r_{\rm in}}{R_0} \approx Q^{7/9}, \tag{2.80}$$

$$\frac{H}{R_0} \approx Q. \tag{2.81}$$

Therefore, the pulsars with $Q > 1$, in which the directivity pattern is a rather narrow cone, mostly have a double mean profile of the radio emission. It is in these pulsars that various irregularities, such as the full radio emission termination (nulling), mode switching, are detected. Conversely, the pulsars with $Q \ll 1$ ($r_{\rm in} \ll R_0$) are characterized by stable radio emission, and their mean profiles are mostly of a single type.

Finally, some properties of radio pulsars (for example, subpulse drift) indirectly confirm the existence of the potential drop and the particle acceleration over the magnetic poles of the neutron star (Ruderman and Sutherland, 1975). Indeed, if in the vicinity of the pulsar surface there is a longitudinal electric field region on the open field lines, an additional potential difference develops between the central and periphery domains over the acceleration region so that the additional electric field is directed to or from the magnetic axis (see Fig. 2.9). As a result, besides the general motion around the rotation axis, the additional electric drift results in the plasma

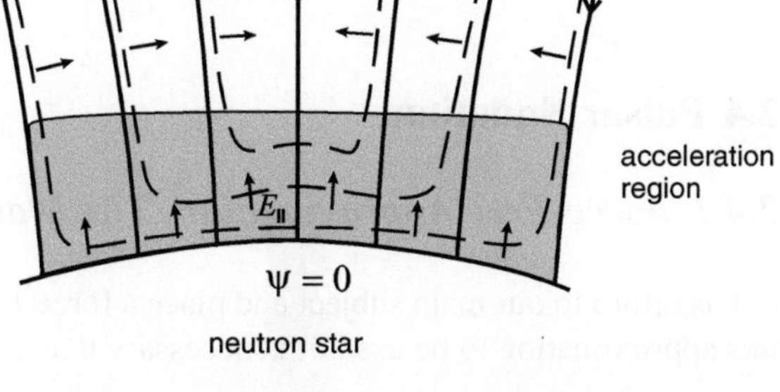

Fig. 2.9 Equipotential surfaces ψ = const (*dashed lines*) in the region of the open field lines. The potential drop in the acceleration region gives rise to an additional potential difference between the magnetic surfaces. The electric drift produced by the additional electric field (*fine arrows*) results in an additional plasma rotation around the magnetic axis

rotation around the magnetic axis, which, in turn, can be observed as the regular drift of radiating regions within the mean pulse (see Fig. 2.6). About 200 radio pulsars with drifting subpulses are known now (Lyne and Graham-Smith, 1998; Weltevrede et al., 2007).

2.3.7 Secondary Plasma Generation—"Outer Gap"

Finally, we should point to another particle generation mechanism that can occur already far from the neutron star. As seen from Fig. 2.3, on some open field lines, where $\boldsymbol{\Omega} \cdot \mathbf{B} = 0$, the charge density, according to (2.39), changes the sign. Clearly, the charge-separated plasma outflowing from the star could not ensure the fulfillment of the condition $\rho_{\rm e} = \rho_{\rm GJ}$. Therefore, the hypothesis for the existence of an "outer gap" in the vicinity of the line $\rho_{\rm GJ}=0$ was put forward, in which the emerging longitudinal electric field also produces the secondary plasma. However, since, because of a weak magnetic field, the one-photon conversion becomes impossible, the main particle generation mechanism is the two-photon conversion process $\gamma + \gamma \rightarrow e^{+} + e^{-}$ (Cheng et al., 1986). At present, the thorough computations of cascade processes in the outer gap were carried out and their aim was to explain the high-energy radiation of radio pulsars (Chiang and Romani, 1994; Zhang and Cheng, 1997; Cheng et al., 2000; Hirotani and Shibata, 2001). The chain of processes is the following:

1. The occurrence of the longitudinal electric field, because the condition $\rho_{\rm e} = \rho_{\rm GJ}$ cannot be satisfied.
2. The acceleration of primary particles.
3. The emission of curvature photons.
4. The IC scattering of thermal X-ray photons emitted from the neutron star surface.
5. The secondary particles generated by the collision of high-energy IC γ-quanta with soft X-ray photons.

Certainly, in the real conditions, plasma outflowing from the magnetosphere contains particles of both signs so that, in principle, the condition $\rho_{\rm e} = \rho_{\rm GJ}$ could be satisfied by slightly changing the longitudinal particle velocities. However, this problem, which requires, generally speaking, kinetic analysis, has not been solved yet (see, e.g., Lyubarskii, 1995).

2.4 Pulsar Equation

2.4.1 Force-Free Approximation. The Magnetization Parameter

Let us return to our main subject and place a force-free limit to the GS equation. For this approximation to be used, it is necessary that

1. the plasma energy density ϵ_{part} is much smaller than the energy density of the electromagnetic field ϵ_{em};
2. the amount of plasma is enough to screen the longitudinal electric field $E_{\parallel}$.

The force-free approximation must be valid in the radio pulsar magnetosphere with large margin, because the plasma filling the magnetosphere is secondary with respect to the magnetic field. Following Michel (1969), for a quantitative estimate, one can introduce the magnetization parameter

$$\sigma = \frac{e\Omega\Psi_{\text{tot}}}{4\lambda m_e c^3}, \tag{2.82}$$

where Ψ_{tot} is the total magnetic flux and $\lambda = n/n_{\text{GJ}}$ (2.74) is the multiplicity of particle generation. One should, however, stress that in Michel (1969), the case of the monopole magnetic field was considered for simplicity. Therefore, we must be careful when determining this value for concrete astrophysical objects. In particular, for radio pulsars

$$\Psi_{\text{tot}} \approx \pi B_0 R_0^2 \approx \pi B_0 R^2 \frac{\Omega R}{c}, \tag{2.83}$$

which corresponds to the magnetic flux only in the region of open field lines. Therefore, for the radio pulsar magnetosphere

$$\sigma = \frac{e B_0 \Omega^2 R^3}{4\lambda m_e c^4}. \tag{2.84}$$

As a result, the smallness condition of the particle contribution to the energy–momentum tensor $T^{\alpha\beta}_{\text{part}} \ll T^{\alpha\beta}_{\text{em}}$ up to the light cylinder can be written as

$$\sigma \gg \gamma_{\text{in}}. \tag{2.85}$$

Here $\gamma_{\text{in}} \sim 10^2$–$10^4$ is the characteristic Lorentz factor of the plasma near the star surface.

Problem 2.11 Using definitions (2.74) and (2.84), check that relation (2.85) really corresponds to the smallness condition of the particle contribution (up to the light cylinder!) for the component T^{00}, i.e., for the energy density.

The magnetization parameter is one of the key dimensionless parameters characterizing the relativistic plasma moving in the magnetic field. As we see, up to the factor γ_{in}, it coincides with the ratio of the electromagnetic energy flux to the particle energy flux. In particular, the large value of σ shows that the main contribution to the energy flux in the interior regions of the magnetosphere is made by the electromagnetic flux. For the characteristic parameters of radio pulsar ($P \sim 1$ s, $B_0 \sim 10^{12}$

G), we have $\sigma \sim 10^4$–10^5, and only for the youngest ones ($P \sim 0.1$ s, $B_0 \sim 10^{13}$ G) the value $\sigma \sim 10^6$. Nevertheless, the condition $\sigma \gg \gamma_{\rm in}$ turns out to be satisfied. As to the screening of longitudinal electric field, this condition must also be satisfied with large margin by relation $\lambda \gg 1$ (2.74).

Thus, in the zero order with respect to the parameters σ^{-1} and λ^{-1}, the radio pulsar magnetosphere can actually be described by the force-free approximation. The force-free approximation implies that in the general equation—the energy–momentum conservation law $\nabla_\alpha T^{\alpha\beta} = 0$—we can now disregard the particle contribution. Using the explicit form of the energy–momentum tensor of the electromagnetic field (Landau and Lifshits, 1989)

$$T^{\alpha\beta}_{\rm em} = \begin{pmatrix} \dfrac{(E^2 + B^2)}{8\pi} & \dfrac{c}{4\pi}\mathbf{E}\times\mathbf{B} \\ \dfrac{c}{4\pi}\mathbf{E}\times\mathbf{B} & -\dfrac{1}{4\pi}(E^i E^k + B^i B^k) + \dfrac{1}{8\pi}(E^2 + B^2)\,\delta^{ik} \end{pmatrix}, \tag{2.86}$$

we obtain for the space components the known equation

$$\frac{1}{c}\mathbf{j}\times\mathbf{B} + \rho_{\rm e}\mathbf{E} = 0, \tag{2.87}$$

or

$$[\nabla\times\mathbf{B}]\times\mathbf{B} + (\nabla\cdot\mathbf{E})\,\mathbf{E} = 0. \tag{2.88}$$

Equation (2.87) in the nonrelativistic limit naturally reduces to zero of Ampére's force $\mathbf{F}_{\rm A} = \mathbf{j}\times\mathbf{B}/c$. Therefore, the approximation studied is called the force-free approximation.

2.4.2 Integrals of Motion

Recall now that we are, first of all, interested in axisymmetric stationary configurations. In this case, it is convenient to take, as an unknown variable, the magnetic flux function $\Psi(r,\theta)$. Strictly, it was just the method first successfully used by H. Grad (1960) and V.D. Shafranov (1958).

Thus, we write the magnetic field as

$$\mathbf{B} = \frac{\nabla\Psi\times\mathbf{e}_\varphi}{2\pi\varpi} - \frac{2I}{c\varpi}\mathbf{e}_\varphi, \tag{2.89}$$

dependent on two scalar functions $\Psi(r,\theta)$ and $I(r,\theta)$. Here the numerical coefficient in the first term is chosen so that the function $\Psi(r,\theta)$ coincides with the magnetic flux passing through the circle r, θ, $0 < \varphi < 2\pi$ (see Fig. 2.10).

Indeed, the definition of the magnetic flux function is quite analogous to that of the stream function $\Phi(r,\theta)$ (1.90) introduced in Sect. 2.4.2. Therefore, all the basic properties retain.

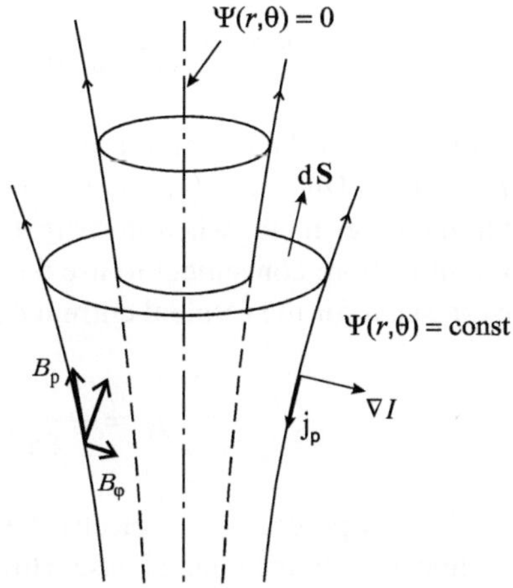

Fig. 2.10 Axisymmetric magnetic surfaces $\Psi(r,\theta) =$ const. For the case $\Psi > 0$, the GJ charge density $\rho_{GJ} < 0$. Therefore, in the vicinity of the north polar cap, I is positive and the current $\mathbf{j}_p$ is antiparallel to the magnetic field $\mathbf{B}$

- The condition $\mathrm{d}\Psi = \mathbf{B}\cdot\mathrm{d}\mathbf{S}$ is always satisfied ($\mathrm{d}\mathbf{S}$—an area element). Therefore, the function $\Psi(r,\theta)$ has the meaning of a magnetic flux.
- The condition $\nabla\cdot\mathbf{B} = 0$ is satisfied automatically. Therefore, three magnetic field components are fully specified by two scalar functions $\Psi(r,\theta)$ and $I(r,\theta)$.
- The condition $\mathbf{B}\cdot\nabla\Psi = 0$ is also satisfied. Therefore, the lines $\Psi(r,\theta) =$ const prescribe the form of the magnetic surfaces.

As to $I(r,\theta)$, it is the total electric current passing through the same circle. We can easily verify this fact by the obvious relation $\int B_\varphi \mathrm{d}\varphi = -(4\pi/\varpi c)I$. The minus sign in this expression and in the toroidal magnetic field expression (2.89) is chosen from the condition that the value I is positive for the electric current connected with the GJ charge density outflow. For the case $\Psi > 0$ shown in Fig. 2.10, the GJ charge density is negative, viz., $\rho_{GJ} < 0$ (and, conversely, $\rho_{GJ} > 0$ for $\Psi < 0$). Therefore, in the vicinity of the north polar cap, the current $\mathbf{j}_p$ is always antiparallel to the magnetic field $\mathbf{B}$. Having written the definition of the poloidal density of the electric current as

$$\mathbf{j}_p = -\frac{\nabla I \times \mathbf{e}_\varphi}{2\pi\varpi}, \tag{2.90}$$

we obtain the same set of properties as for the magnetic flux function.

- The condition $\mathrm{d}I = -\mathbf{j}\cdot\mathrm{d}\mathbf{S}$ is satisfied. Therefore, the function $I(r,\theta)$ has the meaning of the total electric current inflowing into the magnetosphere.
- The continuity condition $\nabla\cdot\mathbf{j} = 0$ is satisfied automatically (recall that we consider the stationary configurations only).
- The condition $\mathbf{j}\cdot\nabla I = 0$ is satisfied. Therefore, the lines $I(r,\theta) =$ const prescribe the form of the current surfaces in the magnetosphere.

Finally, the toroidal electric current can easily be determined from the φ-component of Maxwell's equation $\nabla\times\mathbf{B} = (4\pi/c)\mathbf{j}$. Thus, using the definition (2.89), we have

$$j_\varphi = -\frac{c}{8\pi^2 r \sin\theta}\left[\frac{\partial^2\Psi}{\partial r^2} + \frac{\sin\theta}{r^2}\frac{\partial}{\partial\theta}\left(\frac{1}{\sin\theta}\frac{\partial\Psi}{\partial\theta}\right)\right]. \tag{2.91}$$

As we see, in the definition of the toroidal current density j_φ, the known operator $\hat{\mathcal{L}} = \varpi^2\nabla_k\left(\varpi^{-2}\nabla^k\right)$ (1.119) written in the spherical coordinates is available again. On the other hand, when investigating the radio pulsar magnetosphere, as we will see, it is more convenient to use the cylindrical coordinates (ϖ, z). In this case, the expression for the toroidal current density looks like

$$j_\varphi = -\frac{c}{8\pi^2\varpi}\left[\nabla^2\Psi - \frac{2}{\varpi}\frac{\partial\Psi}{\partial\varpi}\right]. \tag{2.92}$$

We now proceed to the electric field definition. Naturally, it has three independent components in the general case. However,

1. Maxwell's equation $\nabla \times \mathbf{E} = 0$, in the axisymmetric case, yields the condition $E_\varphi = 0$;
2. the full screening assumption yields $E_\parallel = 0$.

Thus, it is convenient to write the electric field as

$$\mathbf{E} = -\frac{\Omega_{\rm F}}{2\pi c}\nabla\Psi, \tag{2.93}$$

i.e., express it in terms of one scalar function $\Omega_{\rm F}(r,\theta)$.

This expression yields the following important properties:

- The condition $\mathbf{E}\cdot\mathbf{B} = 0$ is satisfied automatically.
- From Maxwell's equation $\nabla \times \mathbf{E} = 0$, it follows that $\nabla\Omega_{\rm F} \times \nabla\Psi = 0$. In the axisymmetric case, where all the values depend only on two variables, this implies that

$$\Omega_{\rm F} = \Omega_{\rm F}(\Psi), \tag{2.94}$$

  i.e., the surfaces $\Omega_{\rm F}(r,\theta) = \text{const}$ are to coincide with the magnetic surfaces $\Psi(r,\theta) = \text{const}$.
- The drift velocity $\mathbf{U}_{\rm dr} = c\,\mathbf{E}\times\mathbf{B}/B^2$, as was mentioned, is now written as

$$\mathbf{U}_{\rm dr} = \mathbf{\Omega}_{\rm F} \times \mathbf{r} + j_\parallel\mathbf{B}, \tag{2.95}$$

  where again $j_\parallel$ is some scalar function. As we see, the introduced function $\Omega_{\rm F}$ has the meaning of the angular velocity of particles moving in the magnetosphere. The condition (2.94) is the known Ferraro isorotation law (Ferraro, 1937; Alfven and Fälthammar, 1963) according to which the particle angular velocity is to be constant on the axisymmetric magnetic surfaces.

Finally, using definitions (2.89) and (2.90) for $\mathbf{B}$ and $\mathbf{j}_\mathrm{p}$, we can write the toroidal component of Eq. (2.88) as $[\nabla I \times \mathbf{e}_\varphi] \times [\nabla \Psi \times \mathbf{e}_\varphi] = \nabla I \times \nabla \Psi = 0$. Consequently, the total current inside the magnetic surface is also an integral of motion:

$$I = I(\Psi). \tag{2.96}$$

Problem 2.12 Show that in the force-free limit the total energy and angular momentum losses are now defined as

$$W_\mathrm{tot} = \frac{1}{c}\int E(\Psi)\mathrm{d}\Psi, \qquad K_\mathrm{tot} = \frac{1}{c}\int L(\Psi)\mathrm{d}\Psi, \tag{2.97}$$

where

$$E(\Psi) = \frac{\Omega_\mathrm{F} I}{2\pi}, \tag{2.98}$$

$$L(\Psi) = \frac{I}{2\pi}. \tag{2.99}$$

2.4.3 Grad–Shafranov Equation

We are now ready to formulate the GS equation describing the poloidal structure of the magnetic field. As in the hydrodynamical case, we write the poloidal component of Eq. (2.87) as

$$\frac{j_\varphi}{c}\nabla\Psi + \frac{B_\varphi}{c}\nabla I - \frac{\nabla\cdot\mathbf{E}}{4\pi}\frac{\Omega_\mathrm{F}}{2\pi\varpi}\nabla\Psi = 0. \tag{2.100}$$

This vector equation, under the condition $\nabla I = (\mathrm{d}I/\mathrm{d}\Psi)\nabla\Psi$ resulting from (2.96), can again be reduced to the scalar equation multiplied by $\nabla\Psi$. In the cylindrical coordinates, it has the form

$$-\left(1 - \frac{\Omega_\mathrm{F}^2\varpi^2}{c^2}\right)\nabla^2\Psi + \frac{2}{\varpi}\frac{\partial\Psi}{\partial\varpi} - \frac{16\pi^2}{c^2}I\frac{\mathrm{d}I}{\mathrm{d}\Psi} + \frac{\varpi^2}{c^2}(\nabla\Psi)^2\,\Omega_\mathrm{F}\frac{\mathrm{d}\Omega_\mathrm{F}}{\mathrm{d}\Psi} = 0, \tag{2.101}$$

where ∇^2 is the Laplace operator. It is just the pulsar equation obtained in dozens of papers in the 1970s (see, e.g., Mestel (1973); Scharlemann and Wagoner (1973); Michel (1973a); Mestel and Wang (1979); the final version containing the latter term was deduced by Okamoto (1974)). The nonrelativistic version of the force-free GS equation is formulated in Appendix B.

The pulsar equation has the following properties:

- As any GS equation, it comprises only the stream function $\Psi(\varpi, z)$ and the invariants $\Omega_{\rm F}(\Psi)$ and $I(\Psi)$.
- On the other hand, the force-free equation does not contain any additional parameters associated with the plasma properties; therefore, it must not be supplemented with Bernoulli's equation.
- Equation (2.101) remains elliptic over the entire space where it is defined; this observation, as we will see, is very important. Indeed, the force-free equation (2.87) has meaning only if the condition $|\mathbf{E}| < |\mathbf{B}|$ is satisfied, whereas Eq. (2.101) can formally be extended to the nonphysical domain $|\mathbf{E}| > |\mathbf{B}|$.
- The differential operator

$$\hat{\mathcal{L}}_{\rm psr} = \left(1 - \frac{\Omega_{\rm F}^2 \varpi^2}{c^2}\right) \nabla^2 \Psi - \frac{2}{\varpi} \frac{\partial \Psi}{\partial \varpi} \tag{2.102}$$

  is linear in the derivatives Ψ; for $\Omega_{\rm F} = {\rm const}$, all nonlinearity of the pulsar equation is only in the last two terms associated with the integrals of motion.
- The differential operator (2.102) does not explicitly contain the coordinate z.
- At small distances, as compared to the light cylinder radius $\varpi \ll R_{\rm L}$, the differential operator $\mathcal{L}_{\rm psr}$ coincides with $\hat{\mathcal{L}}$ (1.119).
- The equation contains one critical surface—the light cylinder $\varpi_{\rm L} = c/\Omega_{\rm F}$.
- For known flow structure (i.e., given $\Psi(\varpi, z)$, $\Omega_{\rm F}(\Psi)$, and $I(\Psi)$), the electric field and the toroidal component of the magnetic field are specified from the algebraic relations.
- According to the general formula $b = 2 + i - s'$ for the number of boundary conditions, we have $b = 3$, i.e., the problem requires three boundary conditions.

For example, within the analytical approach, it is convenient to take, as such boundary conditions, two integrals of motion $\Omega_{\rm F} = \Omega_{\rm F}(\Psi)$ and $I = I(\Psi)$, as well as the normal component of the magnetic field on the neutron star surface $r = R$ or, what is the same, the magnetic flux $\Psi = \Psi(R,\theta)$. Thus, for example, for the dipole magnetic field

$$\Psi(R, \theta) \approx |\mathbf{m}| \frac{\sin^2 \theta}{R}. \tag{2.103}$$

Here $\mathbf{m}$ is the magnetic moment of the neutron star. But in this case, it is not clear whether the solution can be extended to infinity. Therefore, in numerical simulations, one generally uses another set of boundary conditions, viz., the angular velocity $\Omega_{\rm F} = \Omega_{\rm F}(\Psi)$ and the magnetic flux Ψ both on the neutron star surface and "at infinity" (i.e., on the outer boundary of the computational domain). Then the current $I(\Psi)$ is to be determined from the solution.

It is very important that Eq. (2.101) contains two key values—the longitudinal current I and the angular rotational velocity $\Omega_{\rm F}$, the latter is directly associated with the voltage drop in the inner gap. Indeed, as shown in the following section, the electric and magnetic fields for the arbitrary inclination angle χ must be connected by the relation

$$\mathbf{E} + \frac{\boldsymbol{\Omega} \times \mathbf{r}}{c} \times \mathbf{B} = -\nabla \psi, \tag{2.104}$$

where ψ at small distances $\varpi \ll R_{\rm L}$ has the meaning of the electric potential in the rotating coordinate system. In particular, since in the interior of a perfectly conducting star $\mathbf{E}_{\rm in} + (\boldsymbol{\Omega} \times \mathbf{r}/c) \times \mathbf{B}_{\rm in} = 0$, we have $\psi_{\rm in} = 0$. On the other hand, for the case of the zero longitudinal electric field $E_{\parallel} = 0$, we have $\mathbf{B} \cdot \nabla \psi = 0$. Thus, in the domain, where the condition $E_{\parallel} = 0$ is satisfied, the potential ψ must be constant on the magnetic surfaces

$$\psi = \psi(\Psi). \tag{2.105}$$

Hence, in the region of the closed magnetic field lines (i.e., the field lines not outgoing beyond the light cylinder), we simply have $\psi = 0$. On the other hand, in the region of the open field lines, which are separated from the neutron star by the longitudinal electric field region, the potential ψ is different from zero (see Fig. 2.9). Its value coincides with the electric potential drop in the particle generation region. The occurrence of the nonzero potential ψ in the region of the open field lines leads to additional plasma rotation around the magnetic axis, which is observed as a subpulse drift (see Fig. 2.6).

Indeed, using the definition of the electric field (2.93), we find that in the axisymmetric case the angular velocity $\Omega_{\rm F}$ can be written as

$$\Omega_{\rm F} = \Omega + 2\pi c \frac{{\rm d}\psi}{{\rm d}\Psi}. \tag{2.106}$$

It is easy to verify that the derivative ${\rm d}\psi/{\rm d}\Psi$ is always negative, so the plasma angular velocity $\Omega_{\rm F}$ is always smaller than the angular velocity of the neutron star Ω. The value $\psi(P, B_0)$ is determined by the concrete particle generation mechanism. In the following, it is convenient to introduce the dimensionless accelerating potential

$$\beta_0 = \frac{\psi(P, B_0)}{\psi_{\rm max}}, \tag{2.107}$$

where $\psi_{\rm max}$ (2.43) is the maximum potential drop in the acceleration region. As a result, the angular velocity $\Omega_{\rm F}$ over the acceleration region, where the secondary plasma screens the longitudinal electric field (and, therefore, the GS equation method can be used), is simply determined by $\Omega_{\rm F} = (1 - \beta_0)\Omega$. As to the longitudinal currents, it is convenient to normalize them to the GJ current density $j_{\rm GJ} = c\rho_{\rm GJ}$. As a result, we can write

$$I(\Psi_{\rm tot}) = i_0 I_{\rm GJ}, \tag{2.108}$$

where

$$I_{\mathrm{GJ}} = \frac{B_0 \Omega^2 R^3}{2c} \tag{2.109}$$

is the characteristic total current across the polar cap surface.

2.4.4 Mathematical Intermezzo—Quasistationary Formalism

In this section, we call attention to some relations involving the quasistationary generalization of the above equations describing the magnetosphere of an inclined rotator. The assumption of quasistationarity implies that we consider the electromagnetic fields that depend on time t and angular coordinate φ only in $\varphi - \Omega t$ combination. Note that the condition for quasistationarity is wider than the condition for time independence of all values in the reference frame rotating with angular velocity Ω, because the quasistationarity condition can be extended beyond the light cylinder where the rotation with angular velocity Ω is impossible. In particular, the spherical wave (2.15), (2.16), (2.17), (2.18), (2.19), and (2.20) emitted by the rotating neutron star in vacuum satisfies the quasistationarity condition.

When the time dependence is available in all equations only in the $\varphi - \Omega t$ combination, all time derivatives can be replaced by derivatives with respect to the coordinates using the relations (Mestel, 1973)

$$\frac{\partial}{\partial t} Q = -\Omega \frac{\partial}{\partial \varphi} Q, \tag{2.110}$$

$$\frac{\partial}{\partial t} \mathbf{V} = -(\boldsymbol{\Omega} \times \mathbf{r}, \nabla)\mathbf{V} + \boldsymbol{\Omega} \times \mathbf{V} \tag{2.111}$$

for the arbitrary scalar $Q(\varpi, \varphi - \Omega t, z)$ and the vector $\mathbf{V}(\varpi, \varphi - \Omega t, z)$ fields. Using now the known vector relation $\nabla \times [\mathbf{U} \times \mathbf{V}] = -(\mathbf{U}\nabla)\mathbf{V} + (\mathbf{V}\nabla)\mathbf{U} + (\nabla \cdot \mathbf{V})\mathbf{U} - (\nabla \cdot \mathbf{U})\mathbf{V}$, we can rewrite the condition (2.111) as

$$\frac{1}{c}\frac{\partial}{\partial t} \mathbf{V} = \nabla \times [\boldsymbol{\beta}_{\mathrm{R}} \times \mathbf{V}] - (\nabla \cdot \mathbf{V})\boldsymbol{\beta}_{\mathrm{R}}. \tag{2.112}$$

Hereafter, by definition,

$$\boldsymbol{\beta}_{\mathrm{R}} = \frac{\boldsymbol{\Omega} \times \mathbf{r}}{c} \tag{2.113}$$

is the corotation vector. As is easily checked, $\nabla \cdot \boldsymbol{\beta}_{\mathrm{R}} = 0$.

Problem 2.13 Check relations (2.110), (2.111), and (2.112).

Using relations (2.110), (2.111), and (2.112), we can rewrite Maxwell's equation as

$$\nabla \cdot \mathbf{E} = 4\pi \rho_{\rm e}, \tag{2.114}$$

$$\nabla \times \mathbf{E} = -\nabla \times [\boldsymbol{\beta}_{\rm R} \times \mathbf{B}], \tag{2.115}$$

$$\nabla \cdot \mathbf{B} = 0, \tag{2.116}$$

$$\nabla \times \mathbf{B} = \nabla \times [\boldsymbol{\beta}_{\rm R} \times \mathbf{E}] + \frac{4\pi}{c}\mathbf{j} - 4\pi \rho_{\rm e} \boldsymbol{\beta}_{\rm R}. \tag{2.117}$$

Equation (2.115) just yields relation $\mathbf{E} + \boldsymbol{\beta}_{\rm R} \times \mathbf{B} = -\nabla\psi$ (2.104), where

$$\psi = \Phi_{\rm e} - (\boldsymbol{\beta}_{\rm R} \cdot \mathbf{A}), \tag{2.118}$$

and $\Phi_{\rm e}$ and $\mathbf{A}$ are, respectively, the scalar and vector potentials of the electromagnetic field.

If the $(4\pi/c)\mathbf{j} - 4\pi\rho_{\rm e}\boldsymbol{\beta}_{\rm R}$ combination in (2.117) is also zero (for example, this is the case for the vacuum approximation), this equation can be resolved as

$$\mathbf{B} - \boldsymbol{\beta}_{\rm R} \times \mathbf{E} = -\nabla h, \tag{2.119}$$

where $h(\varpi, \varphi - \Omega t, z)$ is an arbitrary scalar function. In this case, the electric and magnetic fields are expressed in terms of the potentials ψ and h as

$$\mathbf{E}_{\rm p} = \frac{1}{1 - \boldsymbol{\beta}_{\rm R}^2}\left(-\nabla\psi + \boldsymbol{\beta}_{\rm R} \times \nabla h\right), \tag{2.120}$$

$$E_{\varphi} = -\frac{1}{\varpi}\frac{\partial \psi}{\partial \varphi}, \tag{2.121}$$

$$\mathbf{B}_{\rm p} = \frac{1}{1 - \boldsymbol{\beta}_{\rm R}^2}\left(-\nabla h - \boldsymbol{\beta}_{\rm R} \times \nabla\psi\right), \tag{2.122}$$

$$B_{\varphi} = \frac{1}{\varpi}\frac{\partial h}{\partial \varphi}. \tag{2.123}$$

Substituting these expressions in equations $\nabla \cdot \mathbf{E} = 0$ and $\nabla \cdot \mathbf{B} = 0$ valid for the vacuum case, we obtain the system of equations (Beskin et al., 1993)

$$\hat{\mathcal{L}}_2\psi - \frac{2}{1 - x_r^2}\frac{\partial h}{\partial z'} = 0, \tag{2.124}$$

$$\hat{\mathcal{L}}_2 h + \frac{2}{1 - x_r^2}\frac{\partial \psi}{\partial z'} = 0, \tag{2.125}$$

where $x_r = \Omega\varpi/c$, $z' = \Omega z/c$, and the operator $\hat{\mathcal{L}}_2$ is

$$\hat{\mathcal{L}}_2 = \frac{\partial^2}{\partial x_r^2} + \frac{1}{x_r}\frac{1+x_r^2}{1-x_r^2}\frac{\partial}{\partial x_r} + \frac{1-x_r^2}{x_r^2}\frac{\partial^2}{\partial \varphi^2} + \frac{\partial^2}{\partial z'^2}. \qquad (2.126)$$

Problem 2.14 Check that the solutions to system (2.124) and (2.125) for the orthogonal rotator (i.e., if $\sin\chi = 1$)

$$h = |\mathbf{m}| \sin\theta \,\mathrm{Re}\left(\frac{1}{r^2} - \mathrm{i}\frac{\Omega}{c}\frac{1}{r} - \frac{\Omega^2}{c^2}\right)\exp\left(\mathrm{i}\frac{\Omega r}{c} + \mathrm{i}\varphi - \mathrm{i}\Omega t\right) \qquad (2.127)$$

$$\psi = |\mathbf{m}| \sin\theta\cos\theta \,\mathrm{Re}\left(\frac{\Omega}{c}\frac{1}{r} - \mathrm{i}\frac{\Omega^2}{c^2}\right)\exp\left(\mathrm{i}\frac{\Omega r}{c} + \mathrm{i}\varphi - \mathrm{i}\Omega t\right) \qquad (2.128)$$

exactly correspond to the electromagnetic fields (2.15), (2.16), (2.17), (2.18), (2.19), and (2.20) for the rotating magnetic dipole.

Within the quasistationary approximation, we can write the general equation for the magnetic field. Indeed, the condition for constancy of the total current I (2.96) on the magnetic surfaces can be regarded as a consequence of Eq. (2.95) for the drift velocity $\mathbf{U}_{\mathrm{dr}}$. Therefore, the electric current can also be represented as the expansion $\mathbf{j} = \rho_{\mathrm{e}}\, \boldsymbol{\Omega} \times \mathbf{r} + i_{\parallel}\mathbf{B}$. Substituting this condition in the general equation (2.117), we readily see that $\nabla \cdot (i_{\parallel}\mathbf{B}) = 0$ and, hence, the function $i_{\parallel}$ must also be constant along the magnetic field lines, viz., $\mathbf{B} \cdot \nabla i_{\parallel} = 0$. In particular, if the longitudinal current is zero near the neutron star surface, it is to be zero in the entire magnetosphere. As a result, Eq. (2.117), with account taken of (2.104), can be rewritten as (Beskin et al., 1983)

$$\nabla \times \{(1-\boldsymbol{\beta}_{\mathrm{R}}^2)\mathbf{B} + \boldsymbol{\beta}_{\mathrm{R}}(\boldsymbol{\beta}_{\mathrm{R}}\cdot\mathbf{B}) + [\boldsymbol{\beta}_{\mathrm{R}} \times \nabla\psi]\} = \frac{4\pi}{1-\boldsymbol{\beta}_{\mathrm{R}}^2 + \boldsymbol{\beta}_{\mathrm{R}}[\nabla\psi\times\mathbf{B}]/B^2}\left[\frac{i_{\parallel}}{c}\left((1-\boldsymbol{\beta}_{\mathrm{R}}^2)\mathbf{B} + [\boldsymbol{\beta}_{\mathrm{R}}\times\nabla\psi]\right) + \frac{[\nabla\psi\times\mathbf{B}]}{B^2}\left(\frac{\boldsymbol{\Omega}\cdot\mathbf{B}}{2\pi c} + \frac{1}{4\pi}(\nabla^2\psi - (\boldsymbol{\beta}_{\mathrm{R}}\nabla)(\boldsymbol{\beta}_{\mathrm{R}}\nabla\psi))\right)\right]. \qquad (2.129)$$

Along with the equation $\nabla\cdot\mathbf{B} = 0$ (given the scalar functions $i_{\parallel}$ and ψ), it specifies the quasistationary magnetic field structure.

The quasistationary approximation is a natural generalization to axisymmetrical stationary configurations studied here. On the other hand, the possibility to use it seems unlikely. The point is that in the quasistationary case, it is impossible to introduce the analogue of the unique function Ψ describing the magnetic surfaces. As a result, one fails to reduce Maxwell's equations to a single scalar equation for the stream function by formalizing the constancy condition of the potential ψ and the current $i_{\parallel}$ along the given magnetic field line. Therefore, Eq. (2.129) was not essentially analyzed and its solutions were found only in the exceptional cases (Beskin et al., 1983; Mestel et al., 1999), where it was actually reduced to the system of equations (2.124), (2.125) for the scalar functions ψ and h.

2.5 Energy Losses of Radio Pulsars

2.5.1 Current Loss Mechanism

Before proceeding to the discussion of the exact solutions to the pulsar equation, we consider the problem of the energy losses of the rotating neutron star. As was noted, in the vacuum approximation, the only mechanism resulting in the pulsar slowing down is a magnetodipole radiation. However, in the case of the plasma-filled magnetosphere, another slowing-down mechanism connected with the electric currents flowing in the magnetosphere occurs.

Indeed, the total current outflowing from the pulsar surface is to be zero. On the other hand, as was specially noted above, the charges of the same sign are to outflow from both magnetic poles (the charge densities ρ_{GJ} in the vicinity of the magnetic poles are identical). Therefore, an inverse current making up for the charge loss of the neutron star must inevitably flow along the separatrix dividing the open and closed magnetic field lines. As a result, the currents $\mathbf{J}_s$ that close the longitudinal currents in the magnetosphere flow over the pulsar surface (see Fig. 2.11). The ponderomotive action of these currents must result in the slowing down of the radio pulsar rotation (Beskin et al., 1993). It is important that this slowing-down mechanism occurs for the axisymmetric rotator when the magnetodipole losses are obviously zero. Actually, this mechanism was developed even in P. Goldreich and P. Julian's (1969) pioneer paper that was devoted to the axisymmetric magnetosphere.

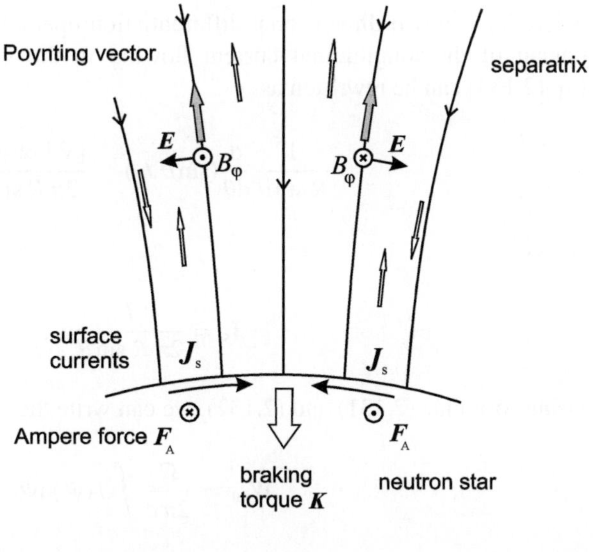

Fig. 2.11 Electric current structure (*contour arrows*) in the magnetic pole region of the neutron star. Ampére's force $\mathbf{F}_A$ connected with the surface current $\mathbf{J}_s$ generates the moment of force $\mathbf{K}$ resulting in the neutron star slowing down. For inclination angles χ not too close to 90°, the slowing-down moment $\mathbf{K}$ is antiparallel to the neutron star magnetic moment. The energy flux over the acceleration region is mainly connected with the Poynting vector (*shaded arrows*)

We first emphasize that if the energy losses of radio pulsars are really connected with the rotational kinetic energy loss of the neutron star, the total energy losses $W_{\text{tot}} = -I_{\text{r}}\Omega\dot{\Omega}$ and the angular momentum losses $K_{\text{tot}} = -I_{\text{r}}\dot{\Omega}$ should be connected by the relation

$$W_{\text{tot}} = \Omega K_{\text{tot}}. \tag{2.130}$$

Hence, the energy and the angular momentum for the outgoing radiation must satisfy the same condition.

To show that relation (2.130) really holds for the current losses, we write the energy losses as

$$W_{\text{tot}} = -\boldsymbol{\Omega} \cdot \mathbf{K}, \tag{2.131}$$

where

$$\mathbf{K} = \frac{1}{c}\int [\mathbf{r} \times [\mathbf{J}_{\text{s}} \times \mathbf{B}]]\mathrm{d}S \tag{2.132}$$

is a slowing-down moment connected with Ampére's force of the current flowing on the surface. Here, for simplicity, we consider the axisymmetric case. The general relations are given in the following section.

It is easy to show that for $\chi = 0°$, the slowing-down moment is exactly antiparallel to the neutron star angular velocity. The surface current $\mathbf{J}_{\text{s}}$ must satisfy the continuity equation

$$\nabla_2 \mathbf{J}_{\text{s}} = j_{\text{n}}, \tag{2.133}$$

where ∇_2 is a two-dimensional differentiation operator and j_{n} is the normal component of the longitudinal current flowing in the magnetosphere. As a result, Eq. (2.133) can be rewritten as

$$\frac{1}{R\sin\theta}\frac{\mathrm{d}}{\mathrm{d}\theta}(\sin\theta\, J_\theta) = \frac{[\nabla I \times \mathbf{e}_\varphi]_n}{2\pi R \sin\theta}. \tag{2.134}$$

It yields

$$\mathbf{J}_{\text{s}} = \frac{I}{2\pi R \sin\theta}\mathbf{e}_\theta. \tag{2.135}$$

Using formulae (2.131) and (2.132), we can write the total energy losses as

$$W_{\text{tot}} = \frac{\Omega}{2\pi c}\int I(\Psi)\mathrm{d}\Psi. \tag{2.136}$$

On the other hand, the total losses of the angular momentum K_{tot} (2.132) are rewritten as

$$K_{\text{tot}} = \frac{1}{2\pi c}\int I(\Psi)\mathrm{d}\Psi. \tag{2.137}$$

As a result, relation (2.130), as was expected, turns out to be identically valid for the current losses.

Besides, we should point out that expression (2.136), as is seen, can be expanded into the sum of two terms

$$W_{\text{tot}} = W_{\text{em}} + W_{\text{part}}. \tag{2.138}$$

Here the first term

$$W_{\text{em}} = \frac{1}{2\pi c}\int \Omega_{\text{F}}(\Psi)I(\Psi)\mathrm{d}\Psi, \tag{2.139}$$

according to definitions (2.89) and (2.93), is just the Poynting vector flux

$$W_{\text{em}} = \frac{c}{4\pi}\int [\mathbf{E}\times\mathbf{B}]\mathrm{d}\mathbf{S}. \tag{2.140}$$

Therefore, W_{em} corresponds to the electromagnetic energy flux flowing away from the neutron star. As is expected, the electromagnetic energy losses are different from zero only in the presence of the longitudinal electric current generating the toroidal magnetic field. Note that the energy is transported at zero frequency; therefore, the electromagnetic field transporting this energy is not an electromagnetic wave in an ordinary sense.

On the other hand, the second term

$$W_{\text{part}} = \frac{1}{2\pi c}\int I(\Psi)[\Omega - \Omega_{\text{F}}(\Psi)]\mathrm{d}\Psi, \tag{2.141}$$

according to relation (2.106), can be rewritten as

$$W_{\text{part}} = -\int \frac{\mathrm{d}\psi}{\mathrm{d}\Psi} I(\Psi)\,\mathrm{d}\Psi = -\int I(\Psi)\mathrm{d}\psi = \int \psi \mathrm{d}I = -\int \psi \mathbf{j}_{\text{e}}\mathrm{d}\mathbf{S}. \tag{2.142}$$

Here, when integrating by parts, we used the zero condition of the potential ψ on the polar cap boundary. As we see, the losses W_{part} correspond to the energy gained by primary particles in the acceleration region.

Problem 2.15 Show that relation (2.138) holds for any inclination angle χ and, in particular, for any form of the polar cap.

(Hint: since the source of both the surface current $\mathbf{J}_s$ and the additional magnetic field $\mathbf{B}_T$ is the longitudinal current $i_\parallel \mathbf{B}$ flowing in the region of open field lines ($\nabla \cdot \mathbf{J}_s = i_\parallel B_n$, $\nabla \times \mathbf{B}_T = (4\pi/c) i_\parallel \mathbf{B}$), as is easily checked, they are connected by the simple relation

$$\mathbf{J}_s = -\frac{c}{4\pi}[\mathbf{B}_T \times \mathbf{n}]. \tag{2.143}$$

As a result, formulae (2.131) and (2.132) valid for any inclination angle χ can be identically rewritten as

$$W_{\rm tot} = \frac{c}{4\pi} \int (\boldsymbol{\beta}_R \cdot \mathbf{B})(\mathbf{B} \cdot d\mathbf{S}). \tag{2.144}$$

Further, it is necessary to use relation (2.104) yielding the identity

$$[\mathbf{E} \times \mathbf{B}] d\mathbf{S} = (\boldsymbol{\beta}_R \cdot \mathbf{B})(\mathbf{B} \cdot d\mathbf{S}) + [\nabla\psi \times \mathbf{B}] d\mathbf{S} \tag{2.145}$$

and the condition $\psi = 0$ on the polar cap boundary.)

Thus, already from the analysis of the axisymmetric case, we can make a number of important conclusions.

1. The compatibility condition $W_{\rm tot} = \Omega K_{\rm tot}$ (2.130) cannot be obtained within the force-free approximation, because in this approximation there is no additional term $W_{\rm part}$ (2.141) corresponding to the energy of particles accelerated in the inner gap. Attempts to solve the loss problem by the force-free approximation, inevitably, lead to misunderstanding (Holloway, 1977; Shibata, 1994).
2. Under the condition $\psi \ll \psi_{\rm max}$, of major importance in the total balance of current losses is the electromagnetic energy flux at zero frequency $W_{\rm em}$ (2.139). But for pulsars located near the "death line" in the P–$\dot{P}$ diagram (for which the condition $\psi \sim \psi_{\rm max}$ is satisfied), the losses $W_{\rm part}$ correspond to the energy gained by primary particles in the acceleration region rather than to the energy of particles flowing along the open field lines. As was shown, a considerable part of the energy loss $W_{\rm part}$ is not used to generate particles but low-energy γ-quanta able to freely escape the neutron star magnetosphere. Therefore, the γ-quanta luminosity of radio pulsars located near the "death line" region is up to a few percent of the total losses $I_r \Omega \dot{\Omega}$. In these pulsars, the efficiency of the rotation energy processing in the high-energy radiation appears much larger than in the radio band. Consequently, the particle energy flux, at least, inside the light cylinder, appears much smaller than the flux $W_{\rm em}$ transported by the electromagnetic field. This fact just corresponds to the condition $\sigma \gg 1$ (2.82) valid for all radio pulsars.

3. On the other hand, for the slowing-down current mechanism discussed, the change in the angular moment $K_{\rm tot}$ is due to the electrodynamic losses (2.137). This must be the case as the angular momentum of photons $\mathcal{L}_{\rm ph}$ emitted in the vicinity of the star surface is much less than $\varepsilon_{\rm ph}/\Omega$. Therefore, the γ-quanta emitted in the vicinity of the neutron star surface cannot play a considerable role in the total balance of the angular momentum losses.

2.5.2 Slowing Down of Inclined and Orthogonal Rotators

We now discuss the problem of the energy losses of neutron stars for the arbitrary inclination angle χ. The necessity to do this is already obvious from an uncertainty in the expression for the energy losses of radio pulsars at the stage of the orthogonal rotator. The point is that the simple assumption based on the analysis of only the longitudinal currents results in a decrease in the factor $(\Omega R/c)^{1/2}$ as compared to the current losses of the axisymmetric rotator (Mestel et al., 1999). Indeed, let us estimate the energy losses by the Poynting vector flux through the light cylinder surface $R_{\rm L} = c/\Omega$

$$W_{\rm tot} = \frac{c}{4\pi}\int [\mathbf{E}\times\mathbf{B}]{\rm d}\mathbf{S} \sim cE(R_{\rm L})B_{\varphi}(R_{\rm L})R_{\rm L}^2. \tag{2.146}$$

The electric field in the vicinity of the light cylinder $E(R_{\rm L})$ is determined only by the value of the poloidal magnetic field $\mathbf{B}_{\rm p}$

$$E(R_{\rm L}) \approx \frac{\Omega R_{\rm L}}{c}B_{\rm p} \approx B_{\rm p}, \tag{2.147}$$

and according to the dependence $B \propto r^{-3}$ for the dipole magnetic field within the light cylinder, we have $B_{\rm p}(R_{\rm L}) \approx (\Omega R/c)^3 B_0$, where B_0 is a magnetic field on the neutron star surface. The toroidal magnetic field B_{φ} is connected with the longitudinal currents flowing in the magnetosphere. Therefore, the charge density of the orthogonal rotator within the polar cap $R_0 \sim (\Omega R/c)^{1/2}R$ is $\varepsilon_{\rm A} = (\Omega R/c)^{1/2}$ times less than that of the axisymmetric rotator. The toroidal magnetic field on the light cylinder can be estimated as

$$B_{\varphi}(R_{\rm L}) \approx \left(\frac{\Omega R}{c}\right)^{1/2} B_{\rm p}(R_{\rm L}), \tag{2.148}$$

which yields the additional factor $\varepsilon_{\rm A}$ in the expression for the energy losses. However, a comprehensive analysis shows that, in reality, a decrease in the factor must have the form $\varepsilon_{\rm A}^2 = (\Omega R/c)$, so that the total losses of the orthogonal rotator should be written as (Beskin et al., 1993; Beskin and Nokhrina, 2004)

$$W_{\rm tot}^{\rm orth} \approx \frac{B_0^2 \Omega^4 R^6}{c^3}\left(\frac{\Omega R}{c}\right). \tag{2.149}$$

To show this, we are to write the most general expression for the surface current $\mathbf{J}_{\rm s}$ in the presence of the strong magnetic field. It can be divided into two components, a parallel and a perpendicular one to the surface electric field $\mathbf{E}_{\rm s}$, i.e., we write the current $\mathbf{J}_{\rm s}$ as

$$\mathbf{J}_{\rm s} = \mathbf{J}_{\rm s}^{(1)} + \mathbf{J}_{\rm s}^{(2)}, \tag{2.150}$$

where

$$\mathbf{J}_{\rm s}^{(1)} = \Sigma_{||}\mathbf{E}_{\rm s}, \tag{2.151}$$

$$\mathbf{J}_{\rm s}^{(2)} = \Sigma_{\perp}\left[\frac{\mathbf{B}_{\rm n}}{B_{\rm n}} \times \mathbf{E}_{\rm s}\right]. \tag{2.152}$$

Here $\Sigma_{||}$ is the Pedersen conductivity and $\Sigma_{\perp}$ is the Hall conductivity. Suppose now that the pulsar surface conductivity perpendicular to the magnetic field is homogeneous and the field $\mathbf{E}_{\rm s}$ has the potential ξ'. Hence, relations (2.151) and (2.152) look like

$$\mathbf{J}_{\rm s}^{(1)} = \nabla\xi', \tag{2.153}$$

$$\mathbf{J}_{\rm s}^{(2)} = \frac{\Sigma_{\perp}}{\Sigma_{||}}\left[\frac{\mathbf{B}_{\rm n}}{B_{\rm n}} \times \nabla\xi'\right]. \tag{2.154}$$

Note at once that since the magnetic field structure in the vicinity of the pulsar surface is symmetric about the plane passing through the vectors of the angular velocity and the magnetic moment of the neutron star, the surface current should have the same symmetry. Thus, the currents proportional to $\Sigma_{\perp}$ do not contribute to the energy losses of the neutron star.

As a result, Eq. (2.133) is now rewritten as

$$\nabla_2^2\xi' = -i_{||}B_0. \tag{2.155}$$

If we make in this equation the substitution $x_{\rm m} = \sin\theta_{\rm m}$ and introduce the dimensionless potential $\xi = 4\pi\xi'/B_0R^2\Omega$ and the current $i_0 = -4\pi i_{||}/\Omega R^2$, we finally get

$$\left(1 - x_{\rm m}^2\right)\frac{\partial^2\xi}{\partial x_{\rm m}^2} + \frac{1 - 2x_{\rm m}^2}{x_{\rm m}}\frac{\partial\xi}{\partial x_{\rm m}} + \frac{1}{x_{\rm m}^2}\frac{\partial^2\xi}{\partial\varphi_{\rm m}^2} = i_0(x_{\rm m}, \varphi_{\rm m}). \tag{2.156}$$

Here again $\theta_{\rm m}$ and $\varphi_{\rm m}$ are spherical coordinates relative to the magnetic axis. Naturally, the solution to Eq. (2.156) substantially depends on the boundary conditions. As is shown below, this boundary condition is the assumption that beyond the polar cap there are no surface currents associated with the bulk longitudinal current flowing in the magnetosphere. In this case, the boundary condition can be written as

$$\xi\,[x_0(\varphi_m), \varphi_m] = \text{const}, \tag{2.157}$$

where the function $x_0(\varphi_m)$ prescribes the form of the polar cap.

We should emphasize that the main uncertainty is just in this assertion. Indeed, the absence of the longitudinal current in the region of the closed field lines $x_m > x_0$, i.e., the fulfillment of the condition $i_0(x_m > x_0, \varphi_m) = 0$, does not imply that the gradient $\nabla\xi$ (and, hence, the surface current $\mathbf{J}_s$) is also zero here. In the case of the inclined rotator, the longitudinal current closure can occur beyond the polar cap, where the equation for the potential ξ has the form $\nabla^2\xi = 0$. The solution to this equation is a set of multipole flows $\xi_n \approx A_n \cos^n \varphi_m / x_m^n$ whose amplitudes A_n could be quite arbitrary. The corresponding jump of the derivative of the potential ξ on the polar cap boundary fixes the value of the surface current flowing along the separatrix dividing the region of closed and open field lines (see Fig. 2.12). Otherwise, this implies that, besides the bulk current flowing along the open field lines, additional surface current must flow in the magnetosphere; the value of the current, at first sight, can be in no way associated with the value of the bulk current.

However, it is easy to show that, in reality, the closing surface currents cannot extend beyond the polar cap. If this were the case, the longitudinal currents would exist in the closed magnetosphere region (see Fig. 2.13). Indeed, as is evident from relations (2.151) and (2.152), the existence of the surface current $\mathbf{J}_s$ must, inevitably, be accompanied by the occurrence of the surface electric field $\mathbf{E}_s$, i.e., the electric potential difference between various points of the neutron star surface, which are

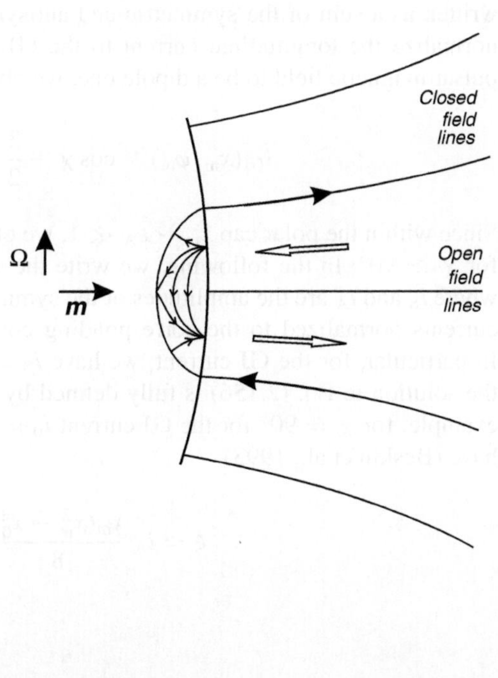

Fig. 2.12 The structure of electric currents flowing in the vicinity of the magnetic poles of the orthogonal rotator. The currents flowing along the separatrix (*bold arrows*) dividing the region of closed and open field lines are compatible with the bulk currents (*contour arrows*), so the closing surface currents (*fine arrows*) are totally concentrated within the polar cap

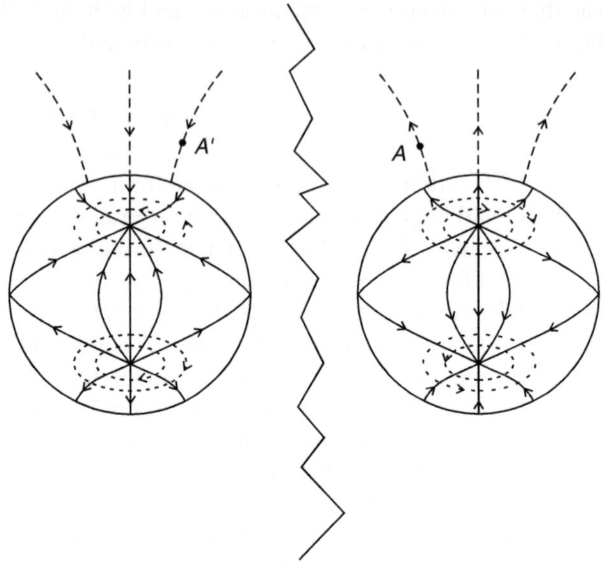

Fig. 2.13 Surface current structure (*fine lines*) and the toroidal magnetic field (*dotted lines*) for two magnetic poles of the orthogonal rotator. If the surface currents flew beyond the polar caps (*dashed lines*), this would give rise to a potential difference between the points A and A' connected by the closed magnetic field line. The surface current structure corresponds to the solution (2.159)

connected by the closed magnetic field lines. But this contradicts the assumption of the absence of longitudinal currents in the closed magnetosphere. Consequently, the current flowing along the separatrix must be compatible with the bulk currents flowing within the open field line region so that the closing surface currents may be totally concentrated within the polar cap. This just leads to the boundary condition (2.157).

On the other hand, for the arbitrary inclination angle χ the current i_0 can be written as a sum of the symmetric and antisymmetric components. It is natural to normalize the longitudinal current to the GJ current $j_{\rm GJ} = c\rho_{\rm GJ}$. Supposing the pulsar magnetic field to be a dipole one, we obtain for the GJ current with $x_{\rm m} \ll 1$

$$i_{\rm GJ}(x_{\rm m}, \varphi_{\rm m}) \approx \cos\chi + \frac{3}{2}x_{\rm m}\cos\varphi_{\rm m}\sin\chi\,. \tag{2.158}$$

Since within the polar cap $x_{\rm m} \sim \varepsilon_{\rm A} \ll 1$, we obtain $i_{\rm GJ} \sim 1$ for $\chi \simeq 0$ and $i_{\rm GJ} \sim \varepsilon_{\rm A}$ for $\chi \simeq 90°$. In the following, we write the current i_0 as $i_0 = i_{\rm S} + i_{\rm A}x_{\rm m}\cos\varphi_{\rm m}$, where $i_{\rm S}$ and $i_{\rm A}$ are the amplitudes of the symmetric and antisymmetric longitudinal currents normalized to the corresponding components of the GJ current (2.158). In particular, for the GJ current, we have $i_{\rm S} = \cos\chi$ and $i_{\rm A} = (3/2)\sin\chi$. Thus, the solution to Eq. (2.156) is fully defined by the bulk longitudinal current i_0. For example, for $\chi = 90°$ for the GJ current $i_0 = i_{\rm A}x_{\rm m}\cos\varphi_{\rm m}$ and for $x_0 = \text{const}$, we have (Beskin et al., 1993)

$$\xi = i_{\rm A}\,\frac{x_{\rm m}(x_{\rm m}^2 - x_0^2)}{8}\,\cos\varphi_{\rm m}. \tag{2.159}$$

Problem 2.16 Show that in this case the total current $I_{\rm sep}$ flowing along the separatrix is 3/4 the total bulk current $I_{\rm bulk}$ flowing in the region of the open field lines:

$$\frac{I_{\rm sep}}{I_{\rm bulk}} = -\frac{3}{4}. \tag{2.160}$$

Further, we expand the slowing-down moment $\mathbf{K}$ (2.132) in terms of the vectors $\mathbf{e}_{\rm m}$, $\mathbf{n}_1$, and $\mathbf{n}_2$, where $\mathbf{e}_{\rm m} = \mathbf{m}/|\mathbf{m}|$, the unit vector $\mathbf{n}_1$ is perpendicular to the magnetic moment $\mathbf{m}$ and lies in the plane formed by the vectors Ω and $\mathbf{m}$ (and $\Omega \cdot \mathbf{n}_1 > 0$), and $\mathbf{n}_2 = \mathbf{e}_{\rm m} \times \mathbf{n}_1$

$$\mathbf{K} = K_{\parallel}\mathbf{e}_m + K_{\perp}\mathbf{n}_1 + K_{\dagger}\mathbf{n}_2. \tag{2.161}$$

As a result, we have (Beskin et al., 1993)

$$K_{\parallel} = -\frac{B_0^2 R^4 \Omega}{c} \int_0^{2\pi} \frac{{\rm d}\varphi_{\rm m}}{2\pi} \int_0^{x_0(\varphi_{\rm m})} {\rm d}x_{\rm m}\, x_{\rm m}^2 \sqrt{1 - x_{\rm m}^2}\, \frac{\partial \xi}{\partial x_{\rm m}}, \tag{2.162}$$

and $K_{\perp} = K_1 + K_2$, where

$$K_1 = \frac{B_0^2 R^4 \Omega}{c} \int_0^{2\pi} \frac{{\rm d}\varphi_{\rm m}}{2\pi} \int_0^{x_0(\varphi_{\rm m})} {\rm d}x_{\rm m} \left(x_{\rm m} \cos\varphi_{\rm m} \frac{\partial \xi}{\partial x_{\rm m}} - \sin\varphi_{\rm m} \frac{\partial \xi}{\partial \varphi_{\rm m}} \right), \tag{2.163}$$

$$K_2 = \frac{B_0^2 R^4 \Omega}{c} \int_0^{2\pi} \frac{{\rm d}\varphi_{\rm m}}{2\pi} \int_0^{x_0(\varphi_{\rm m})} {\rm d}x_{\rm m}\, x_{\rm m}^3 \cos\varphi_{\rm m} \frac{\partial \xi}{\partial x_{\rm m}}, \tag{2.164}$$

and $K_{\dagger}$, as we will see, does not enter the Euler equations at all. Here we also took into account that both magnetic poles contribute to the slowing-down moment.

Since integration over $x_{\rm m}$ in (2.163) and (2.164) is taken to the polar cap boundary $x_0(\varphi_{\rm m}) \sim \varepsilon_{\rm A}$, as an estimate, we could take $K_2 \sim \varepsilon_{\rm A}^2 K_1$, i.e., $K_2 \ll K_1$. However, as is readily checked, when the boundary condition (2.157) is satisfied, the integrand in (2.163) is a complete derivative with respect to $\varphi_{\rm m}$:

$$\int_0^{x_0(\varphi_{\rm m})} {\rm d}x_{\rm m} \left(x_{\rm m} \cos\varphi_{\rm m} \frac{\partial \xi}{\partial x_{\rm m}} - \sin\varphi_{\rm m} \frac{\partial \xi}{\partial \varphi_{\rm m}} \right) =$$
$$\frac{\partial}{\partial \varphi_{\rm m}} \left[-\int_0^{x_0(\varphi_{\rm m})} {\rm d}x_{\rm m}\, \xi \sin\varphi_{\rm m} + \xi\,(x_0, \varphi_{\rm m})\, x_0(\varphi_{\rm m}) \sin\varphi_{\rm m} \right]. \tag{2.165}$$

Therefore, the contribution K_1 appears identically equal to zero. As a result, the expressions for $K_{\parallel}$ and $K_{\perp}$ have the form

$$K_{\parallel} = -\frac{B_0^2 \Omega^3 R^6}{c^3}\left[c_{\parallel} i_{\rm S} + \mu_{\parallel}\left(\frac{\Omega R}{c}\right)^{1/2} i_{\rm A}\right], \tag{2.166}$$

$$K_{\perp} = -\frac{B_0^2 \Omega^3 R^6}{c^3}\left[\mu_{\perp}\left(\frac{\Omega R}{c}\right)^{1/2} i_{\rm S} + c_{\perp}\left(\frac{\Omega R}{c}\right) i_{\rm A}\right], \tag{2.167}$$

where $c_{\parallel}$ and $c_{\perp}$ are factors of the order of unity dependent on the particular profile of the longitudinal current i_0 and the form of the polar cap. As to the coefficients $\mu_{\parallel}$ and $\mu_{\perp}$, they are associated with the polar cap axisymmetry and their contribution proves unessential. In particular, $\mu_{\parallel}(0) = \mu_{\perp}(0) = 0$ and $\mu_{\parallel}(90°) = \mu_{\perp}(90°) = 0$.

We can explain the unavailability of the leading term K_1 (2.163) for the energy losses. As was shown above, the energy losses of radio pulsars $W_{\rm tot}$ can be identically rewritten as (2.144)

$$W_{\rm tot} = \frac{c}{4\pi}\int (\boldsymbol{\beta}_{\rm R}\cdot\mathbf{B})(\mathbf{B}\cdot {\rm d}\mathbf{S}). \tag{2.168}$$

On the light cylinder, expression (2.168) coincides with the estimate (2.146) but can be used in the vicinity of the neutron star surface as well. It is easy to verify that the condition of the current closure within the polar cap (2.157) is equivalent to the condition of the complete screening of the magnetic field $\mathbf{B}_{\rm T}$, which is caused by the longitudinal currents flowing in the region of the open field lines. This fact is obvious for the axisymmetric rotator; however, it needs a substantial additional assumption for the angles $\chi \approx 90°$. As shown in Fig. 2.13, the toroidal magnetic field specifying the value $(\boldsymbol{\beta}_{\rm R}\cdot\mathbf{B})$ must not extend beyond the polar cap. As a result, in the zero approximation, the mean value of the scalar product $(\boldsymbol{\beta}_{\rm R}\cdot\mathbf{B})$ in the region of open field lines turns out to be zero and the energy loss itself is determined by the small corrections $\sim\varepsilon_{\rm A}^2$ associated with the curvature of the neutron star surface. Clearly, the pattern must be the same on the light cylinder. In other words, for the orthogonal rotator, the mean value of the toroidal magnetic field of order $B_{\varphi}(R_{\rm L}) \sim i_0 B_{\rm p}(R_{\rm L})$ is to be zero on the light cylinder. This establishes the difference in the estimates of the energy losses for the orthogonal rotator.

Writing the Euler equations, we can find the change in the angular velocity $\dot{\Omega}$ and the inclination angle $\dot{\chi}$ of the pulsar:

$$I_{\rm r}\frac{{\rm d}\Omega}{{\rm d}t} = K_{\parallel}\cos\chi + K_{\perp}\sin\chi, \tag{2.169}$$

$$I_{\rm r}\Omega\frac{{\rm d}\chi}{{\rm d}t} = K_{\perp}\cos\chi - K_{\parallel}\sin\chi. \tag{2.170}$$

Here we, for simplicity, suppose that the neutron star is spherically symmetric, and its moment of inertia $I_{\rm r}$ is thus independent of the orientation of the rotation axis. As a result, for angles χ not too close to 90°, so that $\cos\chi > \varepsilon_{\rm A}^2$ (i.e., when the symmetric currents are of major importance), we find

$$\frac{\mathrm{d}\Omega}{\mathrm{d}t} = -c_{\parallel}\frac{B_0^2\Omega^3R^6}{I_\mathrm{r}c^3}i_\mathrm{S}\cos\chi, \tag{2.171}$$

$$\frac{\mathrm{d}\chi}{\mathrm{d}t} = c_{\parallel}\frac{B_0^2\Omega^2R^6}{I_\mathrm{r}c^3}i_\mathrm{S}\sin\chi. \tag{2.172}$$

We readily see that Eqs. (2.171) and (2.172) yield the conservation of the invariant

$$\mathcal{I}_\mathrm{cur} = \Omega\sin\chi, \tag{2.173}$$

different from (2.22). This is because, as was mentioned, the slowing-down moment **K** (2.132) for the symmetric currents is opposite to the magnetic dipole **m**, so the projection of the angular velocity Ω onto the axis perpendicular to **m** is an integral of motion. For the orthogonal rotator $\chi \approx 90°$, where $\cos\chi < \varepsilon_\mathrm{A}^2$, we get

$$\frac{\mathrm{d}\Omega}{\mathrm{d}t} = -c_{\perp}\frac{B_0^2\Omega^4R^7}{I_\mathrm{r}c^4}i_\mathrm{A}. \tag{2.174}$$

Because of the dependence $i_\mathrm{S} \approx \cos\chi$, the contribution of the symmetric current can be disregarded here. The comparison of relations (2.171) and (2.174) shows that the energy release of pulsars at the orthogonal rotator stage (and for GJ current $i_\mathrm{A} \approx 1$) is $\Omega R/c$ times less than that of axisymmetric pulsars.

To sum up, we can make the general conclusions:

1. For inclination angles $\chi < 90°$, the slowing-down moment **K** (2.132) is antiparallel to the magnetic moment of the neutron star **m**. Therefore, for the current losses the invariant value is

$$\Omega\sin\chi = \mathrm{const}. \tag{2.175}$$

This conclusion directly follows from the analysis of the Euler equations, viz., the projection of the angular velocity onto the direction perpendicular to the applied moment of forces is an invariant of motion (Landau and Lifshits, 1976). Consequently, unlike the magnetodipole losses, the inclination angle must increase with time. According to the invariant (2.173), the characteristic time of the change in the inclination angle χ ($\tau_\chi = \chi/2\dot{\chi}$) coincides with the dynamical age of the pulsar $\tau_\mathrm{D} = P/2\dot{P}$

$$\tau_\mathrm{D} \approx \frac{I_\mathrm{r}c^3}{2B_0^2\Omega^2R^6} \approx 10\,\mathrm{mln\,years}\left(\frac{P}{1\,s}\right)^2\left(\frac{B_0}{10^{12}\,\mathrm{G}}\right)^{-2}. \tag{2.176}$$

2. The current losses W_tot can be rewritten as $W_\mathrm{tot} = VI$. Here

$$V \sim EL \sim \left(B_0\frac{\Omega R_0}{c}\right)R_0 \tag{2.177}$$

is the characteristic potential drop within the polar cap and I is the total current circulating in the magnetosphere. Using now the definition $i_0 = I/I_{GJ}$ and the fact that for χ not too close to 90°, we can take $V \approx \psi_{max}$ to obtain

$$W_{tot} = c_{\parallel} \frac{B_0^2 \Omega^4 R^6}{c^3} i_0 \cos\chi . \tag{2.178}$$

The coefficient $c_{\parallel} \sim 1$, as seen from relation (2.164), depends on the longitudinal current profile. One should stress here that, besides the factor $\cos\chi$ connected with the scalar product in (2.131), the substantial dependence of the current losses W_{tot} on the inclination angle is in the factor $i_0 \approx i_S$. The point is that in the definition of the dimensionless current, there is the GJ current for the axisymmetric case, whereas for nonzero χ the GJ charge density in the vicinity of the magnetic poles substantially depends on the angle χ, viz., $\rho_{GJ} \approx -(\boldsymbol{\Omega} \cdot \mathbf{B})/2\pi c \propto \cos\chi$. Therefore, it is logical to expect that for the inclined rotator the dimensionless current i_0 is bounded from above

$$i_0^{(max)}(\chi) \sim \cos\chi . \tag{2.179}$$

As a result, the current losses decrease as the angle χ increases, at least, as $\cos^2\chi$.

3. As to radio pulsars, in which the inclination angle χ is close to 90°, for the antisymmetric longitudinal currents i_A the energy losses can be written as

$$W_{tot} = c_{\perp} \frac{B_0^2 \Omega^4 R^6}{c^3} \left(\frac{\Omega R}{c}\right) i_A . \tag{2.180}$$

Here the coefficient $c_{\perp} \sim 1$ already depends not only on the antisymmetric longitudinal current profile but also on the form of the polar cap. Consequently, the current losses for the orthogonal rotator (and for $i_A \sim 1$) turn out to be $(\Omega R/c)$ times less than in the axisymmetric case. Certainly, if the current density can be much larger than the local GJ current $\rho_{GJ,90}c$, then $i_A \gg 1$, the energy losses can be large enough. We discuss this possibility in Sect. 2.6.3.

Problem 2.17 Show that for the constant current density i_S = const within the polar cap (Beskin et al., 1993)

$$c_{\parallel} = \frac{f_*^2}{4}, \tag{2.181}$$

where f_* is the dimensionless area of the polar cap: $S = f_* \pi (\Omega R/c) R^2$.

Problem 2.18 Using relation (2.159), show that for the orthogonal rotator

$$c_{\perp} = \frac{f_{*}^{3}}{64}. \tag{2.182}$$

Thus, the important conclusion is that for currents $I \sim I_{\rm GJ}$ (i.e., for $i_0 \sim 1$) characteristic of the radio pulsar magnetosphere, the current losses (2.178) in this expression coincide with the magnetodipole losses (2.5). On the other hand, the current and magnetodipole losses have a number of considerable differences.

- The magnetodipole losses (2.5) are absent in the axisymmetric case, whereas the current losses are maximal for $\chi = 0°$.
- The magnetodipole losses result in a decrease in the inclination angle with time ($\Omega \cos \chi$ = const), whereas for the current losses the angle χ, on the contrary, is to increase ($\Omega \sin \chi$ = const) approaching 90°. However, in both cases, the evolution of the angle χ is in the range of parameters, where the energy losses of the neutron star become minimal.
- For the magnetodipole losses, the braking index $n_{\rm br}$ is larger than three (see (2.24)), whereas for the current losses, it can be less than three (see Beskin et al. (1993) for details).
- The magnetodipole losses are universal i.e., they are independent of the additional parameters. On the other hand, the current losses (2.178) are proportional to the electric current i_0 circulating in the magnetosphere.

Otherwise, the difference between the current and magnetodipole losses is rather substantial. Theoretically, this brings up the question of the relative role of these two slowing-down mechanisms in the total balance of the energy losses. The answer to this question can be given only together with the solution to the complete problem of the neutron star magnetosphere. On the other hand, one should note that for most radio pulsars the dimensionless current is $i_0 \sim 1$, so that the simplest magnetodipole formula (2.5) yields, in the large, a reliable estimate for the total energy losses of the rotating neutron star. As a result, both the magnetodipole and the current losses give similar results when analyzing the statistical characteristics of radio pulsars (Michel, 1991; Beskin et al., 1993). The direct determination of the sign of the derivative $\dot{\chi}$ different for the two slowing-down mechanisms is now beyond the sensitivities of the present-day receivers. Therefore, up to now, the observations do not allow one to choose between these two slowing-down mechanisms (see Appendix C as well).

2.6 Magnetosphere Structure

2.6.1 Exact Solutions

We again return to our main topic and consider the structure of the radio pulsar magnetosphere. It was shown that in the zero order with respect to the small parameters

σ^{-1} and λ^{-1}, the magnetosphere structure can be described by the force-free equation (2.101). As was noted, this equation contains only one singular surface and, therefore, needs three boundary conditions. As such boundary conditions, one can take the values of the invariants $\Omega_{\rm F}(\Psi)$ and $I(\Psi)$, as well as the normal component of the magnetic field on the neutron star surface (or, what is the same, the stream function $\Psi(R,\theta)$ on its surface).

Equation (2.101) is of a nonlinear type. However, unlike the hydrodynamical GS equation version, the whole nonlinearity is now associated with the integrals of motion. In particular, in the absence of the longitudinal current and for the constant angular velocity $\Omega_{\rm F}(\Psi) = \Omega$, it becomes linear

$$-\left(1-\frac{\Omega^2\varpi^2}{c^2}\right)\nabla^2\Psi + \frac{2}{\varpi}\frac{\partial\Psi}{\partial\varpi} = 0. \tag{2.183}$$

On the other hand, unlike the hydrodynamical case, for the constant value of the angular velocity $\Omega_{\rm F}$, the location of the singular surface $\Omega_{\rm F}\varpi/c = 1$ is known beforehand. Since Eq. (2.183) does not explicitly comprise the cylindrical coordinate z, its solution can be sought by the method of separation of variables (Michel, 1973a; Mestel and Wang, 1979)

$$\Psi(\varpi, z) = \frac{|\mathbf{m}|}{R_{\rm L}}\int_0^\infty R_\lambda(\varpi)\cos(\lambda z)\,{\rm d}\lambda. \tag{2.184}$$

These properties made it possible to obtain the solution to Eq. (2.101) for a number of the simplest cases.

2.6.1.1 Axisymmetric Magnetosphere with the Zero Longitudinal Current for the Dipole Magnetic Field of the Neutron Star

In the absence of the longitudinal currents, the only currents in the magnetosphere are corotation currents $\Omega_{\rm F}\varpi\rho_{\rm GJ}\mathbf{e}_\varphi$. Recall that we assume here $\Omega_{\rm F} = $ const. Therefore, the range of applicability of Eq. (2.183) extends only to the light cylinder which coincides with the light surface. Substituting expansion (2.184) in Eq. (2.183), we obtain for the radial function $R_\lambda(x_r)$ (Michel, 1973a; Mestel and Wang, 1979)

$$\frac{{\rm d}^2R_\lambda(x_r)}{{\rm d}^2x_r} - \frac{(1+x_r^2)}{x_r(1-x_r^2)}\frac{{\rm d}R_\lambda(x_r)}{{\rm d}x_r} - \lambda^2R_\lambda(x_r) = 0. \tag{2.185}$$

Hereafter, we again use the dimensionless variables $x_r = \Omega\varpi/c$ and $z' = \Omega z/c$.

The boundary conditions for Eq. (2.185) are

1. the dipole magnetic field in the vicinity of the star surface $\mathbf{B} = [3(\mathbf{nm})\mathbf{n} - \mathbf{m}]/r^3$, i.e.,

$$\Psi(x_r, z') = \frac{|\mathbf{m}|}{R_{\rm L}}\frac{x_r^2}{(x_r^2+z'^2)^{3/2}} \tag{2.186}$$

for $x_r \to 0$ and $z' \to 0$;

2. the absence of a singularity on the light cylinder $x_r = 1$.

According to the known expansion

$$\frac{x_r^2}{(x_r^2 + z'^2)^{3/2}} = \frac{2}{\pi} \int_0^\infty \lambda x_r \, K_1(\lambda x_r) \cos(\lambda z') \, \mathrm{d}\lambda, \tag{2.187}$$

where $K_1(x)$ is the Macdonald function of the first order, the first condition implies that for $x_r \to 0$ the relation

$$R_\lambda(x_r) \to \frac{2}{\pi} \lambda x_r \, K_1(\lambda x_r) \tag{2.188}$$

must hold. As we see, the situation is absolutely equivalent to the hydrodynamical limit when one of the boundary conditions for the ordinary differential equation is connected with a field source and the second one corresponds to the absence of a singularity on the critical surface.

Problem 2.19 Show that the solution to Eq. (2.185) can be constructed in the form of the series

$$R_\lambda(x_r) = \mathcal{D}(\lambda) \sum_{n=0}^{\infty} a_n (1 - x_r^2)^n, \tag{2.189}$$

where the expansion coefficients a_n satisfy the recurrent relations

$$a_0 = 1, \quad a_1 = 0, \quad a_{n+1} = \frac{4n^2}{4(n+1)^2} a_n + \frac{\lambda^2}{4(n+1)^2} a_{n-1}. \tag{2.190}$$

The value $\mathcal{D}(\lambda)$ can be determined from the boundary condition (2.188) near the neutron star surface. Indeed, using the asymptotic behavior $K_1(x) = x^{-1}$ for $x \to 0$, we get

$$\mathcal{D}(\lambda)^{-1} = \frac{\pi}{2} \sum_{n=0}^{\infty} a_n. \tag{2.191}$$

Figure 2.14 shows the magnetic field structure obtained from the solution to Eq. (2.183) (Michel, 1973a). As was expected, the dipole magnetic field is disturbed only in the vicinity of the light cylinder; at small distances the magnetic field remains dipole. Note also that the magnetic field on the light cylinder appears orthogonal to its surface. This fact can be directly checked by definition (2.89) in the form of expansion (2.189)—the z-component of the magnetic field on the light cylinder $B_z(x_r = 1)$ turns out to be automatically equal to zero. This is, by the way, the solution to the singularity problem in expression (2.41)—the charge density remains finite on the light cylinder. At the equator of the light cylinder ($\varpi = R_\mathrm{L}$, $z = 0$), the magnetic field is zero. Finally, it turned out that the total

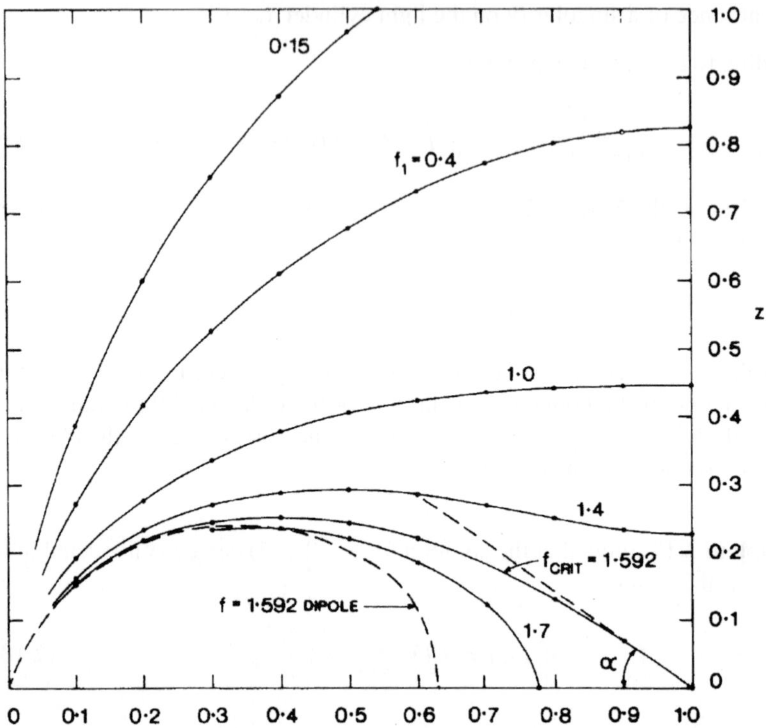

Fig. 2.14 Magnetic field structure for the zero longitudinal current and the accelerating potential ($i_0 = 0$, $\beta_0 = 0$) for the dipole axisymmetric magnetic field of the neutron star (Michel, 1973a). The numbers indicate the values of the dimensionless magnetic field function f ($\Psi = \pi B_0 R^2(\Omega R/c)\, f$) (Reproduced by permission of the AAS, Fig. 1 from Michel, F.C.: Rotating magnetosphere: a simple relativistic model. ApJ **180**, 207–226 (1973))

magnetic flux crossing the light cylinder is about 1.592 times larger than that in the vacuum case. This result implies that the area of the polar cap increases in the same proportion (Michel, 1973a)

$$S_{\rm cap} \approx 1.592\, \pi R_0^2. \tag{2.192}$$

As to the toroidal magnetic field, since the longitudinal electric currents are absent, it is identically equal to zero in the whole magnetosphere.

Problem 2.20 Having written the expression for the magnetic flux through the light cylinder surface, show that the coefficient $f_* \approx 1.592$ (so-called Michel number) is connected with the function $\mathcal{D}(\lambda)$ by the relation

$$f_* = \int_0^\infty \mathcal{D}(\lambda) \mathrm{d}\lambda. \tag{2.193}$$

It is also interesting to note that when receding from the equatorial plane, within the light cylinder, the electric and magnetic fields decrease exponentially fast rather than by a power law: $B \propto \exp(-pz/R_{\mathrm{L}})$, where $p \approx 3.0$. This property is associated with the structure of expansion (2.184) and the existence of the pole of the function R_λ for $\lambda = ip$ (Beskin et al., 1993). This fast decrease in the fields is possible because the magnetic moment of the corotation currents almost fully screens the magnetic moment of the neutron star.

Further, the electric field on the light cylinder is compared in magnitude with the magnetic one, but its direction is along the rotation axis of the neutron star. Since the normal component of the electric field vanishes on the light cylinder, one can conclude that the total charge of the neutron star and the magnetosphere turns out to be zero. Otherwise, part of the charge Q_* (2.12) located, in the vacuum case, on the neutron star surface passes into the pulsar magnetosphere. On the other hand, the equality of the electric and magnetic fields on the light cylinder shows that for the zero longitudinal current the light cylinder coincides with the light surface. Therefore, the constructed solution cannot be extended beyond the light cylinder, though, formally, the pulsar equation does not have any singularities here.

2.6.1.2 Axisymmetric Magnetosphere with the Zero Longitudinal Current for the Monopole Magnetic Field

At first sight, there is no sense to consider this case, because the monopole magnetic field does not occur in reality. However, as we see in the following, the analysis of the rotating monopole magnetosphere proves very fruitful, especially, for the case of the black hole magnetosphere.

The solution of the problem for the monopole magnetic field is analogous to the previous one (Michel, 1973a). There is a difference only in boundary condition 1 on the star surface and, hence, only in the explicit form of the function $\mathcal{D}(\lambda)$. As a result, as for the dipole magnetic field, the magnetic field on the light cylinder appears orthogonal to its surface and also decreases exponentially with distance from the equatorial plane, and at small distances from the star the monopole field perturbations prove small (see Fig. 2.15). On the other hand, as in the previous example, the electric field on the light cylinder is compared with the magnetic one and, therefore, the solution of the pulsar equation cannot be extended beyond the light cylinder.

2.6.1.3 Magnetosphere with the Zero Longitudinal Current for the Inclined Rotator

The exact solution for the zero longitudinal currents (and in the absence of the accelerating potential $\psi = 0$) can be constructed at an arbitrary inclination angle of χ (Beskin et al., 1983). This becomes possible because for $i_{\parallel} = 0$ and $\psi = 0$ the quasistationary GS equation (2.129) also becomes linear

$$\nabla \times \left[(1 - \boldsymbol{\beta}_{\mathrm{R}}^2)\mathbf{B} + \boldsymbol{\beta}_{\mathrm{R}}(\boldsymbol{\beta}_{\mathrm{R}} \cdot \mathbf{B})\right] = 0. \tag{2.194}$$

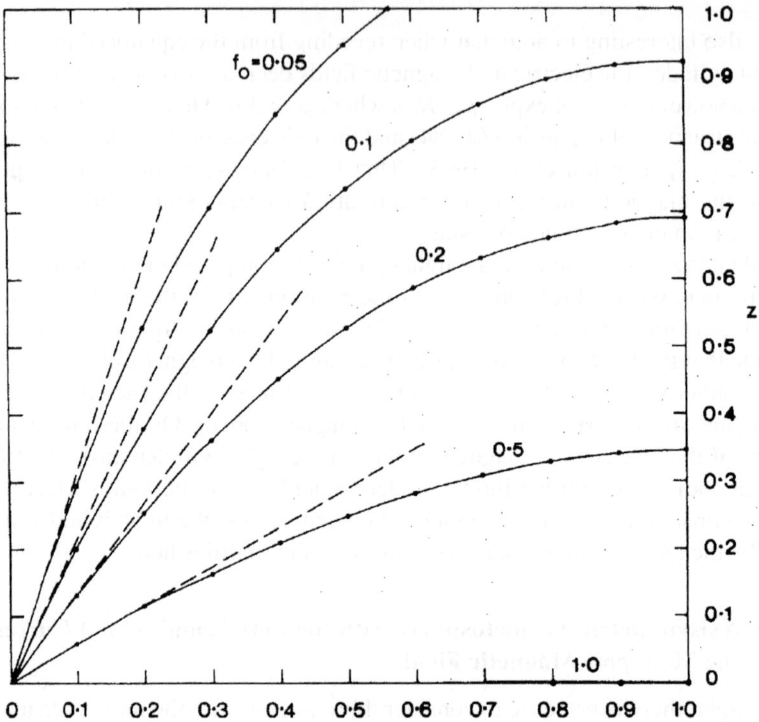

Fig. 2.15 Magnetic field structure for the zero longitudinal current and the accelerating potential ($i_0 = 0$, $\beta_0 = 0$) for the monopole magnetic field of the compact object (Michel, 1973a) [Reproduced by permission of the AAS, Fig. 2 from Michel, F.C.: Rotating magnetosphere: a simple relativistic model. ApJ **180**, 207–226 (1973)]

The solution to Eq. (2.194) [so-called Mestel equation (Mestel, 1973)] can be written as

$$(1 - \boldsymbol{\beta}_{\mathrm{R}}^2)\mathbf{B} + \boldsymbol{\beta}_{\mathrm{R}}(\boldsymbol{\beta}_{\mathrm{R}} \cdot \mathbf{B}) = -\nabla h, \tag{2.195}$$

while, for the zero accelerating potential ψ, Maxwell's equation $\nabla \cdot \mathbf{B} = 0$ looks like $\hat{\mathcal{L}}_2 h = 0$:

$$\frac{\partial^2 h}{\partial x_r^2} + \frac{(1 + x_r^2)}{x_r(1 - x_r^2)}\frac{\partial h}{\partial x_r} + \frac{(1 - x_r^2)}{x_r^2}\frac{\partial^2 h}{\partial \varphi^2} + \frac{\partial^2 h}{\partial z'^2} = 0. \tag{2.196}$$

On the other hand, the electric field for $\psi = 0$ can be found from the condition $\mathbf{E} + \boldsymbol{\beta}_{\mathrm{R}} \times \mathbf{B} = 0$, because relation (2.104) must hold for any quasistationary configurations. Therefore, the electric and magnetic fields can again be specified by equalities (2.120), (2.121), (2.122), and (2.123) in which we must take $\psi = 0$

$$\mathbf{E}_\mathrm{p} = \frac{[\boldsymbol{\beta}_\mathrm{R} \times \nabla h]}{1 - \boldsymbol{\beta}_\mathrm{R}^2}, \tag{2.197}$$

$$E_\varphi = 0, \tag{2.198}$$

$$\mathbf{B}_\mathrm{p} = -\frac{\nabla h}{1 - \boldsymbol{\beta}_\mathrm{R}^2}, \tag{2.199}$$

$$B_\varphi = -\frac{1}{\varpi}\frac{\partial h}{\partial \varphi}. \tag{2.200}$$

To construct the solution to Eq. (2.196), we see that in the studied linear case, the magnetic field of the neutron star can be expanded into axisymmetric and orthogonal parts. In other words, the potential $h(x_r, \varphi - \Omega t, z')$ can be represented as

$$h(x_r, \varphi - \Omega t, z') = h_0(x_r, z') \cos\chi + h_1(x_r, z') \cos(\varphi - \Omega t) \sin\chi, \tag{2.201}$$

and now the potentials $h_0(x_r, z')$ and $h_1(x_r, z')$ satisfy the equations

$$\frac{\partial^2 h_0}{\partial x_r^2} + \frac{(1 + x_r^2)}{x_r(1 - x_r^2)}\frac{\partial h_0}{\partial x_r} + \frac{\partial^2 h_0}{\partial z'^2} = 0, \tag{2.202}$$

$$\frac{\partial^2 h_1}{\partial x_r^2} + \frac{(1 + x_r^2)}{x_r(1 - x_r^2)}\frac{\partial h_1}{\partial x_r} + \frac{\partial^2 h_1}{\partial z'^2} - \frac{(1 - x_r^2)}{x_r^2} h_1 = 0. \tag{2.203}$$

Therefore, as in the case of the axisymmetric rotator, the solution to Eqs. (2.202) and (2.203) can be found in the form

$$h_0(x_r, z') = \frac{|\mathbf{m}|}{R_\mathrm{L}^2}\int_0^\infty R_\lambda^{(0)}(x_r)\sin(\lambda z')\,\mathrm{d}\lambda, \tag{2.204}$$

$$h_1(x_r, z') = \frac{|\mathbf{m}|}{R_\mathrm{L}^2}\int_0^\infty R_\lambda^{(1)}(x_r)\cos(\lambda z')\,\mathrm{d}\lambda, \tag{2.205}$$

where the radial functions $R_\lambda^{(0)}(x_r)$ and $R_\lambda^{(1)}(x_r)$ must satisfy the equations

$$\frac{\mathrm{d}^2 R_\lambda^{(0)}(x_r)}{\mathrm{d}^2 x_r} + \frac{(1 + x_r^2)}{x_r(1 - x_r^2)}\frac{\mathrm{d}R_\lambda^{(0)}(x_r)}{\mathrm{d}x_r} - \lambda^2 R_\lambda^{(0)}(x_r) = 0, \tag{2.206}$$

$$\frac{\mathrm{d}^2 R_\lambda^{(1)}(x_r)}{\mathrm{d}^2 x_r} + \frac{(1 + x_r^2)}{x_r(1 - x_r^2)}\frac{\mathrm{d}R_\lambda^{(1)}(x_r)}{\mathrm{d}x_r} - \left(\lambda^2 + \frac{1 - x_r^2}{x_r^2}\right) R_\lambda^{(1)}(x_r) = 0. \tag{2.207}$$

The boundary conditions for Eqs. (2.206) and (2.207), as before, are

1. the dipole magnetic field $\mathbf{B} = [3(\mathbf{nm})\mathbf{n} - \mathbf{m}]/r^3$ in the vicinity of the star surface, i.e.,

$$h_0(x_r, z') \to \frac{|\mathbf{m}|}{R_\mathrm{L}^2} \frac{z'}{(x_r^2 + z'^2)^{3/2}}, \tag{2.208}$$

$$h_1(x_r, z') \to \frac{|\mathbf{m}|}{R_\mathrm{L}^2} \frac{x_r}{(x_r^2 + z'^2)^{3/2}}, \tag{2.209}$$

for $x_r \to 0$ and $z' \to 0$;

2. the absence of a singularity on the light cylinder $x_r = 1$:

$$\left.\frac{\mathrm{d}R_\lambda^{(0)}}{\mathrm{d}x_r}\right|_{x_r=1} = 0, \tag{2.210}$$

$$\left.\frac{\mathrm{d}R_\lambda^{(1)}}{\mathrm{d}x_r}\right|_{x_r=1} = 0. \tag{2.211}$$

Using expansion (2.187) again, we find that for $x_r \to 0$, the following relations must hold:

$$R_\lambda^{(0)}(x_r) \to \frac{2}{\pi}\lambda K_0(\lambda x_r), \tag{2.212}$$

$$R_\lambda^{(1)}(x_r) \to \frac{2}{\pi}\lambda K_1(\lambda x_r). \tag{2.213}$$

Here $K_0(x)$ and $K_1(x)$ are the Macdonald functions of zero and the first order.

Besides (and it is very important), it is also necessary that the magnetic field should decrease at infinity along the rotation axis for $z \to \infty$. The necessity to introduce an "additional" boundary condition is that the magnetic field line extending to infinity along the rotation axis does not intersect the light cylinder and, hence, there is no additional regularity condition for it. When this condition is not satisfied, we have the nonphysical solution (Endean, 1983)

$$h_\mathrm{E}(x_r, \varphi, z', t) = h_*[x_r J_0(x_r) - J_1(x_r)]\cos(\varphi - \Omega t) \tag{2.214}$$

(h_*—an arbitrary constant) independent of z and, hence, not decreasing at infinity.

Problem 2.21 Show that the solution to Eqs. (2.206) and (2.207) can be constructed in the form of the formal series (Beskin et al., 1983; Mestel et al., 1999)

$$R_\lambda^{(0)}(x_r) = \mathcal{D}_0(\lambda)\sum_{n=2}^{\infty} b_n(1 - x_r^2)^n, \tag{2.215}$$

$$R_\lambda^{(1)}(x_r) = \mathcal{D}_1(\lambda)\sum_{n=2}^{\infty} c_n(1 - x_r^2)^n, \tag{2.216}$$

where the expansion coefficients b_n and c_n satisfy the recurrent relations

$$b_{n+1} = \frac{n}{n+1} b_n + \frac{\lambda^2}{4(n^2-1)} b_{n-1}, \tag{2.217}$$

$$c_{n+1} = \frac{n(2n-3)}{n^2-1} c_n - \frac{4(n-1)(n-2)-\lambda^2}{4(n^2-1)} c_{n-1} - \frac{\lambda^2-1}{4(n^2-1)} c_{n-2}, \tag{2.218}$$

where $b_0 = b_1 = c_0 = c_1 = 0$ and $b_2 = c_2 = 1$.

Problem 2.22 Using the asymptotic behavior $K_0(x) \rightarrow -\ln x$ and $K_1(x) \rightarrow x^{-1}$ for $x \rightarrow 0$, show that

$$\mathcal{D}_0(\lambda)^{-1} = -\frac{\pi}{2\lambda} \lim_{x_r \to 0} \frac{1}{\ln x_r} \sum_{n=2}^{\infty} b_n (1 - x_r^2)^n, \tag{2.219}$$

$$\mathcal{D}_1(\lambda)^{-1} = \frac{\pi}{2} \lim_{x_r \to 0} x_r \sum_{n=2}^{\infty} c_n (1 - x_r^2)^n. \tag{2.220}$$

Problem 2.23 Using definitions (2.197), (2.198), (2.199), and (2.200) and (2.215) and (2.216), show that the magnetic field and the charge density on the light cylinder are defined as

$$B_{\varpi}(R_{\rm L}, \varphi', z') = \tag{2.221}$$
$$4\frac{|\mathbf{m}|}{R_{\rm L}^3} \left[\cos\chi \int_0^{\infty} \mathcal{D}_0(\lambda) \sin(\lambda z') d\lambda + \sin\chi \cos\varphi' \int_0^{\infty} \mathcal{D}_1(\lambda) \cos(\lambda z') d\lambda \right],$$

$$\rho_{\rm e}(R_{\rm L}, \varphi', z') = \tag{2.222}$$
$$\frac{\Omega|\mathbf{m}|}{2\pi c R_{\rm L}^3} \left[\cos\chi \int_0^{\infty} \mathcal{D}_0(\lambda) \lambda \cos(\lambda z') d\lambda - \sin\chi \cos\varphi' \int_0^{\infty} \mathcal{D}_1(\lambda) \lambda \sin(\lambda z') d\lambda \right],$$

where $\varphi' = \varphi - \Omega t$.

Problem 2.24 Show that in the axisymmetric case the singular solution independent of z has a singularity on the light cylinder and, hence, must be abandoned automatically.

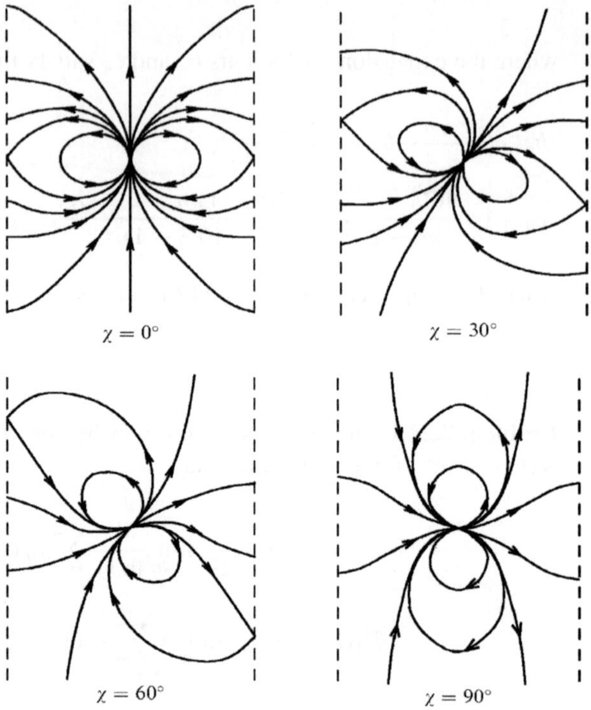

Fig. 2.16 Magnetic field structure for the zero longitudinal current and the accelerating potential ($i_0 = 0$, $\beta_0 = 0$) for the inclined dipole magnetic field of the neutron star (Beskin et al., 1983)

As shown in Fig. 2.16, for the case of the inclined rotator, the basic properties valid for the axisymmetric magnetosphere fully retain. In the absence of the longitudinal currents, the boundary of the region of applicability is the light cylinder, where the corotation currents begin to distort the dipole magnetic field. The magnetic field itself becomes orthogonal to the light cylinder here. On the other hand, the electric and magnetic fields exponentially decrease with distance from the equatorial plane. Finally, the total electric charge in the magnetosphere is zero.

Relations (2.215) and (2.216) allow us to get the complete information concerning the magnetosphere structure. Thus, Table 2.2 gives the values of the magnetic field (in $|\mathbf{m}|/R_{\rm L}^3$ units) and the charge density (in $\Omega B/2\pi c$ units) on the light cylinder for four different inclination angles χ. Besides, Fig. 2.17 shows the change in the polar cap form as the inclination angle χ increases. Its area varies from $1.592\,\pi\,R_0^2$ for $\chi = 0°$ to $1.96\,\pi\,R_0^2$ for $\chi = 90°$.

Problem 2.25 Using the nonphysical solution (2.214), show that the dimensionless area of the polar cap surface $f_*(90) \approx 1.96$ for $\chi = 90°$ is expressed in terms of the Bessel functions J_0 and J_1

$$f_*(90) = \frac{2}{\pi[J_0(1) - J_1(1)]}. \tag{2.223}$$

Table 2.2 The magnetic field B_x (2.221) and the charge density ρ_e (2.222) on the light cylinder at different inclination angles of χ

	$\chi = 0°$		$\chi = 30°$		$\chi = 60°$		$\chi = 90°$	
z/R_L	B_x	ρ_e	B_x	ρ_e	B_x	ρ_e	B_x	ρ_e
1.5	0.16	−0.13	0.17	−0.14	0.12	−0.11	0.05	−0.05
1.4	0.22	−0.17	0.23	−0.19	0.18	−0.16	0.07	−0.08
1.3	0.30	−0.23	0.32	−0.26	0.25	−0.22	0.11	−0.12
1.2	0.41	−0.30	0.44	−0.35	0.36	−0.31	0.18	−0.19
1.1	0.54	−0.38	0.61	−0.48	0.51	−0.44	0.27	−0.29
1.0	0.71	−0.48	0.83	−0.63	0.72	−0.61	0.41	−0.43
0.9	0.93	−0.58	1.11	−0.81	1.00	−0.84	0.62	−0.63
0.8	1.17	−0.65	1.48	−1.02	1.39	−1.11	0.93	−0.90
0.7	1.44	−0.67	1.93	−1.20	1.89	−1.41	1.36	−1.24
0.6	1.70	−0.59	2.43	−1.32	2.52	−1.70	1.93	−1.62
0.5	1.89	−0.35	2.96	−1.30	3.24	−1.90	2.65	−1.99
0.4	1.95	0.08	3.44	−1.05	4.01	−1.90	3.50	−2.23
0.3	1.81	0.67	3.77	−0.53	4.72	−1.59	4.41	−2.22
0.2	1.41	1.31	3.84	0.22	5.23	−0.92	5.23	−1.82
0.1	0.78	1.82	3.58	1.06	5.42	0.01	5.81	−1.04
0.0	0.00	2.01	3.01	1.74	5.22	1.01	6.02	0.00
−0.1	−0.78	1.82	2.23	2.09	4.64	1.81	5.81	1.04
−0.2	−1.41	1.31	1.39	2.05	3.82	2.23	5.23	1.82
−0.3	−1.81	0.67	0.64	1.69	2.91	2.25	4.41	2.22
−0.4	−1.95	0.08	0.06	1.18	2.06	1.97	3.50	2.23
−0.5	−1.89	−0.35	−0.31	0.69	1.35	1.55	2.65	1.99
−0.6	−1.70	−0.59	−0.51	0.30	0.82	1.11	1.93	1.62
−0.7	−1.44	−0.67	−0.57	0.04	0.45	0.74	1.36	1.24
−0.8	−1.17	−0.65	−0.55	−0.11	0.22	0.46	0.93	0.90
−0.9	−0.92	−0.58	−0.49	−0.18	0.08	0.26	0.62	0.63
−1.0	−0.71	−0.48	−0.41	−0.20	0.00	0.14	0.41	0.43
−1.1	−0.54	−0.38	−0.33	−0.19	−0.04	0.06	0.27	0.29
−1.2	−0.41	−0.30	−0.26	−0.16	−0.05	0.02	0.18	0.19
−1.3	−0.30	−0.23	−0.20	−0.13	−0.05	−0.01	0.11	0.12
−1.4	−0.22	−0.17	−0.16	−0.11	−0.05	−0.01	0.07	0.08
−1.5	−0.16	−0.13	0.12	−0.08	−0.04	−0.01	0.05	0.05

(Hint: it is necessary to determine the Poynting vector flux through the light cylinder surface and the neutron star surface.)

On the other hand, from the analysis of the above solutions, we conclude that over the entire surface of the light cylinder, the toroidal magnetic field is zero though, unlike the axisymmetric case, it is not zero in the interior regions of the magnetosphere. Indeed, since expansions (2.215) and (2.216) begin with the second powers $(1 - x_r^2)$, definitions (2.199) and (2.200) show that at small distances from the light cylinder, the magnetic field components behave as

$$B_z \propto (1 - x_r^2), \quad B_\varphi \propto (1 - x_r^2)^2. \tag{2.224}$$

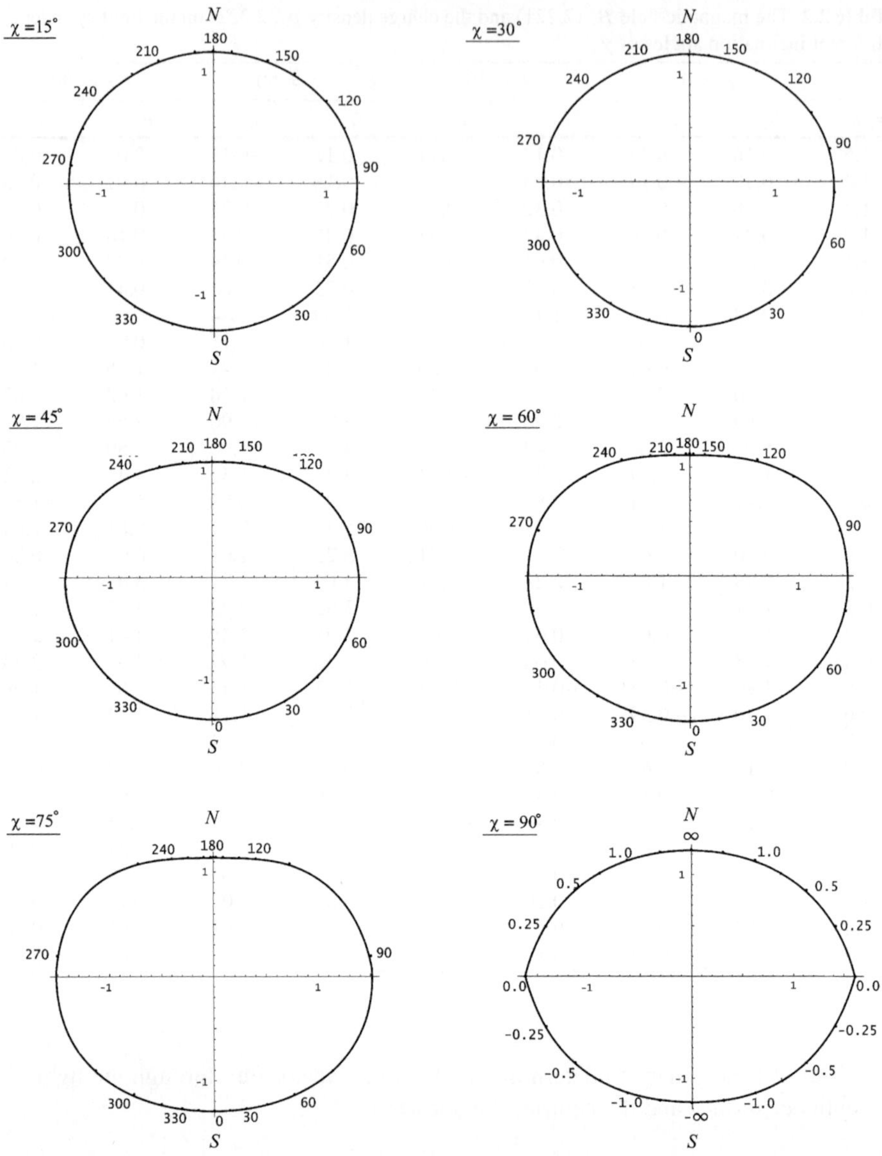

Fig. 2.17 The change in the polar cap form with increasing inclination angle χ. The numbers indicate the values of the angles φ (in degrees) and for $\chi = 90^{\circ}$ the values of $z/R_{\rm L}$ for which the field line coming out from the given point intersects the light cylinder

It is, at first sight, a purely mathematical property but, actually, is of fundamental importance and one of the key conclusions in this chapter. Therefore, the solution of the pulsar equation is thoroughly derived and given in the theorem:

Theorem 2.1 *In the absence of the longitudinal currents and the accelerating potential the Poynting vector flux through the surface of the light cylinder is zero. Otherwise, the corotation currents flowing in the magnetosphere completely screen the magnetodipole radiation of the neutron star. Therefore, in the case of the inclined rotator, all energy losses are connected with the longitudinal current circulating in the magnetosphere* (Beskin et al., 1983; Mestel et al., 1999).

One should note that the conclusion that there are no losses is, in no way, connected with the quasistationary approximation used here. Indeed, as was shown above, the magnetodipole radiation can be produced within this formalism. The point is that in the vacuum case we have two second-order equations (2.124) and (2.125) for the functions ψ and h, which can be rewritten as a single fourth-order equation for one of these values. Therefore, in the vacuum case, two independent solutions corresponding to retarded and advanced potentials are possible. The choice of only the retarded potentials involves an additional physical assumption in the absence of the confluence energy flux from infinity. In the case of the plasma-filled magnetosphere, Eq. (2.196) has the unique solution in the form of a standing wave that does not transport energy to infinity.

2.6.1.4 Axisymmetric Magnetosphere with a Nonzero Longitudinal Current for the Monopole Magnetic Field

F.C. Michel found another remarkable analytical solution for the monopole magnetic field of the star (Michel, 1973b). It turned out that for the special choice of the longitudinal current

$$I(\Psi) = I_{\rm M} = \frac{\Omega_{\rm F}}{4\pi}\left(2\Psi - \frac{\Psi^2}{\Psi_0}\right) \tag{2.225}$$

and for $\Omega_{\rm F} = \text{const}$, the monopole magnetic field

$$\Psi(r,\theta) = \Psi_0(1-\cos\theta) \tag{2.226}$$

is the exact solution to the pulsar equation (2.101), beyond the light cylinder as well. Otherwise, for the current $I = I_{\rm M}$ (2.225), the effects of the longitudinal currents and the corotation currents are fully compensated. It is easy to check that the current I takes the form $I(\theta) = I_{\rm M}^{({\rm A})}\sin^2\theta$, where

$$I_{\rm M}^{({\rm A})} = \frac{\Omega_{\rm F}\Psi_0}{4\pi}, \tag{2.227}$$

which, actually, corresponds to the GJ current density. As is evident from relations (2.225) and (2.226), in the Michel solution the electric field **E** having only the

θ-component is equal in magnitude to the toroidal component of the magnetic field

$$B_{\varphi} = E_{\theta} = -B_0 \left(\frac{\Omega R}{c}\right) \frac{R}{r} \sin\theta, \tag{2.228}$$

which at distances larger than the light cylinder radius becomes larger than the poloidal magnetic field $B_{\rm p} = B_0(R/r)^2$. On the other hand, in this solution the full magnetic field remains larger than the electric one everywhere, which makes the light surface extend to infinity.

As was already noted, the Michel solution, in spite of its artificial character, is of great importance in the black hole magnetosphere theory. Therefore, we return to this solution in the next chapter. We note here that the Michel solution proves useful for the radio pulsar magnetosphere theory as well, because this structure of the magnetic field can be realized beyond the light cylinder in the pulsar wind region. Therefore, we should emphasize at once that under the real conditions we, of course, deal with the so-called split monopole

$$\Psi(r,\theta) = \Psi_0(1 - \cos\theta), \quad \theta < \pi/2, \tag{2.229}$$

$$\Psi(r,\theta) = \Psi_0(1 + \cos\theta), \quad \theta > \pi/2, \tag{2.230}$$

rather than with a monopole when the magnetic flux converges in the lower hemisphere and diverges in the upper one, as shown in Fig. 2.18. In other words, for this solution to exist it is necessary to introduce the current sheet in the equatorial plane dividing the convergent and divergent magnetic fluxes. One should remember that in this geometry topologically equivalent to the dipole magnetic field both in the northern and in the southern parts of the magnetosphere, there is a charge outflow of the same sign. Therefore, the poloidal surface currents closing the bulk currents and ensuring the electric current conservation must flow along the sheet. This sheet is possible in the presence of the accretion disk in which the studied force-free approximation becomes inapplicable.

Problem 2.26 Show by direct substitution in Eq. (2.101) that the monopole magnetic field remains an exact solution for the arbitrary profile of the angular velocity $\Omega_{\rm F}(\Psi)$ if the electric current is still connected with it by the relation (Blandford and Znajek, 1977; Beskin et al., 1992a)

$$4\pi I(\Psi) = \Omega_{\rm F}(\Psi)\left(2\Psi - \frac{\Psi^2}{\Psi_0}\right). \tag{2.231}$$

Later Bogovalov (2001) demonstrated that in the force-free approximation (when massless charged particles can move radially with the velocity of light), the inclined split monopole field

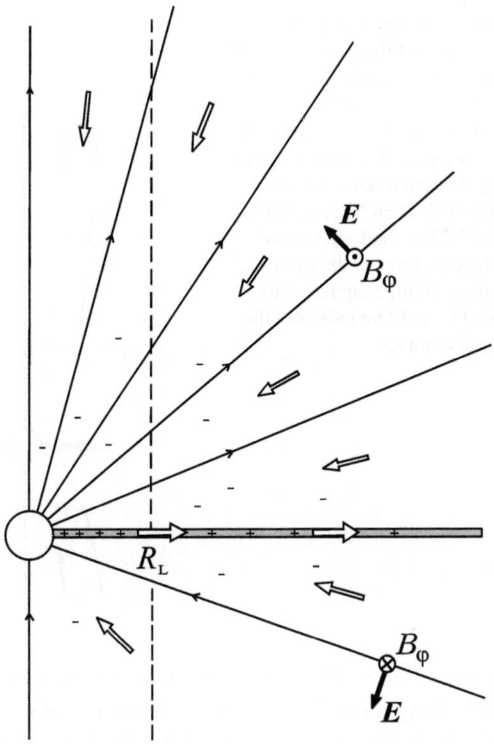

Fig. 2.18 The Michel monopole solution in which the electric field E_θ is exactly equal to the toroidal magnetic field B_φ. In the real conditions, this solution can be realized in the presence of the conducting disk in the equatorial plane along which the electric current closure occurs (*contour arrows*)

$$\Psi(r, \theta, \varphi, t) = \Psi_0(1 - \cos\theta), \quad \theta < \pi/2 - \chi\cos(\varphi - \Omega t + \Omega r/c), \quad (2.232)$$

$$\Psi(r, \theta, \varphi, t) = \Psi_0(1 + \cos\theta), \quad \theta > \pi/2 - \chi\cos(\varphi - \Omega t + \Omega r/c), \quad (2.233)$$

is the solution of the problem as well. In this case, within the cones $\theta < \pi/2 - \chi$, $\pi - \theta < \pi/2 - \chi$ near the rotation axis, the electromagnetic field is not time dependent, while in the equatorial region the electromagnetic fields change the sign at the instant the processing current sheet intersects the given point.

2.6.1.5 Axisymmetric Magnetosphere with a Nonzero Longitudinal Current for the Parabolic Magnetic Field

It turned out that the exact solution can be constructed by the "nonphysical" parabolic field $\Psi \propto r(1 - \cos\theta)$ (1.127) shown in Fig. 2.19 (Blandford, 1976). Certainly, this structure of the magnetic field can again be realized only in the presence of the conducting disk so that the magnetic field lines in the lower and upper hemispheres specularly repeat one another. The jump of the tangential component of the magnetic field is connected with the electric currents flowing within the disk. One should stress at once that in the studied solution only the form of the magnetic

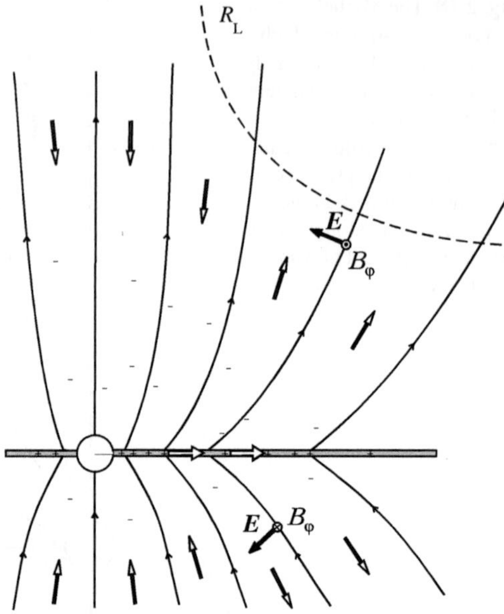

Fig. 2.19 The parabolic structure of the magnetic field and the longitudinal currents for the "nonphysical" solution (2.237). The angular velocity $\Omega_F(\Psi)$ is determined by the rotational velocity of the disk. Therefore, according to (2.235), the longitudinal current closes at the finite values of the magnetic flux Ψ. *Dashed line* indicates the light cylinder

surfaces coincides with the vacuum magnetic field. The density of the magnetic field lines should differ from that of the magnetic field in vacuum. Otherwise, the magnetic flux $\Psi(r, \theta)$ should have the form $\Psi(r, \theta) = \Psi(X)$, where for $\theta < \pi/2$

$$X = r(1 - \cos\theta). \tag{2.234}$$

As in the previous case, this structure of the magnetic field can occur only if there is a certain connection between the angular velocity Ω_F and the current I, viz., when the following relation holds:

$$I(\Psi) = \frac{C\Omega_F(X)X}{2\left[1 + \dfrac{\Omega_F^2(X)X^2}{c^2}\right]^{1/2}}, \tag{2.235}$$

where C is an integration constant. In this case, the magnetic flux can be found from the condition

$$\frac{d\Psi}{dX} = \frac{\pi C}{\left[1 + \dfrac{\Omega_F^2(X)X^2}{c^2}\right]^{1/2}}. \tag{2.236}$$

As we see, here the solution also exists for any profile $\Omega_F(X)$. In particular, for the constant angular velocity, it has the form (Lee and Park, 2004)

$$\Psi(r,\theta) = \frac{\pi C c}{\Omega_{\rm F}} \ln\left[\frac{\Omega_{\rm F} X}{c} + \sqrt{1 + \frac{\Omega_{\rm F}^2 X^2}{c^2}}\right]. \tag{2.237}$$

We should emphasize that though the above solution is formally valid for any value $\Omega_{\rm F}(X)X/c$, in reality, only the configurations in which

$$\frac{\Omega_{\rm F}(X)X}{c} < 1 \tag{2.238}$$

can be realized. The point is that, as shown in Fig. 2.19, all magnetic surfaces must intersect the region of the accretion disk that must determine the value of the angular velocity $\Omega_{\rm F}$. But the accretion disk cannot rotate with the velocity larger than the velocity of light. Since in the equatorial plane $X = \varpi$, the condition (2.238) is to be satisfied over the entire space. As a result, the magnetic field structure does not differ too much from the vacuum solution.

On the other hand, for a fast decrease in the angular velocity $\Omega_{\rm F}(\Psi)$ with increasing Ψ, so that $\Omega_{\rm F}(\Psi)\varpi/c \to 0$, the longitudinal current, according to (2.235), is concentrated only in the region $\Omega_{\rm F}(0)X/c \sim 1$, so that the characteristic magnetic flux, within which the current closure occurs, can be estimated as

$$\Psi_0 = \frac{\pi C c}{\Omega_{\rm F}(0)}. \tag{2.239}$$

This relation defines the connection between the integration constant C and the flow Ψ_0 involved, for example, in the definition of the magnetization parameter σ (2.82).

The "nonphysical" solution was not as known as the Michel monopole solution though it, in many respects, much better describes the structure of the magnetized wind outflowing from compact objects. In particular, it adequately models the jet formation process. On the other hand, one should remember that for the existence of the magnetic field decreasing with distance as r^{-1} (and it is exactly how the magnetic field corresponding to the potential $X = r(1 - \cos\theta)$ is constructed), the toroidal currents flowing in the equatorial plane are needed. In the absence of these currents, the parabolic magnetic field cannot be realized.

Problem 2.27 Show that for the parabolic solution, as in the Michel monopole solution, at large distances $r \to \infty$ the electric field is compared in magnitude with the magnetic one so that $B_{\rm p} \ll B_\varphi$ and $B_\varphi \approx |\mathbf{E}|$, where

$$B_\varphi = -\frac{C\Omega_{\mathrm{F}}}{c\left[1+\dfrac{\Omega_{\mathrm{F}}^2(X)X^2}{c^2}\right]^{1/2}}\frac{(1-\cos\theta)}{\sin\theta}, \tag{2.240}$$

$$|\mathbf{E}| = \frac{C\Omega_{\mathrm{F}}}{c\left[1+\dfrac{\Omega_{\mathrm{F}}^2(X)X^2}{c^2}\right]^{1/2}}\left(\frac{1-\cos\theta}{2}\right)^{1/2}. \tag{2.241}$$

2.6.1.6 Perturbation of the Monopole Magnetic Field

In conclusion, we consider another model problem of the small perturbation of the Michel monopole solution (Beskin et al., 1998). As was already mentioned, Eq. (2.101) needs three boundary conditions. We suppose that the angular velocity $\Omega_{\mathrm{F}} = \mathrm{const}$ remains the same as in the Michel solution. As to the longitudinal current $I(R,\theta)$, it is assumed to differ little from the Michel current (2.228)

$$I(R,\theta) = I_{\mathrm{M}}(\theta) + l(\theta) = I_{\mathrm{M}}^{(\mathrm{A})}\sin^2\theta + l(\theta), \tag{2.242}$$

so that $l(\theta) \ll I_{\mathrm{M}}^{(\mathrm{A})}$. Since the perturbations are assumed to be small, relation (2.242) defines the value of the current as a function of the stream function Ψ.

Writing now the solution to Eq. (2.101) as $\Psi(r,\theta) = \Psi_0[1-\cos\theta+\varepsilon f(r,\theta)]$, we obtain in the first order with respect to the small parameter $\varepsilon = l/I_{\mathrm{M}}^{(\mathrm{A})}$

$$\varepsilon(1-x^2\sin^2\theta)\frac{\partial^2 f}{\partial x^2} + \varepsilon(1-x^2\sin^2\theta)\frac{\sin\theta}{x^2}\frac{\partial}{\partial\theta}\left(\frac{1}{\sin\theta}\frac{\partial f}{\partial\theta}\right) - 2\varepsilon x\sin^2\theta\frac{\partial f}{\partial r}$$
$$-2\varepsilon\sin\theta\cos\theta\frac{\partial f}{\partial\theta} + 2\varepsilon(3\cos^2\theta-1)f = -\frac{1}{I_{\mathrm{M}}^{(\mathrm{A})}\sin\theta}\frac{\mathrm{d}}{\mathrm{d}\theta}(l\sin^2\theta). \tag{2.243}$$

Here $x = \Omega_{\mathrm{F}}r/c$. Equation (2.243), as was expected, has a singularity on the light cylinder $x\sin\theta = 1$.

In the general case, the solution to Eq. (2.243) is extremely cumbersome. However, for the special choice of the perturbation

$$l(\theta) = \varepsilon_* I_{\mathrm{M}}^{(\mathrm{A})}\sin^2\theta, \tag{2.244}$$

where $|\varepsilon_*| = \mathrm{const} \ll 1$, the analytical solution can be found. It has the form

$$\Psi(r,\theta) = \Psi_0\left[1-\cos\theta+\varepsilon_*\left(\frac{\Omega_{\mathrm{F}}r}{c}\right)^2\sin^2\theta\cos\theta\right]. \tag{2.245}$$

The solution (2.245) shows that for $I < I_M$ ($\varepsilon_* < 0$), the magnetic field lines are concentrated near the equator ($\delta\Psi < 0$ for $\theta < \pi/2$). In this case, the light surface is located at finite distance from the light cylinder. It has the form of a cylinder with the radius

$$\varpi_C = |2\varepsilon_*|^{-1/4} R_L, \tag{2.246}$$

on which the monopole field perturbation can still be considered to be small. Accordingly, for $I > I_M$ ($\varepsilon_* > 0$), the magnetic field lines turn to the rotation axis ($\delta\Psi > 0$ for $\theta < \pi/2$), and the light surface is reached only at infinity.

Problem 2.28 Find relation (2.246).

The above exact solutions of the pulsar equation lead to the following general conclusions:

1. The solution to the force-free equation (2.101) can be constructed only within the light surface $|\mathbf{E}| = |\mathbf{B}|$, which, for the zero longitudinal currents, coincides with the light cylinder $\varpi = c/\Omega_F$. Beyond the light surface, the electric field becomes larger than the magnetic one, which results in violation of the frozen-in condition $\mathbf{E} + \mathbf{v} \times \mathbf{B}/c = 0$. In the general case, the light surface does not coincide with the light cylinder but is located at larger distances. As we will see, the presence or the absence of the light surface is of crucial importance for the discussion of the particle acceleration problem (within the force-free approximation the particle Lorentz factor on the light surface, formally, is infinite).
2. In the case of zero longitudinal currents, regardless of the inclination angle χ, the magnetic field on the light cylinder must be perpendicular to its surface (Henriksen and Norton, 1975; Beskin et al., 1983). This mathematical result leads to the most important physical conclusion—the Poynting vector does not have a normal component here and, hence, the electromagnetic energy flux through the light cylinder surface is zero. Consequently, in the absence of the longitudinal currents, the secondary plasma filling the magnetosphere must fully screen the magnetodipole radiation of the neutron star (Beskin et al., 1983; Mestel et al., 1999). Therefore, all the energy losses of the rotating neutron star are to be associated with the ponderomotive action of the surface currents closing the longitudinal currents flowing in the magnetosphere. Thus, formula (2.178) fully defines the slowing down of radio pulsars.
3. In the absence of the longitudinal currents, the magnetic field lines are concentrated in the vicinity of the equator. Otherwise, the toroidal currents $\mathbf{j} = \rho_{GJ}\, \boldsymbol{\Omega} \times \mathbf{r}$ connected with the corotation of the GJ charge density ρ_{GJ} do not collimate the magnetic field lines but, on the contrary, make them diverge and concentrate near the equator. As a result, the magnetic field along the rotation axis decreases exponentially fast rather than by the power law.

2.6.2 Magnetosphere Structure with Longitudinal Currents

We proceed to the key part of this chapter, viz., to the discussion of the magnetosphere structure in the presence of the longitudinal current I and the accelerating potential ψ. The importance of this problem is obvious—as was shown above, the energy losses of radio pulsars are fully specified by the longitudinal electric currents circulating in the magnetosphere. Therefore, the question of the value of the longitudinal currents (and, hence, the presence or the absence of the light surface) is the key one the neutron star magnetosphere theory is to answer.

At the same time, there are two important circumstances. First, as was already noted, within the force-free approximation, the longitudinal current is a free parameter. Second, and it was also mentioned, the particle acceleration problem cannot be solved within this approximation. Therefore, we can analyze a limited set of problems only. A more comprehensive analysis is made in Chap. 5 on the basis of the full magnetohydrodynamic version of the GS equation.

On the other hand, in the presence of the longitudinal current even in the force-free approximation, Eq. (2.129) becomes essentially nonlinear. It is not surprising, therefore, that in most papers the analysis of only the axisymmetric magnetosphere was made. Indeed, since the total current within the polar cap is to be zero, expression $I\,\mathrm{d}I/\mathrm{d}\Psi$ cannot be a linear function Ψ on all open field lines. Except for the Michel and Blandford remarkable solutions (Michel, 1973b; Blandford, 1976), only some analytical solutions were obtained (Beskin et al., 1983; Lyubarskii, 1990; Sulkanen and Lovelace, 1990; Fendt et al., 1995; Beskin and Malyshkin, 1998). Therefore, the problem of construction of magnetosphere with nonzero longitudinal currents is still to be solved. As to the case of an inclined rotator, there are only preliminary results here (Mestel and Wang, 1982; Bogovalov, 1999, 2001).

Technically, the reason is that Eq. (2.101) contains a critical surface—a light cylinder, the passage of which requires the expansion of the solution into eigenfunctions that have no singularity on this surface. Exactly this method of solution was described above when analyzing the magnetosphere with zero longitudinal current. Therefore, in most cases, the similar problem was solved only analytically, which, in turn, could be done only for a certain class of functions $I(\Psi)$, viz., when the current density is constant in the whole region of open magnetic field lines (i.e., when $I(\Psi) = k\Psi$), and the current closure occurs along the separatrix dividing the open and closed field lines. In this statement, Eq. (2.101) appears linear in the region of not only closed but also open magnetic field lines, and the main problem reduces to matching the solutions in these two regions. It is in this direction that the main results of the magnetosphere structure with longitudinal electric field were obtained.

Consider now the analytical method for constructing the solution in more detail, since it allows us to formulate the main problems that arise when trying to construct the self-consistent model of the magnetosphere of radio pulsars containing longitudinal currents. Thus, we consider the axisymmetric force-free magnetosphere of the rotating neutron star. As was already mentioned, with the special choice of the longitudinal current I and the potential ψ, Eq. (2.101) can be reduced to a linear one. This is possible if we take the values of $\Omega_{\mathrm{F}}(\Psi)$ and $I(\Psi)$ in the form

$$\Omega_{\rm F}(\Psi) = \Omega(1 - \beta_0), \tag{2.247}$$

$$I(\Psi) = \frac{\Omega}{2\pi} i_0 \Psi, \tag{2.248}$$

where i_0 and β_0 are constant. Recall that their physical meaning is defined by relations (2.107) and (2.108).

The pulsar equation in the region of open field lines in the dimensionless variables $x_r = \Omega\varpi/c$, $z' = \Omega z/c$ takes the form

$$-\nabla^2\Psi\left[1 - x_r^2(1-\beta_0)^2\right] + \frac{2}{x_r}\frac{\partial\Psi}{\partial x_r} - 4i_0^2\Psi = 0. \tag{2.249}$$

In the region of closed field lines, where, as was already noted, the potential $\psi = 0$ (i.e., $\beta_0 = 0$), we simply have

$$-\nabla^2\Psi\left(1 - x_r^2\right) + \frac{2}{x_r}\frac{\partial\Psi}{\partial x_r} = 0. \tag{2.250}$$

As a result, all nonlinearity is enclosed in a thin transition layer in the vicinity of the separatrix, the very position of which must be found from the solution. Note that, unlike the case of the zero longitudinal current, the zero point of the magnetic field must not necessarily lie on the light cylinder surface $x_r = 1$.

Problem 2.29 Show that, in this case, the solution to Eq. (2.249) that has no singularity on the surface $x_r = (1-\beta_0)^{-1}$ can be constructed in the form of the series (Beskin et al., 1983)

$$R_\lambda(x_1) = \mathcal{D}(\lambda)\sum_{n=0}^{\infty} a_n(1 - x_1^2)^n, \tag{2.251}$$

where $x_1 = (1-\beta_0)x_r$, $\alpha_1 = 4i_0^2/(1-\beta_0)^2$, and the expansion coefficients a_n satisfy the recurrent relations

$$a_0 = 1, \quad a_1 = -\frac{\alpha_1}{4}, \quad a_{n+1} = \frac{4n^2 - \alpha_1}{4(n+1)^2}a_n + \frac{\alpha_1 + \lambda^2}{4(n+1)^2}a_{n-1}. \tag{2.252}$$

Here $\mathcal{D}(\lambda)^{-1} = (\pi/2)(1-\beta_0)^{-1}\sum_{n=0}^{\infty} a_n$.

We now specify the boundary conditions for the system of equations (2.249) and (2.250). In the region of open field lines, Eq. (2.249), according to (1.64), requires three boundary conditions. These conditions are, first of all, the values i_0 and β_0 determined on the star surface. The third boundary condition is not only the value of the stream function $\Psi(R,\theta)$ on the star surface (2.103) but also the value of the function Ψ on the surface of the separatrix $z'_*(x_r)$ dividing the region of the open and closed magnetospheres (Okamoto, 1974)

$$\Psi^{(1)}\big|_{z'=z'_*(x_r)} = \Psi^{(2)}\big|_{z'=z'_*(x_r)} . \tag{2.253}$$

Finally, the regularity condition (2.211) on the light cylinder $x_r = x_{\rm L}$ is written as

$$\frac{2}{x_r}\frac{\partial \Psi}{\partial x_r}\bigg|_{x_r=(1-\beta_0)^{-1}} - 4i_0^2\Psi\big|_{x_r=(1-\beta_0)^{-1}} = 0. \tag{2.254}$$

Clearly, in the presence of the longitudinal current (i.e., for $B_\varphi \neq 0$), the light surface no longer coincides with the light cylinder. Relation (2.254) also shows that in the studied statement of the problem, the magnetic field lines on the light cylinder must be directed from the equator ($B_z > 0$ for $\boldsymbol{\Omega} \cdot \mathbf{m} > 0$).

As to the region of closed field lines, which, in the general case, does not reach the light cylinder, exactly the conditions of matching the regions of closed and open field lines must act as additional boundary conditions for it. These conditions should be, first of all, the coincidence of the location of the separatrix field line $z' = z'_*(x_r)$ for both the regions (2.253) and, besides, the continuity of the value $\mathbf{B}^2 - \mathbf{E}^2$:

$$\{\mathbf{B}^2 - \mathbf{E}^2\} = 0. \tag{2.255}$$

The latter condition is easy to deduce by integrating the force-free equation written as $(\nabla\cdot\mathbf{E})\mathbf{E}+[\nabla\times\mathbf{B}]\times\mathbf{B} = 0$ over a thin transition layer (Okamoto, 1974; Lyubarskii, 1990). It is important that the condition (2.255) is obtained if the curvature of the magnetic field lines is disregarded and, therefore, cannot be used in the vicinity of singular points.

Problem 2.30 Find the condition (2.255) for the Cartesian coordinate system in which the transition layer coincides with the xy-plane (Lyubarskii, 1990).

We can now mention the main papers concerned with the force-free magnetosphere of radio pulsars (in which, in particular, the system of equations (2.249) and (2.250) was analyzed) for the real dipole field of the neutron star.

1. In Beskin et al. (1983), the case $i_0 \neq 0$, $\beta_0 \neq 0$ was studied. Only relation (2.253) was used; the equilibrium condition (2.255) was not taken into account. Besides, the region of closed field lines was supposed to remain the same as in the absence of the longitudinal current.
2. In Lyubarskii (1990), for $\beta_0 = 0$, both the equilibrium conditions (2.253) and (2.255) were taken into account. Incidentally, the additional assumption was that the last open field line, as in the Michel monopole solution, coincides with the equator beyond the light cylinder. Finally, in the paper, the absence of an inverse

current along the separatrix was implicitly assumed, which substantially changed the magnetic field structure in the vicinity of the zero point.

3. In Sulkanen and Lovelace (1990), for $\beta_0 = 0$, the case of the strong longitudinal current $i_0 > 1$ was studied. As was expected, with these longitudinal currents, thc magnetic surface collimates to the rotation axis. The equilibrium conditions with the region of closed field lines were not used at all. As a result, there occurred a region, in which the poloidal magnetic field is absent, between the regions of open and closed field lines.
4. In Beskin and Malyshkin (1998), both the two equilibrium conditions and the perturbation of the region of closed field lines were taken into account. It was also shown that the zero point can be located inside the light cylinder: $x_r^{(*)} < 1$. However, the magnetic field structure in the equatorial region beyond the zero point was not discussed in the paper.

Figure 2.20 shows, as an example, the structure of the magnetic surfaces for the nonzero longitudinal current i_0 and the accelerating potential β_0 obtained numerically by solving Eqs. (2.249) and (2.250) (Beskin and Malyshkin, 1998). It was shown that the solution of the problem cannot be constructed for any values of i_0 and β_0. The point is that, for certain parameters i_0, β_0, the solution to Eq. (2.249) in the region of open field lines shows that the zero line of the magnetic field is located beyond the light cylinder $x_L = 1$. Clearly, in this case, the solution cannot be matched to the closed magnetosphere region because the solution with $i_0 = 0$ cannot be extended to the region $x_r > 1$. As shown in Fig. 2.21, on the plane of the parameters $i_0 - \beta_0$, the forbidden region corresponds to rather small values of i_0.

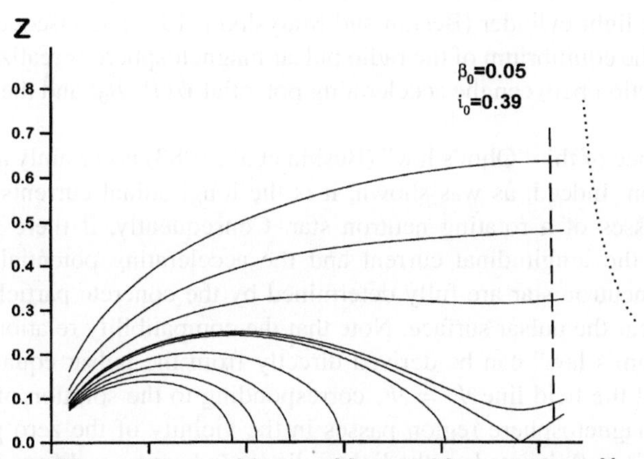

Fig. 2.20 The magnetosphere structure of the axisymmetric rotator for $i_0 = 0.39$ and $\beta_0 = 0.05$. The values of i_0 and β_0 do not correspond to "Ohm's law" (2.256) and, therefore, the zero point is within the "light cylinder" $x_r = 1$. The real light cylinder (*dashed line*) is at a distance of $x_r = 1/(1 - \beta_0)$ from the rotation axis. The *dotted line* indicates the light surface (in this paper, its location was not established) (Beskin and Malyshkin, 1998)

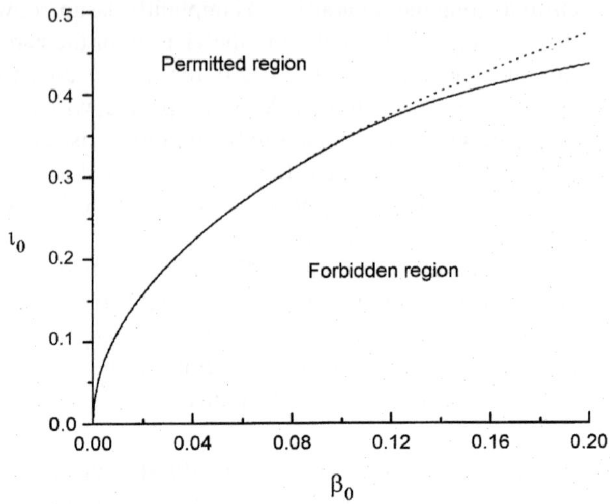

Fig. 2.21 The range of parameters i_0–β_0, for which the construction of the solution is possible. The *dotted line* indicates "Ohm's law" (2.256) (Beskin and Malyshkin, 1998)

Thus, the important conclusion is that the existence in the neutron star magnetosphere of the closed magnetic field lines that do not intersect the light cylinder can impose a certain constraint on the longitudinal currents circulating in the magnetosphere. The thorough computations show that the total energy of the electromagnetic field proves minimal exactly in the vicinity of the boundary line $\beta_0 = \beta_0(i_0)$, when, by the way, the zero point of the magnetic field lies in the vicinity of the light cylinder (Beskin and Malyshkin, 1998). Consequently, we can suppose that the equilibrium of the radio pulsar magnetosphere is realized only for a certain connection between the accelerating potential $\psi(P, B_0)$ and the longitudinal current I.

The existence of this "Ohm's law" (Beskin et al., 1983) is certainly a very important conclusion. Indeed, as was shown, it is the longitudinal currents that specify the energy losses of a rotating neutron star. Consequently, if there is a connection between the longitudinal current and the accelerating potential, the energy losses of the neutron star are fully determined by the concrete particle generation mechanism near the pulsar surface. Note that the compatibility relation prescribing nonlinear "Ohm's law" can be derived directly from the pulsar equation. Indeed, supposing that the field line $\Psi = \Psi_*$ corresponding to the solution of Eq. (2.249) in the open magnetosphere region passes in the vicinity of the zero point (where $B_z \propto (\partial\Psi/\partial x_r) = 0$) located on the light cylinder (where $x_r = 1$), we have directly from (2.249)

$$\beta_0(i_0) = 1 - \left(1 - \frac{i_0^2}{i_{\max}^2}\right)^{1/2}, \tag{2.256}$$

where $i_{max} = \sqrt{(\nabla^2\Psi)_*/4\Psi_*} \approx 0.79$. As seen from Fig. 2.21, the analytical estimate (2.256) is in good agreement with the numerical computations. Relation (2.256), in the large, remains valid for the inclined rotator (Beskin et al., 1993).

On the other hand, as shown in Fig. 2.21, relation (2.256) actually yields only the lower bound for the longitudinal current. Accordingly, the conclusion of the small value of the longitudinal current was not confirmed independently in other papers. Therefore, the question of the value of the longitudinal current remains open. What can be stated with assurance is that the longitudinal current circulating in the radio pulsar magnetosphere does not, evidently, exceed the GJ current. Thus, the problem of the exact value of the energy losses W_{tot} and the existence of the light surface, on which, as we will see, the extra acceleration of particles is possible, remains unsolved. However, for most applications, the estimate $I \approx I_{GJ}$ appears adequate, so that relation (2.5) is a good approximation to W_{tot}. In any event, the problem of the value of the longitudinal current cannot be fully solved by the force-free approximation.

As was already noted, the analytical approach is restricted by the choice of the homogeneous longitudinal current density ($I(\Psi) = k\Psi$). Only a quarter of a century later, after the pulsar equation was formulated, Contopoulos et al. (1999) first studied the system of equations (2.249) and (2.250) numerically. In particular, they succeeded in (by an iterative procedure) passing the singularity on the light cylinder for the arbitrary current $I(\Psi)$. For the case $\beta_0 = 0$, the additional assumption that the last open field line coincides with the equator was also made there (see Fig. 2.22). It is not surprising, therefore, that the longitudinal current (which in the presence of the additional condition is no longer a free parameter) appeared close to

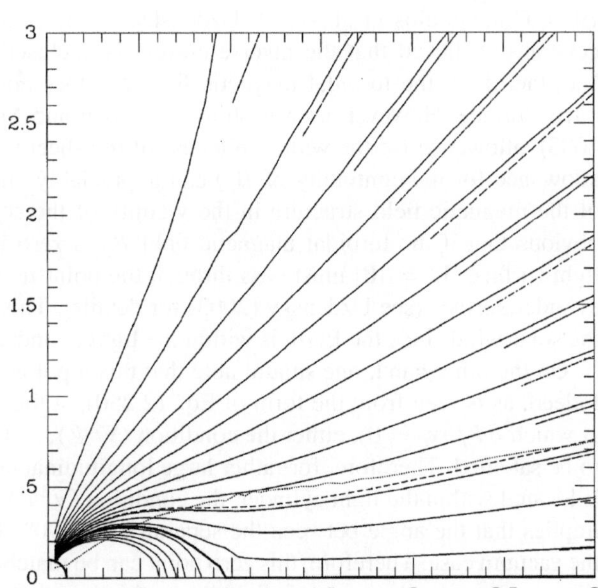

Fig. 2.22 The magnetosphere structure in the model (Contopoulos et al., 1999). The additional assumption that the last open field line coincides with the equator was made (Reproduced by permission of the AAS, Fig. 3 from Contopoulos, I., Kazanas, D., Fendt, C.: The axisymmetric pulsar magnetosphere. ApJ **511**, 351–558 (1999))

the current $I_{\rm M}$ (2.225) for the Michel monopole solution which does not correspond to the GJ current density $j_{\rm GJ} = \rho_{\rm GJ} c \approx$ const. At the same time, the equilibrium condition (2.255) was not taken into account in the paper. This statement of the problem was later discussed in Ogura and Kojima (2003), Goodwin et al. (2004), Gruzinov (2005), Contopoulos (2005), Komissarov (2006), McKinney (2006a), and Timokhin (2006), and, in a number of papers, the case, in which the zero point of the magnetic field can be located within the light cylinder, was also analyzed.

Let us briefly enumerate the main difficulties the force-free magnetosphere theory encounters. First of all, it turned out that the analytical solution method discussed above does not, actually, allow us to uniquely specify the magnetic field structure. The point is that the dipole magnetic field in the vicinity of the neutron star corresponds to the high harmonics λ in expansion (2.184), whereas the visible magnetic field structure on scales comparable with those of the light cylinder is specified by the small values of λ. As a result, the solution

$$\Psi(\varpi, z) = \frac{|\mathbf{m}|}{R_{\rm L}} \int_0^\infty Q(\lambda) R_\lambda(\varpi) \cos \lambda z \mathrm{d}\lambda, \qquad (2.257)$$

where $Q(\lambda) \to 1$ for $\lambda \to \infty$, still corresponds to the dipole magnetic field for $\mathbf{r} \to 0$. This is because on the background of the large dipole magnetic field near the neutron star surface, one fails to control the harmonics with the small value of λ, which is crucial at large distances from the star.

The problem of the magnetic field structure in the equatorial region beyond the zero point is not solved either. As was mentioned, in most papers it was supposed in the example of the solar wind that a current sheet is to develop here, which separates the oppositely directed flows of the magnetic field (see Fig. 2.22) (Lyubarskii, 1990; Contopoulos et al., 1999; Uzdensky, 2003; Goodwin et al., 2004). It was, generally, believed that the inverse current is enclosed in an infinitely thin sheet and, therefore, the toroidal magnetic field B_φ does not disappear up to the separatrix surface. However, as was shown (Beskin and Malyshkin, 1998; Uzdensky, 2003), allowance for the width finiteness of the sheet with the inverse current (i.e., allowance for the continuity of B_φ) can appreciably change the main conclusions of the magnetic field structure in the vicinity of the separatrix. In particular, it is obvious that if the toroidal magnetic field B_φ is zero in the equatorial plane, the light surface $|\mathbf{E}| = |\mathbf{B}|$ must pass through the point $\varpi = c/\Omega$, $z = 0$ on the light cylinder surface (see Uzdensky (2003) for details). This problem does not arise for the solar wind since the Earth is within the light cylinder.

On the other hand, one should note that this topology is not the only possibility. Indeed, as is seen from the form of Eq. (2.250), at the zero point (i.e., at the point at which $\partial\Psi/\partial x_r = 0$), either the condition $(\nabla^2\Psi)_* = 0$ or the condition $x_r^2 = 1$ is to be satisfied. Therefore, for rather large longitudinal currents when the zero point is located within the light cylinder, the condition $(\nabla^2\Psi)_* = 0$ is to be satisfied. This implies that the angle between the separatrices is 90°. There is the same angle for the vacuum case. Therefore, this zero point can be matched to the outer region that is not connected by the magnetic field lines with the neutron star surface, for example, with the chain of magnetic islands located in the equatorial plane (see Fig. 3.12b). Only in the limiting case, in which the zero point lies on the light cylinder $x_r = 1$ (as, for example, is the case for the solution with the zero longitudinal current), the value $(\nabla^2\Psi)_*$ remains finite at the zero point (the angle between the separatrices, as shown in Fig. 2.14, is 70°).

Finally, one should remember that most of the solutions, which influence our viewpoint on the radio pulsar magnetosphere structure, referred to the axisymmetric case. For the inclined rotator, quite new effects can occur, which completely change the entire pattern involved. Unfortunately, in this region (except for the case of the above zero longitudinal current) no reliable results that would allow us to confidently judge the magnetosphere properties of the inclined rotator were obtained (Mestel and Wang 1982; Bogovalov 1999, 2001; Spitkovsky, 2006).

Nevertheless, let us try to point out the general properties following from the analysis of Eq. (2.101) describing the force-free neutron star magnetosphere.

1. In the case of zero longitudinal currents independent of the inclination angle χ, the secondary plasma filling the magnetosphere fully screens the magnetodipole radiation (Beskin et al., 1983; Mestel et al., 1999). Therefore, the energy losses of the rotating radio pulsar can be caused only by the ponderomotive action of the surface currents closing the longitudinal currents flowing in the pulsar magnetosphere. Consequently, formula (2.178) fully defines the slowing down of radio pulsars.
2. When the longitudinal current coincides with the Michel current $I_{\rm M}$, the full compensation of two opposite processes occurs, viz., the decollimation connected with the toroidal current and the collimation due to the longitudinal currents. As a result, the monopole magnetic field, which is an exact vacuum solution, turns out to be an exact solution to Eq. (2.101) in the presence of plasma. Certainly, the exact value of the critical current depends on the concrete geometry of the poloidal magnetic field. However, we can confidently state that $j_{\rm cr} \approx \rho_{\rm GJ} c$.
3. For $j_{\parallel} > j_{\rm cr}$, the light surface (which, in the general case, does not coincide with the light cylinder) extends to infinity. This implies that for sufficiently large longitudinal currents the solution can be really extended to infinity. The magnetic surfaces are collimated to the rotation axis (Sulkanen and Lovelace, 1990).
4. If there are any physical constraints from above on the value of the longitudinal current so that $j_{\parallel} < j_{\rm cr}$, the magnetosphere has a "natural boundary"—the light surface. In this case, the complete problem comprising the outer regions cannot be solved within one-fluid magnetic hydrodynamics because, in this case, multiple flow regions occur.

2.6.3 Magnetosphere Models

As was mentioned, the pulsar wind problem is impossible to solve by the force-free approximation. Therefore, we briefly discuss here only the common features of the most developed models of the radio pulsar magnetosphere. The particle acceleration problems are discussed in Chap. 5.

Recall first that the existence of the light surface depends on the value of the longitudinal current. The point is that, as was noted, the presence of light surface at a finite distance from the neutron star must result in the efficient particle acceleration in the pulsar wind. In particular, in the nonfree particle escape models (in which the electric current in the plasma generation region can be arbitrary), the longitudinal current i_0 is to be determined from relation (2.256). For sufficiently small values of the potential drop $\beta_0 < 1$, the longitudinal current should also be small. This implies that the light surface, on which the additional particle acceleration, inevitably, occurs, should be at a finite distance from the neutron star. Certainly, the existence of the light surface leads to the substantial complication of the theory—in fact, not a single, at least, somewhat reliable result of the plasma behavior beyond the light surface has been obtained yet.

Besides, one should not think that the existence of the light surface can be realized only within the model of the nonfree particle escape from the neutron star surface. Indeed, as is evident from the example of the force-free approximation, the light surface extends to infinity only for rather large values of the longitudinal current. As shown in Chap. 4, this conclusion remains valid for the MHD flows as well. Therefore, for any additional constraints from above on the value of the longitudinal electric current, the occurrence of the light surface at a finite distance from the pulsar can be expected. However, within the particle generation model with free particle escape from the star surface, the value of the longitudinal electric current $4\pi I(\Psi) = 2\Omega_{\rm F}\Psi$ ($j_{\parallel} = j_{\rm GJ}$) is fixed and, what is especially important, substantially differs from the Michel current $4\pi I_{\rm M} = \Omega_{\rm F}(2\Psi - \Psi^2/\Psi_0)$. Therefore, it is not improbable that in the real dipole geometry of the pulsar magnetic field this current is not strong enough for a continuous (in particular, transonic) plasma outflow to exist up to large distances as compared to the light cylinder radius. In any case (and it is very important), in the numerically obtained solutions, the value of the longitudinal current $I(\Psi)$ is smaller than that of the limit current $I_{\rm M}$ (2.225) corresponding to the Michel monopole solution. Therefore, the light surface for this solution can be at a finite distance from the neutron star (see, e.g., Ogura and Kojima, 2003). Certainly, the exact proof of this fact invites further investigation.

Indeed, the analysis of the axisymmetric magnetosphere produced up to now did not clarify this point. As was demonstrated, exact analytical solutions (having the longitudinal current $j_{\parallel} \approx$ const within open magnetic field lines) contain the light surface at a finite distance. But their behavior is irrational near the equatorial plane outside the light cylinder. On the other hand, the numerical calculations postulating the existence of the current sheet outside the light cylinder demand the presence of the longitudinal current $I \approx I_{\rm M}$, which is inconsistent with any particle generation mechanism ($j_{\parallel} \to 0$ near the separatrix).

The above arguments for the existence of the light surface were brought forward for the axisymmetric magnetosphere. It turned out that in the case of the inclined rotator, the situation is much more obvious. Indeed, for the orthogonal rotator the GJ charge density in the vicinity of the magnetic pole should be $\varepsilon_{\rm A} = (\Omega R/c)^{1/2}$ times less than that in the axisymmetric magnetosphere. Accordingly, one can expect that the longitudinal current flowing along the open field lines is weaker in the same proportions. Then in the vicinity of the light cylinder, the toroidal magnetic field appears much smaller than the poloidal magnetic field. On the other hand, as we saw in the example of the Michel solution, for the light surface to extend to infinity, it is necessary that the toroidal magnetic field on the light cylinder be of the order of the poloidal field. Therefore, if the longitudinal current $j_{\parallel}$, in reality, is not $\varepsilon_{\rm A}^{-1/2}$ times higher than $\rho_{\rm GJ,90}c$, where $\rho_{\rm GJ,90}$ is the mean charge density on the polar cap for $\chi \sim 90°$ (and for ordinary pulsars this factor is 10^2), the light surface for the orthogonal rotator must, inevitably, be in the immediate vicinity of the light cylinder.

Thus, the presence or the absence of the light surface must be the basic element when constructing the radio pulsar magnetosphere model. Therefore, we will try to classify the magnetosphere models with this in mind. *The first class* of models suggests the presence of the light surface in the vicinity of the light cylinder, which can be realized for rather weak longitudinal currents flowing in the magnetosphere (Beskin et al., 1983; Chiueh et al., 1998). Within this approach, it is supposed that

- the energy losses of the rotating neutron star are fully defined by the current losses;
- the small value of the longitudinal current $i_0 < 1$ results in the occurrence of the light surface;
- in the vicinity of the light surface almost the total electromagnetic flux is transferred to the particle energy flux;
- accordingly, the full closure of the longitudinal current circulating in the magnetosphere really occurs here (see Fig. 2.23).

The problems of the particle acceleration in the vicinity of the light surface are beyond the scope of our discussion. Therefore, we only point to the main features of this process. In the simplest cylindrical geometry when solving the two-fluid hydrodynamical equations (describing exactly the difference in the electron and positron motion), it was shown (Beskin et al., 1983) that a considerable part of the energy carried within the light surface by the electromagnetic field in the thin transition layer

$$\Delta r \sim \lambda^{-1} R_{\rm L} \tag{2.258}$$

in the vicinity of the light surface is transferred to the particle energy flux ($\lambda \sim 10^4$ is the multiplicity parameter). Here, as shown in Fig. 2.23, the total closure of the longitudinal current circulating in the magnetosphere really occurs. As a result, the high efficiency of the particle acceleration has its logical explanation.

Note, however, that the presence of the light surface leads to a considerable complication of the whole problem of the neutron star magnetosphere structure. In this case, it is possible to somewhat reliably describe only the interior regions of the magnetosphere. The problems of the future destiny of the accelerated particles, the energy transport at large distances, and also the current closure are still to be solved. As was noted, these problems are beyond the scope of one-fluid hydrodynamics; evidently, they cannot be solved at all within the analytical approach.

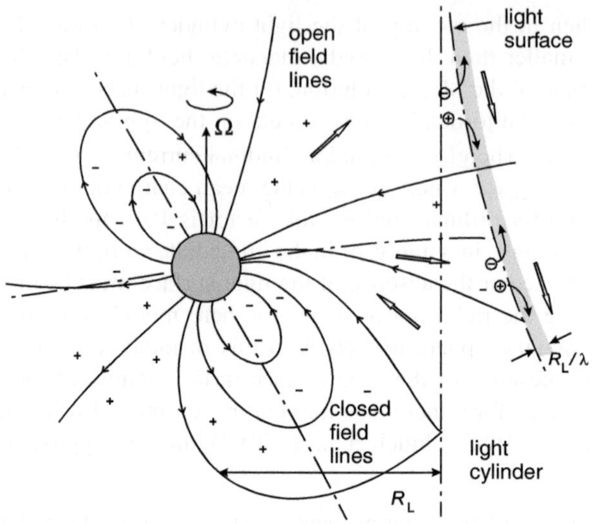

Fig. 2.23 Magnetosphere structure in the model by Beskin et al. (1993). If there are some physical constraints on the value of the longitudinal current (*contour arrows*) so that $j_{\parallel} < j_{\rm cr}$, the magnetosphere has a "natural boundary"—the light surface $|\mathbf{E}| = |\mathbf{B}|$, where the frozen-in condition becomes inapplicable. Therefore, electrons and positrons begin to accelerate in different directions along the electric field and a strong poloidal electric current is generated. As a result, in the thin layer $\Delta r \approx R_{\rm L}/\lambda$, the full closure of the electric current really occurs and the particle energy flux becomes comparable with the total energy flux

The analogous result was later obtained on the basis of the solutions of the two-fluid hydrodynamical equations for more realistic geometry when the poloidal magnetic field is close to the monopole one (Beskin and Rafikov, 2000). It was shown that all results obtained for the cylindrical case remain valid for the more realistic two-dimensional geometry. In particular, it was confirmed that the particles can be accelerated up to energy

$$\varepsilon_{\rm e} \sim eB_0R\frac{1}{\lambda}\left(\frac{\Omega R}{c}\right)^2 \sim 10^4\,{\rm MeV}\left(\frac{\lambda}{10^3}\right)^{-1}\left(\frac{B_0}{10^{12}{\rm G}}\right)\left(\frac{P}{1{\rm s}}\right)^{-2}, \qquad (2.259)$$

but not more than 10^6 MeV, when the radiation friction effects become appreciable. However, as in the one-dimensional case, the problem of constructing the solution beyond the light surface remains unsolved.

The second class of models also suggests the existence of the "dissipation domain" in the vicinity of the light cylinder (see Fig. 2.24). However, only the insignificant energy transfer from the electromagnetic field to particles is postulated here (Mestel and Shibata, 1994; Mestel, 1999). Otherwise, within this model, it is assumed that

- the longitudinal current is close to the critical current ($i_0 \approx 1$);
- in the vicinity of the light surface, only a small amount of the electromagnetic energy flux is transferred to the particle energy flux;

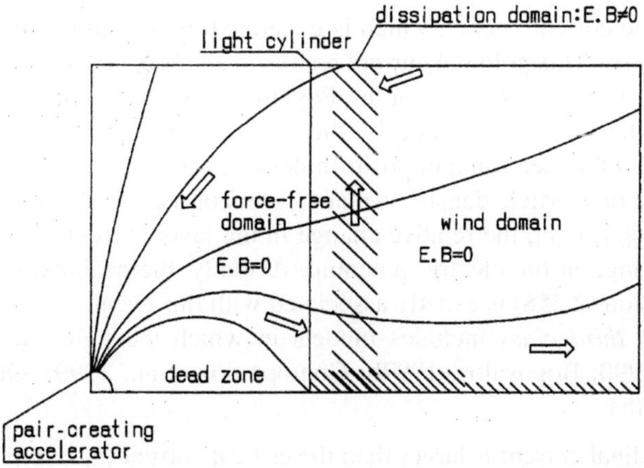

Fig. 2.24 The magnetosphere structure in the Mestel model (Mestel and Shibata, 1994). The existence of the particle acceleration region in the vicinity of the light surface is also supposed. However, only a small change in the longitudinal current (*contour arrows*) is assumed, whereas the potential drop along the magnetic field lines (and, hence, the change in the angular velocity $\Omega_{\rm F}$) was assumed to be significant but insufficient for the particle energy to change appreciably. Therefore, at large distances from the neutron star, the main energy flux is still connected with the Poynting flux

- accordingly, there is only the partial closure of the longitudinal currents circulating in the magnetosphere;
- at large distances from the neutron star, the main energy flux is still connected with the Poynting flux.

Note that in this model the properties of the transition layer were only postulated. In particular, it was assumed that in the transition layer only a small change in the longitudinal current occurs, whereas the relative change in the electric potential along the magnetic field lines (and, hence, the change in the angular velocity $\Omega_{\rm F}$) was assumed to be significant. As a result, the light surface again extended to infinity. Therefore, at large distances from the neutron star, the main energy flux was still connected with the Poynting flux.

One should stress that the basic property of the transition layer studied—the large change in the angular velocity $\Omega_{\rm F}$ with a relatively small longitudinal current—is in contradiction with the properties of the acceleration region in the vicinity of the light surface. As the analysis of the two-fluid MHD equations showed (Beskin et al., 1983; Beskin and Rafikov, 2000), it is the longitudinal current rather than the electric potential that should change most rapidly in the direction perpendicular to the transition layer.

This result can be readily explained. The point is that in the vicinity of the light surface, as was already mentioned, the particle energy formally tends to infinity. As a result, the frozen-in equation is violated, which requires transition to the more exact two-fluid equations. Physically, the result is that electrons and positrons begin to accelerate in different directions along the electric field. Consequently, a strong poloidal electric current occurs, which is generated by the entire electron–positron density $\lambda|\rho_{GJ}|/|e|$. This poloidal current results in an abrupt decrease in the toroidal magnetic field, i.e., in a decrease in the Poynting flux. As to the electric potential, its change in the layer is specified by the electric charge density proportional to the difference in the electron and positron densities only. Since in the radio pulsar magnetosphere the particle density is many orders of magnitude higher than the GJ density $n_{GJ} = |\rho_{GJ}|/|e|$, the relative change in the layer current must considerably exceed the change in the electric potential. Actually, the availability of the factor $1/\lambda$ in expression (2.258) is exactly associated with this event.

Finally, *the third class* includes models in which the light surface is absent (Lyubarskii, 1990; Bogovalov, 1997b; Contopoulos et al., 1999). Otherwise, it is assumed here that

- the longitudinal current is larger than the critical current ($i_0 > 1$);
- the light surface extends to infinity;
- the longitudinal current is closed at large distances from the light cylinder;
- at large distances from the neutron star the main energy flux is still connected with the Poynting flux.

This class of models has presently been studied quite thoroughly, though mainly for the axisymmetric case only (Goodwin et al., 2004; Gruzinov, 2005; Contopoulos, 2005; Komissarov, 2006; McKinney, 2006a; Timokhin, 2006). Only a few years ago, the new and rather fruitful efforts have been made in constructing the force-free model of the inclined rotator (Spitkovsky and Arons, 2003; Spitkovsky, 2006) (see Fig. 2.25). In particular, the existence of the surface currents flowing along the separatrix in the direction opposite to the bulk current in the region of open field lines was confirmed. It was also confirmed that for the existence of the outflowing

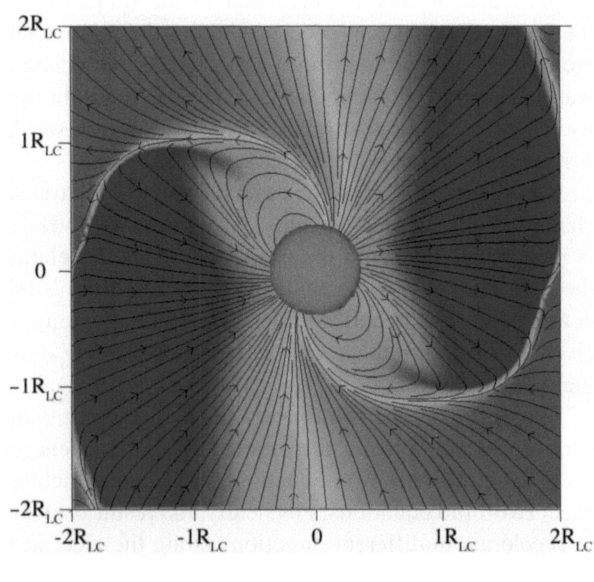

Fig. 2.25 The magnetosphere structure of the orthogonal rotator in which the light surface is absent (Spitkovsky, 2006). At large distances from the neutron star, the main energy flux is connected with the Poynting flux. Rotation axis is perpendicular to the figure plane [Reproduced by permission of the AAS, Fig. 2a from Spitkovsky, A.: Time-dependent force-free pulsar magnetospheres: axisymmetric and oblique rotators. ApJ **648**, L51–L54 (2006)]

wind, the longitudinal current density for the inclined rotator must be much larger than the local GJ one ($i_A \gg 1$). For this reason, it is not surprising that the energy losses even increase with the inclination angle χ

$$W_{tot} = \frac{1}{4}\frac{B_0^2 \Omega^4 R^6}{c^3}\left(1 + \sin^2\chi\right). \tag{2.260}$$

On the other hand, since there is no restriction to the value of the longitudinal current, one fails to confirm or refute the hypothesis for the existence of the light surface in the vicinity of the light cylinder, where the efficient acceleration of particles is possible. Moreover, within this approach, it was impossible to effectively transfer the electromagnetic energy to the particle energy flux. This problem will be studied in more detail in Chap. 5.

2.7 Conclusion

As we see, the consistent theory of the radio pulsar magnetosphere is now still far from completion. One of the main problems is the insufficient potentialities of the analytical methods that fail, in the general case, to construct the solution even in the rather simple force-free approximation. Evidently, only a dozen papers dealing with this set of problems appeared in the 1990s. Attempts to formulate, in general form, the problem of the magnetosphere structure due to the particle motion in the self-consistent electromagnetic field were long beyond the available computating resources (Krause-Polstorff and Michel, 1984, 1985; Petri et al., 2002; Smith et al., 2001).

To sum up, the situation with the existence of the light surface in the pulsar magnetosphere remains unclear. The behavior of the exact analytical solutions corresponding to reasonable longitudinal currents $j_\parallel \approx \text{const}$ within the open magnetic field lines is irrational in the equatorial region outside the light cylinder. On the other hand, the numerical solutions postulating the reasonable quasispherical outflow at large distances are in disagreement with the longitudinal current that can be generated in the polar regions of the neutron star.

Thus, within the force-free approximation, it is impossible to determine the longitudinal current flowing in the magnetosphere and, hence, find the energy losses. Therefore, the force-free statement of the problem, inevitably, calls for the concretization of the medium properties on the boundary of the force-free region, be it infinity or the current sheet, which is to be included in the equatorial region in most models. As we will see, this flaw will be naturally eliminated in the full GS equation version, which takes into account that the particle mass is finite.

Chapter 3
Force-Free Approximation—The Black Hole Magnetosphere

Abstract This chapter mainly deals with the magnetosphere structure of supermassive black holes for which the magnetic fields are of greater importance than for galactic solar mass black holes. Therefore, it is advisable to discuss the problems of the magnetic field generation in accretion disks in this chapter. We also briefly discuss the effect of the magnetic field on the disk accretion and matter outflow processes. Since the black hole itself cannot have a self-magnetic field (so-called "no hair" theorem), the large-scale magnetic field in the vicinity of the black hole can be generated only in the accretion disk. The missing link that helps one understand the energy release mechanism, which effectively transfers the energy from the rotating black hole and/or the inner parts of the disk to the active regions, is, in most cases, exactly associated with this regular magnetic field and the collimation mechanism responsible for the jet formation. In other words, to explain the galactic nuclei activity we use the pulsar idea of a unipolar inductor, the work of which is just provided by the regular poloidal magnetic field, the central object rotation (giving rise to the induction electric field **E**), and the longitudinal electric current (giving rise to the toroidal magnetic field B_{φ}). Within this model, the energy flux, as in the case of the radio pulsar magnetosphere, is fully connected with the electromagnetic energy flux, and the jet formation is, ultimately, due to the known property of the parallel current attraction. Thus, the main constituent elements of the central engine are a supermassive black hole, the accretion disk, and the regular magnetic field. On the basis of this model, the physical nature of the Blandford–Znajek mechanism is discussed in detail. Finally, the exact analytical solutions obtained by the force-free approximation are discussed.

3.1 Astrophysical Introduction—The Central Engine in Active Galactic Nuclei

3.1.1 Possible Mechanisms of Black Hole Formation

There is little doubt about the existence of black holes now. Several dozens of space sources, in which these unusual objects predicted by General Relativity are supposed to exist, have so far been discovered (Camenzind, 2007). Black holes can be

V.S. Beskin, *MHD Flows in Compact Astrophysical Objects*, Astronomy and Astrophysics Library, DOI 10.1007/978-3-642-01290-7_4,

© Springer-Verlag Berlin Heidelberg 2010

generated at the final stage of the evolution of massive stars, for example, due to the supernova explosion or in close binary systems in which, because of the accretion onto the neutron star, its mass becomes larger than the stability limit $\sim 3\,M_{\odot}$. It is clear that the characteristic masses of these black holes cannot be larger than the values of 3–30 $M_{\odot}$. It is these masses that are observed in the galactic candidates for black holes.

If we try to point to the main features of the central engine in the active galactic nuclei (AGN), we can say that most of the astrophysicists now believe that in the center of a host galaxy there is a supermassive black hole (its mass is 10^6–$10^9\,M_{\odot}$) and the surrounding matter accretes onto it (Rees, 1984). It is now the only way to explain both the extremely high efficiency of the energy release and the central engine compactness. The energy source of the galactic nuclei activity can be both the black hole rotation energy

$$E_{\rm tot} = \frac{I_{\rm r}\Omega^2}{2} \sim 10^{63}\,{\rm erg}\left(\frac{M}{10^9\,M_{\odot}}\right)\left(\frac{a}{M}\right)^2 \qquad (3.1)$$

and the accreting matter energy. In both cases, it is possible to assure the extremely high efficiency of the accreting matter reprocessing in the radiation observed. Besides, the indirect evidence for the black hole existence is that the Eddington luminosity

$$L_{\rm Edd} \approx 10^{47}\,{\rm erg/s}\left(\frac{M}{10^9\,M_{\odot}}\right) \qquad (3.2)$$

is close to the AGN luminosity (Zel'dovich and Novikov, 1971).

Further, it is generally (but not by everybody) believed that the matter accretion is of a disk type (Lynden-Bell, 1969). Thus, the preferential direction is naturally in the space, viz., it is the rotation axis along which jets are generated. Generally speaking, it is not improbable that this outflow can be produced in the absence of the regular magnetic field, for example, by the corona heating (Galeev et al., 1979) or the intrinsic energy excess in the accreting matter (Blandford and Begelman, 1999). However, most preferable is a model in which the regular poloidal magnetic field acts as a belt drive between the central engine and the jets (Blandford, 1976; Lovelace, 1976). Indeed, both the acceleration of particles and their efficient emission, for example, due to the synchrotron mechanism can now be most naturally explained by exactly the electrodynamical processes.

A lot of observational data really point to the presence of a supermassive compact object in the galactic nuclei centers (in both active and "quiet" galaxies) (Begelman et al., 1984; Krolik, 1999a). At present, there are several views of the problem of the formation of supermassive black holes, viz., whether they appeared already after the formation of galaxies, in the very process of their formation, or they existed long before the formation of galaxies and thus contributed to their generation (Rees, 1984; Dokuchaev, 1991b).

The massive black holes in the pregalactic era could have been produced by the gravitational instability of the primordial density fluctuations resulting in the collapse of the most developed (and, hence, most dense) regions of dark matter (Zel'dovich and Novikov, 1967a,b; Hawking, 1971; Polnarev and Khlopov, 1985; Gurevich and Zybin, 1995; Gurevich et al., 1997; Niemeyer and Jedamzik, 1999). The formation of black holes can be induced by the phase transitions associated with the diverse topological defects—for example, domain walls (Hawking et al., 1982; Moss, 1994; Khlopov et al., 2000) or cosmological strings (Hawking, 1989; Polnarev and Zembowicz, 1991). One should stress here that only small-mass black holes can be generated in this way. The supermassive black holes can be generated only if the spectrum of initial fluctuations is a nonstandard one, which has a rather extended plateau. Besides, all these mechanisms are, to a large extent, controversial, and it is too early to speak about them in the context of the black hole formation in the galactic nuclei.

Further, the favorable conditions for the formation of a central massive object can be achieved at the earlier stages of the galaxies formation (Rees, 1997) when a young galaxy has a lot of gas that is not condensed yet to form stars. The gas accreting onto the galactic center first generates the spherical component of a stellar population. However, part of the gas cannot participate in the star formation process; moreover, the massive first-generation stars quickly lose part of their mass in the evolution process. Therefore, at a certain stage of the stellar component formation, the free gas can no longer take part in the star formation—the radiation pressure, in particular, hinders the gas condensation (Rees, 1993; Haehnelt and Rees, 1993). Then, when the gas loses a greater part of its angular momentum and kinetic energy due to the heat dissipation, it inevitably condenses to form a black hole as the gas disk is dynamically unstable. By then the first-generation gas of the evolved stars could have taken part in the black hole formation. The black hole is either directly generated from the gas as a whole, while part of the gas already passed through one cycle, or its formation passes through the intermediate stage of a supermassive star $M \sim 10^3$–$10^6\, M_{\odot}$ (Rees, 1997) with its relatively short age of $\sim 10^7$ years and this star inevitably collapses to become a black hole. In the following the black hole mass can increase due to the accretion and the coalescence of black holes at the collision of galaxies (Haehnelt and Kauffmann, 2000).

As an argument for this scenario of the black hole formation, one should often consider the fact that there is a strong correlation between the black hole mass and the mass of the bulge (the old spherical component of the stellar population—a massive one in elliptic galaxies and a low-mass one in spiral galaxies), viz., $M_{\mathrm{c}} \simeq 0.1\%\, M_{\mathrm{buldge}}$ (Kormendy and Richstone, 1995; Faber et al., 1997). In the general case, the mass resulting from the above black hole formation process depends not only on the spherical stellar component mass but also, for example, on the angular momentum of the galaxy. Indeed, in the absence of the efficient withdrawal mechanism of the angular momentum excess the black hole formation process is suppressed. Therefore, the brightest quasars are observed exactly in the elliptic galaxies in which the specific angular momentum is much less than that of the spiral ones.

Finally, the black holes can be generated already after the galaxy formation due to the dynamic evolution of their central parts. The gravitational collapse passes through the evolutionary compression of the central star cluster (Begelman and Rees, 1978; Rees, 1984; Quinlan and Shapiro, 1990; Berezinsky and Dokuchaev, 2001). When in the dynamic evolution process of the central cluster the velocity dispersion of the stars reaches the critical value $\sim 620\,\mathrm{km/s}\,(M_*/M_\odot)^{1/2}(R_*/R_\odot)^{-1/2}$, the mean kinetic energy of the stars is compared with their gravitational binding energy. If two ordinary stars collide or move at a close distance from each other with this kinetic energy, they would inevitably be disrupted by the tidal interaction (Spitzer and Saslaw, 1966; Colgate, 1967; Sanders, 1970; Spitzer, 1971; Dokuchaev, 1991a). Only compact objects such as neutron stars or black holes of small (stellar) mass can survive. At this stage, the central galaxy region is a cluster of compact stars, which is submerged in the gas resulting from the star disruption at the previous evolution stage. In the following, in the dynamic process of the cluster evolution the velocity dispersion of compact stars becomes larger and when it is of the order of the light velocity, the cluster is sure to collapse to form a black hole (Bisnovatyi-Kogan, 1978; Rees, 1984; Berezinsky and Dokuchaev, 2001). What process—the gas condensation or the compact star coalescence—proves more efficient depends on the concrete parameters of the initial stellar cluster.

3.1.2 Nature of Activity and Variability

According to the present-day notions, the massive central objects are present in most galaxies and become active when a sufficient amount of matter ("fuel") falls onto them. Regardless of the interstellar gas composition in the vicinity of the central object, there is always a process that supplies its neighborhood with matter—it is the tidal disruption of the stars moving in the vicinity of the black hole (Lacy et al., 1982; Rees, 1988; Evans and Kochanek, 1989; Khokhlov et al., 1993; Frolov et al., 1994). The estimates show that, for example, in galaxy M31 this event occurs every 10^4 years. Apart from the fact that this process enriches the accretion disk around the central object, an individual event (lasting about several months) can be detected, and the observations could be used to study the nearest neighborhood of the central object. At present, a search for these events continues (Renzini et al., 1995), but the potentialities of this method are limited because of the difficulties associated with gas-dynamic computations. Therefore, it is hardly possible to calculate the parameters of the central object (angular velocity of the black hole, its inclination angle, disk mass, etc.) even if the corresponding observations are available.

In several active galaxies, short events lasting from a few minutes to a day were observed (Papadakis and Lawrence, 1993, 1995; Halpern and Marshall, 1996; Tagliaferri et al., 1996). This "intraday" variability is, possibly, associated with the accretion disk instabilities, for example, the occurrence of a hot spot in the disk (Iwasawa et al., 1998). The mechanism of these flares has not been adequately studied yet. However, in this case, we cannot determine the characteristics of the central object by the observed variability, because the flare is not directly connected with the orbital motion around the black hole but is rather a manifestation of some local instability. This variability also naturally occurs in the same model of the accretion disk, where the iron line is emitted by cold (10^6 K) gas clouds. Indeed, if this cloud in its motion accidentally turns out to be between the observer and the central object, the X-ray flux must change considerably (Abrassart and Czerny, 2000).

Another type of variability is the so-called quasiperiodic oscillations (QPO). Here the characteristic time of the process is from a day for Seyfert galaxies to a month for the most active quasars; therefore, the data collection process becomes more difficult. But, at least, one QPO example has already been found, viz., Seyfert galaxy IRAS 18325-5926 with the variability period of about 16 hours (Iwasawa et al., 1998). This time just coincides with the orbital period around the black hole with mass $M \sim 10^7$–$10^8\,M_\odot$ at a distance of 10–20 $r_{\rm g}$. Possibly, in this case, we directly observe the eigenmodes of accretion disk oscillations—diskoseismic modes which were first used to describe the accreting sources in the interior of our Galaxy (Kato and Fukui, 1980; Nowak and Wagoner, 1992, 1993; Nowak et al., 1997). It is also possible that QPOs result from the geometric effect, viz., if the disk axis does not coincide with the rotation axis of the black hole, due to the GR effects at distances less than 10 $r_{\rm g}$ there occurs a force that tends to change the disk orientation (Bardeen and Petterson, 1975; Rees, 1984). Thus, instabilities and oscillations occur and, as a consequence, radiation variability. All these processes take place in the strong gravitational field of the central black hole and, therefore, are of great interest for the study of its nearest neighborhood. Finally, the disks, in which there are regions of oppositely directed rotational velocities, have been studied lately (Kuznetsov et al., 1999). This can be the case if the accretion disk is fed from the tidal collapse of stars that can have an arbitrary angular momentum.

3.1.3 Magnetized Accretion Disk

When studying accretion disks in the vicinity of supermassive black holes, besides the purely hydrodynamical models discussed in Chap. 1, much attention was given to the study of magnetized flows when there is a rather strong regular poloidal magnetic field in the disk (Bisnovatyi-Kogan and Ruzmaikin, 1973; Campbell et al., 1998; Stone and Pringle, 2001; Krolik, 1999a). We separately discuss below both the nature of occurrence of this field and its role in the jet formation. Here we discuss the role of the regular magnetic field in the matter accretion onto the central black hole and the magnetic field effect on the process of the plasma outflow from the accretion disk surface.

Let us first discuss the main effects for which the regular magnetic field is responsible. First of all, as we will see, the open field lines are very good conductors along which not only the energy but also, what is not less important, the angular momentum can be removed. Therefore, even if the energy density of the magnetic field in the disk is much lower than the plasma energy density, the role of the magnetic field can be significant. In particular, the accretion (i.e., the occurrence of the nonzero radial velocity v_r) is possible even in the absence of viscosity. Besides, the occurrence of Ampére's additional forces connected with the strong regular magnetic field in the disk can change the regular rotational plasma velocity. Note that the magnetic field can be substantial in slightly ionized disks (for example, in their outer and, hence, slightly heated regions), because the ambipolar diffusion effects are able to effectively connect neutrals with the ionized component (Königl, 1989).

There are different versions of magnetized flows from slightly ionized disks, in which the violation of the frozen-in condition is associated with the ambipolar diffusion of charged particles (Wardle and Königl, 1993; Königl, 1989; Li, 1996), to systems, where Ohm's diffusion is of major importance (Li, 1995; Ferreira and Pelletier, 1993a,b, 1995; Ogilvie, 1997; Ogilvie and Livio, 1998; Kaburaki, 2000). In the large, the computations showed that the magnetic field weakly affects the angular gas velocity (this would be possible only if the energy density of the magnetic field $\epsilon_{\rm em}$ were higher than the plasma energy density $\epsilon_{\rm part}$). A more significant effect is a considerable increase in the radial velocity due to which the accretion can become supersonic already at large distances from the black hole (Li, 1995; Shalybkov and Rüdiger, 2000). Finally, the existence of the strong electric currents in the accretion disk is capable of distorting the magnetic field lines of the poloidal magnetic field already on the scale of the disk itself (Ogilvie, 1997; Ogilvie and Livio, 1998). This, in turn, must result in a substantial increase in the plasma outflow velocity.

The point is that, as (Blandford and Payne, 1982) first showed, the magnetic field can act as a sling for the particle ejection from the accretion disk surface. Indeed, let us consider the motion of a particle rotating along the Keplerian orbit in the presence of the regular poloidal magnetic field inclined to the vertical axis at an angle of $\alpha_{\rm m}$. For a rather strong magnetic field, a charged particle, as a "bead on a wire," can move only along the magnetic field lines and, therefore, has to preserve its initial angular velocity. It is easy to verify that for $\alpha_{\rm m} > 30°$, the effective potential

$$\varphi_{\rm eff}(\varpi, z) = -\frac{GM}{\varpi_0}\left[\frac{\varpi_0}{(\varpi^2+z^2)^{1/2}} + \frac{1}{2}\left(\frac{\varpi}{\varpi_0}\right)^2\right], \tag{3.3}$$

which is connected with both the gravitational force of the central body and the centrifugal force, yields the acceleration of particles from the rotation axis. As a result, the outflow from the disk surface is possible even for the cold plasma.

Problem 3.1 Show that a particle moving in the effective potential (3.3) really begins to accelerate from the rotation axis for $\alpha_m > 30°$.

Certainly, the studied process is not universal, because, otherwise, the disk could not exist at all (Ogilvie and Livio, 1998). The outflow is possible only in the corona, where the energy density of the magnetic field exceeds the plasma energy density. In this case, the (poloidal) motion of particles must, as a whole, occur along the magnetic field lines. Besides, in R. Blandford and D. Payne's paper, the effects of the finite disk thickness, in which, in particular, the inclination of the magnetic field lines must change, were not taken into account. Nevertheless, the paper outlined the direction of investigation. Later, attempts were made to construct a unified model describing both the accretion disk, where the magnetic field weakly affects the matter motion, and the corona, in which the outflow along the magnetic field lines is possible (see, e.g., Ferreira and Pelletier, 1995).

Finally, the plasma outflow becomes even more intensive if the particle ejection is accompanied by the generation of a strong longitudinal electric current and, hence, by the occurrence of a toroidal magnetic field (in the previous example, the outflow was for $B_\varphi = 0$). As we see, exactly this model proves most advantageous for interpreting the intensive matter outflow from the accretion disk surface.

In conclusion, note that even if the electrodynamic processes play the main role in the work of the central engine, the significance of the accretion disk should not be underrated. First of all, as is discussed in detail in the following section, it is in the accretion disk that the regular magnetic field, in which the black hole turns out to be submerged, is to be generated. Further, it is possible that the electron–positron plasma generation needed for the work of the central engine is connected with hard γ-quanta emitted from the inner (and, therefore, hottest) regions of the accretion disk. Finally, it is not improbable that the fast-rotating inner parts of the accretion disk are the central engine rather than the black hole (Livio et al., 1999). On the other hand, it is the properties of the accretion disk that are responsible for the high-energy radiation detected in the central source region.

3.1.4 Regular Magnetic Field Generation

As was mentioned, the external magnetic field can not only greatly change the accretion regime (and thus affect the rate of the energy release of the accreting matter) but also facilitate the jet formation. Therefore, the magnetic field generation problem is certainly one of the key problems arising from the construction of the central engine model in AGN.

Theoretically, the magnetic field problem splits into two problems—the problem of generation of the regular magnetic field and the problem of its structure beyond the accretion disk. The point is that, as was noted, the black hole cannot have its self-magnetic field. In particular, any loop of the magnetic field (for example, generated by the magnetic dipole freely falling onto the black hole) disappears for the distant observer in the dynamic time $\tau \sim r_g/c$ (Thorne et al., 1986). The exception is only the monopole magnetic field; however, a large number of magnetic monopoles are needed for its existence, which is, undoubtedly, unrealistic.

On the other hand, it is hardly possible that the strong magnetic field can be connected with the amplification of the external magnetic field, because the magnetic field is frozen in the accreting matter. In any event, it is clear that the magnetic field energy density cannot be larger than the energy density of the accreting plasma, which is also true for the process of the magnetic field generation in the disk. The point is that both the nature of the ordinary viscosity ν resulting in the matter accretion and, hence, the amplification of the magnetic field and the magnetic viscosity $\nu_{\rm m}$ resulting in the violation of the frozen-in condition and, hence, the magnetic field diffusion, are evidently identical, viz., it is the magnetic turbulence (Balbus and Hawley, 1998). Consequently, the Prandtl magnetic number $\mathcal{P} = \nu/\nu_{\rm m}$ cannot greatly differ from unity. It is possible, directly from the analysis of the leading terms in MHD equations (Lubow et al., 1994), to considerably amplify the external (for example, homogeneous) magnetic field attracted to the accretion center only if the condition $(r/H)\mathcal{P} \gg 1$ is satisfied (H is the half-thickness of the accretion disk). Therefore, this process cannot be of crucial importance. It was possible to amplify the magnetic field only by an additional generation mechanism in the disk (Campbell, 1999), but it is a different story.

Thus, the source of the regular magnetic field should be sought in the accretion disk itself rather than in the environment. Indeed, the accretion disk has all the necessary properties to generate a regular magnetic field. Both the differential rotation (the matter rotates at velocities close to the Keplerian velocity $v_{\rm K} = (GM/r)^{1/2}$, and, hence, the angular velocity depends on the distance from the center) and the inhomogeneous matter density in a vertical direction are present in it, which give rise to a reflecting nonsymmetric turbulence. As a result, the accretion disk can work as a dynamo amplifying the stochastic magnetic field (Balbus and Hawley, 1998).

Nevertheless, until recently, the magnetic field generation mechanism was not known. In particular, it is not yet clear if it is possible to generate a sufficiently strong regular magnetic field of order 10^4 G, which is needed, as we will see, for the efficient work of the electrodynamic mechanism of energy release in AGN. Recall that this estimate of the magnetic field is associated with the simple assumption that the magnetic field energy density is compared with the energy density of the accreting plasma giving rise to the Eddington luminosity (3.2). As a result, we have

$$B_{\rm Edd} \approx 10^4\,{\rm G}\left(\frac{M}{10^9 M_{\odot}}\right)^{-1/2}. \tag{3.4}$$

Problem 3.2 Find the formula (3.4).

It is clear that the Eddington magnetic field $B_{\rm Edd}$ (3.4) can be regarded as an upper limit only. Indeed, since the magnetic field amplification mechanism is of a turbulent character, for the generation of the critically strong magnetic field the turbulent motion of the accreting matter should contain, in any case, the energy comparable with its energy. Besides, the dynamo mechanism itself should be quite effective to transfer a considerable part of the turbulent motion energy to the magnetic field. Therefore, in reality, the magnetic field must be much smaller than is known from the estimate (3.4). Finally, it is very difficult to generate the quasidipole magnetic field in the disk systems—the main growing mode is of a quadrupole character (Ruzmaikin et al., 1988).

Nevertheless, in the past years much progress has been made in this field. It is due to the high speed of the present-day computers that provide the direct simulation of the turbulent motion and thus determine the key transport coefficients and, in particular, the model parameter $\alpha_{\rm SS}$ (1.9) by directly averaging the corresponding correlators [it was believed earlier that the machine viscosity fails to realize the turbulence in the numerical simulation (Balbus and Hawley, 1998)].

The magnetorotational instability studied by Velikhov (1959) in the context of the laboratory plasma instability is now regarded as the main source of turbulence needed for the efficient dynamo work. However, as is often the case, it was long unknown in astrophysics and rediscovered by Balbus and Hawley for the case of accretion disks (1991). It is connected with the instability of slow magnetosonic waves in a shear flow realized in the Keplerian motion. Its important property is the absence of the threshold and, therefore, there is an instability in the arbitrarily small magnetic field. In other words, even the small magnetic field gives rise to new unstable degrees of freedom.

As a result, the numerical simulation of the magnetic turbulence showed that the effective amplification of the regular magnetic field can really occur in accretion disks (Brandenburg and Sokoloff, 2002; von Rekowski et al., 2003). The direct computations yielded some unexpected results. It turned out that in accretion disks the dynamo parameter $\alpha_{\rm dyn}$ entering, for example, into the phenomenological equation

$$\frac{\partial < \mathbf{B} >}{\partial t} = \nabla \times (\alpha_{\rm dyn} < \mathbf{B} >) \tag{3.5}$$

can prove negative (Brandenburg et al., 1995; Ziegler and Rüdiger, 2000). In this case, the field amplification due to the low-order process—the so-called $\alpha\omega$-dynamo—is impossible, since the higher-order process—the $\alpha^2\omega$-dynamo—can only be realized (Ruzmaikin et al., 1988). But, in this case, the dipole regular magnetic field rather than the quadrupole one is to be amplified, which is necessary for the most efficient work of the central engine. Finally, much larger values of the viscous parameter $\alpha_{\rm SS}$ (1.9) up to the values of $\alpha_{\rm SS} \sim 0.01$ were obtained by direct calculation (Stone et al., 1996; Arlt and Rüdiger, 2001).

In conclusion, one cannot but mention the number of papers which dealt with the possibility to analyze the magnetic field generation in the disk by the GR effects. The point is that due to the frame-dragging associated with the Lense–Thirring effect the rotating black hole submerged in the magnetic field becomes a source of extra electromagnetic voltage. As seen from the induction equation written in the reference frame of local nonrotating observers (ZAMO)

$$\frac{\partial \mathbf{B}}{\partial t} = -\nabla \times (\alpha \mathbf{E} + \boldsymbol{\beta} \times \mathbf{B}), \tag{3.6}$$

the GR effects give rise to the additional term $\nabla \times [\boldsymbol{\beta} \times \mathbf{B}]$ acting as an electromotive force (EMF). In particular, the Cowling theorem (Alfven and Fälthammar, 1963) known in dynamo theory and prohibiting the existence of an axisymmetric stationary dynamo is no longer valid (Khanna and Camenzind, 1994). In fact, it is due to the differential ZAMO rotation, so that toroidal electric fields can be generated by poloidal electric fields.

Therefore, the hypothesis was put forward (Khanna and Camenzind, 1994, 1996a) that this process can really lead to magnetic field amplification in the vicinity of rotating black holes. However, in numerical calculations of the kinematical problem (Brandenburg, 1996; Khanna and Camenzind, 1996b), the poloidal magnetic field amplification produced by the Lense–Thirring effect is always suppressed by the direct process of the induction action in the variable magnetic field. Nevertheless, the idea of magnetic field generation in accretion disks by GR effects seems rather fruitful and calls for further investigation (Khanna, 1997, 1998; Tomimatsu, 2000). In any event, GR effects should be taken into account when constructing a consistent theory of magnetic field generation in the inner regions of accretion disks.

To sum up, we can say that now we have no direct evidence that in the nuclei of active galaxies there are strong regular magnetic fields that significantly affect the accreting flow dynamics. Nevertheless, even if the effect of the magnetic field can be disregarded when describing the gas accretion, the magnetic field generated in the accretion disk, in view of its long-distance action, can be of crucial importance beyond the accretion disk, viz., in the neighborhood of the black hole and in the corona, in particular, it can specify the outflowing plasma dynamics.

3.2 Basic Equations

3.2.1 (3 + 1)-Splitting for the Electromagnetic Field

We now proceed to the discussion of the theory of force-free black hole magnetospheres. Note at once that in the following we, as before, consider the flows in the Kerr metric rather than in the Kerr–Newman one. Otherwise, we believe that the total charge of black holes is rather small so that it cannot affect the space–time curvature near the black hole. As shown below, in the case of stationary plasma-filled magnetospheres, this condition is satisfied with large margin. The effect of the black hole charge on the electrodynamic processes in its magnetosphere was discussed, for example, in Ruffini et al. (1999) and Punsly (2001).

As was already demonstrated by the example of hydrodynamical flows, a very convenient language to describe the processes in strong gravitational fields is the $(3+1)$-splitting. Thus, Maxwell's equations

$$\nabla_\alpha F_{\beta\gamma} + \nabla_\beta F_{\gamma\alpha} + \nabla_\gamma F_{\alpha\beta} = 0, \tag{3.7}$$

$$\nabla_\alpha F^{\alpha\beta} = 4\pi j^\beta, \tag{3.8}$$

for the stationary flows ($\partial/\partial t = 0$) have the form (Macdonald and Thorne, 1982)

$$\nabla \cdot \mathbf{E} = 4\pi\rho_e, \tag{3.9}$$

$$\nabla \cdot \mathbf{B} = 0, \tag{3.10}$$

$$\nabla \times (\alpha\mathbf{E}) = \hat{\mathcal{L}}_{\boldsymbol{\beta}}\mathbf{B}, \tag{3.11}$$

$$\nabla \times (\alpha\mathbf{B}) = -\hat{\mathcal{L}}_{\boldsymbol{\beta}}\mathbf{E} + 4\pi\alpha\mathbf{j}. \tag{3.12}$$

Here the Lie derivative $\hat{\mathcal{L}}_{\boldsymbol{\beta}}$ acting by the rule

$$\hat{\mathcal{L}}_{\boldsymbol{\beta}}\mathbf{A} = (\boldsymbol{\beta}\nabla)\mathbf{A} - (\mathbf{A}\nabla)\boldsymbol{\beta} \tag{3.13}$$

results from the differential ZAMO rotation (that, as we remember, defines the three-dimensional quantities in their laboratory). We further use the system of units $G = 1$ and $c = 1$ again.

The physical meaning of first two equations (3.9) and (3.10) is quite transparent. Regardless of the space curvature, the magnetic field is a curl one and the electric field sources can be electric charges only. On the other hand, in the case of the rotating black hole the gravitomagnetic field $H_{ik} = 1/\alpha\, \nabla_i \beta_k$ (1.222) yields the Lie derivatives on the right-hand sides of Eqs. (3.11) and (3.12). The Lie derivatives are similar to the additional terms appearing in the quasistationary formalism on the right-hand sides of Eqs. (2.115) and (2.117), so that Maxwell's equations (3.9), (3.10), (3.11), and (3.12) appear close in form to Eqs. (2.114), (2.115), (2.116), and (2.117). As a result, as we will see, it is due to the action of the gravitomagnetic forces that in the induction equation (3.11) there appears an electromotive force resulting in the energy losses of the rotating black hole. In flat space, EMF can be connected only with the time-depending magnetic flux through the given circuit. In the case of the rotating black hole, the change in the magnetic flux is connected with the "space flow" moving with the Lense–Thirring angular velocity ω.

Further, as in flat space, in the axisymmetric case it is convenient to express the magnetic field in terms of the scalar function $\Psi(r, \theta)$ having the meaning of the magnetic flux:

$$\mathbf{B} = \frac{\nabla\Psi \times \mathbf{e}_{\hat{\varphi}}}{2\pi\varpi} - \frac{2I}{\alpha\varpi}\mathbf{e}_{\hat{\varphi}}. \tag{3.14}$$

Here again $I(r, \theta)$ is the total electric current within the magnetic tube $\Psi < \Psi(r, \theta)$ and the indices with caps correspond to the physical components of the vectors. By definition (3.14), Maxwell's equation $\nabla \cdot \mathbf{B} = 0$ holds automatically. It is readily seen that the condition $\mathbf{B} \cdot \nabla\Psi = 0$ is satisfied, and the relation $\Psi(r, \theta) = \text{const}$ thus prescribes the magnetic surfaces. The proportionality factor in (3.14) is again chosen so that Ψ coincides with the magnetic flux inside the tube $\Psi = \text{const}$, i.e., $\mathrm{d}\Psi = \mathbf{B} \cdot \mathrm{d}\mathbf{S}$.

3.2.2 "No Hair" Theorem

Before proceeding further, we formulate an important statement that is generally called the "no hair" theorem (see, e.g., Frolov and Novikov, 1998). We prove it here for the simplest case.

Theorem 3.1 *Nonrotating black holes cannot have magnetic fields themselves; the electric field in the exterior of a nonrotating black hole must coincide with the field of a point charge located in its center.*

Indeed, for a nonrotating (Schwarzschild) black hole, Eqs. (3.11) and (3.12) can be rewritten as

$$\nabla \times (\alpha \mathbf{E}) = 0, \tag{3.15}$$

$$\nabla \times (\alpha \mathbf{B}) = 0. \tag{3.16}$$

It is logical to assume here that the black hole is in vacuum. Therefore, Eqs. (3.15) and (3.16) must be supplemented with the relations $\nabla \cdot \mathbf{E} = 0$ and $\nabla \cdot \mathbf{B} = 0$. As we see, in the case of the nonrotating black hole, the equations for the electric and magnetic fields are separated and appear absolutely identical. As a result, to prove it we can consider the magnetic field structure only. Using now the definition (3.14) and the relation (1.234), to write the equation $\nabla \times (\alpha \mathbf{B}) = 0$, we obtain for $I = 0$

$$r^2 \frac{\partial}{\partial r}\left[\left(1 - \frac{r_{\mathrm{g}}}{r}\right)\frac{\partial \Psi}{\partial r}\right] + \sin\theta \frac{\partial}{\partial \theta}\left(\frac{1}{\sin\theta}\frac{\partial \Psi}{\partial \theta}\right) = 0. \tag{3.17}$$

Problem 3.3 Find Eq. (3.17).

As we see, the differential operator in Eq. (3.17) differs from the GS operator $\hat{\mathcal{L}}$ (1.119) only by the additional factor $(1 - r_{\mathrm{g}}/r)$. Therefore, we can again seek the solution in the form (see, e.g., Ghosh, 2000)

$$\Psi(r, \theta) = \sum_{m=0}^{\infty} g_m(r) Q_m(\theta), \tag{3.18}$$

and now the radial functions $g_m(r)$ are to be found from the solution to the equations

$$x(x-1)\frac{\mathrm{d}^2 g_m}{\mathrm{d}x^2} + \frac{\mathrm{d}g_m}{\mathrm{d}x} - m(m+1)g_m = 0. \tag{3.19}$$

Here $x = r/r_{\mathrm{g}}$.

We consider the solutions to Eqs. (3.19) in more detail. The first family of solutions has the form

$$g_m^{(1)}(x) = x^2 F(1-m, m+2, 3, x), \tag{3.20}$$

where $F(a, b, c, x)$ is a hypergeometric function (see Appendix D). Since $(1-m)$ is an integer less than or equal to zero for $m \neq 0$, the functions $g_m^{(1)}(x)$ for $m \neq 0$ reduce to polynomials. As a result, up to a dimensional factor we have for the full magnetic flux function $\Psi(r, \theta)$

1. $m = 1$
   - $\Psi_1^{(1)}(r, \theta) = r^2 \sin^2\theta$ is a homogeneous field (see Fig. 1.4a),
2. $m = 2$
   - $\Psi_2^{(1)}(r, \theta) = \left(r^3 - \dfrac{3}{4} r_g r^2\right) \sin^2\theta \cos\theta$—a zero point (see Fig. 1.4c),
3. ...
   - $\Psi_m^{(1)}(r, \theta) \propto r^{m+1} Q_m(\theta)$ for $r \to \infty$.

The solution with $m = 0$

$$\Psi_0^{(1)}(r, \theta) = \left[r + r_g \ln\left(1 - \frac{r}{r_g}\right)\right](1 - \cos\theta) \tag{3.21}$$

has a singularity for $r = r_g$. On the other hand, there exists the regular solution

$$\Psi_0^{(1)}(r, \theta) = r(1 - \cos\theta) + r_g(1 + \cos\theta)[1 - \ln(1 + \cos\theta)] - 2r_g(1 - \ln 2), \tag{3.22}$$

corresponding to the asymptotic behavior $\Psi \propto r(1 - \cos\theta)$ for $r \gg r_g$ (the latter term is added to satisfy the condition $\Psi(r, 0) = 0$) (Blandford and Znajek, 1977; Ghosh and Abramowicz, 1997). As for the hydrodynamical case, it is a "nonphysical solution," i.e., it can be realized only in the presence of currents beyond the black hole.

The second family of solutions can be written as

$$g_m^{(2)}(x) = x^{-m}\left[F\left(m, m+2, 1, 1 - \frac{1}{x}\right) \ln\left(1 - \frac{1}{x}\right) + P_m(x)\right], \tag{3.23}$$

where $P_m(x)$ are some polynomials regular on the horizon. Thus, for example,

$$\begin{aligned} P_1(x) &= x^2 + \frac{x}{2}, \\ P_2(x) &= 4x^4 - x^3 - \frac{x^2}{6}, \\ P_3(x) &= 15x^6 - \frac{25x^5}{2} + x^4 + \frac{x^3}{12}. \end{aligned} \tag{3.24}$$

As a result, the asymptotic behavior of the second family for $m \neq 0$ is

- $\Psi_m^{(2)}(r, \theta) \to r^{-m} Q_m(\theta)$ for $r \to \infty$,

- $\Psi_m^{(2)}(r,\theta) \to \ln\left(1 - \frac{r_g}{r}\right) Q_m(\theta)$ for $r \to r_g$.

In particular, for the dipole harmonic $m = 1$, we have

$$\Psi_1^{(2)}(r,\theta) = 2\pi\,|\mathbf{m}|\frac{\sin^2\theta}{r} f(r), \tag{3.25}$$

where (Ginzburg, 1964)

$$f(r) = -3\frac{r^3}{r_g^3}\left[\ln\left(1-\frac{r_g}{r}\right) + \frac{r_g}{r} + \frac{1}{2}\frac{r_g^2}{r^2}\right] \approx 1 + \frac{3}{4}\frac{r_g}{r} + \cdots . \tag{3.26}$$

Since $f(r) \to 1$ for $r \to \infty$, at large distances the stream function $\Psi_1^{(2)}(r,\theta)$ (3.25) really corresponds to the dipole magnetic field. As to a zero harmonic, it again describes the monopole field

$$\Psi_0^{(2)} = \Psi_0(1 - \cos\theta), \tag{3.27}$$

having no singularity on the horizon.

As we see, the first family of solutions does not vanish at infinity and, therefore, this magnetic field can be realized in the presence of external currents only. Otherwise, the fields that have no singularity on the horizon are not the self-magnetic fields of the black hole. As to the second family with multipole behavior r^{-m} at large distances, all of them, except for the case $m = 0$, have a logarithmic singularity on the horizon. This implies that if the electric field in the vicinity of electric or magnetic dipoles (a quadrupole, etc.) located at the point $r_g + \delta r$ is given, at large distances the observer detects a field which is $\ln(r_g/\delta r)$ times less than in the absence of the black hole. The exception is only the monopole solution (3.27), but it, in the absence of the accretion disk, can be realized for the electric field only.

Indeed, let us consider the point charge q located in vacuum in the vicinity of the black hole horizon at the point $r_0 = r_g + \delta r$, $\theta = 0$. Its electric field resulting from the solution to the equations $\nabla\cdot\mathbf{E} = 0$ and $\nabla\times(\alpha\mathbf{E}) = 0$ has the form (Linet, 1976)

$$\mathbf{E} = \frac{q}{r_0 r^2}\left[M\left(1 - \frac{r_0 - M + M\cos\theta}{\mathcal{D}}\right)\right.$$
$$\left. + \frac{r((r-M)(r_0-M) - M^2\cos\theta)(r - M - (r_0 - M)\cos\theta)}{\mathcal{D}^3}\right]\mathbf{e}_{\hat{r}}$$
$$+ \frac{\alpha q(r_0 - 2M)\sin\theta}{\mathcal{D}^3}\mathbf{e}_{\hat{\theta}}, \tag{3.28}$$

where

$$\mathcal{D}^2 = (r-M)^2 + (r_0-M)^2 - M^2 - 2(r-M)(r_0-M)\cos\theta + M^2\cos^2\theta. \tag{3.29}$$

As shown in Fig. 3.1, the electric field is deformed so that the distant observer detects the field of the electric charge located near the center of the black hole

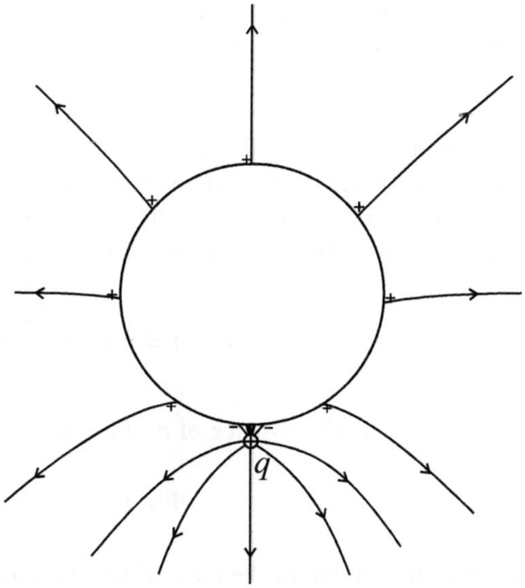

Fig. 3.1 The electric field of the charge q in the vicinity of the black hole horizon. The distant observer detects the field of the electric charge located near the center of the black hole. The fictitious charges appearing in the vicinity of the horizon are also shown

rather than in the vicinity of the horizon (see Thorne et al. (1986) for details). This implies that two opposite charges located in the vicinity of the horizon at a fixed distance from each other are in the vicinity of the black hole center for the distant observer. The closer the charges to the event horizon, the smaller the seeming distance between them. This shows a decrease in the dipole moment for the distant observer.

In the foregoing we dealt with the stationary configurations. The "no hair" theorem was just first proved for the static magnetic field (Ginzburg, 1964). However, this theorem has a clear dynamical meaning. It shows that when the magnetic dipole approaches the black hole horizon, in the characteristic time $\tau \sim r_g/c$ the magnetic field at large distances from the black hole vanishes. Later, the "no hair" theorem was generalized to all other physical fields (see, e.g., Frolov and Novikov, 1998).

Finally, we should remember that for the distant observer any process in the vicinity of the black hole lasts a finite time (in any case, it is limited by the lifetime of the Universe) and, therefore, is not stationary. This implies that in the Boyer–Lindquist coordinates in the vicinity of the horizon, there is always a nonstationary region (a switching-on wave) propagating to the horizon. This switching-on wave due to a relativistic time delay for the distant observer never reaches the event horizon. It is this property that called for the introduction of a "stretched horizon" in the membrane approach, which is located over the transition region where the fields can be regarded as stationary ones. Exactly this stationary field is shown in Fig. 3.1.

3.2.3 *Vacuum Approximation*

Thus, the "no hair" theorem shows that the black hole magnetosphere can occur only in the presence of external electric currents. As was mentioned, an obvious candidate for this "external winding" is the accretion disk. In this section, we discuss the simplest vacuum solutions that, as in the case of the radio pulsar magnetosphere, help us clarify some features of the more realistic plasma-filled magnetosphere.

We first consider the black hole submerged in the external homogeneous magnetic field B_0. The exact solution for the arbitrary angular velocity of the black hole parallel to the external magnetic field was obtained by Wald (1974) (see also Bičák and Dvořák, 1976). In view of the axisymmetric character of the problem, it is convenient to write the solution for the magnetic flux function $\Psi(r, \theta)$ as

$$\Psi(r, \theta) = \pi B_0 \frac{\Sigma^2 - 4a^2 M r}{\rho_{\mathrm{K}}^2} \sin^2\theta. \tag{3.30}$$

In particular, in the absence of rotation we again return to the solution

$$\Psi(r, \theta) = \pi B_0 r^2 \sin^2\theta. \tag{3.31}$$

On the other hand, in the case of the rotating black hole, as seen from Maxwell's equations (3.11) and (3.12), the Lie derivatives act as field sources even for the stationary case. As a result, the rotating black hole submerged in the external magnetic field generates the electric field (Thorne et al., 1986)

$$\begin{aligned}
\mathbf{E} = &-\frac{B_0 a}{\rho_{\mathrm{K}}^2 \Sigma}\left[2\Delta r + 2\rho_{\mathrm{K}}^2(r-M) - 2\frac{\rho_{\mathrm{K}}^2\Delta}{\Sigma^2}(2r^3 + 2a^2 r - ra^2\sin^2\theta + Ma^2\sin^2\theta)\right. \\
&\left. + \frac{M\sin^2\theta}{\rho_{\mathrm{K}}^2}(\Sigma^2 - 4a^2 Mr)\left(1 + 2\frac{ra^2\sin^2\theta}{\Sigma^2}(r-M) - 4\frac{r^2}{\Sigma^2}(r^2+a^2)\right)\right]\mathbf{e}_{\hat{r}} \\
&- \frac{B_0 a^3 \Delta^{1/2}}{\rho_{\mathrm{K}}^2 \Sigma}\sin 2\theta\left[-1 + \frac{\rho_{\mathrm{K}}^2\Delta}{\Sigma^2} + \frac{Mr\sin^2\theta}{\rho_{\mathrm{K}}^2\Sigma^2}(\Sigma^2 - 4a^2 Mr)\right]\mathbf{e}_{\hat{\theta}}.
\end{aligned}$$

Recall once again that we deal with the fields measured by ZAMO.

Problem 3.4 Show that

- the magnetic flux Ψ_* passing through the black hole horizon is

$$\Psi_* = 4\pi M\sqrt{M^2 - a^2}\,B_0; \tag{3.32}$$

  in particular, for the extremely fast-rotating black hole $a = M$ the full magnetic field expulsion occurs so that $\Psi_* = 0$,
- on the other hand, the magnetic flux passing through the black hole ergosphere $r_{\mathrm{erg}} = M + \sqrt{M^2 - a^2\cos^2\theta}$ is independent of the angular velocity Ω_{H} and equal to

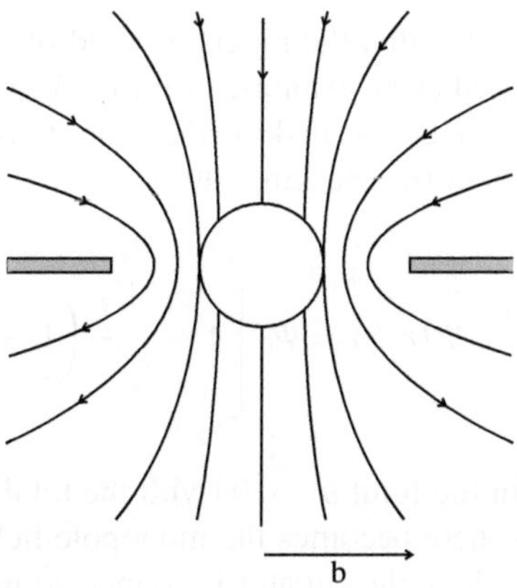

Fig. 3.2 The quasimonopole structure of the magnetic field, which arises in the case of the limited disk of the inner radius b. The jump of the tangential component in the magnetic field is due to the toroidal currents flowing in the accretion disk. All field lines intersecting the black hole horizon extend to infinity

$$\Psi_{\text{erg}} = 4\pi M^2 B_0. \tag{3.33}$$

It is interesting to note that the solution (3.31) for the nonrotating black hole does not formally differ from the homogeneous magnetic field in vacuum. However, this fact is associated only with the convenient choice of a coordinate grid, viz., the Boyer–Lindquist coordinates r, θ. It is easy to verify that, in reality, at the black hole equator there is a singularity, viz., a zero point. To show this we write the physical components of the magnetic field, which, by definition (3.14), have the form

$$B_{\hat{r}} = \frac{1}{2\pi \varpi \sqrt{g_{\theta\theta}}} \frac{\partial \Psi}{\partial \theta}, \tag{3.34}$$

$$B_{\hat{\theta}} = \frac{1}{2\pi \varpi \sqrt{g_{rr}}} \frac{\partial \Psi}{\partial r}. \tag{3.35}$$

As we see, because of the vanishing metric factor $1/\sqrt{g_{rr}}$ for $\alpha = 0$, the θ-component of the magnetic field $B_{\hat{\theta}}$ is to be zero on the entire surface of the black hole horizon. Consequently, the poloidal magnetic field should be orthogonal to the horizon. Since the radial component of the magnetic field at the equator also becomes zero (since $\partial\Psi/\partial\theta = 0$ for $\theta = \pi/2$), here the full magnetic field also turns out to be zero.

Another example is a rotating black hole having the electric charge Q_{H}. In this case, the rotation gives rise to the dipole magnetic field (Thorne et al., 1986)

$$\mathbf{E} = \frac{Q_{\text{H}}}{\rho_{\text{K}}^4 \Sigma} \left[(r^2 + a^2)(r^2 - a^2 \cos^2\theta)\mathbf{e}_{\hat{r}} - 2a^2 r \sqrt{\Delta} \sin\theta \cos\theta \mathbf{e}_{\hat{\theta}} \right], \tag{3.36}$$

$$\mathbf{B} = \frac{Q_{\text{H}} a}{\rho_{\text{K}}^4 \Sigma} \left[2r(r^2 + a^2)\cos\theta \mathbf{e}_{\hat{r}} + (r^2 - a^2\cos^2\theta)\sqrt{\Delta} \sin\theta \mathbf{e}_{\hat{\theta}} \right]. \tag{3.37}$$

Finally, the magnetic field of a thin conducting disk with inner radius b can be used as an example (see Fig. 3.2). If we assume that the magnetic field lines do not intersect the disk surface, the stream function Ψ in the absence of the black hole has the form (Beskin, 1997)

$$\Psi(r,\theta) = \Psi_0 \left[1 - \sqrt{\frac{1}{2}\left(1 - \frac{r^2}{b^2}\right) + \sqrt{\frac{1}{4}\left(1 - \frac{r^2}{b^2}\right)^2 + \frac{r^2\cos^2\theta}{b^2}}} \right]. \tag{3.38}$$

In the limit $b \to 0$ (with the total flow Ψ_0 preserved) formula (3.38) in each hemisphere becomes the monopole field $\Psi = \Psi_0(1 \pm \cos\theta)$. The jump of the magnetic field at the equator is connected with the currents flowing on the disk surface.

Expanding expression (3.38) around $r = 0$ into a series in multipoles and matching each harmonic to the horizon, we easily obtain an expression for the function Ψ in the vacuum approximation (and, hence, for the magnetosphere of the nonrotating black hole) (Beskin et al., 1992a). Thus, for $r_g \ll b$, we have (Beskin, 1997)

$$\Psi_v = \Psi_0 \left[\frac{1}{2}\frac{r^2}{b^2}\sin^2\theta - \frac{1}{120}\frac{r^4}{b^4}\mathcal{F}(r)(4\sin^2\theta\cos^2\theta - \sin^4\theta) + \cdots \right], \tag{3.39}$$

where $\mathcal{F}(r) = F(-4, -2, 1, 1 - r_g/r)$ is a hypergeometric function. Consequently, in the vicinity of the black hole the magnetic field is homogeneous

$$\Psi_v \approx \frac{1}{2}\Psi_0 \frac{r^2\sin^2\theta}{b^2}. \tag{3.40}$$

We emphasize that this procedure is possible, because the second family of particular solutions diverges on the horizon and, therefore, should be dropped. Only in this case, we can uniquely choose the coefficient of the hypergeometric function.

Analogously, solutions for the black hole enclosed in the center of a ring current and in the center of two oppositely directed ring currents were constructed (Chitre and Vishveshwara, 1975; van Putten and Levinson, 2003) and also for the special choice of currents on the disk surface (Tomimatsu and Takahashi, 2001). All of them were constructed by the above procedure. Finally, in the presence of the disk we can use the parabolic solution (3.22). As shown in Fig. 3.3, the solution has the form (Blandford and Znajek, 1977; Ghosh and Abramowicz, 1997)

$$\Psi(r,\theta) = \mathcal{C}(r - r_g)(1 - \cos\theta) - \mathcal{C}r_g(1 + \cos\theta)\ln(1 + \cos\theta), \; \theta < \pi/2, \tag{3.41}$$
$$\Psi(r,\theta) = \mathcal{C}(r - r_g)(1 + \cos\theta) - \mathcal{C}r_g(1 - \cos\theta)\ln(1 - \cos\theta), \; \theta > \pi/2, \tag{3.42}$$

where $\mathcal{C}$ is an arbitrary constant. One should note here that the monopole magnetic field $\Psi = \Psi_0(1 \pm \cos\theta)$, where Ψ_0 is the second constant defining the flow, can be added to solutions (3.41) and (3.42).

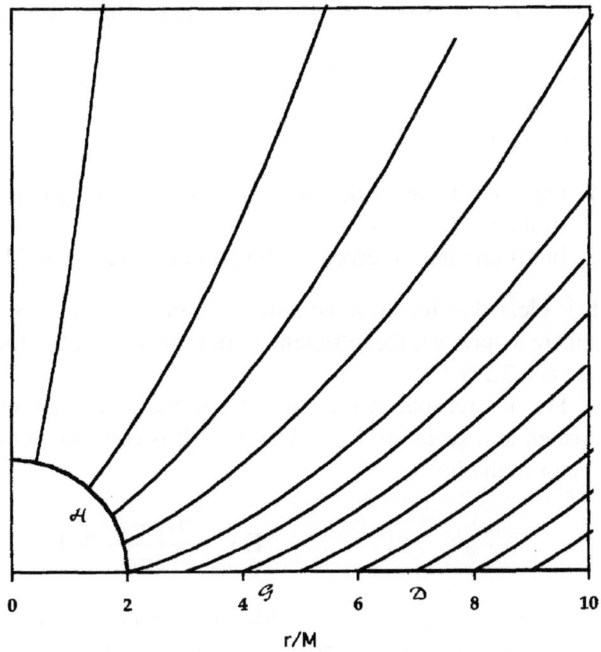

Fig. 3.3 The "nonphysical" parabolic solution that can be realized in the presence of the accretion disk (Ghosh and Abramowicz, 1997). Here the jump of the tangential component on the magnetic field is also provided by the toroidal currents flowing in the equatorial plane

Finally, note that when the magnetic field is generated by two oppositely directed ring currents, the topology of the solution greatly changes (see Fig. 3.4). The field lines passing through the black hole horizon do not extend to infinity now but are closed through the accretion disk surface (or a torus) generating the magnetic field itself. This topology is now actively discussed in the context of the possible sources of cosmological gamma bursts (van Putten and Levinson, 2003).

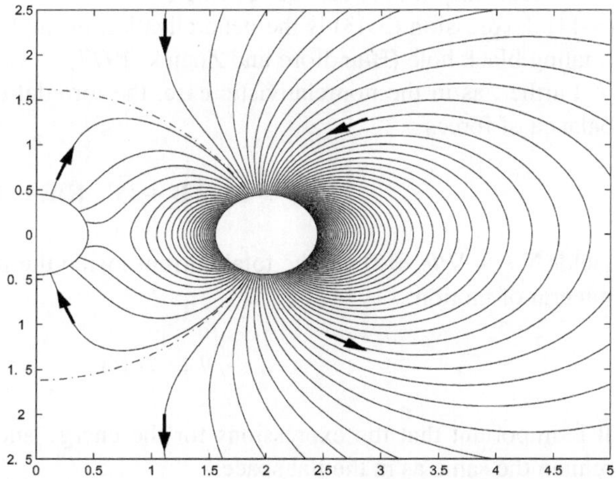

Fig. 3.4 An example of the magnetosphere in which all magnetic field lines passing through the black hole horizon do not extend to infinity but are closed through the torus surface generating the magnetic field (van Putten and Levinson, 2003) (Reproduced by permission of the AAS, Fig. 2 from van Putten, M.H.P.M., Levinson A.: Theory and astrophysical consequences of a magnetized torus around a rapidly rotating black hole. ApJ **584**, 937–953 (2003))

3.2.4 Force-Free Grad–Shafranov Equation in the Kerr Metric

We now proceed to the discussion of the basic equations describing the force-free magnetosphere of the rotating black hole. As in the case of the radio pulsar magnetosphere, we assume that

1. the plasma energy density $\epsilon_{\rm part}$ is much less than the energy density of the electromagnetic field $\epsilon_{\rm em}$;
2. but is enough to screen the longitudinal electric field $E_{\parallel}$.

It is clear that for these conditions to be satisfied it is necessary for the magnetosphere to have a rather efficient particle generation source. This problem is discussed in Sect. 3.2.5.

For this reason, in the studied stationary case the electric field $\mathbf{E}$ can again be written as a scalar multiplied by $\nabla\Psi$. It is convenient to choose the proportionality factor in the form

$$\mathbf{E} = -\frac{\Omega_{\rm F} - \omega}{2\pi\alpha}\nabla\Psi. \tag{3.43}$$

Substituting relation (3.43) in Maxwell's equation (3.11), we see that $\mathbf{B}\cdot\nabla\Omega_{\rm F} = 0$, i.e., $\Omega_{\rm F}$, as in the flat space, is to be constant on the magnetic surfaces:

$$\Omega_{\rm F} = \Omega_{\rm F}(\Psi). \tag{3.44}$$

Indeed, the Lie derivative in Eq. (3.11) under the obvious conditions $\nabla\cdot\mathbf{B} = 0$ and $\nabla\cdot\boldsymbol{\beta} = 0$ can be written as

$$\hat{\mathcal{L}}_{\beta}\mathbf{B} = -\nabla\times[\boldsymbol{\beta}\times\mathbf{B}] = \nabla\times\left(\frac{\omega}{2\pi}\nabla\Psi\right). \tag{3.45}$$

As a result, Eq. (3.11) can be rewritten as $\nabla\times(\Omega_{\rm F}\nabla\Psi) = 0$ and yields relation (3.44). Expression (3.43) is the generalization of the Ferraro isorotation law to the rotating black hole (Blandford and Znajek, 1977).

Further, as in the nonrelativistic case, the toroidal component of the force-free balance of forces

$$[\nabla\times\mathbf{B}]\times\mathbf{B} + (\nabla\cdot\mathbf{E})\mathbf{E} = 0 \tag{3.46}$$

yields $\nabla I\times\nabla\Psi = 0$, so the total current inside the magnetic surface is again an integral of motion:

$$I = I(\Psi). \tag{3.47}$$

It is important that the expressions for the energy and angular momentum fluxes remain the same as in the flat space:

$$E(\Psi) = \frac{\Omega_{\rm F} I}{2\pi}, \qquad L(\Psi) = \frac{I}{2\pi}. \tag{3.48}$$

Finally, the GS equation obtained directly from the poloidal component of Eq. (3.46) can be written as (Macdonald and Thorne, 1982)

$$\frac{1}{\alpha}\nabla_k\left\{\frac{\alpha}{\varpi^2}\left[1-\frac{(\Omega_{\rm F}-\omega)^2\varpi^2}{\alpha^2}\right]\nabla^k\Psi\right\}+\frac{\Omega_{\rm F}-\omega}{\alpha^2}(\nabla\Psi)^2\frac{{\rm d}\Omega_{\rm F}}{{\rm d}\Psi}+\frac{16\pi^2}{\alpha^2\varpi^2}I\frac{{\rm d}I}{{\rm d}\Psi}=0. \tag{3.49}$$

Equation (3.49) has the following properties:

- As any GS equation, it comprises only the stream function $\Psi(\varpi, z)$ and the invariants $\Omega_{\rm F}(\Psi)$, $I(\Psi)$. As in the nonrelativistic case, the force-free equation should not be supplemented with Bernoulli's equation.
- Equation (3.49) is of an elliptic type in the entire region where it is defined. The field of application is limited by the light cylinder on which the electric field becomes equal to the magnetic one: $|\mathbf{E}| = |\mathbf{B}|$. As in the case of the radio pulsar magnetosphere, Eq. (3.49) cannot be extended beyond the light surface. But if the magnetic field is larger than the electric one up to the event horizon, Eq. (3.49) is of an elliptic type up to the horizon.
- In the nonrelativistic limit $\alpha \to 1$, $\omega \to 0$, Eq. (3.49) becomes the pulsar equation (2.101).
- At distances satisfying the condition $\alpha^2 \gg (\Omega_{\rm F}-\omega)^2\varpi^2$, the differential operator $\hat{\mathcal{L}}_{\rm bh} = 1/\alpha\,\nabla_k\{\ldots\nabla^k\}$ coincides with the vacuum operator $\hat{\mathcal{L}}$ (3.17). Hence, for the nonrotating black hole (when $\Omega_{\rm F} = 0$, $I = 0$ as well) the monopole magnetic field $\Psi = \Psi_0(1-\cos\theta)$ is the solution.
- The equation contains two singular surfaces—"light cylinders" defined from the condition $A = 0$, where

$$A = \alpha^2 - (\Omega_{\rm F}-\omega)^2\varpi^2. \tag{3.50}$$

  As we see in the following, it is appropriate to speak about the Alfvén surfaces. One of them corresponding to the outflowing plasma is fully equivalent to the light cylinder in the neutron star magnetosphere and the other inner surface, on which $\alpha \approx (\Omega_{\rm H}-\Omega_{\rm F})\varpi_{\rm H}$, is due to the GR effects.
- Besides, as shown below, the solution can be extended to the horizon only if one more critical condition for $\alpha^2 = 0$ is satisfied.
- According to the general formula $b = 2 + i - s'$ for the number of boundary conditions, we have $b = 1$ for $s' = 3$, i.e., for the force-free magnetosphere regular up to the horizon and extending beyond the outer light cylinder, the problem requires only one boundary condition. If the magnetic field connects the black hole horizon with the accretion disk surface inside the outer light cylinder ($s' = 2$), for the solution regular to the horizon two boundary conditions, for example, the angular velocity $\Omega_{\rm F}(\Psi)$ and the stream function Ψ on the disk surface, are to be given.

- Given the magnetic structure, i.e., the functions $\Psi(\varpi, z)$, $\Omega_F(\Psi)$, and $I(\Psi)$, the electric field and the toroidal component of the magnetic field are determined from the algebraic relations.

Some comments clarifying the properties of Eq. (3.49) are necessary. One should first stress that the occurrence of the inner Alfvén surface due to the GR effects is a property common to the flows in the vicinity of the black hole. This property is demonstrated below for the general MHD case. It is important that, as shown in the following chapter, within ideal magnetohydrodynamics (and for the case of the positive energy flux from the black hole), the plasma can cross the Alfvén surfaces only in one direction; clearly, it has nothing to do with the accretion disk itself in which the viscosity is of vital importance. The outer Alfvén surface can be crossed only in the direction from the black hole ($v^r > 0$) and the inner one in the opposite direction ($v^r < 0$) only. Therefore, on the field lines passing through the horizon there must occur a particle generation region separating the accretion and ejection flows, where the GS equation is no longer used. However, if in the generation region the potential drop is much less than the characteristic potential difference and the surface currents are much weaker than the longitudinal ones, we can assume that $\Omega_F^+ = \Omega_F^-$ and $I^+ = I^-$. As a result, the values of the two integrals of motion can be identical along the entire magnetic field lines connecting the black hole with infinity.

Problem 3.5 Show that on the outer Alfvén surface the radial velocity cannot be negative.
(Hint: in the force-free approximation the velocity is expanded in the drift velocity perpendicular to the magnetic field and the slip velocity along the field, see Fig. 2.4. The total velocity should not be larger than the velocity of light.)

Further, one should mention the problem of the boundary condition on the horizon. Indeed, as was noted above, Eq. (3.49) is of an elliptic type up to the black hole horizon and, therefore, the event horizon can be regarded as some surface on which boundary conditions should be given. On the other hand, the horizon cannot be causally connected with the outer space and, therefore, cannot affect the magnetic field structure beyond the black hole. If we recall that, in reality, the event horizon is separated from the environment by a switching-on wave, prescribing the boundary condition on the horizon becomes problematic (Punsly, 2001). As we will see, this problem can be successfully solved only by the full MHD GS equation version comprising the finite particle mass. We should make some comments on the free-force limit here.

Recall first that by definitions (3.14) and (3.43), the toroidal component of the magnetic field and the θ-component of the electric field diverge on the horizon as α^{-1}. However, this behavior results from the coordinate singularity of the chosen metric and does not correspond to the physical singularity on the horizon. This singularity would occur when freely falling observers rather than ZAMO detect the infinite fields when crossing the horizon. The condition for electric field finiteness on the horizon for the freely falling observer $E'_{\hat{\theta}} = \alpha^{-1}\left(E_{\hat{\theta}} + B_{\hat{\varphi}}\right) < \infty$, i.e.,

$$E_{\hat{\theta}} = -B_{\hat{\varphi}}, \tag{3.51}$$

in the force-free approximation is rewritten as (Znajek, 1977; Macdonald and Thorne, 1982)

$$4\pi I(\Psi) = [\Omega_{\rm H} - \Omega_{\rm F}(\Psi)]\sin\theta \frac{r_{\rm g}^2 + a^2}{r_{\rm g}^2 + a^2\cos^2\theta}\left(\frac{{\rm d}\Psi}{{\rm d}\theta}\right). \tag{3.52}$$

Here we just used definitions (3.14) and (3.43).

The condition (3.52) was generally regarded as an additional boundary condition on the event horizon acting as Ohm's law (Macdonald, 1984). Indeed, for the given stream function on the horizon $\Psi(r_{\rm g}, \theta)$, relation (3.52) connects the longitudinal current $I(\Psi)$ with the angular velocity $\Omega_{\rm F}(\Psi)$ defining the electric field. The problem, however, is that we, initially, do not know the poloidal magnetic field structure in the vicinity of the horizon. Therefore, relation (3.52) cannot be regarded as a coupling between $I(\Psi)$ and $\Omega_{\rm F}(\Psi)$ only. Otherwise, the magnetic flux on the horizon $\Psi(r_{\rm g}, \theta)$ can be found only as a solution to Eq. (3.49), given the values of $I(\Psi)$ and $\Omega_{\rm F}(\Psi)$.

It is easy to check that, in the force-free limit, the "boundary condition on the horizon" (3.52) is really an additional one coinciding with the regularity condition on the event horizon (Beskin, 1997; Uzdensky, 2004). Indeed, the radial derivatives enter into Eq. (3.49) in combinations of $\alpha\partial/\partial r$ and $\alpha^2\partial^2/\partial r^2$ only. The remarkable property of the GS equation is that under the regularity condition of the solution (i.e., if it is possible to drop the terms with the radial derivatives on the horizon) it becomes one-dimensional and can be integrated. As a result of the integration, we again return to relation (3.52). Thus, when the condition (3.52) is satisfied, the singular terms proportional to $\alpha\partial/\partial r$ and $\alpha^2\partial^2/\partial r^2$ are zero on the black hole horizon. This shows the absence of a singularity for $\alpha^2 \to 0$.

However, as will be shown in Chap. 4, in reality, the condition (3.52) is the limit of the critical condition on the fast magnetosonic surface located over the event horizon. In particular, this implies that in finite time the switching-on wave is in the hyperbolic domain of the full GS equation and, hence, cannot affect the flow structure outside the black hole. Therefore, the term "condition on the horizon" is in quotation marks indicating that the full MHD GS equation does not have any additional regularity condition on the horizon. We have already dealt with this property when analyzing the hydrodynamical flows.

Problem 3.6 Find the "boundary condition" (3.52) by the direct integration of the force-free equation (3.49) on the horizon.

Problem 3.7 Show that in the case of the plasma-filled magnetosphere of the black hole, its electric charge $Q_{\rm H}$ is equal to (see, e.g., Lee et al., 2001)

$$Q_{\rm H} = \frac{1}{4\pi}\int_0^{\pi} [\Omega_{\rm H} - \Omega_{\rm F}]\frac{\Sigma^2}{\rho_{\rm K}^2}\left(\frac{\partial \Psi}{\partial r}\right)\sin\theta {\rm d}\theta. \tag{3.53}$$

Evaluate the magnetic field condition in order for the charge $Q_{\rm H}$ (3.53) to begin to disturb the Kerr metric.

As an illustration, show that the additional condition (3.52) really fixes the value of the invariant $L(\Psi) = I(\Psi)/2\pi$ for $\Psi \to 0$. We consider the magnetic field structure problem in the neighborhood of the rotating black hole. We believe that the solution can be extended to the horizon and all magnetic field lines are frozen in the external rotating shell or in the accretion disk within the outer light cylinder. In this case, the formula $b = 2 + i - s'$ shows that the problem requires two boundary conditions only. As such conditions, it is logical to choose the value of the function $\Psi(\mathbf{r})$ on the shell (disk) surface and the angular velocity $\Omega_{\rm F}(\Psi)$. The value $L(\Psi)$ must be found from the condition of the smooth passage of the solution through the critical surfaces.

We now consider the behavior of the invariant $L(\Psi)$ in the vicinity of the rotation axis. It is convenient to write it as

$$L(\Psi) = k\frac{\Omega_{\rm F}}{4\pi^2}\Psi, \tag{3.54}$$

where $\Omega_{\rm F} = \Omega_{\rm F}(0)$ below. Here the proportionality of $L(\Psi)$ and Ψ results from the assumption of the constant density of the electric current in the vicinity of the rotation axis, and the case $k = 1$ corresponds to the GJ current density. Substituting expression (3.54) in the "condition on the horizon" (3.52), we obtain at small angles of θ

$$\frac{{\rm d}\Psi}{\Psi} = 2k\frac{\Omega_{\rm F}}{\Omega_{\rm H} - \Omega_{\rm F}}\frac{{\rm d}\theta}{\sin\theta}, \tag{3.55}$$

i.e., $\Psi(\theta) \propto \theta^w$, where

$$w = 2k\frac{\Omega_{\rm F}}{\Omega_{\rm H} - \Omega_{\rm F}}. \tag{3.56}$$

Formally, k can be arbitrary. But the finiteness condition of the magnetic field on the rotation axis yields $w = 2$, i.e., $k = (\Omega_{\rm H} - \Omega_{\rm F})/\Omega_{\rm H}$. Hence, we conclude that for small Ψ, the invariant $L(\Psi)$ should have the form

$$L(\Psi) = \frac{(\Omega_{\rm H} - \Omega_{\rm F})}{4\pi^2}\,\Psi. \tag{3.57}$$

In the next chapter, we see that the relation (3.57) is really the limit of the critical condition on the inner fast magnetosonic surface as it approaches the event horizon.

Finally, note that the additional condition (3.52) is available only if we deal with the solutions that can be extended to the horizon. If the longitudinal currents are rather small, the field of application of the GS equation is confined to the light surface beyond the horizon. In this case, for the magnetic field lines intersecting both the inner and outer Alfvén surface, we have $s' = 2$, so that the problem requires two boundary conditions. In particular, this implies that the solutions with zero longitudinal current ($I = 0$) and arbitrary angular velocity $\Omega_{\rm F}$ can be constructed. At large distances, the solution coincides with the Michel monopole solution for the zero current (see Fig. 2.15) and, at small distances, extends only to the inner light cylinder $\alpha^2 \approx (\Omega_{\rm F} - \Omega_{\rm H})^2\varpi^2$. This problem is not solved yet.

3.2.5 Particle Generation

In order for the force-free approximation to be used it is necessary that the black hole magnetosphere be filled with plasma screening the longitudinal electric field. On the other hand, as was mentioned, within ideal magnetohydrodynamics, the charged particle accretion along the magnetic field lines from infinity to the black hole horizon becomes impossible. Thus, on the open field lines the plasma must be generated in the magnetosphere itself between two families of singular surfaces. One of its parts outflows beyond the magnetosphere and the other accretes onto the black hole. There are several mechanisms, in which, however, the plasma is ultimately always generated by two-photon interaction. The one-photon conversion, which is of major importance in the radio pulsar magnetosphere, appears inefficient here because for the magnetic fields $B \sim B_{\rm Edd} \sim 10^4$ G the probability of the pair generation $w(\gamma \rightarrow e^+e^-)$ (2.27) is extremely small.

The efficient mechanism may be the direct two-photon process $\gamma+\gamma \rightarrow e^+ + e^-$ (see., e.g., Svensson, 1984), where the necessary γ-quanta are emitted by the inner accretion disk regions. However, rather high temperatures providing the necessary number of hard γ-quanta with energies higher than $\epsilon_{\rm min} = m_{\rm e}c^2$ corresponding to the pair generation threshold are needed in this case.

There is, however, another mechanism that can supply particles in the region of the field lines passing through the black hole horizon even in the absence of hard γ-quanta. It is analogous to the particle generation mechanism in the outer gap of the radio pulsar magnetosphere (Cheng et al., 1986). Indeed, according to (3.9) and (3.43), the exact relativistic expression for the GJ charge density $\rho_{\rm GJ}$ has the form

$$\rho_{\rm GJ} = -\frac{1}{8\pi^2}\nabla_k\left(\frac{\Omega_{\rm F} - \omega}{\alpha}\nabla^k\Psi\right). \tag{3.58}$$

In particular, in the vicinity of the rotation axis we just have

$$\rho_{\rm GJ} \approx -\frac{(\Omega_{\rm F} - \omega)B}{2\pi\alpha}. \tag{3.59}$$

It is exactly this expression that was used in Sect. 2.3.4 in the analysis of the GR effects in the polar regions of the neutron star.

As a result, the GJ charge density due to the GR effects becomes equal to zero for $\omega \approx \Omega_{\rm F}$. This is the case when the condition $0 < \Omega_{\rm F} < \Omega_{\rm H}$ is satisfied, under which, as we will see, the black hole loses its rotational energy. For example, for the monopole magnetic field and the condition $\Omega_{\rm F} = \Omega_{\rm H}/2$, the surface $\rho_{\rm GJ} = 0$ is a sphere of the radius

$$r_{\rm inj} = 2^{1/3} r_{\rm g} \approx 1.26\, r_{\rm g}. \tag{3.60}$$

In the general case, to specify the surface $\rho_{\rm GJ} = 0$ we must know both the magnetic field structure $\Psi(r, \theta)$ and the dependence of the angular velocity $\Omega_{\rm F}$ on the stream function Ψ.

Problem 3.8 Find expression (3.60).

Thus, in the black hole magnetosphere there occurs a region quite analogous to the outer gap in the radio pulsar magnetosphere. In this region, longitudinal electric fields may also occur (see, Beskin et al. (1992b); Hirotani and Okamoto (1998) for details, where it is also shown that the acceleration region size in AGN is much smaller than that of the system, so the acceleration region does not affect the global magnetosphere structure). The chain of processes is the following:

1. The primary particle acceleration by the longitudinal electric field.
2. The curvature photon emission at characteristic frequencies $\omega \leq \omega_{\rm cur}$ (2.26).
3. The interaction between relativistic electrons and soft X-ray photons emitted by the accretion disk; the hard γ-quanta generation by the IC effect.
4. The secondary plasma generation $\gamma + X \rightarrow e^{+} + e^{-}$ due to the interaction between the hard γ-quanta, the curvature, and X-ray photons.
5. The screening of the longitudinal electric field by the secondary plasma.

In any event, the black hole magnetosphere may be filled with plasma only in the presence of the accretion disk providing the sufficient number of photons in the black hole magnetosphere.

3.3 Energy Release Mechanism

3.3.1 Blandford–Znajek Process

We now discuss in detail the energy release mechanism of rotating black holes submerged in the external regular magnetic field [so-called Blandford–Znajek (BZ) process (1977)]. Its main idea is the analogy with the energy transfer process in the inner regions of the radio pulsar magnetosphere. Indeed, we assume that in the neighborhood of the rotating black hole there is a regular poloidal magnetic field along which electric currents flow. Then because of the occurrence of the electric field $E_{\hat{\theta}} \sim \Omega_{\mathrm{F}} \varpi B_{\mathrm{p}}$ connected with the induction action of the plasma rotating with the angular velocity Ω_{F} and the toroidal magnetic field $B_{\varphi} \sim -2I/\varpi$ induced by the longitudinal current I, the electromagnetic energy flux (the Poynting vector flux) flowing along the magnetic field lines is generated.

Certainly, in the vicinity of the black hole the GR effects, by definition, become substantial. Therefore, it is not obvious that the pulsar analogy can be useful everywhere. In any case, it is not applied to the slowing-down mechanism (Punsly, 2001). Indeed, as we saw, in the radio pulsar magnetosphere the neutron star slowing-down is produced by the ponderomotive action of the surface currents closing the electric currents flowing in the magnetosphere. Formally, the surface currents can be introduced for the black hole horizon (or "stretched horizon" as was the case with the membrane paradigm) by analogy with the expression (2.135) as

$$\mathbf{J}_{\mathrm{H}} = \frac{I}{2\pi\varpi}\mathbf{e}_{\hat{\theta}}. \tag{3.61}$$

The difference from the neutron star surface is that in the definition of the jump of the tangential magnetic field there is the finite "regularized" value $B_{\mathrm{H}} = \alpha B_{\hat{\varphi}}$ (see Thorne et al. (1986) for details). As a result, the "condition on the horizon" (3.52) has the form

$$\mathbf{J}_{\mathrm{H}} = \frac{c}{4\pi}\mathbf{E}_{\mathrm{H}}, \tag{3.62}$$

where $E_{\mathrm{H}} = \alpha E_{\hat{\theta}}$ and the factor of proportionality corresponds to the universal resistance $\mathcal{R} = 4\pi/c = 377$ Ohm. It is one of the basic relations in the membrane paradigm.

However, the horizon is not a physically separated surface, not to mention it is not causally connected with the outer space (Punsly and Coroniti, 1990a). Therefore, these currents cannot result in the black hole slowing down. Indeed, in the vicinity of the horizon, no toroidal currents can flow. First, as we saw, the gravitational field appears so strong here that particles (in the ZAMO reference frame) can move in a radial direction only. Second, as shown in Fig. 3.1, for the current to be closed it is not necessary that the real charges flow along the horizon. The gravitational field itself distorts the pattern of electric field lines so that the distant observer seems to see that the charge is in the vicinity of the center of the black hole. Accordingly, there can be no Ampére's force $\mathbf{J}_s \times \mathbf{B}$ associated with charges crossing the magnetic field lines (the motion of the fictitious image charges over the horizon cannot do any work).

In reality, as shown in Chap. 5, the slowing-down moment acts in the plasma generation region over the event horizon. The energy release is due to the negative energy falling onto the horizon. Otherwise, the BZ mechanism is, in fact, the electromagnetic realization of the Penrose effect (Takahashi et al., 1990). Therefore, the rotating black hole, as well as the rotating neutron star, can act as a unipolar inductor effectively transporting the rotation energy to the outer magnetosphere (see Fig. 3.5).

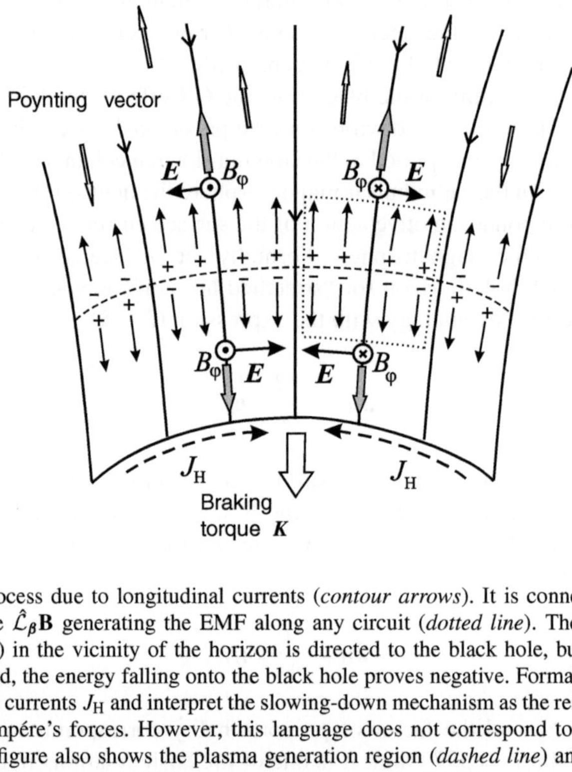

Fig. 3.5 BZ process due to longitudinal currents (*contour arrows*). It is connected with gravitomagnetic force $\hat{\mathcal{L}}_{\beta}\mathbf{B}$ generating the EMF along any circuit (*dotted line*). The Poynting vector (*shaded arrows*) in the vicinity of the horizon is directed to the black hole, but when condition (3.69) is satisfied, the energy falling onto the black hole proves negative. Formally, one can introduce the surface currents $J_{\rm H}$ and interpret the slowing-down mechanism as the result of the braking torque $\mathbf{K}$ of Ampére's forces. However, this language does not correspond to the real physical processes. The figure also shows the plasma generation region (*dashed line*) and the direction of the secondary particle motion. All the vectors are measured by ZAMO

We now define the total energy release that can be produced by the black hole rotation. It is necessary to find the energy flux in the region of magnetic field lines passing through the horizon. In the force-free approximation we have for the energy losses

$$W_{\rm tot} = \int T_{0\beta} {\rm d}S^{\beta} = \int E(\Psi){\rm d}\Psi = \frac{1}{2\pi}\int \Omega_{\rm F}(\Psi) I(\Psi){\rm d}\Psi, \tag{3.63}$$

and for the angular momentum losses

$$K_{\rm tot} = \int L(\Psi)\mathrm{d}\Psi = \frac{1}{2\pi}\int I(\Psi)\mathrm{d}\Psi. \tag{3.64}$$

As we see, the expressions for the energy and the angular momentum fluxes in the case of the Kerr metric are the same as in the flat space. It is the conservation of the values $E(\Psi)$ and $L(\Psi)$ along the magnetic field lines that confirms the existence of the energy flux extending from the black hole horizon to infinity. Finally, using the "condition on the horizon" (3.52), we get

$$W_{\rm tot} = \frac{1}{4\pi}\int \Omega_{\rm F}(\Omega_{\rm H} - \Omega_{\rm F})\sin\theta \frac{r^2 + a^2}{r^2 + a^2\cos^2\theta}\frac{\mathrm{d}\Psi}{\mathrm{d}\theta}\mathrm{d}\Psi. \tag{3.65}$$

Problem 3.9 Show that for $\Omega_{\rm F} = \mathrm{const}$ and the monopole magnetic field $B_n = \mathrm{const}$, the losses $W_{\rm tot}$ (3.65) have the form (Lee et al., 2000)

$$W_{\rm tot} = \frac{\Omega_{\rm F}(\Omega_{\rm H} - \Omega_{\rm F})}{\Omega_{\rm H}^2}\left(\frac{a}{M}\right)^2 B_n^2 M^2 c\, f\left(\frac{a}{r_{\rm g}}\right), \tag{3.66}$$

where

$$f(h) = \frac{1 + h^2}{h^2}\left[\left(h + \frac{1}{h}\right)\mathrm{arctgh} - 1\right], \tag{3.67}$$

so that $f(0) = 2/3$ and $f(1) = \pi - 2$.

We first discuss the above result from a physical viewpoint.

- As seen from relation (3.65), to exactly determine the energy losses one should know not only the magnetic field structure in the vicinity of the black hole horizon, i.e., the magnetic flux function $\Psi(r_{\rm g}, \theta)$, but also the angular velocity $\Omega_{\rm F}(\Psi)$. Thus, in the case of the black hole magnetosphere the main theoretical problem is not to find the longitudinal current $I(\Psi)$, as was the case with the radio pulsar magnetosphere, but the angular plasma velocity $\Omega_{\rm F}(\Psi)$. This problem cannot be successfully solved by the force-free approximation. Therefore, as in the case of radio pulsars, we have to put off the discussion of this problem until the next chapters.
- If we are interested in the estimate of energy losses in order of magnitude, we take $\Omega_{\rm F} = \mathrm{const}$ and $\Psi = \pi r_{\rm g}^2 B_0$. Hence, $W_{\rm tot} \sim W_{\rm BZ}$, where the losses $W_{\rm BZ}$ correspond to the conditions $\Omega_{\rm F} = \Omega_{\rm H}/2$ and $B_n = \mathrm{const}$

$$W_{\rm BZ} \approx 10^{45}\,{\rm erg/s}\left(\frac{a}{M}\right)^2\left(\frac{B_0}{10^4\,{\rm G}}\right)^2\left(\frac{M}{10^9\,M_{\odot}}\right)^2. \tag{3.68}$$

One can check that for an extremely fast-rotating black hole and for $B = B_{\rm Edd}$ the losses $W_{\rm BZ}$ coincide with the Eddington luminosity $L_{\rm Edd} = \dot{M}_{\rm Edd}c^2$. Thus, for the fast-rotating black hole, the theory can really explain the observed energy of the AGN jets (Blandford and Znajek, 1977; Thorne et al., 1986).

- The energy losses occur only under the condition

$$0 < \Omega_{\rm F} < \Omega_{\rm H}, \tag{3.69}$$

and the largest losses are for $\Omega_{\rm F} = \Omega_{\rm H}/2$. It is important that for $\Omega_{\rm F} < \Omega_{\rm H}$, the Poynting vector $\mathbf{S} = (c/4\pi)[\mathbf{E} \times \mathbf{B}]$ on the event horizon is directed to the black hole. Indeed, as shown in Fig. 3.5, according to the dependence $E_{\hat{\theta}} \propto (\Omega_{\rm F} - \omega)$ (3.43), for $0 < \Omega_{\rm F} < \Omega_{\rm H}$, the direction of the electric field in the vicinity of the horizon ($\omega \approx \Omega_{\rm H}$) is opposite to that of the electric field far from the black hole ($\omega = 0$), whereas the direction of the toroidal magnetic field remains unchanged. However, this does not imply that the energy of the rotating black hole increases with time, because Eq. (1.239)

$$\frac{1}{\alpha^2}\nabla\cdot(\alpha^2\mathbf{S}) = H_{ik}T^{ik}, \tag{3.70}$$

besides its own conservation law $\nabla\cdot\mathbf{S} = 0$, contains additional terms responsible for the space curvature. As a result, the flux of the vector $\mathbf{S}$ does not retain along the magnetic field lines. The constant energy flux is given by the invariant expression (3.63).

- The magnetic field B_0 (or, to be exact, the magnetic flux on the horizon $\Psi(r_{\rm g},\theta)$) available in expression (3.65) is not known beforehand and must be defined from the solution to the GS equation. In particular, as we will see, the magnetic field expulsion from the rotating black hole, which takes place in the case of the vacuum magnetosphere, can be compensated by the currents flowing in the magnetosphere of the black hole.

Recall once again that the expression for the energy losses $W_{\rm tot}$ (3.63) by analogy with the radio pulsar magnetosphere can, formally, be written under the action of the electromagnetic forces on the "stretched horizon" (Thorne et al., 1986) as

$$W_{\rm tot} = \int [\mathbf{E}_{\rm H}\mathbf{J}_{\rm H} - \boldsymbol{\beta}_{\rm H}\cdot(\sigma_{\rm H}\mathbf{E}_{\rm H} + \mathbf{J}_{\rm H}\times\mathbf{B}_{\rm n})]\,{\rm d}S. \tag{3.71}$$

Here $\boldsymbol{\beta}_{\rm H} = \boldsymbol{\beta}(r_{\rm g})$, and the "surface charge" $\sigma_{\rm H}$ and the "surface current" $\mathbf{J}_{\rm H}$ are again defined as

$$\sigma_{\mathrm{H}} = \frac{E_{\mathrm{H}}}{4\pi}, \tag{3.72}$$

$$4\pi \mathbf{J}_{\mathrm{H}} \times \mathbf{n} = B_{\mathrm{H}} \mathbf{e}_{\varphi} = \alpha B_{\varphi} \mathbf{e}_{\varphi}. \tag{3.73}$$

As was already mentioned, this analogy cannot be interpreted literally.

It is interesting to note that the condition (3.69) of the energy extraction from the rotating black hole can be given as a theorem.

Theorem 3.2 *The energy flux passing through the event horizon is negative (and, hence, the black hole loses its rotational energy) if the inner Alfvén surface is within the ergosphere (Takahashi et al., 1990).*

Indeed, as is evident from relations (1.218) and (3.50), for $\Omega_{\mathrm{F}} = 0$, the Alfvén surface coincides with the ergosphere surface. In the other limiting case, $\Omega_{\mathrm{F}} = \Omega_{\mathrm{H}}$, the condition $\alpha^2 = (\Omega_{\mathrm{F}} - \omega)^2 \varpi^2$ has the solution $\alpha = 0$, i.e., the Alfvén surface coincides with the event horizon. Therefore, in the whole range of parameters, in which the condition (3.69) is satisfied, the Alfvén surface is located within the ergosphere.

We emphasize once again that the regular magnetic field is only a connecting link that allows one to effectively extract the energy and the angular momentum from the central engine. The latter is especially important, because, within the standard model, when the angular momentum transfer is associated with the viscous stresses in the disk, the angular momentum losses needed for the rather large accretion rate are not quite large. As for astrophysical applications, they will be considered in Sect. 5.1.

3.3.2 *Physical Intermezzo—Black Hole Thermodynamics*

There is another substantial difference between the black hole magnetosphere and the neutron star magnetosphere. The point is that the energy store of the black hole is not only in the rotational energy but also in the so-called irreducible mass M_{irr}, which depends on the area of the black hole horizon S

$$M_{\mathrm{irr}} = \left(\frac{S}{16\pi}\right)^{1/2} = \frac{1}{2}\left(r_{\mathrm{g}}^2 + a^2\right)^{1/2}. \tag{3.74}$$

Otherwise, for the black hole there is one more degree of freedom associated with the area of its surface, and, therefore, relation $W_{\mathrm{tot}} = \Omega_{\mathrm{H}} K_{\mathrm{tot}}$ equivalent to the condition (2.130) for radio pulsars may not be satisfied for black holes. Since this problem is closely connected with black hole thermodynamics, makes it possible to clarify the process of the energy and angular momentum losses, it seems advisable to mention the main assertions of this theory here (see Thorne et al. (1986); Frolov and Novikov (1998) for details).

We first determine the area of the black hole surface

$$S = 4\pi(r_g^2 + a^2) = 8\pi M(M + \sqrt{M^2 - a^2}). \tag{3.75}$$

If we consider the area S as functions of the mass M and the angular momentum $J = a/M$, we can write the relation for increments as

$$\delta M = \frac{1}{8\pi} g_H \delta S + \Omega_H \delta J. \tag{3.76}$$

Here $g_H = (r_g - M)/2Mr_g$ is the so-called surface gravity.

Since the mass M is naturally associated with the energy, Eq. (3.76) can be represented in the thermodynamic form

$$\frac{d\mathcal{E}}{dt} = T_H \frac{dS_H}{dt} + \Omega_H \frac{dJ}{dt}. \tag{3.77}$$

Here $d\mathcal{E}/dt = -W_{tot}$, $dJ/dt = -K_{tot}$, so that the change in the mass is really associated with the energy loss connected with the Poynting vector flux. As to the "temperature" T_H and the "entropy" S_H of the black hole, they are represented as (Frolov and Novikov, 1998)

$$T_H = \frac{\hbar}{2\pi} g_H, \tag{3.78}$$

$$S_H = \frac{1}{4\hbar} S. \tag{3.79}$$

The factors are chosen so that the temperature T_H coincides with that of the black hole, which describes the radiation associated with the Hawking effect. The above thermodynamic analogy is based on the important theorem.

Theorem 3.3 *In any classical processes (matter accretion, black hole coalescence, etc.) the black hole surface area cannot become smaller*

$$\delta S \geq 0. \tag{3.80}$$

As we see, this theorem is quite analogous to the second law of thermodynamics, which is the basis for its thermodynamic interpretation. Nevertheless, we used the terms temperature and entropy in quotation marks, because, in spite of the large number of papers devoted to this theme, it is not clear yet whether expression (3.79) corresponds to the real entropy of the black hole.

The problem of the black hole entropy is beyond the scope of our discussion. It is important here that the electromagnetic process of energy release from the rotating black hole does not contradict the basic laws of physics. Indeed, as is seen from the definition of the gravitational radius $r_g = M + \sqrt{M^2 - a^2}$, as the angular velocity Ω_H decreases, the black hole radius increases, so does its area due to this process.

Moreover, because of the thermodynamic relation (3.77), it is possible to clarify the key properties of the energy transfer from the rotating black hole. Indeed, using definitions (3.63) and (3.64), we can rewrite relation (3.77) as

$$-\int E(\Psi)\mathrm{d}\Psi = \frac{1}{2\pi}\int (\Omega_{\mathrm{H}} - \Omega_{\mathrm{F}})I(\Psi)\mathrm{d}\Psi - \Omega_{\mathrm{H}}\int L(\Psi)\mathrm{d}\Psi. \tag{3.81}$$

The expression on the left-hand side of Eq. (3.81) exactly corresponds to the energy losses, and two terms on the right-hand side to the change in the "entropy" and the angular momentum. For example, for $\Omega_{\mathrm{F}} = 0$ (when $E = \Omega_{\mathrm{F}} I/2\pi = 0$) the energy losses of the black hole turn out to be zero and a decrease in the angular momentum is fully compensated by increasing the black hole area

$$\delta S_{\mathrm{H}} = -\frac{\Omega_{\mathrm{H}}}{T_{\mathrm{H}}}\delta J. \tag{3.82}$$

This case (which can take place if the homogeneous magnetic field is frozen in a nonrotating cloud far from the black hole) corresponds to the adiabatic process in thermodynamics. Further, for $\Omega_{\mathrm{F}} = \Omega_{\mathrm{H}}$, i.e., for $I = 0$ (when both the energy and angular momentum fluxes are zero) not only the mass but also the angular momentum of the black hole remain constant. This case corresponds to the full corotation of the radio pulsar magnetosphere. If $\Omega_{\mathrm{F}} = \Omega_{\mathrm{H}}/2$, the variations of all the values are comparable.

To sum up, we once more point to the basic properties of the electrodynamic mechanism of the energy release of the black hole submerged in the external magnetic field.

1. The BZ slowing-down process of the black hole is due to the gravitomagnetic forces beyond the event horizon. The electromagnetic flux passing through the horizon transports the negative energy to the black hole. Otherwise, the BZ mechanism is the electromagnetic realization of the Penrose process. Recall that this effect allows one to release the energy from the rotating black hole due to the particle decay in the ergosphere when one of the secondary particles extends to infinity and the other (having the negative energy) falls onto the event horizon (Frolov and Novikov, 1998). Exactly this pattern, as we saw, is realized in the black hole magnetosphere. The electric field and the currents connected with the particles escaping the black hole magnetosphere generate an electromagnetic energy flux extending to infinity. As a result, the black hole submerged in the external magnetic field acts as an unipolar inductor losing the rotational energy due to the electric currents flowing in its magnetosphere.
2. Formally, it is possible to introduce the surface charges and the currents and interpret the slowing-down mechanism of the black hole by analogy with radio pulsars, i.e., by the slowing-down mechanism of Ampére's forces acting on the membrane [see (3.71)]. However, this language does not correspond to the real physical processes in the black hole magnetosphere.
3. Technically, the problem of determining the energy losses of the rotating black hole is associated not so much with the problem of the value of the longitudi-

nal current $I(\Psi)$, as in the case of the radio pulsar magnetosphere, as with the problem of the value of the angular velocity of the magnetic field lines $\Omega_F(\Psi)$. Indeed, in the case of the black hole magnetosphere, the angular velocity $\Omega_F(\Psi)$ available in expression (3.65), generally speaking, is in no way connected with the angular velocity of the black hole Ω_H but must be defined from the solution of the complete problem. In particular, Ω_F is to depend on the properties of the plasma source located between two Alfvén surfaces on the field lines passing through the black hole. This important conclusion following from the analysis of the basic properties of the GS equation is associated with the occurrence of additional critical surfaces due to the GR effects.

4. As in the case of the radio pulsar magnetosphere, the problem of the energy losses cannot be successfully solved by the force-free approximation. This is because the force-free equation remains elliptic up to the black hole horizon.
5. Within the full MHD version (i.e., for the finite particle mass), as shown in the following chapter, the horizon is in the hyperbolic domain of the GS equation. Therefore, as was expected, the physical conditions on the horizon cannot affect the flow in the outer regions of the magnetosphere.

3.4 Black Hole Magnetosphere Structure

3.4.1 General Properties

We now discuss the main theoretical results of the magnetosphere structure of the rotating black hole, i.e., the region over the accretion disk, where the magnetic fields are of crucial importance. At present, the MHD model of the central engine in AGN and gamma-bursters seems the most realistic now (see Sect. 5.1). But one should recognize that, in spite of the numerous efforts of theorists over the past 30 years, they have failed to construct the consistent model of the black hole magnetosphere, which would help one answer the key questions, i.e., determine the value of the energy losses and the matter ejection rate as a function of the natural physical parameters characterizing the central engine (such as the mass M and the angular velocity Ω_H of the black hole and also the accretion rate $\dot{M}$), and, besides, show the possibility of the efficient collimation of the magnetic field lines in the direction of the rotation axis. This is because this problem is of an inherently two-dimensional type. In this sense, it turned out to be much more complicated than the disk accretion theory when, in some cases, the problem can be reduced, with adequate accuracy, to the system of ordinary differential equations or even to an algebraic one.

First of all, as was noted, the GR effects give rise to the second family of singular surfaces corresponding to the plasma accretion onto the black hole (Phinney, 1983). Indeed, from the expression for the poloidal four-velocity measured by the local nonrotating observers (ZAMO) it follows that (1.264)

$$u_p^2 = \frac{(E - \omega L)^2 - \alpha^2 L^2/\varpi^2 - \alpha^2\mu^2}{\alpha^2\mu^2}, \tag{3.83}$$

the physical four-velocity component u_p tends to infinity when approaching the black hole horizon. Therefore, the velocity of particles on the horizon ($\alpha = 0$) approaches that of light; hence, it must, inevitably, exceed the speed of any natural oscillations.

One should also point to some features of the black hole magnetosphere, which substantially reduce the potentialities of the analytical approach. First of all, for the central engine efficiency to be explained the angular velocity must be close to the maximum one, so (unlike, for example, the radio pulsar magnetosphere) the small parameter $\varepsilon_A^2 = (\Omega R/c)$ is not available in the problem. Physically, this implies that the outer light cylinder is to be in the vicinity of the central engine and, therefore, besides the inner surfaces, the problems of passing the inner singular surfaces are to be taken into account in the problem as well. In particular, that is why the magnetic field is not considered to be given. Indeed, as we saw in the example of radio pulsars, the electric currents flowing in the magnetosphere greatly distort the magnetic field in the vicinity of the light cylinder. On the other hand, as was mentioned, the black hole does not have its self-magnetic field. Consequently, when studying the black hole magnetosphere one cannot use the results of the numerous papers devoted to the accretion onto neutron (Ghosh and Lamb, 1979) and young (Bardou and Heyvaerts, 1996; Ustyugova et al., 2000; Agapitou and Papaloizou, 2000) stars, where the magnetic field of the central star is of vital importance.

In conclusion, we should mention another important property of the accretion onto fast-rotating black holes,which, on the contrary, greatly simplifies the problem studied here. The point is that due to the gravitomagnetic forces associated with the Lense–Thirring effect the accretion disk in the vicinity of the black hole is in the equatorial plane of the rotating black hole [so-called Bardeen–Petterson effect Thorne et al. (1986)]. This is the case even if at large distances the disk plane substantially deflects from the equatorial plane. Since the magnetic field in the vicinity of the black hole horizon must be generated exactly in the disk and, hence, exactly copy its geometry, it is logical to suppose that the symmetry axis of the regular poloidal magnetic field is to coincide with the black hole rotation axis. Therefore (at least, for rather cold disks) the black hole magnetosphere can be considered to be axisymmetric so that, in this sense, the magnetosphere structure of the black hole appears simpler than that of radio pulsars and accreting neutron stars, where the magnetic axis can be at an arbitrary angle to the rotation axis.

3.4.2 Exact Solutions

3.4.2.1 Slowly Rotating Black Hole with the Quasimonopole Magnetic Field

The first example of the exact solution to the force-free GS equation was constructed by Blandford and Znajek (1977) who, as a zero approximation, considered the magnetosphere of the nonrotating black hole with the split monopole magnetic field (see Fig. 3.6). This geometry, as was mentioned, can be realized in the presence of the thin accretion disk in which there are necessary electric currents. The monopole

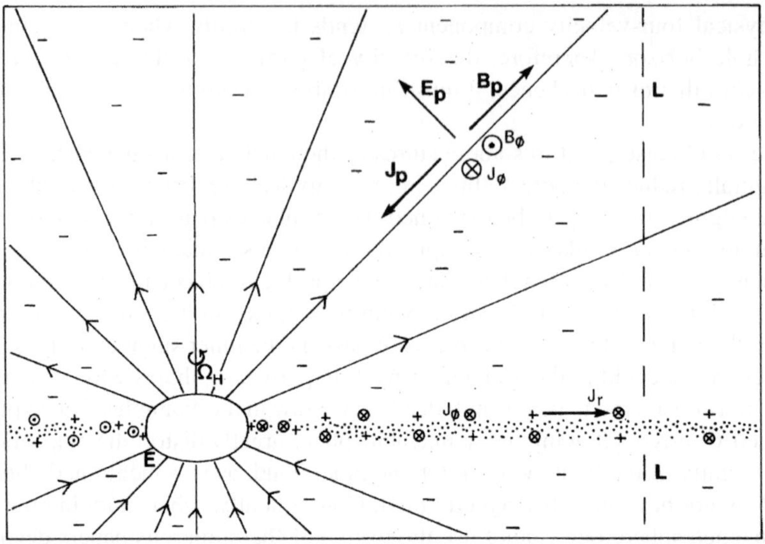

Fig. 3.6 The structure of electromagnetic fields in the case of the quasimonopole magnetic field in the neighborhood of the slowly rotating black hole (Blandford and Znajek, 1977). The currents flowing in the well-conducting disk located in the equatorial plane provide both the jump of the toroidal magnetic field and the closure of bulk currents in the upper and lower hemispheres. *Dashed line* indicates the light cylinder

magnetic field $\Psi = \Psi_0(1 - \cos\theta)$, $\theta < \pi/2$, as was noted, is the exact solution to the GS equation for the nonrotating black hole. Here Ψ_0 is again the total magnetic flux in the upper hemisphere.

We now consider the case of the slowly rotating black hole. As was formulated above, if the solution to Eq. (3.49) can be extended to the black hole horizon, the problem requires only one boundary condition. For example, it could be possible to fix the angular velocity $\Omega_F(\Psi)$ or the longitudinal current $I(\Psi)$. However, in Blandford and Znajek's paper, the inverse problem was actually studied when the values of the longitudinal current $I(\Psi)$ and the angular velocity $\Omega_F(\Psi)$ were determined for the concrete structure of the poloidal magnetic field at infinity. These values of $I(\Psi)$ and $\Omega_F(\Psi)$, for which the solution to the GS equation remains close to the monopole one, were found. As was noted, in a more complete statement (i.e., when the finite particle mass is taken into account) these two values can no longer be free and are to be found from the solution of the problem.

Thus, we again seek the solution to the GS equation for $\theta < \pi/2$ in the form

$$\Psi = \Psi_0[1 - \cos\theta + \varepsilon_3^2 f(r, \theta)], \tag{3.84}$$

where again, $\varepsilon_3 = a/M$. Since in the broad domain $r_g(\Omega_H r_g)^2 \ll r - r_g \ll 1/\Omega_H$ the GS equation up to the small value of ε_3^2 coincides with the vacuum equation, the problem of determining the disturbances of the vacuum magnetic field, in fact, is separated from the problem of determining the longitudinal current I and the angular velocity $\Omega_{\rm F}$. Indeed, as was shown above, at large distances the monopole magnetic field $\Psi = \Psi_0(1 - \cos\theta)$ remains the exact solution to the full GS equation (and, in particular, has no singularity on the light cylinder $\varpi_{\rm L} = c/\Omega_{\rm F}$) under the condition (2.231) $4\pi I(\Psi) = \Omega_{\rm F}(\Psi)(2\Psi - \Psi^2/\Psi_0)$, which for the monopole magnetic field looks like

$$4\pi I(\theta) = \Omega_{\rm F}(\theta)\Psi_0 \sin^2\theta. \tag{3.85}$$

On the other hand, if we assume that in the vicinity of the horizon the magnetic field is close to the monopole one, we can use the condition (3.52) which yields

$$4\pi I(\theta) = [\Omega_{\rm H} - \Omega_{\rm F}(\theta)]\Psi_0 \sin^2\theta. \tag{3.86}$$

As a result, combining relations (3.85) and (3.86), we find

$$\Omega_{\rm F} = \frac{\Omega_{\rm H}}{2}, \tag{3.87}$$

$$I(\Psi) = I_{\rm M} = \frac{\Omega_{\rm F}}{4\pi}\left(2\Psi - \frac{\Psi^2}{\Psi_0}\right). \tag{3.88}$$

Thus, for the problem studied the longitudinal current automatically appears equal to the critical one, and the angular velocity $\Omega_{\rm F}$ exactly corresponds to the case of the most efficient energy release. We emphasize once again that in the force-free approximation the values $\Omega_{\rm F}$ and I are not fixed and the obtained results correspond to only one of the infinite numbers of the possible force-free solutions.

Using relations (3.87) and (3.88), we obtain for the linearized GS equation (Blandford and Znajek, 1977)

$$r^2\frac{\partial}{\partial r}\left[\left(1 - \frac{r_{\rm g}}{r}\right)\frac{\partial f}{\partial r}\right] + \sin\theta\frac{\partial}{\partial\theta}\left(\frac{1}{\sin\theta}\frac{\partial f}{\partial\theta}\right) = -\frac{1}{2}\frac{r_{\rm g}}{r}\left(1 + \frac{r_{\rm g}}{r}\right)\sin^2\theta\cos\theta. \tag{3.89}$$

As is seen, it can again be solved by the separation of variables. We, however, should point to one important difference between Eq. (3.89) and the corresponding hydrodynamical equations discussed in detail in Sect.1.4. The point is that the linearized equation (3.89) has a singularity on the event horizon, more exactly, for $r = r_{\rm g}$. This is because in the method of successive approximations, in the zero order with respect to ε_3^2 the inner Alfvén surface coincides with the black hole horizon. Exactly the singularity on the critical surface led to the singularity of Eq. (3.89) for $r = r_{\rm g}$. In reality, the Alfvén surface $\alpha^2 = (\Omega_{\rm F} - \omega)^2\varpi^2$ is beyond the event horizon, where, in the general case, the regularity condition should be formulated. As to the "condition on the horizon" (3.52), it was used when defining the right-hand side of Eq. (3.89). Otherwise, the factor α^{-2} would appear here and Eq. (3.89) would have a double singularity for $\alpha^2 = 0$.

Indeed, all terms in the GS equation (3.49) of small order ε_3^2 (the terms proportional to $\Omega_{\rm F}^2$ and I^2) comprise the factor α^{-2}. Therefore, in the general case, on the right-hand side of the linearized equation (3.89) the same factor was to appear. And only upon substituting the value given by the condition (3.86) in the expression for the current I, the singularities in the numerator and the denominator are analytically cancelled and, hence, the right-hand side of Eq. (3.89) is finite for $r = r_{\rm g}$.

As a result, the solution to Eq. (3.89) for $\theta < \pi/2$ can be written as

$$\Psi(r,\theta) = \Psi_0\left[1 - \cos\theta + \varepsilon_3^2 g_2(r)\sin^2\theta\cos\theta\right], \tag{3.90}$$

and the equation for the radial function $g_2(r)$ has the form

$$r^2\frac{\rm d}{{\rm d}r}\left[\left(1-\frac{r_{\rm g}}{r}\right)\frac{{\rm d}g_2}{{\rm d}r}\right] - 6g_2 = -\frac{1}{2}\frac{r_{\rm g}}{r}\left(1+\frac{r_{\rm g}}{r}\right). \tag{3.91}$$

In the case discussed here, the boundary conditions for Eq. (3.91) are

1. the absence of the singularity for $r = r_{\rm g}$ (but not on the horizon!)

$$g_m(r) < \infty \qquad \text{for} \qquad r = r_{\rm g}, \tag{3.92}$$

2. the monopole field at infinity

$$g_2 \to 0 \qquad \text{for} \qquad r \to \infty. \tag{3.93}$$

Thus, the solution to Eq. (3.91) has the form (Blandford and Znajek, 1977)

$$\begin{aligned} g_2(r) = {} & 2\frac{r^2}{r_{\rm g}^2} - 3\frac{r}{r_{\rm g}} + \frac{7}{24} + \frac{1}{36}\frac{r_{\rm g}}{r} + \frac{1}{2}\frac{r^2}{r_{\rm g}^2}\left(1-\frac{r_{\rm g}}{r}\right)\left(4\frac{r}{r_{\rm g}}-3\right)\ln\left(1-\frac{r_{\rm g}}{r}\right) \\ & + \frac{1}{2}\left(\frac{r^2}{r_{\rm g}^2}\right)\left(4\frac{r}{r_{\rm g}}-3\right) I_1\left(\frac{r}{r_{\rm g}}\right) \\ & - \frac{1}{2}\left[4\frac{r^3}{r_{\rm g}^3} - \frac{r^2}{r_{\rm g}^2} - 3\frac{r}{r_{\rm g}} + \frac{r^2}{r_{\rm g}^2}\ln\frac{r}{r_{\rm g}}\right] \times \left[4 - \frac{r_{\rm g}}{r} - \frac{1}{6}\frac{r_{\rm g}^2}{r^2} + \left(4\frac{r}{r_{\rm g}}-3\right)\ln\left(1-\frac{r_{\rm g}}{r}\right)\right]. \end{aligned} \tag{3.94}$$

Here

$$I_1(x) = \int_x^\infty \frac{{\rm d}t}{t}\ln\left(\frac{t}{t-1}\right), \tag{3.95}$$

$g_2(r_{\rm g}) = (\pi^2/12 - 49/72)$, and $g_2(r) \to r_{\rm g}/8r$ for $r \to \infty$. The constructed solution is a generalization of the Michel solution (1973a) to the case of the slowly rotating black hole. Recall once more that it contains the electromagnetic energy flux as a consequence of the problem posed.

3.4.2.2 Slowly Rotating Black Hole in the Parabolic Magnetic Field

Another solution was constructed for the case in which the stream function Ψ corresponds to the parabolic solution (3.41) and (3.42) (see Fig. 3.7) (Blandford and Znajek, 1977). Indeed, as we saw, for the slow rotation $\Omega_{\rm F} r_{\rm g} \ll 1$ (and in the flat space), the parabolic solution (2.237) actually coincides with the vacuum solution (3.41). Therefore, for the slow rotation of the black hole, the magnetic flux function can again be specified as

$$\Psi_0^{(1)}(r,\theta) = \pi C X(r,\theta), \tag{3.96}$$

where now for $\theta < \pi/2$

$$X(r,\theta) = r(1-\cos\theta) + r_{\rm g}(1+\cos\theta)\,[1-\ln(1+\cos\theta)] - 2r_{\rm g}(1-\ln 2). \tag{3.97}$$

At large distances from the black hole but at small distances as compared to the light cylinder radius, it becomes the vacuum solution

$$\Psi(r,\theta) = \pi C r(1-\cos\theta). \tag{3.98}$$

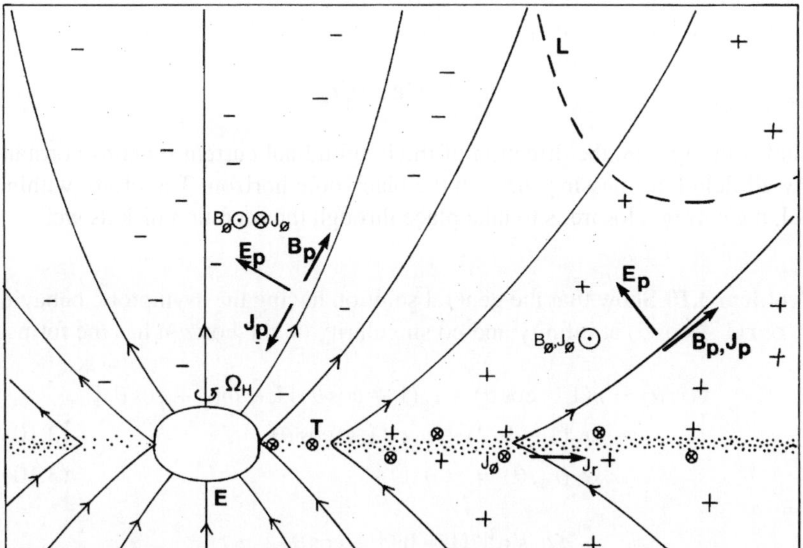

Fig. 3.7 Electromagnetic field structure in the case of the parabolic magnetic field in the neighborhood of the slowly rotating black hole (Blandford and Znajek, 1977). *Dashed line* indicates the light cylinder

However, as was shown in Sect. 2.6.1, under the condition $\Omega_{\rm F}(X)X \ll 1$, the magnetic flux (3.98) remains the solution to the GS equation beyond the light cylinder as well. As a result, we can use the procedure for constructing the solution described above. In particular, for the slow rotation, we can also seek the solution here as a small perturbation of the vacuum magnetic field.

In order for the longitudinal electric current I and the angular velocity $\Omega_{\rm F}$ to be determined, it is again enough to use algebraic relations (3.52) and (2.235). Supposing that the poloidal magnetic field at infinity remains the same as in the vacuum case, we obtain for the slow rotation (Blandford and Znajek, 1977)

$$4\pi I(\Psi) = 2\Omega_{\rm F}(\Psi)\Psi. \tag{3.99}$$

In turn, the "condition on the horizon" again has the form

$$4\pi I(\Psi) = [\Omega_{\rm H} - \Omega_{\rm F}(\Psi)] \sin\theta \frac{{\rm d}\Psi}{{\rm d}\theta}. \tag{3.100}$$

As a result, analyzing the system of two equations (3.99) and (3.100) again, we readily find that the profile of the angular velocity has the form (Blandford and Znajek, 1977)

$$\Omega_{\rm F}(r_{\rm g}, \theta) = \frac{\Omega_{\rm H}\sin^2\theta[1+\ln(1+\cos\theta)]}{4\ln 2 + \sin^2\theta + [\sin^2\theta - 2(1+\cos\theta)]\ln(1+\cos\theta)}, \tag{3.101}$$

the angular velocity varies from $\Omega_{\rm H}/2$ for $\Psi = 0$ to $\Omega_{\rm H}/(4\ln 2 + 1) \approx 0.265\,\Omega_{\rm H}$ on the last field line $\Psi = \Psi_*$ passing through the black hole horizon. In this case,

$$\Psi_* = 2\ln 2\pi C r_{\rm g}. \tag{3.102}$$

As shown in Fig. 3.8, the direction of the longitudinal current remains constant on nearly all field lines passing through the black hole horizon. Therefore, within this model, the current closure is to take place through the accretion disk as well.

Problem 3.10 Show that the general solution having the asymptotic behavior $\Psi \propto r(1-\cos\theta)$ at infinity and no singularity on the horizon has the form

$$X(r,\theta) = r(1-\cos\theta) + r_{\rm g}(1+\cos\theta)[1-\ln(1+\cos\theta)] - 2r_{\rm g}(1-\ln 2) + a_k r_{\rm g}(1-\cos\theta), \tag{3.103}$$

$$\Omega_{\rm F}(r_{\rm g},\theta) = \tag{3.104}$$

$$\frac{\Omega_{\rm H}\sin^2\theta[1+\ln(1+\cos\theta)+a_k]}{4\ln 2+\sin^2\theta+[\sin^2\theta-2(1+\cos\theta)]\ln(1+\cos\theta)+a_k(2-2\cos\theta+\sin^2\theta)},$$

where a_k is an arbitrary constant. In the general case, $\Psi_* = \pi C r_{\rm g}(2\ln 2 + a_k)$.

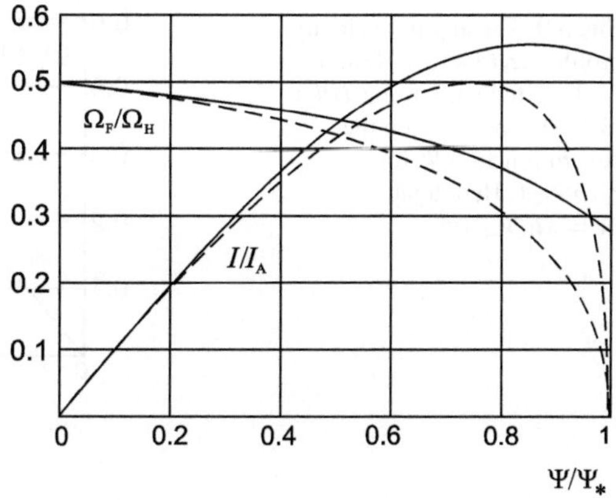

Fig. 3.8 The profiles of the angular velocity $\Omega_F(\Psi)$ (3.101) and the longitudinal current $I(\Psi)$ (3.99) in the vicinity of the horizon for the case of the parabolic magnetic field. *Dashed lines* indicate the solutions (3.103) and (3.104) for $a_k = -1$. Here $I_A = \Omega_H \Psi_* / 4\pi$

Problem 3.11 Show that for $a_k = -1$, the total current within the field lines passing through the black hole is zero (see Fig. 3.8).

3.4.2.3 Slowly Rotating Black Hole Located in the Center of the Limited Disk

Unlike the monopole magnetic field, the more realistic case in which the black hole is in the center of the well-conducting disk of the inner radius b (see Fig. 3.10) was studied in Beskin et al. (1992a) (see also Beskin, 1997). It was also assumed that the magnetic field lines do not penetrate the accretion disk. Then the stream function Ψ for the nonrotating black hole is described by formula (3.39). In particular, in the vicinity of the black hole the magnetic field is homogeneous

$$\Psi_v \approx \frac{1}{2}\Psi_0 \frac{r^2 \sin^2\theta}{b^2}, \tag{3.105}$$

and at large distances $r \gg b$ still remains a monopole one.

If we again consider the slowly rotating black hole and assume that for $a \ll M$ the perturbation of the magnetic field is small, we can find the values of the angular velocity $\Omega_F(\Psi)$ and the longitudinal current $I(\Psi)$, which satisfy this assumption. The only change here is that in the "boundary condition on the horizon" we should substitute the homogeneous magnetic field (3.105) for the monopole one. As a result, instead of relation (3.86), the condition (3.52) yields

$$4\pi I(\Psi) = 2\,\Psi\sqrt{1 - \Psi/\Psi_*}\,[\Omega_H - \Omega_F(\Psi)]. \tag{3.106}$$

Here $\Psi_* = 0.5\Psi_0 r_g^2/b^2$ is a magnetic flux passing through the black hole horizon. The condition at large distances is again defined by relation (3.85). Combining (3.85) and (3.106), we get for $\Psi_* \ll \Psi_0$ (Beskin et al., 1992a) (see Fig. 3.9)

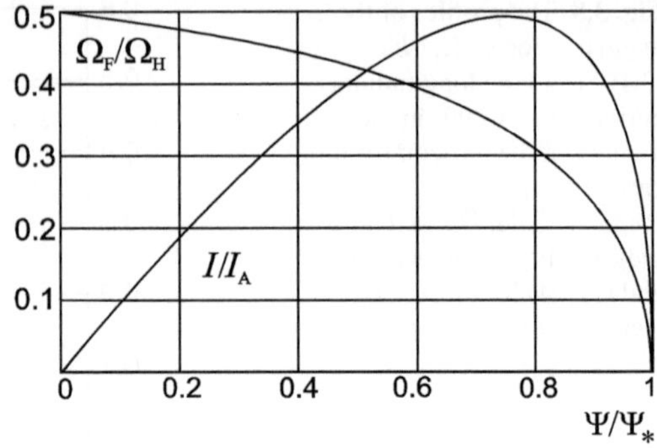

Fig. 3.9 The angular velocity profiles $\Omega_{\rm F}(\Psi)$ (3.107) and the longitudinal current $I(\Psi)$ (3.108) depending on the stream function Ψ for $r_{\rm g}/b \ll 1$. Here again $I_{\rm A} = \Omega_{\rm H}\Psi_*/4\pi$

$$\Omega_{\rm F}(\Psi) = \frac{\sqrt{1-\Psi/\Psi_*}}{1+\sqrt{1-\Psi/\Psi_*}}\,\Omega_{\rm H}, \tag{3.107}$$

$$4\pi\, I(\Psi) = 2\,\Psi\,\frac{\sqrt{1-\Psi/\Psi_*}}{1+\sqrt{1-\Psi/\Psi_*}}\,\Omega_{\rm H}. \tag{3.108}$$

Relations (3.107) and (3.108) specify the structure of the electric field and the longitudinal currents in the black hole magnetosphere. To determine the magnetic field perturbation we would have to solve the partial differential equation in which we fail to carry out the separation of variables. This is because of the complex form of the zero approximation, which is not as simple as in the case of the monopole magnetic field. It is important, however, that, as in the monopole magnetic field, the magnetic surface perturbation problem is separated from the problem of determining the angular velocity $\Omega_{\rm F}(\Psi)$ and the longitudinal current $I(\Psi)$.

We see that in the case studied, $I(\Psi_*) = 0$. Consequently, in the above model, within the magnetic field lines passing through the black hole horizon, an inverse bulk current inevitably occurs, and the total current flowing in the domain $\Psi < \Psi_*$ is thus automatically equal to zero. Accordingly, the condition $\Omega_{\rm F}(\Psi_*) = 0$ shows that the total electric charge in this region is zero as well. Further, as seen from Fig. 3.10, the longitudinal current region generates a "jet" at an angle of

$$\theta_{\rm jet} = \arccos\left(1 - \frac{\Psi_*}{\Psi_0}\right) \approx \frac{r_{\rm g}}{b}. \tag{3.109}$$

The energy transported by the electromagnetic field $W_{\rm tot} = (1/2\pi)\int \Omega_{\rm F}(\Psi)I(\Psi){\rm d}\Psi$ within the "jet" is close to the maximum possible one ($W_{\rm tot} = 0.489\, W_{\rm BZ}$), because the angular velocity (3.107) is close to $\Omega_{\rm H}/2$. Besides, Fig. 3.10 shows the form of the surfaces $\rho_{\rm GJ} = 0$, in the vicinity of which, as shown above, the effective plasma generation is possible. This surface is always located between the inner and outer Alfvén surfaces, which is necessary for the plasma to effectively fill the black hole magnetosphere.

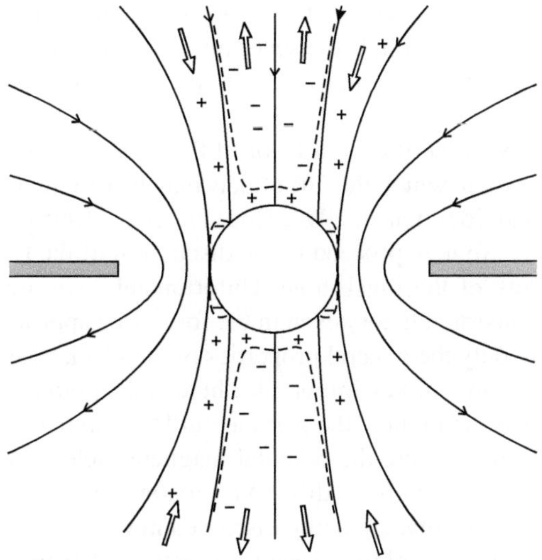

Fig. 3.10 Magnetosphere structure and the form of the surfaces $\rho_{GJ} = 0$ (*dashed lines*) for the black hole located in the center of the limited disk. The total electric current (*contour arrows*) within "jets" is zero

3.4.3 Magnetosphere Models

Thus, for the black hole magnetosphere it has now become possible to obtain the analytic solutions of the force-free equation for several model cases only. For example, the solutions with zero longitudinal currents have not been constructed yet. On the other hand, a number of numerical solutions (Fendt, 1997a; Fendt and Geiner, 2001; Komissarov, 2001; Uzdensky, 2004; Komissarov, 2004b) make it possible to point to some general properties typical of the black hole magnetosphere. One should note that for the strongly magnetized flows (when the energy density of the electromagnetic field is much higher than the particle energy density), the magnetosphere structure should be close to the force-free one. In particular, if the longitudinal current is much weaker than the critical one, with the particle mass taken into account, there is light surface in the magnetosphere beyond which the GS equation method cannot be applied. Therefore, the force-free solutions allow one to judge the general case corresponding to the full GS equation version.

We emphasize that when discussing the concrete astrophysical objects the problem of the substantial role of the GR effects is not always obvious. There are, for example, indications that the jets from young stellar objects are connected with the accretion disk rather than the central rotating star (Pelletier and Pudritz, 1992). If the jet generation mechanism in AGN is of the same character as that of the young stars, it is not improbable that the black hole plays only a passive role in the ejection process and the GR effects are not of vital importance for understanding the nature of their generation.

On the other hand, the GR effects may play an important role in most compact objects. First of all, this is evident from the hard spectra and the annihilation lines in the galactic X-ray sources that are candidates for solar mass black holes (Camenzind, 2007). These properties are never observed in X-ray sources in which the accretion is known to occur onto the neutron star rather than the black hole. These may also include the superluminal motion in quasars (Begelman et al., 1984), which are, possibly, connected with the relativistic electron–positron plasma outflow along with the weakly relativistic wind (Sol et al., 1989; Henri and Pelletier, 1991; Rees, 1997). All this is in favor of the existence of the extra particle acceleration mechanism in which the GR effects may be of vital importance (Okamoto, 1992; Hirotani and Tomimatsu, 1994; Horiuchi et al., 1995).

We now proceed to the discussion of the two-dimensional magnetosphere structure of the black hole. Unfortunately, we are still far from the construction of a consistent theory even in the force-free approximation. Therefore, we can now point to only the general properties of the black hole magnetosphere. It is clear, for example, that of key importance here is the poloidal magnetic field topology, because, as was mentioned, the particle outflow and the electromagnetic energy transport occur exactly along the poloidal magnetic field. There is no complete clarity just at this point. Therefore, different versions, in which the magnetic field structure differs appreciably, are being discussed now.

The simplest geometry is present if there was, initially, the quasihomogeneous magnetic field (see Fig. 3.11a). One can assume that even if the field is amplified in the accretion disk, its geometry, as a whole, retains and the magnetic field lines extend to infinity. As a result, the plasma can freely outflow beyond the magnetosphere. The exact solutions studied in the previous section do have this structure. On the other hand, it is obvious that as the magnetic field is generated in the accretion disk, there can exist another family of field lines passing through the black hole horizon and crossing the accretion disk in the vicinity of the black hole rather

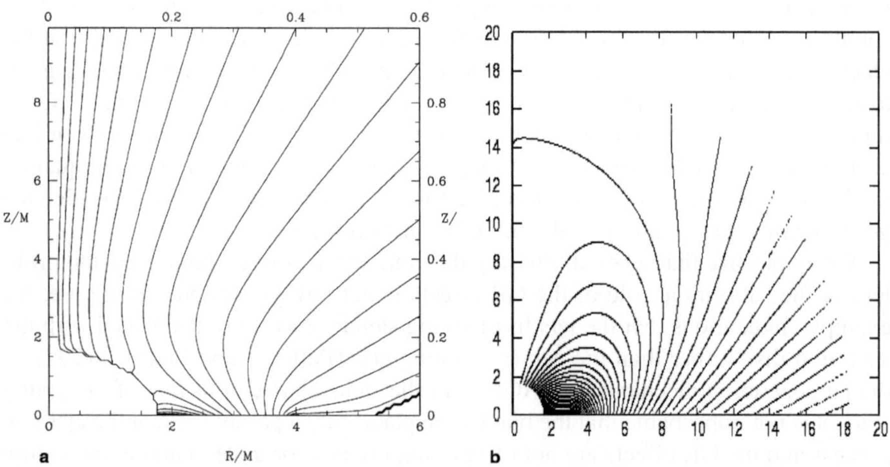

Fig. 3.11 Numerical solutions obtained for the force-free black hole magnetosphere. (**a**) The magnetosphere with quasimonopole magnetic field at infinity (Fendt, 1997a). (**b**) An example of the case in which magnetic field lines passing through the horizon do not extend to infinity (Uzdensky, 2005) [Reproduced by permission of the AAS, Fig. 5 from Uzdensky, D.A.: Force-free magnetosphere of an accretion disk-black hole system. II. Kerr geometry. ApJ **620**, 889–904 (2005)]

than extending to infinity (see Fig. 3.11b). In particular, these field lines no longer pass through the outer singular surfaces. This situation was actively studied in the numerical computations of the accretion onto the neutron star (Ghosh and Lamb, 1979; Bardou and Heyvaerts, 1996) and the young stars (Ustyugova et al., 2000; Fendt and Elstner, 2000). In the nonstationary regime, this case corresponds to a "magnetic tower" (Lynden-Bell, 2003) when the disk rotation generates the toroidal magnetic field. Recently, this geometry not infrequently has been discussed for the magnetosphere of rotating black holes as well (Uzdensky, 2004). As a result, the complete problem of the magnetosphere structure of the rotating black hole is to be constructed by taking into account the existence of both groups of magnetic field lines whose properties, as we saw, greatly differ from one another. Here, however, there are a lot of problems for the closed field lines connecting various parts of the accretion disk, or the accretion disk with the black hole horizon (Fendt, 1997a; Tomimatsu and Takahashi, 2001; Li, 2003; van Putten and Levinson, 2003), or the central star (Bardou and Heyvaerts, 1996).

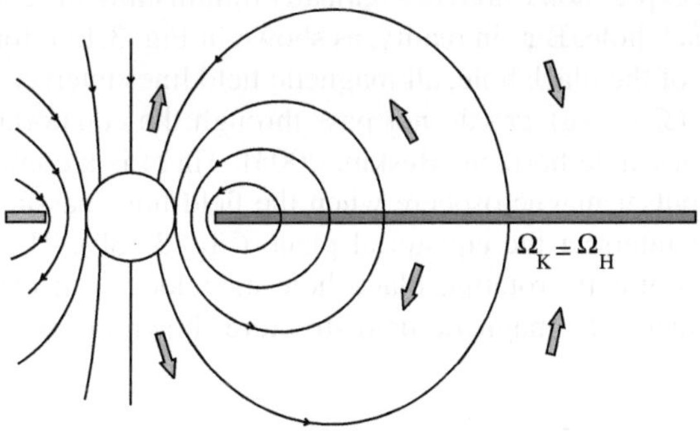

Fig. 3.12 Interaction between the black hole and the accretion disk due to magnetic field lines. In those regions, where the disk rotates with the angular velocity exceeding that of the black hole, the electromagnetic energy flux (*shaded arrows*) is from the disk to the black hole. In the domain, where the condition $\Omega_F < \Omega_K$ is satisfied, the black hole rotational energy is transported to the accretion disk

Indeed, let us consider, for example, the magnetic field lines passing through both the black hole horizon and the accretion disk (see Fig. 3.12). Clearly, for the well-conducting disk the angular velocity Ω_F is to be close to the Keplerian angular velocity Ω_K in the domain of intersection of the field line and the disk. The EMF is only due to an insignificant incompatibility between the angular velocities Ω_F and Ω_K. As a result, if the disk rotates with the angular velocity exceeding that of the black hole, the electromagnetic energy flux is always directed from the disk to the black hole. In this case, the angular velocity of the black hole increases.

However, if the black hole rotates rather fast, it is possible to satisfy the condition $\Omega_F < \Omega_H$ for the field lines passing through the accretion disk in the areas where the Keplerian velocity is rather small. This implies that the rotational energy of the black hole along these field lines is transported from the black hole to the accretion disk (Li, 2003). This, in turn, should result in an extra energy release which, in principle, can be detected. This heating mechanism of the inner regions of the accretion disk was used when discussing the extremely broad iron line in the bright Seyfert galaxy MCG–6–30–15 (Wilms et al., 2001). Accordingly, the reverse energy flux is to flow inside the "magnetic tower" as well (Sherwin and Lynden-Bell, 2007).

Some more comments on the black hole magnetosphere structure are necessary. It is clear, first of all, that for the zero longitudinal currents the solution cannot be extended not only beyond the outer (as was the case in the neutron star magnetosphere) but also inside the inner light cylinder. As a result, in the vicinity of the black hole, there must occur a region, in which the electric field is larger than the magnetic one (see, e.g., Komissarov, 2003). Unfortunately, with account taken of the GR effects, the separation of variables in the GS equation is impossible and, therefore, the above-discussed methods are not applied here.

Further, for the fast rotation of the black hole the Wald solution (3.30) and (3.32) for the vacuum magnetosphere demonstrates the magnetic field expulsion into the ergosphere (Thorne et al., 1986). This could result in the additional factor $(1 - a^2/M^2)$ in expression (3.68) dramatically diminishing the energy losses for the fast-rotating black hole. But, in reality, as shown in Fig. 3.13a, for the plasma-filled magnetosphere of the black hole, all magnetic field lines intersecting the inner light cylinder $\alpha^2 = (\Omega_{\rm F} - \omega)^2\varpi^2$ do not pass through the equatorial plane but are to intersect the black hole horizon (Beskin, 2003). There is a complete analogy here with the radio pulsar magnetosphere when the field lines passing beyond the light cylinder do not intersect the equatorial plane (Fig. 3.13b). Therefore, the energy release for the critically rotating black hole coincides with expression (3.65) in order of magnitude. The magnetic field structure shown in Fig. 3.13a was recently

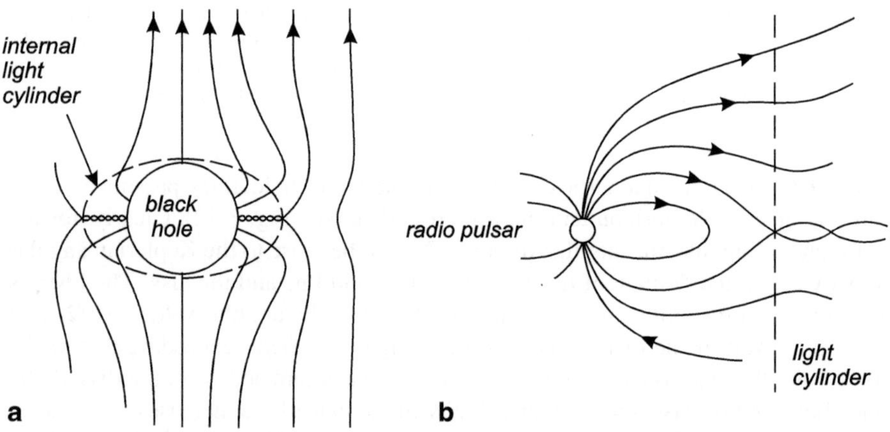

Fig. 3.13 **a** The magnetosphere structure of the fast-rotating black hole submerged in the external homogeneous magnetic field. Magnetic field lines passing through the inner "light cylinder" do not intersect the equatorial plane but bend to the black hole. **b** Radio pulsar magnetosphere structure. The field lines beyond the light cylinder do not cross the equatorial plane but extend to infinity (Beskin, 2003)

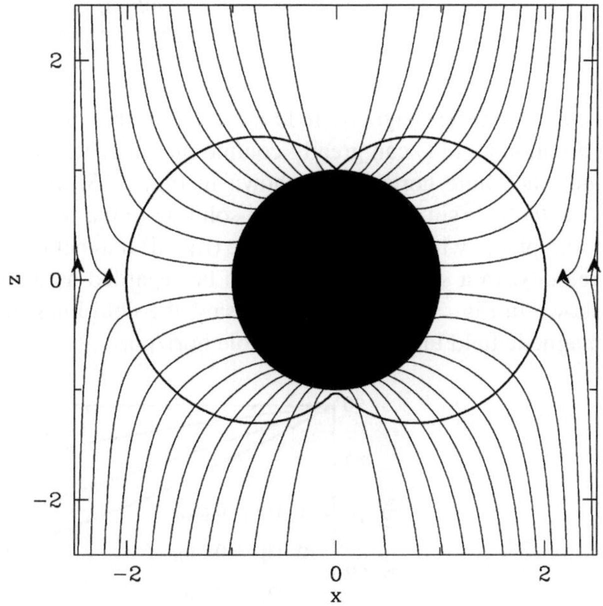

Fig. 3.14 The structure of the force-free magnetosphere of the fast-rotating black hole submerged in the external homogeneous magnetic field (Komissarov and McKinney, 2007). It contains the equatorial current sheet inside the ergosphere

obtained numerically (Komissarov, 2005; Komissarov and McKinney, 2007) (see Fig. 3.14).

Finally, it is important that the GS equation, as was noted, holds true for the black hole horizon only if the electric field remains smaller than the magnetic one. On the other hand, relation (3.52) shows that on the horizon itself the electric field is equal to the magnetic one. Therefore, one can use the force-free approach up to the black hole horizon if the following condition is satisfied (Beskin and Kuznetsova, 2000a):

$$\frac{\mathrm{d}}{\mathrm{d}r}(B^2 - E^2)\bigg|_{r=r_\mathrm{g}} > 0. \tag{3.110}$$

Using the definitions of the electric and magnetic fields, we can rewrite the condition (3.110) as (Hirotani et al., 1992)

$$-2\frac{\cos\theta}{\sin\theta}P_\mathrm{J} - 4\frac{a^2}{\rho_\mathrm{g}^2}\sin\theta\cos\theta\, P_\mathrm{J} + 2\frac{\partial P}{\partial\theta} - 2\left(1 - \frac{M}{r_\mathrm{g}}\right)P_\mathrm{J}^2$$
$$+\frac{r_\mathrm{g}}{(\Omega_\mathrm{H} - \Omega_\mathrm{F})^2\varpi_\mathrm{g}^2}\left(\frac{\partial\alpha^2}{\partial r}\right)_{r_\mathrm{g}} - 2\frac{r_\mathrm{g}}{\Omega_\mathrm{H} - \Omega_\mathrm{F}}\left(\frac{\partial\omega}{\partial r}\right)_{r_\mathrm{g}}$$
$$-2\frac{r_\mathrm{g}}{\varpi_\mathrm{g}}\left(\frac{\partial\varpi}{\partial r}\right)_{r_\mathrm{g}} + 2\frac{r_\mathrm{g}^2}{\rho_\mathrm{g}^2} > 0, \tag{3.111}$$

where the parameter

$$P_{\mathrm{J}} = -r_{\mathrm{g}} \frac{\partial \Psi / \partial r|_{r=r_{\mathrm{g}}}}{\partial \Psi / \partial \theta|_{r=r_{\mathrm{g}}}} \tag{3.112}$$

depends on the magnetic field structure in the vicinity of the horizon. Hence, the condition (3.111) can greatly confine the magnetic field structure in the vicinity of the black hole horizon. As shown in Fig. 3.15, the homogeneous magnetic field $\Psi(r, \theta) \approx r^2 \sin^2 \theta$ cannot be the solution to the force-free equation near the equatorial plane, whereas the monopole (radial) magnetic field can occur at any angular velocity. In a sense, this result can be regarded as the confirmation of the pattern shown in Fig. 3.14, where, as we saw, the field lines of the external homogeneous magnetic field turn to the black hole horizon.

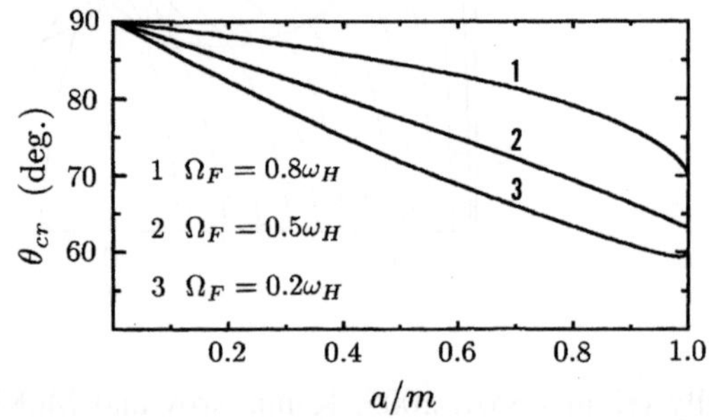

Fig. 3.15 The critical polar angle θ_{cr} on the black hole horizon within which the homogeneous magnetic field can be the solution to the GS equation in the vicinity of the black hole (Hirotani et al., 1992) (Reproduced by permission of the AAS, Fig. 1 from Hirotani, K., Takahashi, M., Nitta, S., Tomimatsu, A.: Accretion in a Kerr black hole magnetosphere – Energy and angular momentum transport between the magnetic field and the matter. ApJ **386**, 455–463 (1992))

On the other hand, one should remember that, with the finite particle mass taken into account, as in the hydrodynamical case, the hyperbolic domain of the full GS equation inevitably occurs, which, actually, cannot be described by the force-free approximation. Therefore, it would make no sense to analyze in detail the behavior of the solution to the force-free equation (3.49) in the vicinity of the horizon. Moreover, it is easy to show that attempts to describe the magnetic field by the force-free approximation can lead to an incorrect result. The appropriate example will be given in Sect. 4.4.6.

3.5 Conclusion

To sum up, we emphasize that all the above examples are certainly too simplified. Unfortunately, they are so far the only exact solutions describing the inner regions of the rotating black hole magnetosphere. Nevertheless, the solutions obtained are capable of formulating several key statements that allow one to judge the basic properties of the central engine.

First of all, as we saw, in all the cases, in which the field lines extend to infinity, the angular velocity Ω_F proved close to half the angular velocity of the black hole. This implies that the energy release efficiency of the rotating black hole is close to a maximum one. Since, according to the papers dealing with the analysis of the secular evolution of accreting black holes (Moderski and Sikora, 1996; Moderski et al., 1998; Wang et al., 2002), their angular velocity can really be close to the limit value $a \sim M$, one can conclude that the BZ process can play an appreciable role in the energy release mechanism of the central engine.

On the other hand, the above exact solutions, as was noted, correspond to the simplest magnetic field topology when there are no field lines connecting the event horizon of the black hole with the accretion disk (a torus, etc.). And in the presence of these field lines there are new interesting possibilities. Finally, recall once again that the studied solutions corresponded to the force-free approximation in which the longitudinal current I is, in fact, a free variable. Therefore, here, within the force-free statement of the problem, the value of the current had to be chosen from some extra condition on the external load and the "surface conductivity on the horizon" (Thorne et al., 1986; Li, 1995). In the next chapter, we will see how this problem is solved by the full GS equation version, with the finite particle mass taken into account.

Chapter 4
Full MHD Version—General Properties

Abstract The full MHD version of the Grad–Shafranov equation combines a well-conducting hydrodynamical medium and an electromagnetic field. Within this version, we can rather simply describe both the transformation of the energy of the electromagnetic field into particles and the whole magnetic field structure. As a result, one of the main problems in theory—the construction of the current system and the determination of the energy losses—can be posed in a mathematically rigorous way. This chapter deals with the general properties of the full MHD version of the GS equation. As in the presence of the magnetic field, as the number of normal modes increases, the structure of the GS equation becomes much more complicated than that in the hydrodynamical case; the properties of the critical surfaces are to be described in detail. This makes it possible to formulate some important theorems on the general properties in the magnetosphere of compact objects. In particular, it is shown that, within the full MHD approach, the electric current circulating in the magnetosphere is no longer a free parameter but is to be determined from the critical conditions on the singular surfaces. Asymptotic behavior of the flows at infinity and in the vicinity of the black hole horizon is considered in detail as well.

4.1 Physical Introduction—Magnetohydrodynamic Waves

Before formulating the basic equations related to the full version of the GS equation, let us recall the basic properties of MHD waves that can propagate in the magnetized plasma. Indeed, as was demonstrated in Chap. 1, the structure of the hydrodynamical GS equation greatly depends on the location of the sonic surface connected with the ordinary sound—the only normal mode capable of propagating in the non-magnetized medium. In the presence of the magnetic field, as the number of normal modes increases, the GS equation structure becomes much more complicated. In this section, we consider only the nonrelativistic case for simplicity, the corresponding general formulae are given below.

Within ideal one-fluid magnetohydrodynamics, we deal with eight unknowns—two thermodynamic functions and two vector quantities, viz., the velocity $\mathbf{v}$ and the magnetic field $\mathbf{B}$. All other characteristics, i.e., the electric field $\mathbf{E} = -\mathbf{v} \times \mathbf{B}/c$, the charge density $\varrho_{\mathrm{e}} = \nabla \cdot \mathbf{E}/4\pi$, etc., can be expressed in terms of these quantities.

V.S. Beskin, *MHD Flows in Compact Astrophysical Objects*, Astronomy and Astrophysics Library, DOI 10.1007/978-3-642-01290-7_5,
© Springer-Verlag Berlin Heidelberg 2010

However, when considering the wave processes there are, actually, only seven independent variables. Indeed, let us consider a wave propagating along the z-axis at an angle of θ with the external magnetic field $\mathbf{B}^{(0)}$. Choosing the corresponding coordinate system, we can always assume that the external homogeneous magnetic field lies in the xz-plane ($B_y^{(0)} = 0$) and the undisturbed velocity is zero ($\mathbf{v}^{(0)} = 0$). Further, we, as usual, consider the small perturbations in the form

$$\rho = \rho^{(0)} + \rho' \exp(-\mathrm{i}\omega t + \mathrm{i}kz), \tag{4.1}$$

$$s = s^{(0)} + s' \exp(-\mathrm{i}\omega t + \mathrm{i}kz), \tag{4.2}$$

$$\mathbf{v} = \mathbf{v}' \exp(-\mathrm{i}\omega t + \mathrm{i}kz), \tag{4.3}$$

$$\mathbf{B} = \mathbf{B}^{(0)} + \mathbf{b} \exp(-\mathrm{i}\omega t + \mathrm{i}kz), \tag{4.4}$$

etc., i.e., we assume that all the disturbed quantities depend only on the time t and the coordinate z as $\exp(-\mathrm{i}\omega t + \mathrm{i}kz)$, where k is a scalar wave vector (the index "0" is omitted in the following). As a result, Maxwell's equation $\nabla \cdot \mathbf{B} = 0$ gives the condition $kb_z = 0$. This implies that the longitudinal disturbance of the magnetic field can be eliminated by the proper choice of the coordinate system.

The other seven equations, i.e.,

- the continuity equation

$$\frac{\partial \rho'}{\partial t} + \rho \frac{\partial v_z'}{\partial z} = 0, \tag{4.5}$$

- the energy equation

$$\frac{\partial s'}{\partial t} = 0, \tag{4.6}$$

- three components of the Euler equation

$$\frac{\partial v_x'}{\partial t} - \frac{B_z}{4\pi\rho}\frac{\partial b_x}{\partial z} = 0, \tag{4.7}$$

$$\frac{\partial v_y'}{\partial t} - \frac{B_z}{4\pi\rho}\frac{\partial b_y}{\partial z} = 0, \tag{4.8}$$

$$\frac{\partial v_z'}{\partial t} + \frac{(\partial P/\partial \rho)_s}{\rho}\frac{\partial \rho'}{\partial z} + \frac{(\partial P/\partial s)_\rho}{\rho}\frac{\partial s'}{\partial z} + \frac{B_x}{4\pi\rho}\frac{\partial b_x}{\partial z} = 0, \tag{4.9}$$

- and two components of Maxwell's equation $c\nabla \times \mathbf{E} = -\partial \mathbf{B}/\partial t$

$$\frac{\partial b_x}{\partial t} - B_z\frac{\partial v_x'}{\partial z} + B_x\frac{\partial v_z'}{\partial z} = 0, \tag{4.10}$$

$$\frac{\partial b_y}{\partial t} - B_z\frac{\partial v_y'}{\partial z} = 0, \tag{4.11}$$

in which we used the conditions $\mathbf{v}^{(0)} = 0$ and $\mathbf{E} = -\mathbf{v} \times \mathbf{B}/c$, can be rewritten as one matrix equation (Akhiezer et al., 1975)

$$\frac{\partial d_i}{\partial t} + A_{ik}\frac{\partial d_k}{\partial z} = 0. \tag{4.12}$$

Here the vector d_i and the matrix of the coefficients A_{ik} have the form

$$d_i = \begin{pmatrix} \rho' \\ s' \\ v'_x \\ v'_y \\ v'_z \\ b_x \\ b_y \end{pmatrix}, \; A_{ik} = \begin{pmatrix} 0 & 0 & 0 & 0 & \rho & 0 & 0 \\ 0 & 0 & 0 & 0 & 0 & 0 & 0 \\ 0 & 0 & 0 & 0 & 0 & -B_z/4\pi\rho & 0 \\ 0 & 0 & 0 & 0 & 0 & 0 & -B_z/4\pi\rho \\ P_\rho/\rho & P_s/\rho & 0 & 0 & 0 & B_x/4\pi\rho & 0 \\ 0 & 0 & -B_z & 0 & B_x & 0 & 0 \\ 0 & 0 & 0 & -B_z & 0 & 0 & 0 \end{pmatrix}, \tag{4.13}$$

where $P_\rho = (\partial P/\partial \rho)_s$ and $P_s = (\partial P/\partial s)_\rho$.

Problem 4.1 Explain why the three-dimensional equation $c\nabla \times \mathbf{E} = -\partial \mathbf{B}/\partial t$ yields only two equations for the disturbed values available in d_i.

Substituting in (4.12) the wave dependence $\exp(-i\omega t + ikz)$, we get

$$V d_i = A_{ik} d_k, \tag{4.14}$$

where, by definition, $V = \omega/k$ is the phase velocity of the normal modes. Thus, the dispersion equation defining the dispersion $\omega = \omega(k, \theta)$ and the normal wave polarization is written as

$$\det(A_{ik} - V I_{ik}) = 0, \tag{4.15}$$

where I_{ik} is the unit matrix. Analyzing Eq. (4.15), we readily show that four different types of disturbances $V_{(i)}$, $i = 1, 2, 3, 4$, can propagate in the magnetized plasma. These normal waves are

- the fast and slow magnetosonic waves

$$V^2_{(1,2)} = \frac{1}{2}\left(V^2_\mathrm{A} + c^2_\mathrm{s}\right) \pm \frac{1}{2}\left[\left(V^2_\mathrm{A} + c^2_\mathrm{s}\right)^2 - 4V^2_\mathrm{A}c^2_\mathrm{s}\cos^2\theta\right]^{1/2}, \tag{4.16}$$

- the Alfvén wave

$$V_{(3)} = V_\mathrm{A}\cos\theta, \qquad \omega_{(3)} = \pm\frac{\mathbf{kB}}{\sqrt{4\pi\rho}}, \tag{4.17}$$

- and, finally, the entropy wave

$$V_{(4)} = 0. \tag{4.18}$$

Here

$$V_{\mathrm{A}} = \frac{B}{\sqrt{4\pi\rho}} \tag{4.19}$$

is the so-called Alfvén velocity and the value $c_{\mathrm{s}}^2 = (\partial P/\partial\rho)_s$ again corresponds to the velocity of sound. As we see, three first normal waves turn to be doubly degenerated, because for $\mathbf{v}^{(0)} = 0$ the properties of two waves propagating in opposite directions appear identical. In the general case, we have seven different values for the phase velocities. In particular, for $\theta = 0$

$$v_{(1,2,\ldots,7)} = v^{(0)} \pm V_{(1,2,3,4)}. \tag{4.20}$$

It is convenient to show the properties of the normal waves on phase and group polars which are the dependence of the phase $V_{(i)}$ and group $\mathbf{v}_{\mathrm{gr}}(\partial\omega/\partial\mathbf{k})_i$ velocities on the angle between the external magnetic field $\mathbf{B}$ (it is directed along the horizontal axis in all figures) and the wave vector $\mathbf{k}$ and group velocity $\mathbf{v}_{\mathrm{gr}}$, respectively (see Fig. 4.1). We see that the fast magnetosonic wave ($i = 1$) is a generalization of the ordinary sonic wave to the case of the nonzero external magnetic field. On all polars, it intersects the axes at the points with the coordinates

$$V_{(1)}(\theta = 0, \pi) = \max(V_{\mathrm{A}}, c_{\mathrm{s}}), \tag{4.21}$$

$$V_{(1)}(\theta = \pi/2, 3\pi/2) = \sqrt{V_{\mathrm{A}}^2 + c_{\mathrm{s}}^2}. \tag{4.22}$$

Therefore, for the small magnetic fields $V_{\mathrm{A}} \ll c_{\mathrm{s}}$, its properties do not essentially differ from those of ordinary sound. However, as we see, in the external magnetic field, besides a fast wave, the slow magnetosonic wave $i = 2$ occurs, the properties of which prove much more nontrivial. This is especially true for the group polar that has the form of a swallow tail and is characterized not only by the outer point of intersection with the X-axis

$$V_{(2)}(\theta = 0, \pi) = \min(V_{\mathrm{A}}, c_{\mathrm{s}}), \tag{4.23}$$

but also by the so-called cusp velocity

$$V_{\mathrm{cusp}} = \frac{V_{\mathrm{A}} c_{\mathrm{s}}}{\sqrt{V_{\mathrm{A}}^2 + c_{\mathrm{s}}^2}}, \tag{4.24}$$

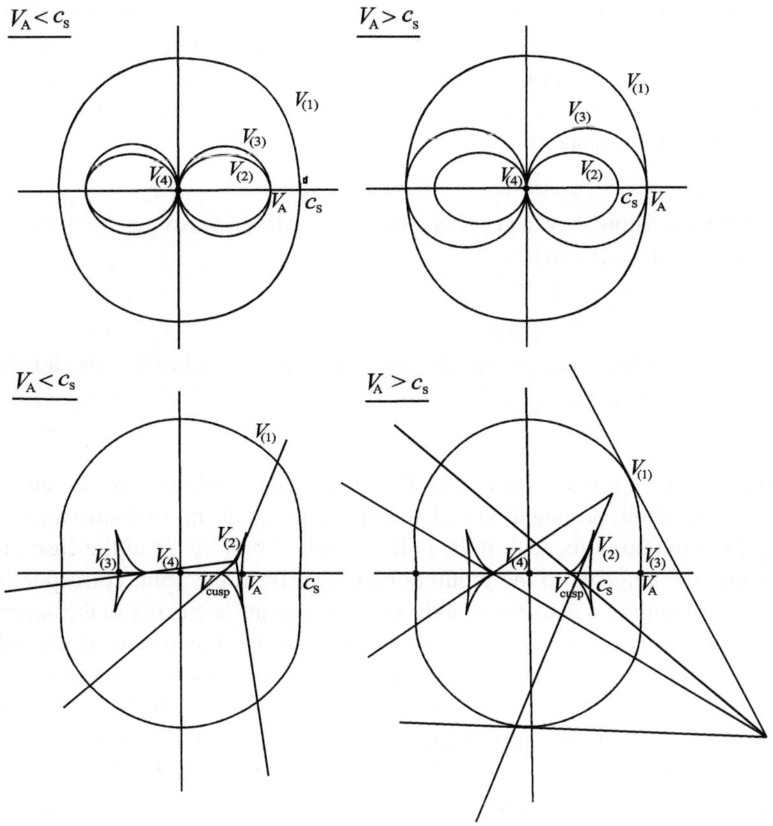

Fig. 4.1 Phase (*top*) and group (*bottom*) polars of magnetohydrodynamic waves for weak (*left*) and strong (*right*) magnetic fields. There are also shown Mach cones, the number of which depends on the relation between the velocity of a disturbing body and the proper velocities of the MHD waves. The magnetic field is directed along the horizontal axis

corresponding to the smallest velocity of this wave propagation. Both in the fast and in the slow waves the disturbances affect both the density and the velocity of matter and, therefore, it is generally difficult to excite them separately. Note also that at zero temperature the slow wave is absent.

Further, the properties of the Alfvén waves $i = 3$ are close to those of an ordinary string whose role here is played by the external magnetic field. As is evident from their relation (4.17), the group velocity of the Alfvén waves

$$\mathbf{v}_{\mathrm{gr}} = \pm \frac{\mathbf{B}}{\sqrt{4\pi\rho}} \tag{4.25}$$

is always directed along the external magnetic field. Therefore, it is shown on group polars by two points. Only the magnetic field b_y and velocity v'_y components perpendicular to the plane, which contains the vectors **B** and **k**, are disturbed in this wave. The thermodynamic values are not disturbed. It is not surprising, therefore, that the Alfvén wave exists at zero temperature. Finally, the entropy wave $i = 4$, in which there are oscillations of only the density and the entropy (temperature) at constant pressure, does not disturb the velocity and the magnetic field. In the system of plasma at rest it is a standing wave ($\omega_{(4)} = 0$).

Problem 4.2 Show that the eigenvector of the Alfvén wave has the form $\mathrm{d_i} = (0, 0, 0, -1, 0, 0, \sqrt{4\pi\rho})$.

Problem 4.3 Show that at any angles θ, the three conditions are satisfied: $V_{(2)} \leq V_{(3)} \leq V_{(1)}$, $V_{(1)} \geq c_{\mathrm{s}}$, $V_{(2)} \leq c_{\mathrm{s}}$.

In the following it is important that the group polars indicate the velocity of the energy (information) propagation and, therefore, are the fronts of disturbances propagating from the point located in the polar center. Therefore, as in the case of ordinary sound, the tangents to the group polar drawn from the point corresponding to the proper velocity of a disturbing body are Mach cones occurring in the supersonic motion. For the ordinary sound, these tangents can be drawn only if the velocity of body is larger than the sound velocity c_{s}. In the presence of the magnetic field, these tangents can be drawn at any arbitrarily small velocity of bodies. However, the number of Mach cones depends on the velocity of the disturbing body.

Indeed, as shown in Fig. 4.1, at large velocities, in the general case, there are two Mach cones, one of which is tangent to the fast wave and the second one to the slow wave. At smaller velocities, which, however, lie beyond the group polar corresponding to the slow wave, there is only one Mach cone. However, at velocities within the slow group polar, the number of cones is again equal to two. Thus, one can make the unexpected conclusion that for the nonzero magnetic field there is no minimum velocity of the disturbing body for which the Mach cones are absent.

One should stress that the case, in which the body velocity is directed along the magnetic field, is isolated. Here the Mach cone (and only one) exists only at rather large velocities $V > V_{(1)}$ and also in the domain $V_{\mathrm{cusp}} < V < V_{(2)}$. At velocities satisfying the conditions $V_{(2)} < V < V_{(1)}$ and $V < V_{\mathrm{cusp}}$, the Mach cones are absent. As we will see, this property is needed when specifying the structure of the GS equation.

4.2 Relativistic Flows in the Kerr Metric

4.2.1 Integrals of Motion

We consider the full MHD version of the GS equation. This implies that in the general energy–momentum conservation law $\nabla_\beta T^{\alpha\beta} = 0$ the energy–momentum tensor should be written as a sum of the hydrodynamical and electromagnetic parts. We are at once ready to write all equations in the Kerr metric (1.209), i.e., in the most general axisymmetric stationary metric, because all the necessary values were already given in Chaps. 2 and 3. In particular, the expressions for the electric and magnetic fields remain the same as in the force-free approximation

$$\mathbf{B} = \frac{\nabla\Psi \times \mathbf{e}_{\hat\varphi}}{2\pi\varpi} - \frac{2I}{\alpha\varpi}\mathbf{e}_{\hat\varphi}, \tag{4.26}$$

$$\mathbf{E} = -\frac{\Omega_{\rm F} - \omega}{2\pi\alpha}\nabla\Psi \tag{4.27}$$

(for the generally relativistic expressions we again take $c = 1$ and $G = 1$). It is important that, as shown above, for the plasma-filled magnetosphere the electric charge of the black hole is no longer a free parameter, but its value is so small that the Kerr metric disturbance can be disregarded. As to the flat space, the corresponding equations are obtained by the trivial substitution $\alpha \to 1$, $\omega \to 0$. Hereafter we do not write these relations separately.

We first show how for the axisymmetric stationary flows, in the general case, five integrals of motion on the magnetic surfaces occur. We assume, as before, that in the magnetosphere there is enough plasma to screen the longitudinal electric field. However, now instead of the relation $E_{\parallel} = 0$, we can write the more informative frozen-in condition

$$\mathbf{E} + \mathbf{v} \times \mathbf{B} = 0, \tag{4.28}$$

which fixes the electric field perpendicular to the magnetic field lines. Substituting the expressions for the electric and magnetic fields in Maxwell's equation (3.11), we again conclude that $\Omega_{\rm F}$ must be constant on the magnetic surfaces (Ferraro isorotation law):

$$\Omega_{\rm F} = \Omega_{\rm F}(\Psi). \tag{4.29}$$

Further, by virtue of the frozen-in condition (because in the stationary case the electric field $\mathbf{E}$ has no toroidal component), we can conclude that the poloidal component of the hydrodynamical velocity $\mathbf{v}$ and, hence, the four-velocity $\mathbf{u}$ must be parallel to the magnetic field. It is convenient to take the factor of proportionality in the form

$$\mathbf{u}_{\rm p} = \frac{\eta}{\alpha n}\mathbf{B}_{\rm p}, \tag{4.30}$$

where η is a scalar function that has the meaning of the particle-to-magnetic flux ratio. As a result, using the frozen-in condition again, we can write the four-velocity of matter $\mathbf{u}$ as

$$\mathbf{u} = \frac{\eta}{\alpha n}\mathbf{B} + \gamma(\Omega_{\rm F} - \omega)\frac{\varpi}{\alpha}\mathbf{e}_{\hat\varphi}, \tag{4.31}$$

where γ is the Lorentz factor of the medium (as measured by ZAMO). According to relation $\nabla \cdot (\eta \mathbf{B}) = 0$ resulting from the continuity equation (1.247) and Maxwell's equation (3.10), the function η must also be constant on the magnetic surfaces $\Psi = \text{const}$

$$\eta = \eta(\Psi). \tag{4.32}$$

Two integrals of motion are associated with the axisymmetric and stationary character of the flows studied, which leads to the conservation of the total energy E and the z-component of the angular momentum L

$$E = E(\Psi) = \frac{\Omega_{\rm F} I}{2\pi} + \mu\eta(\alpha\gamma + \omega\varpi u_{\hat{\varphi}}), \tag{4.33}$$

$$L = L(\Psi) = \frac{I}{2\pi} + \mu\eta\varpi u_{\hat{\varphi}}. \tag{4.34}$$

We see that both the energy and angular momentum fluxes consist of the contribution of the electromagnetic field and particles, the electromagnetic contribution fully coinciding with the force-free limit, and the hydrodynamical contribution differs from the hydrodynamical case only by the additional factor $\eta(\Psi)$. This factor results from the different normalization in the expressions for the energy and momentum fluxes, since expressions (1.251) and (1.252) in Chap. 1 correspond to the normalization to the flow of matter $\mathrm{d}\Phi$. It is convenient to define the losses of the energy $W_{\rm tot}$ and the angular momentum $K_{\rm tot}$ in the general MHD case by the relations

$$W_{\rm tot} = \int_0^{\Psi_{\max}} E(\Psi)\mathrm{d}\Psi, \tag{4.35}$$

$$K_{\rm tot} = \int_0^{\Psi_{\max}} L(\Psi)\mathrm{d}\Psi, \tag{4.36}$$

i.e., normalize to the unit magnetic flux. Finally, in the axisymmetric case, the isentropy condition yields $s = s(\Psi)$, so that the entropy on one particle $s(\Psi)$ is, in fact, the fifth integral of motion.

Problem 4.4 Find expression (4.31) for the hydrodynamical four-velocity **u**.

Problem 4.5 Show that

$$\eta(\Psi) = \frac{\mathrm{d}\Phi}{\mathrm{d}\Psi}. \tag{4.37}$$

We show now that for the given five integrals of motion $\Omega_{\rm F}(\Psi)$, $\eta(\Psi)$, $s(\Psi)$, $E(\Psi)$, and $L(\Psi)$ and also the poloidal magnetic field $\mathbf{B}_{\rm p}$, we can restore the toroidal magnetic field $B_{\hat{\varphi}}$ and all the other plasma parameters. To do this, we use the conservation laws (4.33) and (4.34), which, along with the φ-component of Eq. (4.31), yield (Camenzind, 1986)

$$\frac{I}{2\pi} = \frac{\alpha^2 L - (\Omega_F - \omega)\varpi^2(E - \omega L)}{\alpha^2 - (\Omega_F - \omega)^2\varpi^2 - \mathcal{M}^2}, \tag{4.38}$$

$$\gamma = \frac{1}{\alpha\mu\eta}\frac{\alpha^2(E - \Omega_F L) - \mathcal{M}^2(E - \omega L)}{\alpha^2 - (\Omega_F - \omega)^2\varpi^2 - \mathcal{M}^2}, \tag{4.39}$$

$$u_{\hat{\varphi}} = \frac{1}{\varpi\mu\eta}\frac{(E - \Omega_F L)(\Omega_F - \omega)\varpi^2 - L\mathcal{M}^2}{\alpha^2 - (\Omega_F - \omega)^2\varpi^2 - \mathcal{M}^2}, \tag{4.40}$$

where now

$$\mathcal{M}^2 = \frac{4\pi\eta^2\mu}{n}. \tag{4.41}$$

We readily see that the value $\mathcal{M}^2$ up to the coefficient α^2 is the square of the Mach number of the poloidal velocity u_p with respect to the poloidal component of the Alfvén velocity

$$u_{A,p} = \frac{B_p}{\sqrt{4\pi n\mu}}, \tag{4.42}$$

i.e., $\mathcal{M}^2 = \alpha^2 u_p^2/u_{Ap}^2$. In the following it is convenient to use exactly the value $\mathcal{M}^2$, since it remains finite on the black hole horizon.

Since the relativistic enthalpy μ can be expressed in terms of two other thermodynamic functions, it is convenient to write it as $\mu = \mu(n, s)$. Thus, the definition (4.41) allows us to express the concentration n (and, hence, the specific enthalpy μ) as the function η, s, and $\mathcal{M}^2$. The value $\nabla\mu$ by the general thermodynamic relation (1.25) and definition (4.41) must be defined from the relation (Beskin and Pariev, 1993)

$$\nabla\mu = \frac{c_s^2}{1 - c_s^2}\mu\left(2\frac{\nabla\eta}{\eta} - \frac{\nabla\mathcal{M}^2}{\mathcal{M}^2}\right) + \frac{1}{1 - c_s^2}\left[\frac{1}{n}\left(\frac{\partial P}{\partial s}\right)_n + T\right]\nabla s. \tag{4.43}$$

This implies that, along with five integrals of motion, the expressions for I, γ, and $u_{\hat{\varphi}}$ depend only on one additional value, i.e., the Mach number $\mathcal{M}$. To determine the Mach number $\mathcal{M}$ one should use the obvious relation $\gamma^2 - u_{\hat{\varphi}}^2 = u_p^2 + 1$, which, according to expressions (4.39) and (4.40), can be rewritten as relativistic Bernoulli's equation

$$\frac{K}{\varpi^2 A^2} = \frac{1}{64\pi^4}\frac{\mathcal{M}^4(\nabla\Psi)^2}{\varpi^2} + \alpha^2\eta^2\mu^2. \tag{4.44}$$

Here

$$A = \alpha^2 - (\Omega_{\rm F} - \omega)^2\varpi^2 - \mathcal{M}^2 \tag{4.45}$$

is the Alfvén factor and

$$K = \alpha^2\varpi^2(E - \Omega_{\rm F}L)^2\left[\alpha^2 - (\Omega_{\rm F} - \omega)^2\varpi^2 - 2\mathcal{M}^2\right] \\ +\mathcal{M}^4\left[\varpi^2(E - \omega L)^2 - \alpha^2 L^2\right]. \tag{4.46}$$

Relations (4.38), (4.39), (4.40), and (4.44) are algebraic relations permitting us to determine, though in implicit form, all the flow characteristics by the known function field $\mathbf{B}_{\rm p}$ (i.e., the known function Ψ) and the five integrals of motion. One should stress that at nonzero temperature they are extremely cumbersome, primarily, because of the necessity to solve Eq. (4.43). For the cold plasma ($s = 0$, i.e., for $\mu =$ const), Bernoulli's equation (4.44) becomes an algebraic four-order equation by the value $\mathcal{M}^2$. As shown below, this often helps one find asymptotic solutions. Equalities (4.38), (4.39), (4.40), and (4.44) were analyzed in a great number of papers, starting from the stellar (solar) wind (Weber and Davis, 1967; Mestel, 1968; Sakurai, 1990), where, certainly, their nonrelativistic limit was used, to the relativistic pulsar wind (Michel, 1969; Ardavan, 1976; Okamoto, 1978), the hydrodynamical and MHD accretion of matter onto the black holes (Camenzind, 1986; Punsly and Coroniti, 1990b; Takahashi et al., 1990; Chakrabarti, 1990). Below, we give several general assertions that can be derived directly from the analysis of the relations (4.38), (4.39), and (4.40).

- Within the Alfvén surface ($A \ll 1$), we have $I \approx 2\pi L$, i.e., the current I remains constant on the magnetic surfaces. Consequently, the current closure is impossible within the Alfvén surface.
- Accordingly, within the Alfvén surface

$$\gamma \approx \frac{\alpha(E - \Omega_{\rm F}L)}{\mu\eta} \approx \gamma_{\rm in}. \tag{4.47}$$

Hence, the particle acceleration within the Alfvén surface is also impossible.
- The toroidal velocity $v_{\hat{\varphi}} = u_{\hat{\varphi}}/\gamma$ within the Alfvén surface is

$$v_{\hat{\varphi}} \approx \Omega_{\rm F}\varpi. \tag{4.48}$$

Hence, the full plasma corotation is to occur here.
- Conversely, beyond the outer Alfvén surface we have

$$\frac{I}{2\pi} = \frac{\Omega_{\rm F}E}{\Omega_{\rm F}^2 + \mathcal{M}^2\varpi^{-2}}. \tag{4.49}$$

For the magnetically dominated flows, when the contribution $\mathcal{M}^2$ can be disregarded in the denominator, the current I, as was expected, still remains an integral of motion. Therefore, the current closure is possible for the weakly magnetized plasma only.

- Further, since $L \approx \Omega_{\rm F}\varpi^2(r_{\rm A})E$ (see the numerator in (4.38)), beyond the outer Alfvén surface the toroidal velocity behaves as $v_{\hat\varphi} \propto \varpi^{-1}$.
- Finally, in the vicinity of the horizon the physical component of the toroidal four-velocity remains finite, whereas the Lorentz factor behaves as $\gamma \propto \alpha^{-1}$. Thus, as in the hydrodynamical case, the radial component of the particle velocity (as measured by ZAMO) for $\alpha \to 0$ tends to the velocity of light and the other velocity components to zero.

4.2.2 Singular Surfaces

4.2.2.1 Alfvén Surface

Algebraic relations (4.38), (4.39), (4.40), and (4.44) make it possible to define the singular surfaces in the MHD flows studied, on which the poloidal velocities $v_{\rm p}$ become equal to the intrinsic velocities of the axisymmetric disturbances capable of propagating in plasma. The Alfvén surface A is defined from the vanishing condition of the denominator A (4.45) in algebraic relations (4.38), (4.39), and (4.40)

$$A = 0. \tag{4.50}$$

In the force-free limit $\mathcal{M}^2 \to 0$ the Alfvén surface coincides with the light cylinder. By definitions (4.41) and (4.50), we find that on the Alfvén surface

$$u_{\rm p}^2 = u_{\rm A,p}^2 \left[1 - \frac{(\Omega_{\rm F} - \omega)^2\varpi^2}{\alpha^2}\right], \tag{4.51}$$

which in the nonrelativistic limit naturally coincides with the condition $v_{\rm p} = V_A$ (4.17). As shown in Weber and Davis (1967), on the u^r–r-plane, it is a point of higher order than, for example, a saddle or a focus. On the other hand, it turned out that all trajectories with the positive energy square E^2 pass through it. This implies that there is no singularity in algebraic relations (4.38), (4.39), and (4.40) themselves, and the regularity condition (the zero equality of the numerators for the zero denominator) specifies the location of the Alfvén surface only. On the other hand, as will be shown, the GS equation itself has a singularity on the Alfvén surface.

4.2.2.2 Fast and Slow Magnetosonic Surfaces

The fast and slow magnetosonic surfaces F and S are most easily defined as singularities in the expression for the gradient of the Mach number $\mathcal{M}$. Indeed, using relations (4.44), (4.45), and (4.46), which can be rewritten as $(\nabla\Psi)^2 =$

$F(\mathcal{M}^2, E, L, \eta, \Omega_{\rm F}, \mu)$, where

$$F = \frac{64\pi^4}{\mathcal{M}^4}\frac{K}{A^2} - \frac{64\pi^4}{\mathcal{M}^4}\alpha^2\varpi^2\eta^2\mu^2, \tag{4.52}$$

we find

$$\nabla_k\mathcal{M}^2 = \frac{N_k}{D}, \tag{4.53}$$

where now

$$N_k = -\frac{A}{(\nabla\Psi)^2}\nabla^i\Psi\cdot\nabla_i\nabla_k\Psi + \frac{A}{2}\frac{\nabla'_k F}{(\nabla\Psi)^2}. \tag{4.54}$$

Here the operator ∇'_k again acts on all values except for $\mathcal{M}^2$. The denominator D can be rewritten as

$$D = \frac{A}{\mathcal{M}^2} + \frac{\alpha^2}{\mathcal{M}^2}\frac{B_\varphi^2}{B_{\rm p}^2} - \frac{1}{u_{\rm p}^2}\frac{A}{\mathcal{M}^2}\frac{c_{\rm s}^2}{1-c_{\rm s}^2}, \tag{4.55}$$

where $c_{\rm s}^2 = 1/\mu(\partial P/\partial n)_s$ is the velocity of sound. We emphasize that when differentiating $\nabla'_a F$ in (4.54) we should use relation (4.43), since expression (4.52) for F comprises μ. The zero condition of the denominator in expression (4.53),

$$D = 0, \tag{4.56}$$

specifies the fast and slow singular surfaces. Indeed, using definition (4.41), we find that $D = 0$ for

$$(u_{\rm p})^2_{1,2} = \frac{1}{2}\left(\mathcal{W}u^2_{\rm A,p} + \frac{c_{\rm s}^2}{1-c_{\rm s}^2}\right)$$
$$\pm\frac{1}{2}\left[\left(\mathcal{W}u^2_{\rm A,p} + \frac{c_{\rm s}^2}{1-c_{\rm s}^2}\right)^2 - 4\left(1 - \frac{(\Omega_{\rm F}-\omega)^2\varpi^2}{\alpha^2}\right)\frac{c_{\rm s}^2u^2_{\rm A,p}}{1-c_{\rm s}^2}\right]^{1/2}, \tag{4.57}$$

where

$$\mathcal{W} = 1 - \frac{(\Omega_{\rm F}-\omega)^2\varpi^2}{\alpha^2} + \frac{B_\varphi^2}{B_{\rm p}^2}, \tag{4.58}$$

and $u_{\rm A,p} = B_{\rm p}/\sqrt{4\pi n\mu}$, i.e., is again defined by the poloidal component of the magnetic field. In the nonrelativistic approximation, relation (4.57) naturally becomes expression (4.16)

$$(v_{\rm p})^2_{1,2} = \frac{1}{2}\left(V_{\rm A}^2 + c_{\rm s}^2\right) \pm \frac{1}{2}\sqrt{\left(V_{\rm A}^2 + c_{\rm s}^2\right)^2 - 4c_{\rm s}^2 V_{\rm A}^2 \cos^2\theta}, \tag{4.59}$$

where again $V_{\rm A} = B/\sqrt{4\pi\rho}$ (ρ is the mass density) and $\cos\theta = B_{\rm p}/B$. Otherwise, a singularity occurs when the poloidal velocity becomes equal to that of the fast or slow magnetosonic wave. For the cold plasma $c_{\rm s} = 0$, the slow magnetosonic velocity is zero (Akhiezer et al., 1975), so that, in this case, the slow magnetosonic surface is absent. The fast and slow surfaces, unlike the Alfvén one, are saddle points, i.e., the transonic solutions exist only for a certain connection between the integrals of motion. They result from the regularity condition

$$N_r = 0; \quad N_\theta = 0 \tag{4.60}$$

for $D = 0$.

The regularity conditions (4.56) and (4.60), as we already saw in the example of the hydrodynamical flows, play the key role in the construction of analytic solutions to the GS equation. Therefore, we give the basic properties of the denominator D specifying, in particular, the locations of the fast and slow magnetosonic surfaces.

- As seen from (4.55), for the zero toroidal magnetic field $B_{\hat\varphi} = 0$, the condition $D = 0$ coincides with the Alfvén condition $A = 0$. This property immediately follows from the phase polar structure shown in Fig. 4.1. Indeed, as was noted, the axisymmetry condition permits all disturbances to propagate in the poloidal plane only. Hence, the poloidal velocity, according to (4.31), must be parallel to the magnetic field. When a wave propagates along the magnetic field, the Alfvén wave coincides with one of the sonic polars. Therefore, the important conclusion is that on the rotation axis, where the toroidal magnetic field is obviously absent, the Alfvén and one of the sonic surfaces must be tangent to one another.
- Within the hydrodynamical limit $\mathcal{M} \to \infty$ ($\eta \to \infty$), we turn to expression $D = -1 + c_{\rm s}^2/u_{\rm p}^2(1 - c_{\rm s}^2)$ (1.262) and for the small velocities to nonrelativistic expression $D = -1 + c_{\rm s}^2/v_{\rm p}^2$ (1.112).
- As shown in the next section, $D = -1$ on the black hole horizon. This implies that the external subsonic space $D > 0$ is automatically separated from the event horizon by the hyperbolic domain $-1 < D < 0$. Below, this most important property is considered in detail.
- The denominator D can also be rewritten as

$$D = -1 + \frac{\alpha^2(\mathbf{B}^2 - \mathbf{E}^2)}{\mathcal{M}^2 B_{\rm p}^2} - \frac{1}{u_{\rm p}^2}\frac{A}{\mathcal{M}^2}\frac{c_{\rm s}^2}{1 - c_{\rm s}^2}. \tag{4.61}$$

Hence, in any case at zero temperatures, the flow crosses the fast magnetosonic surface in the field of application of the approach $|\mathbf{B}| > |\mathbf{E}|$ studied.

Problem 4.6 Show that on the fast magnetosonic surface the condition $|\mathbf{B}| > |\mathbf{E}|$ is satisfied at nonzero temperatures as well.

4.2.2.3 Cusp Surface

As we will see, the GS equation has another singular surface—the cusp surface C defined from the condition $D = -1$. This surface is associated with the singularity occurring on the group polar as a cusp point for a slow magnetosonic wave (see Fig. 4.1). As a result, we obtain for the corresponding velocity

$$u^2_{\text{cusp, p}} = u^2_{\text{A,p}} \frac{\left[\alpha^2 - (\Omega_{\text{F}} - \omega)^2 \varpi^2\right] c_{\text{s}}^2}{\left[\alpha^2 - (\Omega_{\text{F}} - \omega)^2 \varpi^2 + \alpha^2 B_{\varphi}^2 / B_{\text{p}}^2\right] u^2_{\text{A,p}}(1 - c_{\text{s}}^2) + \alpha^2 c_{\text{s}}^2}. \tag{4.62}$$

Within the nonrelativistic limit, this expression coincides with (4.24). Indeed, as shown in Fig. 4.1, the cusp singularity propagates along the magnetic field. The singularity in the axisymmetric case, as was seen, occurs when the poloidal velocities and the velocities of normal modes are equal, with which the nonrelativistic limit of Eq. (4.62) is associated:

$$u_{\text{cusp,p}} = \frac{u_{\text{A,p}} c_{\text{s}}}{\sqrt{u_{\text{A}}^2 + c_{\text{s}}^2}}. \tag{4.63}$$

The existence of the cusp surface, in the general case, does not provide for additional regularity conditions.

4.2.2.4 Light Cylinder and Light Surface

The light cylinder R_{L} is a surface on which the electric field $|\mathbf{E}|$ is equal to the poloidal component of the magnetic field $|\mathbf{B}_{\text{p}}|$. According to (4.45) and (4.50), far from gravitating bodies it is located at a distance of $\varpi = R_{\text{L}} = 1/\Omega_{\text{F}}$. In the case of the black hole magnetosphere, there is another "light cylinder" located on the surface $\alpha = |\Omega_{\text{F}} - \Omega_{\text{H}}|\varpi$. No additional regularity condition is to be specified on the light cylinder. Note also that since $\mathcal{M}^2 > 0$, the Alfvén singularity does not coincide with the light cylinder. Both for the outer and inner families of critical surfaces, the flow first crosses the Alfvén surface and then the light cylinder.

Finally, as was noted, the characteristic surface is the light surface S_{L} on which the electric field $|\mathbf{E}|$ is equal to the magnetic field $|\mathbf{B}|$. The light surface is similar to a limiting line; in ordinary hydrodynamics it specifies the natural boundary of a continuous flow.

Thus, the plasma with nonzero temperature in its motion first crosses the cusp and the slow magnetosonic surfaces, then the Alfvén surface, the light cylinder, afterward the fast magnetosonic surface. The light surface (if any) is located even at larger distances. On the other hand, the GR effects give rise to the second family of singular surfaces in the vicinity of the black hole horizon. Of great importance is the fact that under the condition of the energy release by the rotating black hole the outer Alfvén surface (through which all trajectories pass) corresponds to the values of $u^r > 0$, i.e., the values of the outflowing plasma, whereas the inner Alfvén surface corresponds to the value of $u^r < 0$, i.e., the value of the accretion (Takahashi et al., 1990) (this theorem is proved below). But this contradicts the assumption of the constancy of the function η on the field lines $\Psi = \text{const}$. Consequently, the plasma flow in the black hole magnetosphere (to be exact, on the field lines passing through the horizon) cannot be continuous and we must consider the plasma generation regions in which the GS equation is not applied. Clearly, the above discussion is not pertinent to the hydrodynamical accretion regime for which the Alfvén surface is absent (the limiting process from the MHD regime to hydrodynamics is discussed in detail in Takahashi (2002)).

4.2.3 Grad–Shafranov Equation

We now proceed to the discussion of the GS equation—the equilibrium equation for the magnetic surfaces. Having written the poloidal component of the energy–momentum conservation law (which requires a lot of efforts, because the total number of terms in the case of the Kerr metric is about a hundred), we again see that this vector equation reduces to a scalar second-order equation multiplied by $\nabla_k\Psi$. In compact form, it can be written as (Nitta et al., 1991; Beskin and Pariev, 1993)

$$\frac{1}{\alpha}\nabla_k\left\{\frac{1}{\alpha\varpi^2}[\alpha^2-(\Omega_{\rm F}-\omega)^2\varpi^2-\mathcal{M}^2]\nabla^k\Psi\right\}+\frac{\Omega_{\rm F}-\omega}{\alpha^2}(\nabla\Psi)^2\frac{{\rm d}\Omega_{\rm F}}{{\rm d}\Psi}$$
$$+\frac{64\pi^4}{\alpha^2\varpi^2}\frac{1}{2\mathcal{M}^2}\frac{\partial}{\partial\Psi}\left(\frac{G}{A}\right)-16\pi^3\mu n\frac{1}{\eta}\frac{{\rm d}\eta}{{\rm d}\Psi}-16\pi^3 nT\frac{{\rm d}s}{{\rm d}\Psi}=0, \quad (4.64)$$

where

$$G=\alpha^2\varpi^2(E-\Omega_{\rm F}L)^2+\alpha^2\mathcal{M}^2L^2-\mathcal{M}^2\varpi^2(E-\omega L)^2. \quad (4.65)$$

Substituting in (4.64) the term $\nabla_k\mathcal{M}^2$, by definitions (4.53) and (4.54), we finally get

$$A\left[\frac{1}{\alpha}\nabla_k\left(\frac{1}{\alpha\varpi^2}\nabla^k\Psi\right)+\frac{1}{\alpha^2\varpi^2(\nabla\Psi)^2}\frac{\nabla^i\Psi\cdot\nabla^k\Psi\cdot\nabla_i\nabla_k\Psi}{D}\right]$$
$$+\frac{1}{\alpha^2\varpi^2}\nabla_k' A\cdot\nabla^k\Psi-\frac{A}{\alpha^2\varpi^2(\nabla\Psi)^2}\frac{1}{2D}\nabla_k' F\cdot\nabla^k\Psi$$
$$+\frac{\Omega_{\rm F}-\omega}{\alpha^2}(\nabla\Psi)^2\frac{{\rm d}\Omega_{\rm F}}{{\rm d}\Psi}+\frac{64\pi^4}{\alpha^2\varpi^2}\frac{1}{2\mathcal{M}^2}\frac{\partial}{\partial\Psi}\left(\frac{G}{A}\right)$$
$$-16\pi^3\mu n\frac{1}{\eta}\frac{{\rm d}\eta}{{\rm d}\Psi}-16\pi^3 nT\frac{{\rm d}s}{{\rm d}\Psi}=0, \qquad (4.66)$$

where the gradient ∇_k' again acts on all values except for $\mathcal{M}^2$ and the derivative $\partial/\partial\Psi$ on the integrals of motion only. Formula (4.66) defines, in general form, the GS equation describing the equilibrium of the magnetic surfaces. Recall once again that Eq. (4.66) comprises only the stream function Ψ and five integrals of motion. Indeed, the thermodynamic values by the equation of state and definitions (4.41) and (4.43) can be expressed in terms of the function $s(\Psi)$, the value $\eta(\Psi)$, and the square of the Mach number $\mathcal{M}^2$. The value $\mathcal{M}^2$ itself, because of Bernoulli's equation (4.44), is expressed, though implicitly, in terms of the gradient $(\nabla\Psi)^2$ and five integrals of motion. The physical root of (4.44), certainly, is to be chosen. As to the classical GS equation

$$r^2\sin^2\theta\nabla_k\left(\frac{1}{r^2\sin^2\theta}\nabla^k\Psi\right)+16\pi^2 I\frac{{\rm d}I}{{\rm d}\Psi}+16\pi^3 r^2\sin^2\theta\frac{{\rm d}P}{{\rm d}\Psi}=0, \qquad (4.67)$$

i.e., the equation describing, in the nonrelativistic case ($\alpha=1$, $\omega=0$), the static ($\mathbf{v}=0$, i.e., $\gamma=1$) axisymmetric configurations is derived from (4.66) in the limit $\Omega_{\rm F}\to 0$ (which corresponds to the infinitely distant light cylinder $R_{\rm L}\to\infty$), $L\to 0$, and $\eta\to 0$. In this case, as seen from definitions (4.41) and (4.55), also $\mathcal{M}^2\to 0$, $D^{-1}\to 0$, and $E\to\mu\eta$, while the current I and the enthalpy μ (and, hence, any other thermodynamic function) become integrals of motion. The hydrodynamical version of the GS equation (1.115) studied in Chap. 1 can be deduced in the limit $\Psi\to 0$, $\eta\to\infty$, when, however, the product $\eta\Psi\sim\Phi$ remains finite. Formally, this limit corresponds to the substitution of $\Psi\to\Phi$ and $\eta\to 1$.

The GS equation (4.66) is a second-order equation linear with respect to higher derivatives. Otherwise, it can be written in canonical form

$$\mathcal{A}\frac{\partial^2\Psi}{\partial r^2}+2\mathcal{B}\frac{\partial^2\Psi}{\partial r\partial\theta}+\mathcal{C}\frac{\partial^2\Psi}{\partial\theta^2}+\mathcal{F}=0, \qquad (4.68)$$

where

$$\mathcal{A} \propto A\left[D + \frac{(\nabla_{\hat{r}}\Psi)^2}{(\nabla\Psi)^2}\right], \tag{4.69}$$

$$\mathcal{B} \propto A\frac{(\nabla_{\hat{r}}\Psi)(\nabla_{\hat{\theta}}\Psi)}{(\nabla\Psi)^2}, \tag{4.70}$$

$$\mathcal{C} \propto A\left[D + \frac{(\nabla_{\hat{\theta}}\Psi)^2}{(\nabla\Psi)^2}\right]. \tag{4.71}$$

The expression $\mathcal{F}$ does not contain the second derivatives of Ψ. As one can check,

$$\mathcal{AC} - \mathcal{B}^2 \propto A^2 D(D+1). \tag{4.72}$$

Therefore, as was expected, the GS equation varies from an elliptic to a hyperbolic type on the singular surfaces on which the poloidal velocity of matter becomes equal either to the fast or slow magnetosonic velocity ($D = 0$) or to the cusp velocity ($D = -1$). On the Alfvén surface $A = 0$, the type of the equation does not change. Nevertheless, the Alfvén surface is still the singular surface of the GS equation, since the regularity conditions should be satisfied on it

$$\begin{aligned}&\frac{1}{\alpha^2\varpi^2}\nabla'_k A \cdot \nabla^k \Psi + \frac{\Omega_{\rm F} - \omega}{\alpha^2}\frac{{\rm d}\Omega_{\rm F}}{{\rm d}\Psi}(\nabla\Psi)^2 + \\ &\frac{64\pi^4}{\alpha^2\varpi^2}\frac{1}{2\mathcal{M}^2}\frac{\partial}{\partial\Psi}\left(\frac{G}{A}\right) - 16\pi^3\mu n\frac{1}{\eta}\frac{{\rm d}\eta}{{\rm d}\Psi} - 16\pi^3 n T\frac{{\rm d}s}{{\rm d}\Psi} = 0,\end{aligned} \tag{4.73}$$

which immediately follows from (4.66). We dealt with this singularity when discussing the pulsar equation.

To conclude, a few words should be said about the number of boundary conditions. As we see, the full MHD version of the GS equation contains five integrals of motion which, generally speaking, must be determined by the boundary conditions. On the other hand, the regularity conditions (4.60) and (4.73) must be satisfied on the singular surfaces. Therefore, at least, for the simplest topologies when all field lines cross all s' singular surfaces, the number of boundary conditions b can again be written as (Beskin, 1997)

$$b = 2 + i - s', \tag{4.74}$$

where i is the number of integrals of motion. We emphasize that since the GS equation can be rewritten as $D + K_1 N_r + K_2 N_\theta = 0$, relations (4.60) prescribe only one regularity condition. The second condition is needed to specify the location of the singular surface.

4.3 Special Cases

4.3.1 Nonrelativistic Flows

In this section we formulate the basic equations for the nonrelativistic version of the GS equation, i.e., for the nonrelativistic velocities $v \ll c$ and the Newtonian gravitational potential $\varphi_{\rm g} \ll c^2$. Unlike the relativistic motion in flat space when all equations can be derived by the obvious limit $\alpha \to 1$, $\omega \to 0$, here the situation should be specially considered, because in all equations the contribution of the rest energy must be eliminated. In particular, as we saw, in the nonrelativistic version the dimensions of the integrals of motion $E_{\rm n}$, $L_{\rm n}$ differ from their dimensions in the relativistic case E and L. For the nonrelativistic case, in the following we restore the dimension and consider the physical components of the vectors only.

We first write the relations defining the electromagnetic fields and the velocity in terms of the integrals of motion:

$$\mathbf{B} = \frac{\nabla\Psi \times \mathbf{e}_\varphi}{2\pi\varpi} - \frac{2I}{\varpi c}\mathbf{e}_\varphi, \tag{4.75}$$

$$\mathbf{E} = -\frac{\Omega_{\rm F}}{2\pi c}\nabla\Psi, \tag{4.76}$$

$$\mathbf{v} = \frac{\eta_{\rm n}}{\rho}\mathbf{B} + \Omega_{\rm F}\varpi\mathbf{e}_\varphi. \tag{4.77}$$

Here again $\rho = m_{\rm p}n$ is the mass density and the nonrelativistic matter-to-magnetic flux ratio $\eta_{\rm n}(\Psi)$ is connected with the corresponding relativistic value as

$$\eta_{\rm n} = m_{\rm p}c\eta. \tag{4.78}$$

The expressions for the energy and the angular momentum fluxes are now written as

$$E_{\rm n}(\Psi) = \frac{\Omega_{\rm F}I}{2\pi c\eta_{\rm n}} + \frac{v^2}{2} + w + \varphi_{\rm g}, \tag{4.79}$$

$$L_{\rm n}(\Psi) = \frac{I}{2\pi c\eta_{\rm n}} + v_\varphi r\sin\theta, \tag{4.80}$$

the relativistic and nonrelativistic expressions connected by the relations

$$E = c\eta_{\rm n} + \frac{\eta_{\rm n}E_{\rm n}}{c}, \tag{4.81}$$

$$L = \frac{\eta_{\rm n}L_{\rm n}}{c}. \tag{4.82}$$

As we see, the nonrelativistic integrals, as in the hydrodynamical case, are normalized to the unit matter flux $\mathrm{d}\Phi$ and, therefore, the total energy and angular momentum losses must be written as (cf. (2.97))

$$W_{\mathrm{tot}} = \int E_{\mathrm{n}}(\Psi)\eta_{\mathrm{n}}(\Psi)\mathrm{d}\Psi, \tag{4.83}$$

$$K_{\mathrm{tot}} = \int L_{\mathrm{n}}(\Psi)\eta_{\mathrm{n}}(\Psi)\mathrm{d}\Psi. \tag{4.84}$$

Two more invariants are again the angular velocity $\Omega_{\mathrm{F}}(\Psi)$ and the entropy $s(\Psi)$.

As a result, algebraic relations (4.38), (4.39), and (4.40) defining the longitudinal current and the toroidal velocity have the form (Weber and Davis, 1967)

$$\frac{I}{2\pi} = c\eta_{\mathrm{n}}\frac{L_{\mathrm{n}} - \Omega_{\mathrm{F}}\varpi^2}{1 - \mathcal{M}^2}, \tag{4.85}$$

$$v_{\varphi} = \frac{1}{\varpi}\frac{\Omega_{\mathrm{F}}\varpi^2 - L_{\mathrm{n}}\mathcal{M}^2}{1 - \mathcal{M}^2}, \tag{4.86}$$

where now

$$\mathcal{M}^2 = \frac{4\pi\eta_{\mathrm{n}}^2}{\rho}. \tag{4.87}$$

As to the value $\mathcal{M}^2$, it must, as in the relativistic case, be determined from Bernoulli's equation (4.79) written as

$$\frac{\mathcal{M}^4}{64\pi^4\eta_{\mathrm{n}}^2}(\nabla\Psi)^2 = 2\varpi^2(E_{\mathrm{n}} - w - \varphi_{\mathrm{g}})$$
$$- \frac{(\Omega_{\mathrm{F}}\varpi^2 - L_{\mathrm{n}}\mathcal{M}^2)^2}{(1 - \mathcal{M}^2)^2} - 2\varpi^2\Omega_{\mathrm{F}}\frac{L_{\mathrm{n}} - \Omega_{\mathrm{F}}\varpi^2}{1 - \mathcal{M}^2}. \tag{4.88}$$

Recall that in Eq. (4.88), as in its relativistic version (4.43), the specific enthalpy w should be regarded as a function of the entropy s, the Mach number $\mathcal{M}^2$, and the integral η_{n}. The corresponding connection has the form

$$\nabla w = c_{\mathrm{s}}^2\left(2\frac{\nabla\eta_{\mathrm{n}}}{\eta_{\mathrm{n}}} - \frac{\nabla\mathcal{M}^2}{\mathcal{M}^2}\right) + \left[\frac{1}{\rho}\left(\frac{\partial P}{\partial s}\right)_n + \frac{T}{m_{\mathrm{p}}}\right]\nabla s. \tag{4.89}$$

Relations (4.85) and (4.86) again lead to a number of general conclusions.

- Within the Alfvén surface $r \ll r_{\mathrm{f}}$, we have $I \approx 2\pi c\eta_{\mathrm{n}}L_{\mathrm{n}}$, i.e., the current I remains constant on the magnetic surfaces. Hence, the current closure within the Alfvén surface is impossible in the nonrelativistic case as well.
- The toroidal velocity v_{φ} within the Alfvén surface is simply

$$v_\varphi \approx \Omega_F \varpi. \tag{4.90}$$

Hence, the full corotation must be available here.

- Conversely, beyond the Alfvén surface we have

$$\frac{I}{2\pi} = c\eta_n \frac{\Omega_F \varpi^2}{\mathcal{M}^2}. \tag{4.91}$$

Therefore, the current closure (in the field of application of the GS equation) is possible only under the strong collimation, i.e., when the condition $\mathcal{M}^2 \propto \varpi^2$ is violated.

- Finally, since

$$L_n \approx \Omega_F \varpi^2(r_A) \tag{4.92}$$

(see the numerator in (4.85)), beyond the Alfvén surface the toroidal velocity behaves as

$$v_\varphi \approx \frac{\Omega_F \varpi_A^2}{\varpi}. \tag{4.93}$$

Thus, most of the features of the nonrelativistic flows prove close to those of the relativistic flows (as to the differences, they are given below).

Problem 4.7 Show that, within the Alfvén surface $r < r_A$, we have

$$v_p^2 \approx v_0^2 + \Omega_F^2 \varpi^2, \tag{4.94}$$

so that for the sufficiently small initial velocity v_0 the relation $v_p \approx v_\varphi$ can hold here.

Further, the Alfvén factor in the nonrelativistic case is

$$A = 1 - \mathcal{M}^2. \tag{4.95}$$

It is easy to check that this singularity really corresponds to the condition of the equality of the poloidal velocity v_p to the poloidal component of the Alfvén velocity $V_{A,p}$. Indeed, using definitions (4.77) and (4.87), we find

$$\mathcal{M}^2 = \frac{4\pi\rho v_p^2}{B_p^2}. \tag{4.96}$$

For this reason, it gives the name of the singularity $A = 0$.

As to the derivative of the Mach number $\mathcal{M}^2$, in the nonrelativistic case, it can still be represented as

$$\nabla_k \mathcal{M}^2 = \frac{N_k}{D}, \tag{4.97}$$

where now

$$N_k = -\frac{(1-\mathcal{M}^2)}{(\nabla\Psi)^2}\nabla^i\Psi \cdot \nabla_i\nabla_k\Psi + \frac{(1-\mathcal{M}^2)}{2}\frac{\nabla_k' F_n}{(\nabla\Psi)^2}. \tag{4.98}$$

In this case,

$$F_n = \frac{64\pi^4\eta_n^2 K_n}{\mathcal{M}^4(1-\mathcal{M}^2)^2}, \tag{4.99}$$

where

$$K_n = 2(1-\mathcal{M}^2)^2(E_n - w - \varphi_g)\varpi^2 + (1-2\mathcal{M}^2)\Omega_F^2\varpi^4$$
$$-2(1-2\mathcal{M}^2)\Omega_F L_n\varpi^2 - \mathcal{M}^4 L_n^2. \tag{4.100}$$

The factor D now looks like

$$D = \frac{1-\mathcal{M}^2}{\mathcal{M}^2} + \frac{1}{\mathcal{M}^2}\frac{B_\varphi^2}{B_p^2} - \frac{c_s^2}{v_p^2}\frac{1-\mathcal{M}^2}{\mathcal{M}^2}. \tag{4.101}$$

Problem 4.8 Check that the condition $D = 0$ really yields expression (4.16) for the fast and slow magnetosonic velocity $V_{(1,2)}$ and the condition $D = -1$ for expression (4.24) for the cusp velocity V_{cusp}.

Finally, the GS equation in compact form can be written as (Heyvaerts and Norman, 1989)

$$\frac{1}{16\pi^3\rho}\nabla_k\left(\frac{1-\mathcal{M}^2}{\varpi^2}\nabla^k\Psi\right) + \frac{dE_n}{d\Psi}$$
$$+\frac{\Omega_F\varpi^2 - L_n}{1-\mathcal{M}^2}\frac{d\Omega_F}{d\Psi} + \frac{1}{\varpi^2}\frac{\mathcal{M}^2 L_n - \Omega_F\varpi^2}{1-\mathcal{M}^2}\frac{dL_n}{d\Psi}$$
$$+\left(2E_n + \frac{1}{\varpi^2}\frac{\Omega_F^2\varpi^4 - 2\Omega_F L_n\varpi^2 + \mathcal{M}^2 L_n^2}{1-\mathcal{M}^2}\right)\frac{1}{\eta_n}\frac{d\eta_n}{d\Psi} - \frac{T}{m_p}\frac{ds}{d\Psi} = 0. \tag{4.102}$$

As was mentioned, the full version of the nonrelativistic GS equation containing all five invariants was formulated by Soloviev (1965). Though little known among the astrophysicists, this equation was often fully reformulated in the following (Okamoto, 1975; Heinemann and Olbert, 1978; Heyvaerts, 1996). In particular, for this reason, there is no unique nomenclature yet. Therefore, it is sometimes very difficult to compare results in various papers with one another.

4.3.2 Anisotropic Pressure

Here we give general expressions for the GS equation for the medium with anisotropic pressure. This is the case, for example, for the solar wind when the free path l_ν is larger than the Larmor radius $r_{\rm L}$ by many orders, so that near the Earth we have $l_\nu/r_{\rm L} \sim 10^9$. We restrict ourselves, for simplicity, to the nonrelativistic case only (Beskin and Kuznetsova, 2000b); the full relativistic version, with account taken of the GR effects, is formulated in Kuznetsova (2005).

In fact, our problem is again to transform the equation of motion

$$\rho(\mathbf{v}\cdot\nabla)\mathbf{v} = -\nabla_k P_{ik} + \frac{1}{4\pi}[\nabla\times\mathbf{B}]\times\mathbf{B} - \rho\nabla\varphi_{\rm g}. \tag{4.103}$$

Following Chew et al. (1956), for collisionless plasma the anisotropic pressure P_{ik} can be written as

$$P_{ik} = P_s\delta_{ik} + (P_n - P_s)\frac{B_i B_k}{B^2}, \tag{4.104}$$

which contains two scalar function $P_{\rm s}$ and $P_{\rm n}$. As a result, instead of the fifth integral, viz., the entropy $s(\Psi)$, we can introduce two additional invariants constant on the magnetic surfaces. This approximation was called the double adiabatic approximation, because it is based on the conservation of two adiabatic invariants

$$s_1(\Psi) = \frac{P_{\rm n}B^2}{\rho^3}, \tag{4.105}$$

$$s_2(\Psi) = \frac{P_{\rm s}}{\rho B}. \tag{4.106}$$

Thus, these invariants correspond to the polytropic equation of state with $\Gamma_{\parallel} = 3$ and $\Gamma_{\perp} = 1$.

Integrating now the toroidal and longitudinal (parallel to $\mathbf{v}_{\rm p}$) components of Eq. (4.103), we readily obtain the expressions for energy and angular momentum (Asséo and Beaufils, 1983; Tsikarishvili et al., 1995)

$$E_{\rm n}(\Psi) = \frac{v^2}{2} + \frac{P_{\rm s}}{\rho} + \frac{3}{2}\frac{P_{\rm n}}{\rho} + \frac{\Omega_{\rm F} I}{2\pi c\eta_{\rm n}}(1-\beta_{\rm a}) + \varphi_{\rm g}, \tag{4.107}$$

$$L_{\rm n}(\Psi) = v_\varphi r\sin\theta + \frac{I}{2\pi c\eta_{\rm n}}(1-\beta_{\rm a}). \tag{4.108}$$

Here

$$\beta_{\rm a} = 4\pi \frac{P_{\rm n} - P_{\rm s}}{B^2} \tag{4.109}$$

is an anisotropy parameter. Along with the angular velocity $\Omega_{\rm F}(\Psi)$ and the integral $\eta_{\rm n}(\Psi)$, these six invariants define all characteristics of the flow, for example (Asséo and Beaufils, 1983; Lovelace et al., 1986)

$$\frac{I}{2\pi} = c\eta_{\rm n} \frac{L_{\rm n} - \Omega_{\rm F}\varpi^2}{1 - \mathcal{M}^2 - \beta_{\rm a}}, \tag{4.110}$$

$$v_\varphi = \frac{1}{\varpi} \frac{\Omega_{\rm F}\varpi^2(1 - \beta_{\rm a}) - \mathcal{M}^2 L_{\rm n}}{1 - \mathcal{M}^2 - \beta_{\rm a}}. \tag{4.111}$$

Here again $\mathcal{M}^2 = 4\pi\eta_{\rm n}^2/\rho$. Hence, in the case of the anisotropic pressure the Alfvén singularity condition has the form

$$A = 1 - \mathcal{M}^2 - \beta_{\rm a} = 0. \tag{4.112}$$

One should stress that condition (4.110) specifies the current I in an implicit way, since the current I in terms of B_φ^2 entering into $\beta_{\rm a}$ is available on the right-hand side of Eq. (4.110). Nevertheless, the standard procedure for determining the flow parameters by the given values of the poloidal field $\mathbf{B}_{\rm p}$ remains unchanged. Bernoulli's equation (4.107) that can be rewritten as

$$\frac{\mathcal{M}^4}{64\pi^4\eta_{\rm n}^2}(\nabla\Psi)^2 = 2\varpi^2\left(E_{\rm n} - \frac{P_{\rm s}}{\rho} - \frac{3}{2}\frac{P_{\rm n}}{\rho} - \varphi_{\rm g}\right)$$
$$-\frac{[\Omega_{\rm F}\varpi^2(1 - \beta_{\rm a}) - L_{\rm n}\mathcal{M}^2]^2}{(1 - \mathcal{M}^2 - \beta_{\rm a})^2} - 2\varpi^2\Omega_{\rm F}(1 - \beta_{\rm a})\frac{L_{\rm n} - \Omega_{\rm F}\varpi^2}{1 - \mathcal{M}^2 - \beta_{\rm a}}, \tag{4.113}$$

relation (4.110), and definitions (4.105) and (4.106) implicitly give the Mach number $\mathcal{M}^2$ and the anisotropy parameter $\beta_{\rm a}$ as the function Ψ (or $\mathbf{B}_{\rm p}$) and all six invariants:

$$\mathcal{M}^2 = \mathcal{M}^2[(\nabla\Psi)^2, E_{\rm n}, L_{\rm n}, \Omega_{\rm F}, \eta_{\rm n}, s_1, s_2], \tag{4.114}$$

$$\beta_{\rm a} = \beta_{\rm a}[(\nabla\Psi)^2, E_{\rm n}, L_{\rm n}, \Omega_{\rm F}, \eta_{\rm n}, s_1, s_2]. \tag{4.115}$$

Relations (4.110) and (4.111) make it possible to determine all the other values as a function of the magnetic flux Ψ and the six integrals of motion.

As to the GS equation, it can again be derived from the unused component in the Euler equation orthogonal to the poloidal magnetic field. In compact form, it looks like (Beskin and Kuznetsova, 2000b)

$$\nabla_k\left[\frac{1}{\varpi^2}\left(1-\mathcal{M}^2-\beta_{\rm a}\right)\nabla^k\Psi\right]+\frac{64\pi^4}{\varpi^2}\frac{1}{2\mathcal{M}^2}\frac{\partial}{\partial\Psi}\left(\frac{G}{A}\right)$$
$$-8\pi^3 P_{\rm n}\frac{1}{s_1}\frac{{\rm d}s_1}{{\rm d}\Psi}-16\pi^3 P_{\rm s}\frac{1}{s_2}\frac{{\rm d}s_2}{{\rm d}\Psi}=0, \qquad (4.116)$$

where

$$\left(\frac{G}{A}\right)=2\varpi^2\eta_{\rm n}^2\left(E_{\rm n}-\frac{P_{\rm s}}{\rho}-\frac{3}{2}\frac{P_{\rm n}}{\rho}-\varphi_{\rm g}\right)$$
$$+\eta_{\rm n}^2\frac{\varpi^4\Omega_{\rm F}^2(1-\beta_{\rm a})-2\varpi^2\Omega_{\rm F}L_{\rm n}(1-\beta_{\rm a})+\mathcal{M}^2L_{\rm n}^2}{1-\mathcal{M}^2-\beta_{\rm a}}. \qquad (4.117)$$

Here again ∇_k is a covariant derivative and the operator $\partial/\partial\Psi$ acts only on the invariants $E_{\rm n}(\Psi)$, $L_{\rm n}(\Psi)$, $\Omega_{\rm F}(\Psi)$, and $\eta_{\rm n}(\Psi)$. By definitions (4.105) and (4.106) and relations (4.114) and (4.115), this equation, as any GS-type equation, comprises only the unknown function of the magnetic flux $\Psi(r,\theta)$ and the six integrals of motion. In particular, in the static case $\mathbf{v}=0$, i.e., when $\Omega_{\rm F}\to 0$, $\eta_{\rm n}\to 0$ (so that $\mathcal{M}^2\to 0$), but $\eta_{\rm n}L_{\rm n}\to$ const, and for $\varphi_{\rm g}=0$, we have

$$\nabla_k\left[\frac{1}{\varpi^2}\left(1-\beta_{\rm a}\right)\nabla^k\Psi\right]+\frac{16\pi^2}{\varpi^2}(1-\beta_{\rm a})I\frac{\partial I}{\partial\Psi}$$
$$+16\pi^3\rho\frac{\rm d}{{\rm d}\Psi}\left(\frac{P_{\rm s}}{\rho}+\frac{3}{2}\frac{P_{\rm n}}{\rho}\right)-8\pi^3 P_{\rm n}\frac{1}{s_1}\frac{{\rm d}s_1}{{\rm d}\Psi}-16\pi^3 P_{\rm s}\frac{1}{s_2}\frac{{\rm d}s_2}{{\rm d}\Psi}=0. \qquad (4.118)$$

This equation is the generalization of the static GS equation (4.67) to the medium with anisotropic pressure. Note that there is the partial derivative $\partial I/\partial\Psi$ here. This implies that in expression (4.110) the integrals of motion should be differentiated.

It is interesting to note that relations (4.107), (4.108), (4.109), (4.110), and (4.111) remain valid for the more general case (Denton et al., 1994)

$$(\mathbf{v}\cdot\nabla)\left(\frac{P_{\rm n}B^2}{\rho^3}\right)=2\mathcal{P}_t\left(\frac{B^2}{\rho^3}\right), \qquad (4.119)$$

$$(\mathbf{v}\cdot\nabla)\left(\frac{P_{\rm s}}{\rho B}\right)=-\mathcal{P}_t\left(\frac{1}{\rho B}\right), \qquad (4.120)$$

where $\mathcal{P}_t$ describes the dissipation-free energy exchange between the longitudinal and transverse particle motions. In this approximation, we have the conservation of the "total entropy" $S=S(\Psi)$ determined from the condition

$$(\mathbf{v}\cdot\nabla)S=\frac{1}{2}\frac{\rho^3}{B^2}(\mathbf{v}\cdot\nabla)s_1+\rho B(\mathbf{v}\cdot\nabla)s_2=0. \qquad (4.121)$$

In this case, the GS equation (4.118) has the form

$$\nabla_k \left[\frac{1}{\varpi^2}(1 - \mathcal{M}^2 - \beta_a)\nabla^k \Psi \right] + \frac{32\pi^4}{\varpi^2 \mathcal{M}^2} \frac{\partial}{\partial \Psi} \left(\frac{G}{A} \right) - 16\pi^3 \frac{\mathrm{d}S}{\mathrm{d}\Psi} = 0. \quad (4.122)$$

If we introduce the efficient pressure $P_{\text{eff}} = P_{\text{eff}}(\Psi)$

$$\mathrm{d}P_{\text{eff}} = \rho \, \mathrm{d} \left(\frac{P_s}{\rho} + \frac{3}{2} \frac{P_n}{\rho} \right) - \mathrm{d}S, \quad (4.123)$$

which is equivalent to the ordinary thermodynamic relation $\mathrm{d}P = \rho \mathrm{d}w - nT\mathrm{d}s$, instead of (4.118), we have for the static configurations

$$\nabla_k \left[\frac{1}{\varpi^2} (1 - \beta_a) \nabla^k \Psi \right] + \frac{16\pi^2}{\varpi^2}(1 - \beta_a) I \frac{\partial I}{\partial \Psi} + 16\pi^3 \frac{\mathrm{d}P_{\text{eff}}}{\mathrm{d}\Psi} = 0. \quad (4.124)$$

For $I = 0$, this equation was obtained in Nötzel et al. (1985).

Equation (4.118), as all the above equations, is a mixed-type equation. Using the implicit relations yielding expressions (4.114) and (4.115), we can write the second-order operator in (4.118) in ordinary expanded form

$$A \left[\nabla_k \left(\frac{1}{\varpi^2} \nabla^k \Psi \right) + \frac{\nabla^i \Psi \cdot \nabla^k \Psi \cdot \nabla_i \nabla_k \Psi}{\varpi^2 (\nabla \Psi)^2 D} \right] + \cdots = 0, \quad (4.125)$$

where

$$D = \frac{N}{d}, \quad (4.126)$$

and

$$\begin{aligned} N = {} & \left(1 - \mathcal{M}^2 - \beta_a + 4\pi \frac{4P_n - P_s}{B^2} \frac{B_\varphi^2}{B^2} \right) \left(1 - 3 \frac{P_n}{\rho v_p^2} \right) \\ & + \frac{B_\varphi^2}{B_p^2} \left(1 - \beta_a - 4\pi \frac{3P_n - P_s}{B^2} + 4\pi \frac{4P_n - P_s}{B^2} \frac{B_\varphi^2}{B^2} \right) \\ & - \left(\mathcal{M}^2 - 4\pi \frac{3P_n - P_s}{B^2} \right) \frac{3P_n - P_s}{\rho v_p^2} \frac{B_\varphi^2}{B^2}, \end{aligned} \quad (4.127)$$

$$d = \left(1 - 3 \frac{P_n}{\rho v_p^2} \frac{B_p^2}{B^2} \right) \left(\mathcal{M}^2 + 4\pi \frac{P_n}{B^2} \right) + \frac{P_s}{\rho v_p^2} \frac{B_p^2}{B^2} \left(\mathcal{M}^2 - 4\pi \frac{3P_n - P_s}{B^2} \right). \quad (4.128)$$

As we see, the form (4.125) of Eq. (4.116) coincides with the canonical form of the GS equation. Hence, it is of an elliptic type in the domain $D > 0$ and $D < -1$ and of a hyperbolic type for $-1 < D < 0$. This implies that the condition $D = 0$

must define the velocities of the fast and slow magnetosonic waves propagating in the medium with anisotropic pressure. Accordingly, the condition $D = -1$ is to fix the cusp velocity. This property can be readily verified. Indeed, for the nonrotating medium ($\Omega_{\rm F} = 0$, $L_{\rm n} = 0$) when according to (4.110) $B_{\varphi} = 0$, we have

$$N = (1 - \mathcal{M}^2 - \beta_{\rm a})\left(1 - 3\frac{P_{\rm n}}{\rho v_{\rm p}^2}\right). \tag{4.129}$$

Using now relations (4.127) and (4.128), we can easily reproduce the expressions for the sonic, Alfvén, and cusp velocities for the medium with anisotropic pressure (Clemmow and Dougherty, 1969):

$$c_{\rm s}^2 = 3\frac{P_{\rm n}}{\rho}, \tag{4.130}$$

$$V_{\rm A}^2 = \frac{B^2}{4\pi\rho} - \frac{P_{\rm n} - P_{\rm s}}{\rho}, \tag{4.131}$$

$$V_{\rm cusp}^2 = \frac{3\dfrac{P_{\rm n}}{\rho}\dfrac{B^2}{4\pi\rho} + 6\dfrac{P_{\rm n}P_{\rm s}}{\rho^2} - \dfrac{P_{\rm s}^2}{\rho^2}}{\dfrac{B^2}{4\pi\rho} + 2\dfrac{P_{\rm s}}{\rho}}. \tag{4.132}$$

Recall that the condition $V_{\rm A}^2 < 0$ corresponds to a fire-hose instability and the condition $V_{\rm cusp}^2 < 0$ to a mirror one.

4.4 General Properties

4.4.1 Some Useful Relations

Before proceeding to the consistent analysis of the exact solutions to the GS equation, we try to point to some common properties of the magnetized flows. To begin with, we formulate several general relations widely used in the following. We first introduce some new notation. As we saw, in most expressions we have the combination of invariants $e' = E - \Omega_{\rm F}L$. Using definitions (4.33) and (4.34), we obtain

$$e' = \mu\eta\left[\alpha\gamma - (\Omega_{\rm F} - \omega)\varpi u_{\hat{\varphi}}\right]. \tag{4.133}$$

Otherwise, the value e' corresponds to the particle contribution only. Obviously, for the estimate we can take

$$e' \approx \mu\eta\gamma_{\rm inj}, \tag{4.134}$$

where $\gamma_{\rm inj}$ is the characteristic Lorentz factor of particles in the plasma generation region.

Further, it is logical to express the square of the Mach number $\mathcal{M}^2$ in the form of the ratio

$$q = \frac{\mathcal{M}^2}{(\Omega_{\rm F} - \omega)^2 \varpi^2}. \tag{4.135}$$

Indeed, since at large distances $r \to \infty$ for the quasispherical flows $\mathcal{M}^2 \propto 1/n \propto r^2$, even for the magnetically dominated flow (i.e., for the flow in which in the vicinity of the compact object $\mathcal{M}^2 \ll 1$), at large distances the value $\mathcal{M}^2$ is much larger than unity. On the other hand, the parameter q for the magnetically dominated flow remains small over the whole space. One should stress at once that the definition (4.135) proves inconvenient in the vicinity of the rotation axis, where $\varpi \to 0$, and also in the region $\Omega_{\rm F} \approx \omega$.

Finally, in the following we often use the dimensionless value:

$$\Sigma_{\rm r}^2 = \frac{(\Omega_{\rm F} - \omega)^4 \varpi^2 (\nabla \Psi)^2}{64\pi^4 E^2}. \tag{4.136}$$

It yields us the information concerning the poloidal magnetic field structure at large distances from the central object. Indeed, when moving along the magnetic field line at large distances from the central source ($\omega = 0$), the dependence $\Sigma_{\rm r}^2$ on the radius r is only in the factor $\varpi^2(\nabla\Psi)^2$ defined by the character of the divergence of the magnetic field lines. On the other hand, the parameter $\Sigma_{\rm r}^2$ depends on the ratio $E/\Omega_{\rm F} \approx L$ and, hence, its value can provide the information concerning the longitudinal currents flowing in the magnetosphere. In particular, it is easy to check that for the Michel monopole solution $\Sigma_{\rm r}^2 = 1$.

Let us see how, using the above values, we can define a number of the key parameters characterizing the magnetized plasma flow. We first use expression (4.39) for the particle Lorentz factor. Far from the Alfvén surface $r \gg R_{\rm L}$ it can be rewritten as

$$\gamma = \frac{q}{1+q}\left(\frac{E}{\mu\eta}\right). \tag{4.137}$$

But $\gamma\mu\eta$ is nothing but the particle energy flux density. Therefore, the asymptotic behavior of q is, in fact, the particle-to-electromagnetic energy flux ratio

$$q = \frac{W_{\rm part}}{W_{\rm em}}. \tag{4.138}$$

As seen from relation (4.137), the maximum particle Lorentz factor is defined by the ratio $E/\mu\eta$ that for the nonrelativistic temperatures ($\mu \approx m_{\rm p}$) is given by the values of the invariants E and η.

Problem 4.9 Show by the direct substitution of the condition $nu_{\mathrm{p}} = \eta B_{\mathrm{p}}$ and asymptotic relations $|\mathbf{E}| \approx B_{\varphi} \approx \Omega_{\mathrm{F}} \varpi B_{\mathrm{p}}$ in the definition $\mathcal{M}^2 = 4\pi\mu\eta^2/n$ that q is really the particle-to-electromagnetic energy flux ratio.

Problem 4.10 Show that

$$\gamma_{\max} = \frac{E}{\mu\eta} = \sigma, \tag{4.139}$$

where σ is the above magnetization parameter (2.84). Otherwise, σ has the meaning of the limiting value of the particle Lorentz factor.

Further, we consider the asymptotic behavior of relativistic Bernoulli's equation $\gamma^2 - \mathbf{u}^2 = 1$ (4.44). At distances of ϖ from the rotation axis, which are much larger than the light cylinder radius R_{L}, it has the form

$$\frac{\mathcal{M}^4 E^2}{\mu^2\eta^2 A^2} - \frac{\mathcal{M}^4(\nabla\Psi)^2}{64\pi^4\mu^2\eta^2\varpi^2} = 1. \tag{4.140}$$

This relation leads to two important conclusions. We first use the fact that for the ultrarelativistic flow we have the condition $\mathbf{u}^2 \approx \gamma^2$. Comparing expression (4.140) with the definition Σ_{r} (4.136), we get

$$\gamma \approx \Sigma_{\mathrm{r}} \left(\frac{E}{\mu\eta}\right) q. \tag{4.141}$$

Hence, the conclusion is that the full transformation of the electromagnetic energy into the particle energy ($\gamma \to E/\mu\eta$, $q \to \infty$) is possible only if the condition $\Sigma_{\mathrm{r}} \to 0$ is satisfied. In other words, there exists a quite definite connection between the poloidal magnetic field structure and the particle energy at infinity. Below, this problem is discussed in more detail. On the other hand, for the magnetically dominated flows $q \ll 1$ (and in the asymptotic domain $r \gg r_{\mathrm{f}}$, where r_{f} is the radius of the fast magnetosonic surface) we find

$$\Sigma_{\mathrm{r}} \approx \frac{1}{1+q}, \tag{4.142}$$

i.e., the parameter Σ_{r} is to be close to unity.

Relation (4.140) leads to one more important assertion. Indeed, using the definitions of the electric and magnetic fields (3.14) and (3.43), we can rewrite condition (4.140) as

$$\left(B_{\hat{\varphi}}^2 - |\mathbf{E}|^2\right) \frac{n^2}{\eta^2} \Omega_{\mathrm{F}}^2 \varpi^2 = 1. \tag{4.143}$$

This implies that for the relativistic flow the condition

$$B_{\hat{\varphi}}^2 - |\mathbf{E}|^2 \approx \frac{B_{\hat{\varphi}}^2}{\gamma^2} \tag{4.144}$$

must be satisfied. On the other hand, in the vicinity of the black hole horizon ($\alpha^2 \to 0$) the asymptotic behavior of Bernoulli's equation looks like

$$\frac{\mathcal{M}^2(E-\omega L)^2}{\alpha^2\mu^2\eta^2A^2} - \frac{1}{64\pi^4}\frac{\mathcal{M}^2(\nabla\Psi)^2}{\alpha^2\mu^2\eta^2\varpi^2} = 1. \tag{4.145}$$

As a result, using definitions (3.14) and (3.43) again, we get

$$B_{\hat{\varphi}}^2 - |\mathbf{E}|^2 \approx \left[\frac{(e')^2}{(\Omega_{\mathrm{F}}-\omega)^2\varpi^2\mu^2\eta^2} + 1\right]\frac{B_{\hat{\varphi}}^2}{\gamma^2}. \tag{4.146}$$

As, according to relation (4.134), we have $e'/\mu\eta \sim \gamma_{\mathrm{inj}} > 1$, the first term in square brackets is much larger than unity.

In both cases, as we see, the usability condition $|\mathbf{B}| > |\mathbf{E}|$ in the studied approach seems, at first sight, self-satisfied. However, this does not imply that, with the finite particle mass taken into account, the flow can always be extended to infinity (or to the event horizon). The point is that for the solution to exist it is necessary that the physical root of Bernoulli's algebraic equation could be extended to the asymptotic domain. And this is the case not for any choice of the integrals of motion. This behavior was already observed when analyzing the hydrodynamical flow. As shown in Fig. 1.1, for rather large accretion rates $\Phi > \Phi_{\mathrm{cr}}$, the trajectory cannot be extended to small distances from the gravity center. There is a singularity in its derivative rather than in the value of the velocity (energy) itself. Accordingly, for rather small electric currents the solution to the force-free equation is confined to the light surface located at a finite distance from the compact object.

Finally, note that relation (4.144) has a simple physical meaning. Having written the expression for drift velocity as

$$U_{\mathrm{dr}}^2 \approx \frac{\mathbf{E}^2}{B_{\varphi}^2} \approx \left(1 - \frac{B_{\hat{\varphi}}^2 - \mathbf{E}^2}{B_{\hat{\varphi}}^2}\right) \approx \left(1 - \frac{1}{\gamma^2}\right), \tag{4.147}$$

we conclude that, provided that relation (4.144) holds, the drift velocity fully defines the particle energy. The drift velocity itself is actually directed nearly almost along the poloidal magnetic field, because at large distances the toroidal magnetic field is much larger than its poloidal component. Hence, one can make one more important conclusion:

Theorem 4.1 *For the relativistic flow beyond the outer Alfvén surface the particle motion is mainly determined by the drift motion in the crossed electric and toroidal magnetic fields along the poloidal magnetic field.*

On the other hand, relation (4.146) shows that this is not the case in the vicinity of the black hole horizon and the particle energy is not fully defined by its drift motion. This must be the case, because in the vicinity of the horizon (where the energy density of the electromagnetic field is not enough to distort the Kerr metric) the gravitational forces should be of vital importance in the determination of the particle energy (Punsly, 2001).

4.4.2 Alfvén Surface

We proceed to the discussion of the general properties of the magnetized flows. Rather general conclusions can be made from the analysis of the critical conditions on the singular surfaces. We first consider the problem of the direction of the particle motion on the Alfvén surface. As shown above, the conclusion that the particles can cross the Alfvén surface only in one direction leads to the most important consequence that plasma must be generated in the black hole magnetosphere. In view of this, it seems advisable to consider this problem in more detail.

Thus, we aim to prove the following theorem:

Theorem 4.2 *The particles can cross the Alfvén surface in one direction only. When the central source loses its rotational energy, the outer Alfvén surface can be crossed only in the direction from the compact object. When the energy flux is directed to the black hole, the surface is crossed in the direction of the event horizon. As to the inner Alfvén surface (which exists only in the black hole magnetosphere), the particles can cross it only in the direction of the horizon.*

To prove it we use expression (4.39) for the particle Lorentz factor which, far from the outer Alfvén surface $r \gg r_{\rm A}$ (or in the vicinity of the horizon $\alpha^2 \to 0$), can be written as

$$\gamma = \frac{1}{\alpha\mu}\,\frac{\mathcal{M}^2}{(\Omega_{\rm F}-\omega)^2\varpi^2+\mathcal{M}^2}\,\frac{E-\omega L}{\eta}. \tag{4.148}$$

It is obvious that the sign of γ must be positive. Hence, this condition can be written as

$$\mathrm{sign}(E-\omega L) = \mathrm{sign}\,\eta. \tag{4.149}$$

We first consider the outer Alfvén surface ($\omega \approx 0$). We suppose that the radial magnetic field is positive ($B_r > 0$). Then from the definition $\mathrm{d}\Psi = \mathbf{B}\cdot\mathrm{d}\mathbf{S}$ it follows that $\Psi > 0$. On the other hand, for the positive energy losses ($W_{\rm tot} = \int E\mathrm{d}\Psi > 0$) the condition $E(\Psi) > 0$ must be satisfied. Hence, according to (4.149), we find that $\eta > 0$. Using definition (4.31) $\alpha n\mathbf{u}_{\rm p} = \eta\mathbf{B}_{\rm p}$, we finally have

$$u_r(r_{\rm A}^{\rm (ext)}) > 0. \tag{4.150}$$

The similar arguments can be used for the inner Alfvén surface. Indeed, for $B_r > 0$ (i.e., for $\Psi > 0$) and for the positive energy flux ($0 < \Omega_{\rm F} < \Omega_{\rm H}$), the condition $E \approx \Omega_{\rm F} L < \Omega_{\rm H} L$ must be satisfied. It yields $E - \Omega_{\rm H} L < 0$. Therefore, in the vicinity of the horizon $\eta < 0$, i.e.,

$$u_r(r_{\rm A}^{\rm (int)}) < 0. \tag{4.151}$$

Problem 4.11 Show that conditions (4.150) and (4.151) do not change for $B_r < 0$.

Problem 4.12 Show that the particles can cross the inner Alfvén surface only in the direction of the event horizon for any energy flux direction.

Here, however, it would be more correct to look at condition (4.149) on the other hand. Indeed, if we do not assume that an additional plasma source is present within the inner Alfvén surface, which generates a particle flow in the direction from the black hole, the flows in the vicinity of the horizon can be directed to the event horizon only. Consequently, for this class of flows (and under the condition $B_r > 0$) the inequality $\eta < 0$ must hold. Hence, the MHD flows can be extended to the horizon only if the condition

$$E - \Omega_{\rm H} L < 0 \tag{4.152}$$

is satisfied. For example, the flow with $E > 0$ for $\Omega_{\rm H} = 0$ (i.e., the extraction of energy from nonrotating black hole) appears impossible. For the magnetically dominated flows ($E \approx \Omega_{\rm F} L$) condition (4.152) coincides with condition (3.69) that the energy loss is positive.

The above proof would seem to be unrelated to the properties of the Alfvén singularity. However, this is not the case. Relation (4.149) fixes the sign of γ when expanding the singularity of the type 0/0 by L'Hospital's rule in expressions (4.38), (4.39), and (4.40). Therefore, the condition on the Alfvén surface exactly confines the class of possible flows.

On the other hand, as was noted, the presence of the singularity on the Alfvén surface in algebraic expressions (4.38), (4.39), and (4.40) does not give rise to additional critical conditions as was the case on the sonic surfaces. It only prescribes the location of the Alfvén surface itself. Indeed, using the numerator in relation (4.38), we find

$$\varpi_{\rm A}^2 = \frac{\alpha_{\rm A}^2 L}{(\Omega_{\rm F} - \omega_{\rm A})(E - \omega_{\rm A} L)}. \tag{4.153}$$

In particular, for the outer Alfvén surface ($\omega \approx 0$, $\alpha^2 \approx 1$) we have

$$\varpi_{\rm A}^2 = \frac{L}{\Omega_{\rm F} E}, \tag{4.154}$$

which in the force-free limit ($E = \Omega_{\rm F} L$), as was expected, yields the expression $\varpi_{\rm A} = R_{\rm L} = c/\Omega_{\rm F}$ for the radius of the light cylinder. Finally, for the nonrelativistic flow we have

$$\varpi_{\rm A}^2 = \frac{L_{\rm n}}{\Omega_{\rm F}}. \tag{4.155}$$

We emphasize that since the invariants themselves are the functions of the magnetic flux Ψ, in the general case the location of the Alfvén surface can be found only after the solution to the GS equation.

As to the position of the inner Alfvén surface, it is convenient here to define $\alpha_{\rm A}$. As a result, we have for $\alpha_{\rm A}^2 \ll 1$

$$\alpha_{\rm A}^2 \approx \frac{(\Omega_{\rm H} - \Omega_{\rm F})(\Omega_{\rm H} L - E)\varpi_{\rm g}^2}{L}, \tag{4.156}$$

where $\varpi_{\rm g}$ corresponds to ϖ on the event horizon. Thus, all trajectories, for which conditions (4.150) and (4.151) are satisfied, can freely cross the Alfvén surface. This is because the Alfvén surface is a higher order singularity than the simplest singularity of a saddle type (see Fig. 5.14).

4.4.3 Fast Magnetosonic Surface—Relativistic Flows

We first consider qualitatively the features of the relativistic flows in the vicinity of the outer fast magnetosonic surface. If it is assumed to be at distances much larger than the Alfvén surface, the denominator D can be rewritten as

$$D = -1 + \frac{B_{\hat\varphi}^2 - \mathbf{E}^2}{\mathcal{M}^2 B_{\rm p}^2} + \frac{1}{\mathcal{M}^2}. \tag{4.157}$$

Here we disregarded the finite temperature contribution that, as shown below, becomes insignificant at large distances from the rotation axis. Using relations (4.137) and (4.144) and the estimate $B_{\hat\varphi} \approx |\mathbf{E}| \approx \Omega_{\rm F}\varpi B_{\rm p}$ resulting from the definition of the electromagnetic fields, we get

$$D = -1 + \frac{1}{\gamma^3}\left(\frac{E}{\mu\eta}\right) + \frac{1}{\gamma\Omega_{\rm F}^2\varpi^2}\left(\frac{E}{\mu\eta}\right). \tag{4.158}$$

When defining the third term $1/\mathcal{M}^2$ here, we again use the estimate $\gamma = q(E/\mu\eta)$. Thus, we see that when the third term in expression (4.158) can be disregarded, the particle Lorentz factor on the fast magnetosonic surface has the universal value (Michel, 1969)

$$\gamma(r_{\rm f}) \approx \left(\frac{E}{\mu\eta}\right)^{1/3} \approx \sigma^{1/3}. \tag{4.159}$$

In the general case, to determine the value $\gamma_{\rm f} = \gamma(r_{\rm f})$ we must study Bernoulli's equation (4.41) in more detail. It can be written as

$$q^4 + 2q^3 + \left[1 - \frac{(E-\omega L)^2}{\Sigma_{\rm r}^2 E^2} - 2\frac{\alpha^2}{(\Omega_{\rm F}-\omega)^2\varpi^2} + \frac{\alpha^2 L^2}{\Sigma_{\rm r}^2 E^2 \varpi^2}\right] q^2 \tag{4.160}$$
$$+\alpha^2\left(\frac{\mu^2\eta^2}{\Sigma_{\rm r}^2 E^2}\right) + \frac{\alpha^2}{\Sigma_{\rm r}^2(\Omega_{\rm F}-\omega)^2\varpi^2}\left(\frac{e'}{E}\right)^2 = 0.$$

Here and wherever possible we disregarded the terms of order $\alpha^2/(\Omega_{\rm F}-\omega)^2\varpi^2$ as compared to unity. This assumption is obviously valid for the outer fast magnetosonic surface at distances much larger than the light cylinder radius. As to the inner surface, the smallness of $\alpha(r_{\rm f})^2/(\Omega_{\rm F}-\omega_{\rm f})^2\varpi_{\rm f}^2$ is proved below. Further, in Eq. (4.160) the terms proportional to the first degree of q are omitted as they also prove small. Finally, when deriving Eq. (4.160) we again disregard the finite temperature contribution.

As was noted, Eq. (4.160) defines q (i.e., $\mathcal{M}^2$) as the function of the integrals of motion and the poloidal magnetic field (the stream function Ψ). The magnetic field structure, as we see, is available only through the parameter $\Sigma_{\rm r}^2$. In particular, Eq. (4.160) contains the solution (4.141) obtained for $q \gg \sigma^{-2/3}$.

We first consider the outer fast magnetosonic surface. As we saw, for the magnetically dominated flows, at least to the fast magnetosonic surface, we should expect the values of $\gamma_{\rm f} \sim \sigma^{1/3}$, i.e., $q \sim \sigma^{-2/3} \ll 1$. In this case, in Eq. (4.160) we can disregard the term q^4 and take $\omega = 0$ and $\alpha = 1$. It gives

$$q^3+\frac{1}{2}\left[1-\frac{1}{\Sigma_{\rm r}^2}-\left(2-\frac{1}{\Sigma_{\rm r}^2}\right)\frac{1}{\Omega_{\rm F}^2\varpi^2}\right]q^2+\frac{1}{2\Sigma_{\rm r}^2}\left(\frac{\mu\eta}{E}\right)^2+\frac{1}{2\Sigma_{\rm r}^2}\frac{1}{\Omega_{\rm F}^2\varpi^2}\left(\frac{e'}{E}\right)^2 = 0. \tag{4.161}$$

Here we use the relation $E \approx \Omega_{\rm F} L$ valid for the magnetically dominated flows.

First of all, a very important conclusion can be made from the definitions of q and $\Sigma_{\rm r}$. Indeed, having solved Eq. (4.161) for $\Sigma_{\rm r}^2$

$$\Sigma_{\rm r}^2 = \frac{1 - \dfrac{L^2}{\varpi^2 E^2} - \dfrac{1}{\Omega_{\rm F}^2\varpi^2}\left(\dfrac{e'}{E}\right)^2\dfrac{1}{q^2} - \left(\dfrac{\mu\eta}{E}\right)^2\dfrac{1}{q^2}}{1 - \dfrac{2}{\Omega_{\rm F}^2\varpi^2} + 2q} \tag{4.162}$$

(this relation generalizes (4.142)), we readily show that in the vicinity of the fast magnetosonic surface the value $\Sigma_{\rm r}$ must be close to unity. This implies that the angular momentum $L \approx E/\Omega_{\rm F}$ fixing the longitudinal electric current must have the quite definite value $L \approx L_{\rm cr}$, where

$$L_{\rm cr} \approx \frac{\Omega_{\rm F}\varpi|\nabla\Psi|}{8\pi^2}. \quad (4.163)$$

Since $\varpi|\nabla\Psi| \approx \Psi$ in the vicinity of the fast magnetosonic surface, the longitudinal current $I_{\rm cr} = 2\pi L_{\rm cr}$ is close to the GJ current $I_{\rm GJ} = \Omega_{\rm F}\Psi/2\pi$. Otherwise, we make the important conclusion:

Theorem 4.3 *For the relativistic flows the smooth passage of the fast magnetosonic surface is possible only if the longitudinal current is close to the GJ current.*

We emphasize that this conclusion naturally follows from the conclusion made in the analysis of the force-free flows. Indeed, as we saw, the flow for small currents could not be extended beyond the light surface and for rather large currents it remains subsonic to infinity. In both cases, the force-free flow cannot cross the fast magnetosonic surface. This is the case for the moderate longitudinal currents $I \approx I_{\rm GJ}$ only.

Problem 4.13 Check that for the magnetically dominated flows (i.e., for $\sigma \gg 1$, $\gamma_{\rm inj} \ll \sigma$), the correction to unity in the numerator and the denominator of expression (4.162) really proves small.

At first sight, the above conclusion is too rigorous and at least unexpected. However, we readily show that for $r \approx r_{\rm f}$ the condition $\Sigma_{\rm r} \approx 1$ must be satisfied for the transonic flows. To show this, it is supposed that, given the angular velocity $\Omega_{\rm F}$ (and, hence, the approximate location of the Alfvén surface $r_{\rm A}$), we try to give the boundary conditions on the surface $r \gg r_{\rm A}$, i.e., in the domain where the conditions needed to obtain Eq. (4.161) are valid. It is obvious that the physically meaningful (i.e., real positive) roots of Eq. (4.161) exist only if the coefficient of q^2 in square brackets is positive. But this is not, obviously, the case for

$$\Sigma_{\rm r}^2 > \frac{1 - \dfrac{2}{x_r^2}}{1 - \dfrac{1}{x_r^2}} \quad (4.164)$$

($x_r = \Omega_{\rm F}\varpi$). These large values of $\Sigma_{\rm r}$ correspond to the small longitudinal currents for which, as was demonstrated in Chap. 2, the solution cannot be extended to large distances from the light cylinder. As to the range of parameters $\Sigma_{\rm r} \ll 1$ (i.e., for large longitudinal currents), they, as we saw, can be realized only for the subsonic flows. The flows with $\Sigma_{\rm r} \approx 1$, which can satisfy the GS equation, invite further investigation.

We now proceed to an extensive analysis of the cubic equation (4.161). Here, we use the fact that the fast magnetosonic surface corresponds to crossing two roots of Eq. (4.161) at the same point. On the other hand, Eq. (4.161) has three real roots only if $Q < 0$, where the discriminant of the cubic equation Q (for $r \approx r_{\rm f}$) is

$$Q = \left[\frac{\mu^2\eta^2}{\Sigma_{\rm r}^2 E^2} + \frac{1}{\Sigma_{\rm r}^2 \Omega_{\rm F}^2 \varpi^2}\left(\frac{e'}{E}\right)^2\right]^2 - \frac{1}{27}\left[\frac{\mu^2\eta^2}{\Sigma_{\rm r}^2 E^2} + \frac{1}{\Sigma_{\rm r}^2 \Omega_{\rm F}^2 \varpi^2}\left(\frac{e'}{E}\right)^2\right]\left[\frac{1}{\Sigma_{\rm r}^2} - 1 + \left(2 - \frac{1}{\Sigma_{\rm r}^2}\right)\frac{1}{\Omega_{\rm F}^2 r^2 \sin^2\theta}\right]^3. \quad (4.165)$$

Therefore, the regularity conditions of the solution in the vicinity of the fast magnetosonic surface $r = r_{\rm f}$ can be written as

$$Q|_{r=r_{\rm f}} = 0, \qquad \left.\frac{\partial Q}{\partial r}\right|_{r=r_{\rm f}} = 0, \qquad \left.\frac{\partial Q}{\partial \theta}\right|_{r=r_{\rm f}} = 0. \quad (4.166)$$

Indeed, as shown in Fig. 4.2, the zero value of Q for curves 1 corresponds to the coincidence of two of the three roots at the stopping point, so that these solutions cannot be extended to the domain $Q > 0$. On the other hand, for the parameters corresponding to curves 3, the roots of the cubic equation do not intersect at all. Only curves 2, for which conditions (4.166) are satisfied, intersect at the saddle point that corresponds to the fast magnetosonic surface.

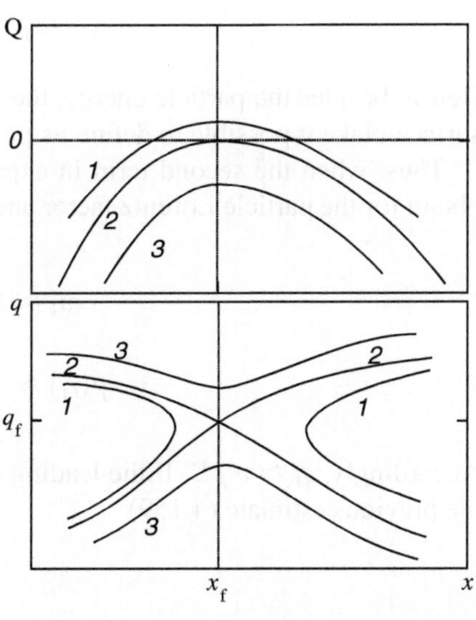

Fig. 4.2 Motion of the roots of the cubic equation in the vicinity of the saddle singular point. The zero value of Q for curves 1 corresponds to the coincidence of two of the three roots at the stopping point, so that these roots cannot be extended to the domain $Q > 0$. For the parameters corresponding to curves 3, the roots of the cubic equation do not intersect at all. Only curves 2, for which conditions (4.166) are satisfied, intersect at the saddle point that corresponds to the fast magnetosonic surface

Problem 4.14 Check that conditions (4.166) exactly coincide with the regularity conditions $D = 0$ (4.56), $N_r = 0$, and $N_\theta = 0$ (4.60).

As a result, the relations $Q = 0$ and $\partial Q/\partial r = 0$ under the condition $\Sigma_\mathrm{r} \approx 1$ can be written as

$$\frac{1}{x_\mathrm{f}^2} + \frac{1}{\Sigma_\mathrm{f}^2} - 1 \approx 3q_\mathrm{f}, \tag{4.167}$$

$$\frac{1}{x_\mathrm{f}^2} + \frac{x_\mathrm{f}\Sigma_\mathrm{f}'}{\Sigma_\mathrm{f}^3} \approx q_\mathrm{f}, \tag{4.168}$$

where again $x_r = \Omega_\mathrm{F}\varpi$, $\Sigma_\mathrm{r}' = \mathrm{d}\Sigma_\mathrm{r}/\mathrm{dr}$, and the indices 'f' everywhere correspond to the quantities on the fast magnetosonic surface. When deriving relations (4.167) and (4.168) we also used the exact expression for q at the moment of the coincidence of the roots

$$q_\mathrm{f} = \left[\left(\frac{\mu\eta}{E}\right)^2 + \frac{1}{\Omega_\mathrm{F}^2\varpi^2}\left(\frac{e'}{E}\right)^2\right]^{1/3}. \tag{4.169}$$

Since, as seen from (4.162), the derivative Σ_f' in order of magnitude can be estimated as

$$x_\mathrm{f}\Sigma_\mathrm{f}' \sim 1 - \Sigma_\mathrm{f}^2, \tag{4.170}$$

we obtain the relation

$$x_\mathrm{f} \approx q_\mathrm{f}^{-1/2}. \tag{4.171}$$

Hence, besides the particle energy, the critical conditions on the fast magnetosonic surface make it possible to define its position.

Thus, when the second term in expression (4.169) is a leading one, we finally obtain for the particle Lorentz factor and the radius of the fast magnetosonic surface

$$\varpi_\mathrm{f} \approx R_\mathrm{L}\left(\frac{E}{e'}\right)^{1/2}, \tag{4.172}$$

$$\gamma(r_\mathrm{f}) \approx \frac{e'}{\mu\eta} = \gamma_\mathrm{inj}. \tag{4.173}$$

Accordingly, $q_\mathrm{f} \approx e'/E$. If the leading term in (4.169) is the first term, we have by the previous estimate (4.159)

$$\varpi_{\rm f} \approx R_{\rm L}\left(\frac{E}{\mu\eta}\right)^{1/3}, \tag{4.174}$$

$$\gamma(r_{\rm f}) \approx \left(\frac{E}{\mu\eta}\right)^{1/3} \sim \sigma^{1/3}, \tag{4.175}$$

and $q_{\rm f} \approx (\mu\eta/E)^{2/3}$. Recall that the exact angular dependence $r_{\rm f}$ on θ can be defined, given the functions $E(\Psi)$ and $\eta(\Psi)$ and the solution to the GS equation. Recall also that all relations are obtained by assuming that $\Omega_{\rm F}\varpi \gg 1$. Therefore, in the vicinity of the rotation axis a more detailed consideration is needed.

Besides, one more important conclusion is that for the strongly magnetized flows ($\sigma \gg 1$, $\gamma_{\rm inj} \ll \sigma^{1/3}$) the particle energy on the fast magnetosonic surface region is merely a small part as compared to the maximum possible energy ($\gamma \approx \sigma$). Indeed, the condition $\gamma_{\rm f} \sim \sigma^{1/3}$ shows that the particle-to-electromagnetic energy flux ratio is only

$$\frac{W_{\rm part}}{W_{\rm em}} \sim \max\left(\sigma^{-2/3}, \frac{\gamma_{\rm inj}}{\sigma}\right). \tag{4.176}$$

Hence, in the relativistic case the substantial transformation of the energy flux from the electromagnetic field into the particle energy is possible only beyond the fast magnetosonic surface.

Thus, in the foregoing we proved the following theorem.

Theorem 4.4 *In the relativistic case, in the vicinity of the outer fast magnetosonic surface the particle Lorentz factor attains the values of*

$$\gamma(r_{\rm f}) = \max\left[\left(\frac{E}{\mu\eta}\right)^{1/3}, \gamma_{\rm inj}\right]. \tag{4.177}$$

Therefore, here for the strongly magnetized flows ($\gamma_{\rm inj} \ll \sigma^{1/3}$) the energy fraction transported by the particles is only a small part ($\sim\sigma^{-2/3}$) as compared to the electromagnetic energy flux. The smooth passage of the fast magnetosonic surface is possible only if the condition $\Sigma_{\rm r}(r_{\rm f}) \approx 1$ is satisfied, which corresponds to the GJ longitudinal current flowing in the magnetosphere.

Estimate (4.175) for the particle energy on the fast magnetosonic surface was first derived by Michel (1969). However, the assertion was made that the Lorentz factor $\gamma = \sigma^{1/3}$ is attained only at an infinite distance from the central body. Afterward this assertion was reproduced in a lot of papers (see, e.g., Okamoto (1978); Kennel et al. (1983); Li et al. (1992); Lery et al. (1998)) and regarded as the general property of the strongly magnetized flows. In reality, the conclusion that the fast magnetosonic surface must be at infinity was based on the self-inconsistent choice of the poloidal magnetic field structure. The point is that in all the above papers the Michel monopole solution, which is not the exact solution for the nonzero particle mass, was used as a poloidal magnetic field. As seen from Eqs. (4.167) and (4.168), they indeed have no solution for finite $x_{\rm f}$ for the Michel monopole magnetic field ($\Sigma_{\rm r}=1$, $\Sigma_{\rm r}'=0$). If the difference between the poloidal magnetic field and the monopole one is considered self-consistently, as we saw, the fast magnetosonic surface transfers at a finite distance from the central body.

Finally, note that the conditions for the weakly and strongly magnetized flows $\gamma_{\rm inj} \ll \sigma^{1/3}$ and $\gamma_{\rm inj} \gg \sigma^{1/3}$ correspond to those of the fast and slow rotation of the central body (Bogovalov, 2001). Indeed, the equality $\gamma_{\rm inj} = \sigma^{1/3}$ can be rewritten as $\Omega_{\rm F} = \Omega_{\rm cr}$, where

$$\Omega_{\rm cr} = 2\pi \left(\frac{\mu\eta\gamma_{\rm inj}^3}{\Psi_0}\right)^{1/2}, \tag{4.178}$$

so that for the fast rotation ($\Omega_{\rm F} \gg \Omega_{\rm cr}$) the particle energy may greatly increase when approaching the fast magnetosonic surface, whereas for the slowly rotating sources $\Omega_{\rm F} \ll \Omega_{\rm cr}$ the particle energy remains actually the same as near the origin. The values $\Omega_{\rm cr}$ for concrete astrophysical objects will be considered in Sect. 5.1.

On the other hand, the leading terms in algebraic equation (4.160) in the vicinity of the inner fast magnetosonic surface have the form

$$q^3 + \frac{1}{2}\left[1 - \frac{(E-\omega L)^2}{\Sigma_{\rm r}^2 E^2} - 2\frac{\alpha^2}{(\Omega_{\rm F}-\omega)^2\varpi^2} + \frac{\alpha^2 L^2}{\Sigma_{\rm r}^2 E^2 \varpi^2}\right] q^2$$
$$+\frac{1}{2}\frac{\alpha^2}{\Sigma_{\rm r}^2(\Omega_{\rm F}-\omega)^2\varpi^2}\left(\frac{e'}{E}\right)^2 = 0. \tag{4.179}$$

Here, as we see, we can always disregard the next-to-last term in Eq. (4.160), which in the vicinity of the horizon is $\Omega_{\rm F}^2\varpi_{\rm g}^2/\gamma_{\rm inj}^2$ times less than the term proportional to $(e'/E)^2$. As a result, this equation can be analyzed in the same way as the outer fast magnetosonic surface. Expression (4.162) for $\Sigma_{\rm r}^2$ can now be rewritten as

$$\Sigma_{\rm r}^2 = \frac{\dfrac{(E-\omega L)^2}{E^2} - \dfrac{\alpha^2 L^2}{\varpi^2 E^2} - \dfrac{\alpha^2}{(\Omega_{\rm F}-\omega)^2\varpi^2}\left(\dfrac{e'}{E}\right)^2\dfrac{1}{q^2}}{1 - 2\dfrac{\alpha^2}{(\Omega_{\rm F}-\omega)^2\varpi^2} + 2q}. \tag{4.180}$$

As in the vicinity of the outer fast magnetosonic surface, all corrections to the first terms in the numerator and the denominator appear much less than unity, we can take

$$\Sigma_{\rm r}^2 \approx \frac{(E-\Omega_{\rm H} L)^2}{E^2}. \tag{4.181}$$

Here we also used the condition for closeness of the inner fast magnetosonic surface to the event horizon.

Further, the discriminant of Eq. (4.179) can be written as

$$Q = \left[\frac{\alpha^2}{\Sigma_{\rm r}^2(\Omega_{\rm F}-\omega)^2\varpi^2}\left(\frac{e'}{E}\right)^2\right]^2 \tag{4.182}$$
$$-\frac{\alpha^2}{27\Sigma_{\rm r}^2(\Omega_{\rm F}-\omega)^2\varpi^2}\left(\frac{e'}{E}\right)^2\left[\frac{(E-\omega L)^2}{\Sigma_{\rm r}^2E^2}-1+2\frac{\alpha^2}{(\Omega_{\rm F}-\omega)^2\varpi^2}-\frac{\alpha^2}{\Sigma_{\rm r}^2}\frac{L^2}{\varpi^2E^2}\right]^3.$$

It yields (Hirotani et al., 1992; Beskin and Kuznetsova, 2000a)

$$q(r_{\rm f}) \approx \frac{1}{\Sigma_{\rm r}}\left(\frac{e'}{E}\right), \tag{4.183}$$

$$\alpha_{\rm f}^2 = \alpha^2(r_{\rm f}) \approx \frac{(\Omega_{\rm H}-\Omega_{\rm F})^2\varpi_{\rm g}^2}{\Sigma_{\rm r}}\left(\frac{e'}{E}\right), \tag{4.184}$$

$$\gamma(r_{\rm f}) \approx \frac{\gamma_{\rm inj}}{\alpha_{\rm f}}. \tag{4.185}$$

Main attention should be given to the latter relation. As we see, the particle Lorentz factor on the inner fast magnetosonic surface $\gamma_{\rm f} = \gamma(r_{\rm f})$ differs from the Lorentz factor in the ejection region only by the coordinate factor $\alpha_{\rm f}$. This implies that no additional electromagnetic acceleration near the black hole horizon actually occurs (Punsly, 2001). The difference in the values γ is just associated with the particle acceleration by the strong gravitational field of the black hole. Thus, the following theorem was proved:

Theorem 4.5 *On the inner fast magnetosonic surface the particle energy differs from the particle energy at large distances from the black hole only by the coordinate factor α: $\gamma_{\rm f} \approx \gamma_{\rm inj}/\alpha(r_{\rm f})$. This implies that there is no additional electromagnetic acceleration of particles in the vicinity of the horizon.*

Besides, we see that the fast magnetosonic surface at angles of θ not too close to zero is located much closer to the horizon than the Alfvén surface

$$\alpha^2(r_{\rm f}) \approx \alpha^2(r_{\rm A})\frac{\gamma_{\rm inj}}{\sigma}, \tag{4.186}$$

so that $\alpha^2(r_{\rm f}) \ll \alpha^2(r_{\rm A})$, where $\alpha^2(r_{\rm A}) \approx (\Omega_{\rm H}-\Omega_{\rm F})^2\varpi_{\rm g}^2$ corresponds to the Alfvén surface. As is evident from (4.184), in the vicinity of the fast magnetosonic surface the ratio

$$\frac{\alpha_{\rm f}^2}{(\Omega_{\rm F}-\omega)^2\varpi^2} \approx \frac{e'}{E} \tag{4.187}$$

really appears small as compared to unity, which justifies dropping the corresponding terms in the general equation (4.160). Finally, comparing solution (4.181) with definition (4.136) (and also taking $E \approx \Omega_{\rm F}L$, which is the case for the strongly magnetized flows), we immediately obtain the relation

$$L \approx \frac{(\Omega_{\rm H} - \Omega_{\rm F})}{8\pi^2} \sin\theta \, \frac{{\rm d}\Psi}{{\rm d}\theta}, \tag{4.188}$$

which, as is readily checked, actually coincides with the "boundary condition on the horizon." It is shown below that this coincidence is not accidental.

4.4.4 Fast Magnetosonic Surface—Nonrelativistic Flows

We now proceed to the analysis of the nonrelativistic flows. It is obvious that in this case we deal with the outer fast magnetosonic surface only. Here, on the contrary, we can quote the following theorem:

Theorem 4.6 *For the nonrelativistic flows the smooth crossing of the fast magnetosonic surface is possible only if the particle energy flux is of the order of one-third of the total energy flux. Otherwise, the transonic nonrelativistic flows cannot be magnetically dominated at large distances* $r \gg r_{\rm f}$.

We first show under what conditions we have the relation $W_{\rm part}(r_{\rm f}) = W_{\rm tot}(r_{\rm f})/3$. We consider nonrelativistic Bernoulli's equation (4.88) assuming that the fast magnetosonic surface is at distances much larger than the Alfvén surface radius. In this case $\mathcal{M}^2(r_{\rm f}) \gg 1$, so that Eq. (4.88) can be roughly rewritten as

$$\frac{1}{2}\frac{\mathcal{M}^4}{64\pi^4\eta_{\rm n}^2\varpi^2}(\nabla\Psi)^2 + \frac{\Omega_{\rm F}^2\varpi^2}{\mathcal{M}^2} = E_{\rm n}. \tag{4.189}$$

Here the first term on the right-hand side corresponds to the kinetic particle energy $v_{\rm p}^2/2$ (the toroidal velocity decreases with the distance as r^{-1} and for $r_{\rm f} \gg r_{\rm A}$ can be dropped). The second term is an asymptotic expression for the Poynting vector flux. Naturally, we also disregarded the contribution of the enthalpy and the gravitational potential which also vanish at large distances. As a result, Eq. (4.189) can be rewritten as

$$q_{\rm n}^3 - 2\frac{v_{\rm in}^2}{\Sigma_{\rm n}^2 E_{\rm n}} q_{\rm n} + 2\frac{v_{\rm in}^4}{\Sigma_{\rm n}^2 E_{\rm n}^2} = 0. \tag{4.190}$$

Here now

$$\Sigma_{\rm n}^2 = \frac{\Omega_{\rm F}^4\varpi^2(\nabla\Psi)^2}{64\pi^4 v_{\rm in}^2\eta_{\rm n}^2 E_{\rm n}^2}. \tag{4.191}$$

The value $q_{\rm n}$ is given by the relation

$$q_{\rm n} = \frac{v_{\rm in}^2}{\Omega_{\rm F}^2\varpi^2}\mathcal{M}^2, \tag{4.192}$$

and the normalization factor $v_{\rm in} = {\rm const}$ in order of magnitude (up to the heat terms dropped here) coincides with the plasma ejection velocity in the source. The first term in (4.190) corresponds to the particle flow contribution, the latter to the electromagnetic flux, and the second term with the opposite sign to the total energy flux $E_{\rm n}$.

Using the expression for the discriminant of the cubic equation (4.190)

$$Q_{\rm n} = -\frac{8}{27}\frac{v_{\rm in}^6}{E_{\rm n}^3\Sigma_{\rm n}^6} + \frac{v_{\rm in}^8}{E_{\rm n}^4\Sigma_{\rm n}^4}, \tag{4.193}$$

and the regularity condition $Q_{\rm n} = 0$ at the fast magnetosonic point, we obtain for $\Sigma_{\rm f}^2 = \Sigma_{\rm n}^2(r_{\rm f})$

$$\Sigma_{\rm f}^2 = \frac{8}{27}\frac{E_{\rm n}}{v_{\rm in}^2}. \tag{4.194}$$

As a result, we have

$$q_{\rm n}(r_{\rm f}) = \frac{3}{2}\frac{v_{\rm in}^2}{E_{\rm n}}. \tag{4.195}$$

Comparing the corresponding terms in (4.190) and relations (4.194) and (4.195), we obtain the condition $W_{\rm part}(r_{\rm f}) = W_{\rm tot}(r_{\rm f})/3$.

As was specially emphasized, we would have the condition $W_{\rm part} = W_{\rm tot}/3$ only if the fast magnetosonic surface is located at the distance much larger than the Alfvén radius $r_{\rm f}$. In reality, as is readily verified, solution (4.195) corresponds to the condition $\mathcal{M}_{\rm f}^2 \sim 1$. Therefore, in the nonrelativistic case, the radius of the fast magnetosonic surface $r_{\rm f}$ is always close to that of the Alfvén radius $r_{\rm A}$ (to be exact, the ratio $r_{\rm f}/r_{\rm A}$ is not larger than several units). Therefore, we must analyze full Bernoulli's equation

$$\mathcal{M}^4\Sigma_{\rm n}^2 + \frac{x_{\rm n}^2}{v_{\rm in}^4}\left(\frac{v_{\rm in}^2}{E_{\rm n}}\right)^2 \frac{(v_{\rm in}^2 x_{\rm n}^2 - \mathcal{M}^2\Omega_{\rm F}L_{\rm n})^2}{(1-\mathcal{M}^2)^2}$$
$$+2\frac{x_{\rm n}^4}{v_{\rm in}^2}\left(\frac{v_{\rm in}^2}{E_{\rm n}}\right)^2 \frac{\Omega_{\rm F}L_{\rm n} - x_{\rm n}^2 v_{\rm in}^2}{1-\mathcal{M}^2} = 2x_{\rm n}^4\left(\frac{v_{\rm in}^2}{E_{\rm n}}\right), \tag{4.196}$$

where the first two terms correspond to the particle flux, the third term to the electromagnetic flux, and there is the total energy flux on the right-hand side. Here now $x_{\rm n} = \Omega_{\rm F}\varpi/v_{\rm in}$, and we, for simplicity, returned to the variable $\mathcal{M}^2$.

Analysis shows that for fast rotation, solutions (4.194) and (4.195) remain approximately valid for the full equation (4.196). The fast rotation is associated with the condition $E_{\rm n} \gg v_{\rm in}^2$, when in the vicinity of the fast magnetosonic surface the flow is strongly magnetized, i.e., the contribution of the electromagnetic energy flux (Poynting vector) proves larger than the particle flux. In this case,

$$\Sigma_{\rm f}^2 \approx \frac{E_{\rm n}}{v_{\rm in}^2}, \tag{4.197}$$

$$x_{\rm f}^2 \approx \frac{E_{\rm n}}{v_{\rm in}^2}, \tag{4.198}$$

$$\mathcal{M}_{\rm f}^2 \approx 1 \tag{4.199}$$

(the third condition just implies that the Alfvén and fast magnetic surfaces are close). Relation (4.198) can be rewritten as

$$r_{\rm f}^2 \approx \frac{E_{\rm n}}{\Omega_{\rm F}^2}. \tag{4.200}$$

Since for the fast rotation we can take $E_{\rm n} \approx \Omega_{\rm F} L_{\rm n}$, the estimate (4.200) is naturally close to estimate (4.155) for the Alfvén radius.

Problem 4.15 Using relations (4.88) and (4.101), show that for the magnetically dominated ($E_{\rm n} \approx \Omega_{\rm F} L_{\rm n}$) cold flow the Mach number on the fast magnetosonic surface $\mathcal{M}_{\rm f}^2 = \mathcal{M}^2(r_{\rm f}, \theta)$ is given by the expression

$$\mathcal{M}_{\rm f}^2 = \frac{1}{1-(1-L_{\rm n}/\Omega_{\rm F}\varpi_{\rm f}^2)^{2/3}}. \tag{4.201}$$

Further, comparing solution (4.194) with definition (4.191) and again taking $\varpi(\nabla\Psi) \approx \Psi_0$, we find

$$E_{\rm n} \approx \frac{\Omega_{\rm F}^{4/3}\Psi_0^{2/3}}{4\pi^{4/3}\eta_{\rm n}^{2/3}}. \tag{4.202}$$

Accordingly, the total energy losses, according to (4.83), can be estimated as

$$W_{\rm tot} \approx E_{\rm n}\eta_{\rm n}\Psi_0 \approx \Omega^{4/3}\dot{M}^{1/3}\Psi_0^{4/3}, \tag{4.203}$$

where $\dot{M} \approx \eta_{\rm n}\Psi_0 \approx 4\pi\rho_{\rm in}v_{\rm in}R_{\rm in}^2$ is the mass ejection rate. Thus, we get

$$W_{\rm tot} \approx \Omega^{4/3}\dot{M}^{1/3}B_{\rm in}^{4/3}R_{\rm in}^{8/3}, \tag{4.204}$$

where again the subscripts "in" correspond to the quantities in the source. Besides, condition (4.202) allows us to estimate the critical angular velocity $\Omega_{\rm crit}$ separating the fast and slow rotations. Indeed, taking in (4.202) $E_{\rm n} = v_{\rm in}^2/2$, we obtain

$$\Omega_{\rm crit} = \left(\frac{v_{\rm in}^3\eta_{\rm n}}{\Psi_0}\right)^{1/2} = \frac{v_{\rm in}}{R_{\rm in}}\left(\frac{4\pi\rho_{\rm in}v_{\rm in}^2}{B_{\rm in}^2}\right)^{1/2}. \tag{4.205}$$

Hereafter we take $\Psi_0 \approx \pi R_{\rm in}^2 B_{\rm in}$ and $\eta_{\rm in} \approx \rho_{\rm in} v_{\rm in}/B_{\rm in}$. We discuss the values $W_{\rm tot}$ and $\Omega_{\rm crit}$ for young stellar objects in Sect. 5.1.3.

Further, the expression for the radius of the fast magnetosonic surface $r_{\rm f}$ can be written as

$$r_{\rm f}^2 \approx R_{\rm in}^2 \left(\frac{B_{\rm in}^2}{4\pi \rho_{\rm in} v_{\rm in}^2}\right) \left(\frac{\Omega_{\rm F}}{\Omega_{\rm cr}}\right)^{-2/3}. \tag{4.206}$$

Evaluating now the longitudinal current as $I \approx 2\pi c \eta_{\rm n} E_{\rm n}/\Omega_{\rm F}$, we make the unexpected conclusion that the dimensionless current $i_0 = I/I_{\rm GJ}$ must be much larger than unity

$$i_0 \approx \frac{c}{v_{\rm in}} \left(\frac{\Omega_{\rm F}}{\Omega_{\rm cr}}\right)^{-2/3} \approx \left(\frac{c^2}{2E_{\rm n}}\right)^{1/2}. \tag{4.207}$$

Thus, in the nonrelativistic case for the smooth passage of the fast magnetosonic surface the longitudinal current must be much larger than the GJ one. As seen from the second equality in (4.207), in the relativistic limit ($E_{\rm n} \to c^2$) the longitudinal current approaches the GJ one.

Finally, comparing the corresponding terms in (4.196), we conclude that in the whole range of parameters $\Omega_{\rm F} \gg \Omega_{\rm crit}$ the particle energy near the fast magnetosonic surface is to be comparable with the total energy $E_{\rm n}$. Otherwise, in the nonrelativistic case the transonic flow cannot be magnetically dominated at large distances. Even if the Poynting flux at the base of the flow is much larger than the particle energy flux, the smooth passage through the fast magnetosonic surface is possible only if there is substantial particle acceleration in the vicinity of this surface.

As to the case of small angular velocities $\Omega_{\rm F} \ll \Omega_{\rm crit}$, we can disregard the second and third terms in (4.196). As a result, we have

$$\Sigma_{\rm f}^2 \approx 2x_{\rm f}^4, \tag{4.208}$$

$$r_{\rm f}^2 \approx \frac{\Psi_0}{\eta_{\rm n}}. \tag{4.209}$$

Thus, we find

$$r_{\rm f}^2 \approx R_{\rm in}^2 \left(\frac{B_{\rm in}^2}{4\pi \rho_{\rm in} v_{\rm in}^2}\right), \tag{4.210}$$

$$i_0 \approx \frac{c}{v_{\rm in}}. \tag{4.211}$$

In this case, the particle energy flux coincides with the total energy flux.

Note that the above expressions have meaning only if the condition $r_{\rm f} \gg R_{\rm in}$ is satisfied. Otherwise, the critical surfaces would be in the immediate vicinity of the compact object and there would be no pronounced particle acceleration. According to (4.206), the condition $r_{\rm f} \gg R_{\rm in}$ is satisfied only for rather small angular velocities $\Omega_{\rm F} \ll \Omega_2$, where

$$\Omega_2 = \frac{B_{\rm in}^2}{4\pi \rho_{\rm in} v_{\rm in}^2} \frac{v_{\rm in}}{R_{\rm in}}. \tag{4.212}$$

Therefore, it is convenient to introduce the nonrelativistic parameter $\mu_{\rm l} = \Omega_{\rm F}/\Omega_2$

$$\mu_{\rm l} = \frac{4\pi \rho_{\rm in} v_{\rm in} \Omega_{\rm F} R_{\rm in}}{B_{\rm in}^2}, \tag{4.213}$$

which is sometimes called the "mass loading" (Anderson et al., 2005). As a result, for $\mu_{\rm l} \ll 1$, the radius of the fast magnetosonic surface $r_{\rm f}$ and the particle velocity $v_{\rm f} \approx \sqrt{2E_{\rm n}/3}$ (which, in fact, is close to the limit velocity v_∞) can be written as (see, e.g., Spruit, 1996)

$$r_{\rm f} \approx \mu_{\rm l}^{-1/3} R_{\rm in}, \tag{4.214}$$

$$v_{\rm f} \approx \mu_{\rm l}^{-1/3} \Omega_{\rm F} R_{\rm in}. \tag{4.215}$$

Accordingly, for $\mu_{\rm l} \gg 1$, we simply have $r_{\rm f} \approx R_{\rm in}$ and $v_\infty \approx \Omega_{\rm F} R_{\rm in}$. Note that the mass loading $\mu_{\rm l}$ can be estimated directly from the observations. Indeed,

$$\mu_{\rm l} = \frac{\dot{M} \Omega_{\rm F}}{B_{\rm in}^2 R_{\rm in}}. \tag{4.216}$$

As will be shown in Sect. 5.1.3, for young stellar objects (YSO) $\mu_{\rm l} \ll 1$ and $P_2 = 2\pi/\Omega_2 \sim 10^2$–$10^3$ s. As a result, the condition $\Omega \ll \Omega_2$ is always satisfied for the young stars.

Thus, the matter can be effectively accelerated in the range of angular velocities $\Omega_{\rm crit} \ll \Omega_{\rm F} \ll \Omega_2$ only. Clearly, it is possible only if the condition $\Omega_{\rm crit} \ll \Omega_2$ is satisfied, which, as is readily verified, corresponds to the condition

$$\frac{B_{\rm in}^2}{4\pi \rho_{\rm in} v_{\rm in}^2} \gg 1. \tag{4.217}$$

This must be the case, since, as inequality (4.217) is violated, the energy density of the outflowing plasma would be larger than the energy density of the electromagnetic field and the electromagnetic acceleration of particles could not take place.

Problem 4.16 Show that, as in the relativistic case, the toroidal magnetic field on the Alfvén surface is compared with the poloidal one.

Problem 4.17 Show that the criterion for the slow rotation $\Omega_{\rm F} \ll \Omega_{\rm crit}$ corresponds to the condition $\omega_{\rm n} \ll 1$ and for the fast rotation $\Omega_{\rm F} \gg \Omega_{\rm crit}$ to the condition $\omega_{\rm n} \approx 1$, where (see, e.g., Lery et al., 1998)

$$\omega_{\rm n} = \frac{\Omega_{\rm F} r_{\rm A}}{{\rm V}_{\rm A}(r_{\rm A})}. \qquad (4.218)$$

Problem 4.18 Show that the value $r_{\rm f} \approx r_{\rm A}$ (4.206) can be obtained directly from the definition $\mathcal{M}(r_{\rm A}) = 1$ under the quasimonopole outflow condition $\rho v r^2 \approx$ const.

Problem 4.19 Show that if we again introduce the magnetization parameter

$$\sigma = \frac{\Omega^2 \Psi_0}{8\pi^2 c^3 \eta_{\rm n}} = \frac{v_{\rm in}}{c}\left(\frac{\Omega R_{\rm in}}{c}\right)^2 \frac{B_{\rm in}^2}{8\pi \rho_{\rm in} V_{\rm in}^2}, \qquad (4.219)$$

we have

- for the nonrelativistic case $\sigma \ll 1$,
- expression (4.207) for the current i_0 can be written in compact form as

$$i_0 = \sigma^{-1/3}, \qquad (4.220)$$

- the "mass loading" $\mu_{\rm l}$ can be written as

$$\mu_{\rm l} = \left(\frac{\Omega_{\rm F} R_{\rm in}}{c}\right)^3 \sigma^{-1}. \qquad (4.221)$$

Note that the condition of the efficient particle acceleration in the vicinity of the singular surfaces ($W_{\rm part} \sim W_{\rm tot}$ for $r \approx r_{\rm f}$) can be derived from the simplest estimate of the value of the toroidal velocity $v_{\varphi}(r_{\rm f}) \approx \Omega_{\rm F} r_{\rm f}$ (4.90). Indeed, for the magnetically dominated flows we can take $E_{\rm n} \approx \Omega_{\rm F} I / 2\pi c \eta_{\rm n}$, which immediately yields the value of $E_{\rm n} \approx \Omega_{\rm F}^2 r_{\rm f}^2$ for $I \approx i_0 I_{\rm GJ}$. Moreover, we readily show that the poloidal velocity $v_{\rm p}$ in the vicinity of the singular surfaces is also of order $\Omega_{\rm F} r_{\rm f}$. Thus, the character of the particle acceleration in the magnetically dominated nonrelativistic wind becomes evident.

Theorem 4.7 *Inside the Alfvén surface (where the energy density of the magnetic field is much larger than that of particles), the magnetic field acts as a sling ensuring the constant angular velocity of plasma. As a result, the particle velocity linearly* *increases with distance from the rotation axis:* $v \sim \Omega_{\rm F}\varpi$. *However, this acceleration ceases at distances of* $r > r_{\rm f}$, *where the plasma energy density is compared with the energy density of the electromagnetic field.*

In this domain, the particle energy is mainly determined by the poloidal velocity component, as the flow has the different asymptotic behavior $v_\varphi \approx \Omega_{\rm F}\varpi_{\rm A}^2/\varpi$ (4.93).

Finally, recall that exactly in the vicinity of the fast magnetosonic surface the structure of characteristic surfaces, which was described in detail for the hydrodynamical flows, can be reconstructed. Of special interest here is, undoubtedly, the moment of the analyticity loss when a shock wave is to occur in the vicinity of the nonstandard singular point as the flow parameters slowly change. Using a concrete example below, we show that this effect, evidently, really happens for the magnetically dominated wind.

4.4.5 Behavior of the Solution at Large Distances

The behavior of the solution to the GS equation in the asymptotically distant region $r \gg r_{\rm f}$ has also been the focus of attention for many years, because the problems of the current closure and the formation of jets observed in the broadest class of astrophysical objects are connected with it. Below we return to this problem when discussing in detail the exact solutions for the magnetized wind. Here we state several general assertions of the particle collimation and acceleration.

Note at once that one must distinguish between the physical and mathematical infinities. As we will see, in the relativistic case the collimation of the magnetic surfaces is weakly pronounced, so that the visible collimation may be exponentially far as compared to the characteristic scales in the problem—the light cylinder radius $R_{\rm L}$ or the fast magnetosonic surface radius $r_{\rm f}$. Clearly, under the real conditions the flow properties change much earlier due to the interaction between the wind and the environment with finite pressure and finite magnetic field. For example, for the characteristic parameters of the pulsar wind ($B_\varphi(R_{\rm L}) \approx B_{\rm p}(R_{\rm L}) \sim 1$ G, $R_{\rm L} \sim 10^{10}$ cm) and the interstellar medium ($B_{\rm ext} \sim 10^{-6}$ G), this change is to occur, at least, at distances where the toroidal magnetic field of a wind becomes equal to the external magnetic field. The corresponding scale

$$R_{\rm t} \sim R_{\rm L}\frac{B_{\rm L}}{B_{\rm ext}} \sim 10^{16}\,\text{cm}, \tag{4.222}$$

where $B_{\rm L} = B(R_{\rm L})$, as we see, is comparable with the transverse size of the jet of young pulsars. However, it is not improbable that the pronounced distortion of the quasimonopole wind occurs already on much smaller scales when the poloidal component of the magnetic field is compared with the external magnetic field. This is the case already at distances

$$R_{\rm t}' \sim R_{\rm L}\left(\frac{B_{\rm L}}{B_{\rm ext}}\right)^{1/2} \sim 10^{13}\,\text{cm}, \tag{4.223}$$

which for the characteristic distances to the radio pulsars of order 1 kpc is within the angular resolution of the present-day telescopes. Thus, under the real conditions the general results of the asymptotic behavior of the solution for $r \to \infty$ should be used with caution.

Let us now try to formulate some general properties the magnetized flows must have at large distances from the central object. It is obvious that we are primarily interested in the transonic flows when the flow is supersonic at large distances. Therefore, the asymptotic condition can be written as

$$\varpi \gg r_{\rm f}. \tag{4.224}$$

Besides, we assume that the solution can be extended to infinity. The longitudinal electric current, as we saw, should be rather large.

We first show that for the polytropic index $\Gamma > 1$ we can disregard the pressure gradient contribution at large distances, i.e., we believe that the outflowing plasma is cold. This conclusion can be readily deduced from both the GS equation and the Bernoulli's equation. Analyzing, for example, the nonrelativistic expressions (4.88) and (4.102), we see how both the enthalpy $w = c_{\rm s}^2/(\Gamma - 1) \propto n^{\Gamma-1}$ in (4.88) and the temperature $T = k(s)n^{\Gamma-1}$ in (4.102) decrease with distance from the compact source, because for any divergent flow the concentration vanishes at large distances, i.e., $n \to 0$ for $r \to \infty$. Therefore, the final temperature contribution (enthalpy, entropy) as compared to the total energy E and its derivative $\mathrm{d}E/\mathrm{d}\Psi$ is negligible at large distances.

Problem 4.20 Show that for $\Gamma > 1$ the finite temperature contribution can be disregarded in the relativistic case, both at large distances (i.e., for $r \gg r_{\rm f}$) and in the vicinity of the black hole horizon ($\alpha^2 \ll \alpha_{\rm f}^2$).

Problem 4.21 Show that, on the contrary, for the cylindrical jets the thermal effects can be substantial.

Analyzing the leading terms in the GS equation (4.66), we can show that at large distances it can be written as (Heyvaerts and Norman, 1989; Bogovalov, 1998; Okamoto, 1999)

$$\frac{1}{2}\mathbf{n}_{\perp} \cdot \nabla(\mathbf{B}_{\rm p}^2) - \frac{B_{\varphi}^2 + 4\pi n m_{\rm p} c^2 \gamma}{R_{\rm c}} + \frac{B_{\varphi}^2 - \mathbf{E}^2}{\varpi}(\mathbf{n}_{\perp} \cdot \mathbf{e}_{\varpi}) = 0. \tag{4.225}$$

Here $R_{\rm c}$ is the curvature radius of the magnetic field line in the poloidal plane and $\mathbf{n}_{\perp} = \nabla\Psi/|\nabla\Psi|$. Otherwise, for $r \gg r_{\rm f}$, the GS equation describes the balance of the bulk centrifugal force

$$\mathcal{F}_{\rm c} = \frac{nmc^2\gamma + S/c}{R_{\rm c}} \tag{4.226}$$

and of the bulk electromagnetic force

$$\mathcal{F}_{\rm em} = \rho_{\rm e}\mathbf{E} + \nabla\left(\frac{B^2}{8\pi}\right). \tag{4.227}$$

As we see, in the numerator of expression (4.226), besides the obvious term associated with the particles, there appears a term due to the electromagnetic energy flux (Poynting vector $S \approx (c/4\pi)E_\theta B_\varphi \approx (c/4\pi)B_\varphi^2$). The point is that, as was noted, the electromagnetic energy, as well as the outflowing matter, propagates along the curved magnetic surfaces. Consequently, an additional force is needed to bend the vector **S** along the particle trajectory. In the nonrelativistic case, this force is provided by Ampére's force associated with the longitudinal electric current. In the relativistic case, it is necessary to include the force $\rho_{\rm e}\mathbf{E}$ associated with the electric field.

Analysis of Eq. (4.225) shows that if the curvature of the magnetic field lines is sufficiently large, the first term due to the bulk force $j_\varphi B_{\rm p}/c$ can be dropped. Thus, in the nonrelativistic case, the GS equation can be written as (Okamoto, 1999)

$$\frac{\rho v_\parallel^2}{R_{\rm c}} = \frac{1}{c} j_\parallel B_\varphi. \tag{4.228}$$

In the relativistic case, using the general asymptotic expression (4.144), we obtain for the weakly magnetized outflow

$$\frac{\mu n u^2}{R_{\rm c}} \sim \frac{1}{c\gamma^2} j_\parallel B_\varphi \tag{4.229}$$

(a more exact expression is obtained in Sect. 5.2.3). For the magnetically dominated flow when the main energy is transferred by the Poynting vector, we find using the general relation (4.144)

$$\frac{|\mathbf{E}| B_\varphi}{4\pi R_{\rm c}} \approx \frac{j_\parallel B_\varphi}{c\gamma^2}. \tag{4.230}$$

Since at large distances $|\mathbf{E}| \approx B_\varphi$ and $(4\pi/c)j_\parallel = \partial B_\varphi/\partial\varpi \sim B_\varphi/\varpi$, relation (4.230) can be rewritten in compact form as (Beskin et al., 2004)

$$\gamma \approx \sqrt{\frac{R_{\rm c}}{\varpi}}. \tag{4.231}$$

Relations (4.229), (4.230), and (4.231) allow us to make a number of rather general conclusions:

1. The sign of the curvature radius depends on the direction of the longitudinal current $j_\parallel$. Therefore, if there is a bulk back current in the magnetosphere, one

should expect the collimation of the magnetic surfaces in the vicinity of the axis and their decollimation in the vicinity of the equator (Okamoto, 1999).

2. In the relativistic case ($\gamma \gg 1$), the curvature radius of the magnetic surfaces R_c must be much larger than the radius r. Hence, the collimation (decollimation) of the magnetic surfaces for the ultrarelativistic outflow must be strongly suppressed.
3. Conversely, in the nonrelativistic case ($\gamma \approx 1$), the curvature radius R_c is comparable with the radius r. This implies that already in the vicinity of the fast magnetosonic surface region the pronounced collimation (decollimation) of the magnetic surfaces must be present.

As was already mentioned, the above relations correspond to the case in which the curvature of the magnetic surfaces is of vital importance. Clearly, for the flows close to the cylindrical ones, the equilibrium is to be established already by the balance of the first and third terms in Eq. (4.225) when the centrifugal force can be disregarded and the force $j_\varphi B_p/c \sim \nabla(B^2/8\pi)$, on the contrary, proves substantial. Using the estimate $B_\varphi \approx (\Omega\varpi/c)B_p$, we obtain for the relativistic flows the simple relation

$$\gamma = \frac{\Omega\varpi}{c}, \tag{4.232}$$

which, as we will see, is also rather universal. As a result, the choice between the asymptotic solutions (4.231) and (4.232) must depend on how much the magnetic surfaces would be curved.

It is easy to show that the parabolic magnetic field, where the field lines at a large distance from the central source are described by $z(\varpi) \propto \varpi^2$, is a terminating configuration. Indeed, the curvature radius R_c for the magnetic surfaces prescribed by $z(\varpi) \propto \varpi^k$ can be defined as (Korn and Korn, 1968)

$$R_c = \frac{[z'^2 + 1]^{3/2}}{z''}, \tag{4.233}$$

where $z' = dz/d\varpi$ and $z'' = d^2z/d\varpi^2$. At large distance from the equatorial plane where $z' \gg 1$ formula (4.231) yields for the particle energy moving along the magnetic field line

$$\gamma \propto \varpi^{k-1}. \tag{4.234}$$

Therefore, for $k = 2$ the acceleration efficiency defined by relations (4.231) and (4.232) appears identical. As a result, if the magnetic surfaces are collimated more strongly than those for the parabolic field (i.e., for $k > 2$), at large distances the curvature of the magnetic surfaces can be disregarded and the particle energy is defined by (4.232). If the flow is weakly collimated (i.e., for $1 < k < 2$), the particle acceleration is less effective and relation (4.231) should be used. The cases $k = 2$ (parabola) and $k = 1$ (monopole field) need a special examination given in Sects. 5.3.4 and 5.3.5. The numerical modeling (Narayan et al., 2007; Barkov and Komissarov, 2008) completely confirms the above pattern.

Recall once again that these conclusions can be applied to not too large distances from the compact object (i.e., only at physical infinity) and to the domain distant from the rotation axis ($\theta \sim 1$). As to the behavior of the solution at true (mathematical) infinity, this problem invites separate investigation. The point is that at large distances the right-hand sides of Eqs. (4.228) and (4.229) must decrease as r^{-3}, whereas the numerator on the left-hand side as r^{-2}. As a result, the curvature radius of the magnetic surfaces R_c must increase as r. This conclusion immediately follows from (4.231), because the particle Lorentz factor is bounded from above by the value of σ. But this behavior cannot be realized at mathematical infinity (Heyvaerts and Norman, 2003a,b,c) and, therefore, for $r \to \infty$, only the right-hand sides of Eqs. (4.228) and (4.229) prove to be leading ones. In particular, for the nonrelativistic flow we simply have

$$j_{\parallel} = 0. \tag{4.235}$$

This conclusion can be formulated as follows:

Theorem 4.8 *At "mathematical infinity" practically all longitudinal current must be concentrated in the vicinity of the rotation axis (Heyvaerts and Norman, 1989). Then the returning current is to flow along the equatorial plane.*

Recall that this hypothesis concerns the absence of the environment.

Problem 4.22 Show that at mathematical infinity the decollimation of magnetic surfaces is impossible (Heyvaerts and Norman, 1989).

On the other hand, as seen from (4.231), for the ultrarelativistic case the curvature radius of the magnetic surfaces R_c must be much larger than the radius r. Therefore, the flow must be radial with adequate accuracy. As shown below, a great difference from the radial flow (i.e., passage from physical to mathematical infinity) can be only exponentially far from the compact object.

Nevertheless, analysis of the relativistic equations makes it possible to get important additional information. Assuming that at mathematical infinity $r \to \infty$

1. we can disregard the radial derivatives in Bernoulli's equation and the GS equation (which implies that the left-hand side in (4.228) and (4.229) is dropped),
2. the square of the Mach number $\mathcal{M}^2$ is proportional to r^2 and $\mathcal{M}^2/r^2$ is thus independent of r,

we reduce the problem to the one-dimensional one. And, as we know, in the one-dimensional case the GS equation can be integrated (Heyvaerts and Norman, 1989). Indeed, for the conical magnetic surfaces $\Psi = \Psi(\theta)$, in the asymptotically distant region $r \to \infty$ the GS equation (4.66) for $\mu = \text{const}$ (and $c = 1$) is rewritten as

$$-\frac{1}{\sin\theta}\frac{\mathrm{d}}{\mathrm{d}\theta}\left[\frac{\Omega_{\rm F}^2\sin^2\theta+m^2(\theta)}{\sin\theta}\frac{\mathrm{d}\Psi}{\mathrm{d}\theta}\right]+\left(\frac{\mathrm{d}\Psi}{\mathrm{d}\theta}\right)^2\Omega_{\rm F}\frac{\mathrm{d}\Omega_{\rm F}}{\mathrm{d}\Psi}$$

$$+32\pi^4\frac{\partial}{\partial\Psi}\left[\frac{E^2}{\Omega_{\rm F}^2\sin^2\theta+m^2(\theta)}\right]-\frac{32\pi^4}{m^2(\theta)}\frac{\mathrm{d}}{\mathrm{d}\Psi}(\mu^2\eta^2)=0, \qquad (4.236)$$

where the value $m^2(\theta) = \mathcal{M}^2/r^2$ is independent of the radius r. We recall that here the derivative $\partial/\partial\Psi$ acts on the integrals of motion only. If we now multiply Eq. (4.236) by $(\Omega_{\rm F}^2\sin^2\theta + m^2(\theta))(\mathrm{d}\Psi/\mathrm{d}\theta)$ and use the asymptotic behavior of the Bernoulli's relativistic equation (4.44)

$$\frac{[\Omega_{\rm F}^2\sin^2\theta+m^2(\theta)]^2}{64\pi^4r^4\sin^2\theta}\left(\frac{\mathrm{d}\Psi}{\mathrm{d}\theta}\right)^2=E^2-\frac{\mu^2\eta^2}{m^4(\theta)}[\Omega_{\rm F}^2\sin^2\theta+m^2(\theta)]^2, \qquad (4.237)$$

we obtain upon the elementary, though cumbersome, transformations (Heyvaerts and Norman, 1989)

$$\frac{\mathrm{d}}{\mathrm{d}\theta}\left(\frac{q^2\Omega_{\rm F}^2}{\mu^2\eta^2}\right)=0, \qquad (4.238)$$

where again $q = \mathcal{M}^2/\Omega_{\rm F}^2\varpi^2$. In particular, for $\eta = \text{const}$, $\Omega_{\rm F} = \text{const}$, we simply have

$$q = \text{const}. \qquad (4.239)$$

We emphasize at once that relation (4.238) is of a universal character, i.e., it holds for both the relativistic and the nonrelativistic cases. Indeed, using the asymptotic behavior of the nonrelativistic expression (4.85) for the electric current $I \approx 2\pi c\eta_{\rm n}\Omega_{\rm F}\varpi^2/\mathcal{M}^2$, we at once conclude that at large distances the condition $I \approx \text{const}$ is to be satisfied. This implies that the longitudinal current density $j_{\parallel}$ should vanish on the conical surfaces. Otherwise, we again return to relation (4.235) obtained by disregarding the centrifugal force ($R_{\rm c} \to \infty$).

Problem 4.23 Show that in the relativistic case condition (4.238) yields relation $I/\gamma \approx \text{const}$ refining expression (4.229).

As a result, taking into account condition (4.238), Eq. (4.236) can be readily integrated. For example, under the condition $\Omega_{\rm F} = \text{const}$ and $\eta = \text{const}$, i.e., for $q = \text{const}$, its solution has the form

$$\Psi(r\to\infty,\theta)=\Psi_0\left[1-\sqrt{1+\frac{\gamma_{\rm in}}{\sigma}}\cdot\frac{(1+\cos\theta)^p-(1-\cos\theta)^p}{(1+\cos\theta)^p+(1-\cos\theta)^p}\right], \qquad (4.240)$$

where the exponent p is

$$p = \frac{\sqrt{1+\gamma_{\rm in}/\sigma}}{1+q}. \tag{4.241}$$

We also used the following assumptions:

1. The angular momentum $L(\Psi)$ can be taken as

$$L(\Psi) = \frac{\mu\eta}{\Omega_{\rm F}}\sigma\left(2\frac{\Psi}{\Psi_0} - \frac{\Psi^2}{\Psi_0^2}\right), \tag{4.242}$$

which corresponds to the Michel monopole solution (2.225).
2. Bernoulli integral $E(\Psi)$ is

$$E(\Psi) = \mu\eta\gamma_{\rm in} + \mu\eta\sigma\left(2\frac{\Psi}{\Psi_0} - \frac{\Psi^2}{\Psi_0^2}\right), \tag{4.243}$$

where the first term corresponds to the particle contribution and the second to the electromagnetic field contribution. The constant $\gamma_{\rm in}$ was supposed to be constant (below we discuss this form of the Bernoulli integral in more detail).
3. We used the boundary condition $\Psi(\pi/2) = \Psi_0$.

The characteristic profile of the stream function $\Psi(\theta)$ is shown in Fig. 4.3. Analysis of the exact solution (4.240) leads to a number of conclusions confirming the above assertions.

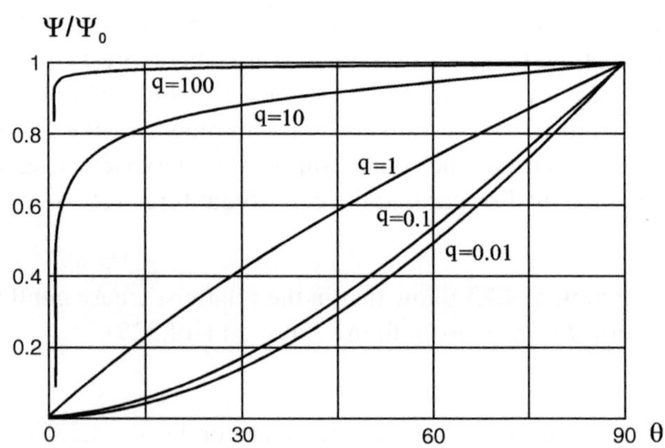

Fig. 4.3 The behavior of the stream function $\Psi(\theta)$ (4.240) in the asymptotically far domain $r \to \infty$ for the different values of q for $\gamma_{\rm in} \ll \sigma$

- When considering the finite particle mass ($\sigma < \infty$) we cannot satisfy the boundary condition $\Psi(0) = 0$. This implies that in the vicinity of the rotation axis the flow must differ from the radial one (Bogovalov, 1995).

- If at large distances the energy is transported by the electromagnetic field (i.e., $W_{\rm em} \gg W_{\rm part}$, so that $q \ll 1$), the flow is to remain actually a radial one, i.e., $\Psi \approx \Psi_0(1 - \cos\theta)$.
- On the other hand, if at large distances practically the total energy is transported by particles (so that $q \gg 1$), there would be the strong collimation of the magnetic surfaces to the rotation axis.

Recall once again that the solution (4.240) allows us to observe only the coupling between the magnetic field structure and the efficiency of the particle acceleration. In reality, the value q is not initially known and its value is to be found along with the solution of the complete problem.

Problem 4.24 Using definition (5.114) for nonrelativistic Bernoulli integral

$$E_{\rm n} = \frac{v_{\rm in}^2}{2} + i_0\sigma c^2\left(2\frac{\Psi}{\Psi_0} - \frac{\Psi^2}{\Psi_0^2}\right), \tag{4.244}$$

and relation (5.123), which are needed to find the values $q = \mathcal{M}^2/\Omega_{\rm F}^2\varpi^2$ (and for $\Omega_{\rm F} = \text{const}$, $\eta = \text{const}$, i.e., for $q = \text{const}$), find the magnetic flux $\Psi(\theta)$ in the asymptotically distant domain $r \to \infty$.

Problem 4.25 Estimate the angular dimension of the central core for the nonrelativistic flow.

4.4.6 Behavior of the Solution in the Vicinity of the Horizon

Finally, we consider the behavior of the solution to the GS equation in the vicinity of the rotating black hole. As shown in the hydrodynamical case, for the physically reasonable transonic flows the radial velocity of matter on the horizon must not be zero. Consequently, the concentration n on the horizon (recall that it is determined in the comoving reference frame) must be finite and, therefore, the Mach number must be different from zero: $\mathcal{M}^2(r_{\rm g}) \neq 0$. As a result, the denominator $D(r_{\rm g}, \theta)$, as is readily verified, can be rewritten as ($c = G = 1$)

$$D(r_{\rm g}, \theta) = -1 + \frac{\alpha^2}{\mathcal{M}^2 B_{\rm p}^2}(B_{\hat\varphi}^2 - E_{\hat\theta}^2). \tag{4.245}$$

Here we dropped all terms that obviously vanish for $\alpha^2 \to 0$. On the other hand, if we take α^2 to be zero in algebraic Bernoulli's equation (4.44), under the condition $\mathcal{M}^2(r_{\rm g}) \neq 0$ on the horizon it can be rewritten as

$$\frac{(E-\Omega_{\rm H}L)^2}{\left[(\Omega_{\rm F}-\Omega_{\rm H})^2\varpi^2+\mathcal{M}^2\right]^2}=\frac{1}{64\pi^4\rho_{\rm K}^2\varpi^2}\left(\frac{{\rm d}\Psi}{{\rm d}\theta}\right)^2. \tag{4.246}$$

By definitions (4.26) and (4.27), this relation can be rewritten as $|B_{\hat{\varphi}}(r_{\rm g})| = |E_{\hat{\theta}}(r_{\rm g})|$. This result is in good agreement with the main assertion of the "membrane paradigm," according to which ZAMO are to detect the φ-component of the magnetic field and the θ-component of the electric field, which diverge as $1/\alpha$ (Thorne et al., 1986). Thus, the second term in expression (4.245) must also vanish for $\alpha^2 \to 0$. Therefore, on the horizon the condition

$$D(r_{\rm g},\theta) = -1 \tag{4.247}$$

must be satisfied. This implies that for $\mathcal{M}^2(r_{\rm g}) \neq 0$, Eq. (4.66) in the vicinity of the horizon must be of a hyperbolic type. Therefore, the full MHD version of the GS equation (4.66) does not need any boundary condition on the horizon. This must be the case, because no signal, by definition, can propagate from the horizon to the outer regions of the magnetosphere (Punsly and Coroniti, 1990b).

Problem 4.26 Show that in the vicinity of the horizon $D = -1+\alpha^2\mathcal{K}$, where

$$\mathcal{K}=\frac{(E-\Omega_{\rm F}L)^2}{(E-\Omega_{\rm H}L)^2}\frac{A^2}{\mathcal{M}^6}+\frac{\mu^2\eta^2A^2}{(E-\Omega_{\rm H}L)^2\mathcal{M}^6(1-c_{\rm s}^2)}\left[\varpi^2(\Omega_{\rm F}-\Omega_{\rm H})^2+\mathcal{M}^2c_{\rm s}^2\right], \tag{4.248}$$

so that $\mathcal{K} > 0$. As we see, in the vicinity of the horizon $D > -1$, which just corresponds to the hyperbolic domain of the GS equation.

Problem 4.27 Show that the estimate of the location of the inner fast magnetosonic surface $\alpha_{\rm f}^2 \approx 1/\mathcal{K}$ derived from the condition $D = 0$ coincides with expression (4.184).

Problem 4.28 Show that the hydrodynamical limit of (4.246) fixes the accretion rate (1.271) for the case $c_* = 1$ ($r_* = r_{\rm g}$).

Hence, within the full MHD version, the hyperbolic domain of the general equation (4.66) must be above the black hole horizon. In particular, according to (4.247), even in the limit $\mathcal{M}^2 \to 0$, the condition $D = -1$ must be satisfied on the horizon. On the other hand, as we saw, the force-free equation (3.49) remains elliptic up to the horizon. Therefore, it is necessary to consider the passage from MHD to the force-free approximation in more detail.

We first consider the limit of the regularity conditions (4.56) and (4.60) on the fast magnetosonic surface for $\mathcal{M}^2 \to 0$. As is obvious from (4.55), the very location of this surface for $\mathcal{M}^2 \to 0$ extends to the black hole horizon. On the other hand, the total singularity in D is in the factor $1/\mathcal{M}^2$. The values $\mathcal{M}^2 D$ and $N_k' = N_k/A$ in the force-free limit remain finite on the horizon. Therefore, as the limit of the regularity conditions (4.60), we can consider the values of N_a' for $\mathcal{M}^2 = 0$ and $r = r_{\rm g}$.

As a result, the condition $N_\theta'(r_{\rm g}) = 0$ can be rewritten as

$$\frac{\mathrm{d}}{\mathrm{d}\theta}\left[64\pi^4 \frac{(E - \Omega_{\rm H} L)^2}{(\Omega_{\rm F} - \Omega_{\rm H})^4 \varpi_{\rm g}^2} - \frac{1}{\rho_{\rm K}^2(r_{\rm g})}\left(\frac{\mathrm{d}\Psi}{\mathrm{d}\theta}\right)^2\right] = 0. \tag{4.249}$$

We readily see that condition (4.249) coincides with the "boundary condition on the horizon" (4.246). Analogously, as is readily verified, the equality $\mathcal{M}^2 D(r_{\rm g}) = 0$ yields condition (4.246) as well. Finally, the condition $N_r'(r_{\rm g}) = 0$ upon elementary, though cumbersome, transformations can be reduced to the form

$$r_{\rm g}\frac{\partial}{\partial r}\left[(\nabla\Psi)^2 - \frac{16\pi^2 I^2}{(\Omega_{\rm F} - \omega)^2 \varpi^2}\right]_{r=r_{\rm g}}$$
$$+\frac{2\xi + (\Omega_{\rm H} - \Omega_{\rm F})\varpi_{\rm g}[\xi^2 + 1/(\alpha\gamma)^2]}{(\Omega_{\rm H} - \Omega_{\rm F})\varpi_{\rm g}\rho_{\rm K}^2}\left(\frac{\mathrm{d}\Psi}{\mathrm{d}\theta}\right)^2 = 0, \tag{4.250}$$

where

$$\xi = \left.\frac{u_{\hat\varphi}}{\alpha\gamma}\right|_{r=r_{\rm g}}, \tag{4.251}$$

and all the values in (4.250) are taken on the black hole horizon.

As we see, condition (4.250), besides the "force-free" values $I(\Psi)$ and $\Omega_{\rm F}(\Psi)$, also comprises the plasma parameters γ and $u_{\hat\varphi}$, while for the relativistic plasma ($\alpha\gamma \gg 1$) is only their ratio ξ (4.251). Evaluating $(\partial/\partial r)(\nabla\Psi)^2 \sim \Psi_0^2/r_{\rm g}^3$, we have

$$u_{\hat\varphi}(r_{\rm g}) \sim \alpha\gamma\Omega_{\rm H}\varpi_{\rm g}, \tag{4.252}$$

in complete agreement with general relations (4.39) and (4.40).

Here we have one of the key points that shed light on the difference between the force-free approximation and the full MHD version of the GS equation. Indeed, as is readily seen, Bernoulli's equation (4.246) for $\alpha^2 = 0$ in the force-free limit $\mathcal{M}^2 = 0$, $E = \Omega_{\rm F} L$ coincides with the "boundary condition on the horizon" (3.52). As a result, the additional relation resulting from Bernoulli's equation, available in the full version, automatically ensures the regularity condition for $\alpha^2 \to 0$ (certainly, provided that the flow is not bounded by the light surface located beyond the event horizon). To show this, it is convenient to use Eq. (4.64). As a result, in the limit $\alpha^2 \to 0$, the GS equation has the form

$$\frac{1}{\alpha}\nabla_k\left[\frac{1}{\varpi^2\alpha}A\nabla^k\Psi\right]+\frac{\Omega_{\rm F}-\Omega_{\rm H}}{\alpha^2}(\nabla\Psi)^2\frac{{\rm d}\Omega_{\rm F}}{{\rm d}\Psi}-\frac{32\pi^4}{\alpha^2}\frac{\partial}{\partial\Psi}\left[\frac{(E-\Omega_{\rm H}L)^2}{A}\right]=0. \tag{4.253}$$

If we suppose now that the solution is regular on the horizon, i.e., if we drop the terms containing $\alpha\partial\Psi/\partial r$ and $\alpha^2\partial^2\Psi/\partial r^2$, we obtain multiplying (4.253) by $2A({\rm d}\Psi/{\rm d}\theta)/\sin^2\theta$

$$2A\varpi_{\rm g}^2(\Omega_{\rm F}-\Omega_{\rm H})\frac{{\rm d}\Omega_{\rm F}}{{\rm d}\theta}\left[\frac{1}{\varpi_{\rm g}^2\rho_{\rm K}^2(r_{\rm g})}\left(\frac{{\rm d}\Psi}{{\rm d}\theta}\right)^2-\frac{64\pi^4(E-\Omega_{\rm H}L)^2}{A^2}\right]$$
$$+\frac{\rm d}{{\rm d}\theta}\left[\frac{A^2}{\varpi_{\rm g}^2\rho_{\rm K}^2(r_{\rm g})}\left(\frac{{\rm d}\Psi}{{\rm d}\theta}\right)^2-64\pi^4(E-\Omega_{\rm H}L)^2\right]=0. \tag{4.254}$$

Clearly, this relation is automatically true if condition (4.246) is satisfied. But, on the other hand, if the "condition on the horizon" resulting from Bernoulli's equation is satisfied, the singular terms $\alpha\partial\Psi/\partial r$ and $\alpha^2\partial^2\Psi/\partial r^2$ in the GS equation are to be zero for $\alpha^2\to 0$. This implies that the solution is regular on the event horizon.

Thus, we are ready to quote the key statement:

Theorem 4.9 *The "boundary condition on the horizon" in the force-free approximation is the rudiment of the critical condition on the fast magnetosonic surface when it tends to the black hole horizon. Hence, this condition is given, in fact, on the surface causally connected with the outer space. Consequently, the force-free approximation has no intrinsic contradiction associated with the necessity to use an additional boundary condition in the causally unconnected domain.*

We can also note here that, as seen from Fig. 4.4, with account taken of the nonzero particle mass, the Alfvén surface in the vicinity of the rotation axis is beyond the ergosphere that, by definition (1.218), crosses the horizon for $\theta=0$. Thus, the hypotheses of Theorem 3.2 advanced in the previous chapter are formally violated. According to this theorem, the energy release in the magnetosphere of the rotating black hole is possible only if the Alfvén surface is within the ergosphere. Actually, the theorem is formulated for the continuous flows whose hypotheses are

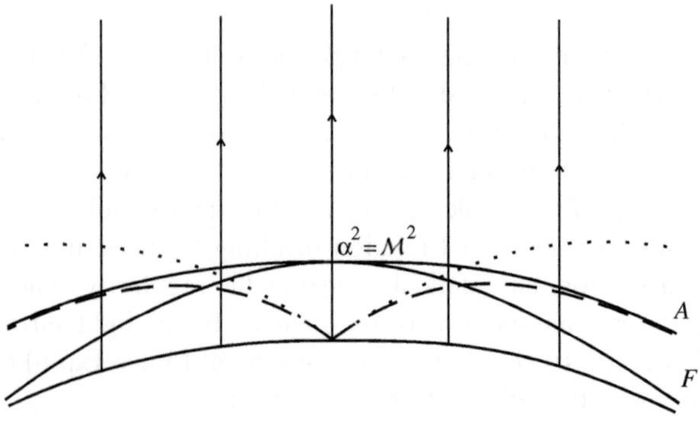

Fig. 4.4 The location of the Alfvén (A) and fast magnetosonic (F) surfaces in the vicinity of the black hole poles. The *dashed line* indicates the Alfvén surface in the force-free approximation and the *dotted line* indicates the ergosphere surface

violated in the plasma generation region. As a result, the particle energy flux on different sides of the generation region, because of the different sign of the value η, is positive both for the outflowing plasma and the matter falling onto the black hole. In particular, in the vicinity of the rotation axis, where the total energy flux is associated with the particles only, the positive energy flux leaving the black hole occurs along with the positive energy flux propagating to the event horizon. The energy source on these field lines contributing to both the increase in the energy of the black hole and the energy transfer at infinity is a photon source contributing to the plasma generation.

Finally, the above-mentioned property that the GS equation and Bernoulli's equation yield the same "condition on the horizon" can be considered differently. As we saw, the "condition on the horizon" is deduced from the GS equation (4.66), when all the terms that do not diverge in the vicinity of the horizon are dropped as α^{-2}. But, since the same asymptotic behavior follows from Bernoulli's equation (without which the GS equation is not closed), all terms of order α^{-2} can be analytically eliminated. As a result, in general form the GS equation in the vicinity of the horizon is

$$\frac{\Sigma^2}{\rho_{\rm K}^2}\frac{\partial^2\Psi}{\partial r^2} + \frac{(D+1)}{\alpha^2}\frac{\partial^2\Psi}{\partial\theta^2} + \mathcal{G} = 0, \tag{4.255}$$

where $\mathcal{G}$ (which is finite for $\alpha^2 \to 0$) does not comprise the second derivatives Ψ. Since always $D+1 \propto \alpha^2$, the full equation (4.66) (as well as the above hydrodynamical equations) is regular, i.e., in the limit $\alpha^2 \to 0$, it has no diverging coefficients of the higher derivatives.

Hence, the structures of the solution to the GS equation in the vicinity of the horizon and at large distances greatly differ from one another. Indeed, though for $\sigma \to \infty$ the fast magnetosonic surface tends to infinity, for any finite value σ beyond the outer fast magnetosonic surface there is always, figuratively speaking, infinitely much space, where the collimation of the magnetic field lines is possible. Therefore, the angular dependence $\Psi(\theta)$ at infinity and in the sonic surface region can differ considerably. In the vicinity of the black hole there are no small parameters of the higher derivatives. Therefore, when the fast magnetosonic surface approaches the event horizon, the substantial reconstruction of the magnetic field structure between these surfaces is impossible. That is why the limit of the critical condition on the fast magnetosonic surface for $r_{\rm f} \to r_{\rm g}$ and the "condition on the horizon" are equivalent.

Finally, one can show that the rudiment of the critical condition on the fast magnetosonic surface coincides with the regularity condition (3.57) in the vicinity of the black hole poles. To show this, recall that for the strongly magnetized flow ($V_{\rm A} \gg c_{\rm s}$) in the vicinity of the rotation axis, where $B_{\hat\varphi} \to 0$, the fast magnetosonic and Alfvén surfaces coincide with each other. It is important that for the finite particle mass these surfaces cross the rotation axis for $\alpha^2 = \mathcal{M}^2 > 0$, i.e., above the black hole horizon (see Fig. 4.4). As a result, for the finite particle mass the critical condition on the fast magnetosonic surface in the vicinity of the rotation axis can be deduced from the analysis of the numerator in algebraic relations (4.40) (in the general case, as was mentioned, the singularities in algebraic relations (4.38), (4.39), and (4.40) do not give rise to additional restrictions to the flow parameters). Since in the vicinity of the axis, as was noted, the energy E is specified only by particle contribution, we find from (4.40)

$$L \approx \frac{\omega(r_{\mathrm{f}}, 0) - \Omega_{\mathrm{F}}(0)}{\mathcal{M}^2(r_{\mathrm{f}}, 0)} (r_{\mathrm{f}}^2 + a^2)(\alpha\gamma)\mu|\eta|\theta^2. \tag{4.256}$$

Using now the definitions $\alpha n u_{\mathrm{p}} = \eta B_{\mathrm{p}}$ and $\mathcal{M}^2 = 4\pi\mu\eta^2/n$ and also the relation $\Psi \approx \pi B_{\mathrm{p}}(r_{\mathrm{f}}^2 + a^2)\theta^2$ valid in the vicinity of the rotation axis, we finally get

$$L \approx \frac{1}{4\pi^2}\left(\frac{\gamma}{u_{\mathrm{p}}}\right)[\omega(r_{\mathrm{f}}, 0) - \Omega_{\mathrm{F}}(0)]\,\Psi. \tag{4.257}$$

In the force-free limit ($r_{\mathrm{f}} \to r_{\mathrm{g}}$, $u_{\mathrm{p}} \to \gamma$), condition (4.257) naturally becomes relation (3.57) ensuring the regularity of the force-free solution on the horizon in the vicinity of the rotation axis.

The above example once again points to the substantial difference of the full MHD version of the GS equation from its force-free limit in which the Alfvén surface $\alpha^2 = (\omega - \Omega_{\mathrm{F}})^2\varpi^2$ crossed the event horizon for $\theta = 0$, which gave rise to the irrational solutions $\Psi(\theta) \propto \theta^q$ with $q \neq 2$. For the finite particle mass, the "condition on the horizon" (4.246) for $\theta \to 0$ has now the different behavior

$$\frac{E(0)}{\mathcal{M}^2}(r_{\mathrm{g}}^2 + a^2)\theta = \frac{1}{8\pi^2}\left(\frac{\mathrm{d}\Psi}{\mathrm{d}\theta}\right). \tag{4.258}$$

Since in the vicinity of the axis, where $u_{\hat{\varphi}} \to 0$ and $I \to 0$, the energy $E(0)$ is specified only by the particle contribution, we obtain the condition

$$2\pi B_{\mathrm{p}}(r_{\mathrm{g}}^2 + a^2)\,\theta = \left(\frac{\mathrm{d}\Psi}{\mathrm{d}\theta}\right), \tag{4.259}$$

which is identically satisfied by definition (3.14). This conclusion once again confirms the general conclusion that the GS equation for the finite particle mass has no singularity on the horizon.

Chapter 5
Full MHD Version—Particle Acceleration and Collimation

Abstract In this chapter, we discuss the analytical solutions obtained for cylindrical jets, quasimonopole outflows, and the black hole magnetosphere. It is demonstrated that the self-consistent analysis makes it possible to determine the main characteristics of the outgoing wind, including the determination of the total energy loss and the poloidal magnetic field structure. The conditions of the effective energy transformation from the electromagnetic flux to the particle flux are given as well. For cylindrical jets both the relativistic and nonrelativistic versions are discussed. It is shown that taking into account the finite ambient pressure, one can determine the magnetic flux within the central core. For nonrelativistic flows, which are magnetically dominated near the origin, the solution can be constructed only in the presence of an oblique shock near the jet base, where additional heating is to take place. The analytical solutions obtained for quasimonopole and parabolic magnetic fields illustrated the general property that the efficient particle acceleration can take place only for the strongly collimated outflows. Two toy solutions for the black hole magnetosphere support the physical nature of the Blandford–Znajek process. Finally, the results obtained by the self-similar approach and in the numerical simulation are briefly discussed. It is demonstrated that there is now a lot of numerical data that confirms the analytical results obtained by the Grad–Shafranov equation method.

5.1 Astrophysical Introduction

Problems of jet collimation and particle acceleration (the internal structure of jets, their stability, the interaction with energy release regions) are so diverse that a separate survey would be needed to discuss them (see, e.g., Krolik (1999b) and Camenzind (2007) and the bibliography therein). Therefore, we would like to discuss only the problems that can be studied by the GS equation method discussed here.

5.1.1 Radio Pulsars

As was already mentioned, the radio pulsar wind theory has not been constructed yet. In other words, there is no consistent model describing in a single language

V.S. Beskin, *MHD Flows in Compact Astrophysical Objects,* Astronomy and Astrophysics Library, DOI 10.1007/978-3-642-01290-7_6,

© Springer-Verlag Berlin Heidelberg 2010

the energy transfer from the neutron star surface to infinity, including the efficient particle acceleration, i.e., the total energy transfer from the electromagnetic flux to the outflowing plasma energy flux. Moreover, there is no reasonable theory of jets observed in the Crab and Vela radio pulsars (Weisskopf et al., 2000; Helfand et al., 2001).

Indeed, as was shown in Chap. 2, if the energy release is fully connected with the current energy losses, the major part of energy within the light cylinder must be transported by the electromagnetic field and the particle contribution is very small ($\sigma \sim 10^4$–10^6). On the other hand, the observations show that at large distances from the neutron star the major part of the energy must be transported by relativistic particles. For example, the analysis of the Crab Nebula radiation from a shock wave located at a distance of $\sim 10^{17}$ cm from the pulsar in the interaction region of the pulsar wind with supernova remnant uniquely shows that the ratio $W_{\rm part}/W_{\rm em}$ is only $\approx 10^{-3}$ here (Kennel and Coroniti, 1984a,b). Therefore, in the asymptotically far region the Poynting vector flux must be fully transferred to the outflowing plasma flux.

However, the transformation, evidently, occurs already at much smaller distances, viz., at distances comparable with the light cylinder. There is evidence that the alternating optical emission from companions in some close binary systems containing radio pulsars was detected (Djorgovski and Evans, 1988). It is logical to associate this optical emission, whose periodicity exactly coincides with the orbital period of the binary system, with the heating of part of the companion star facing the radio pulsar. It turned out that the energy reradiated by the companion actually coincides with the total energy emitted by the radio pulsar at the corresponding solid angle. Clearly, this fact cannot be explained either by the magnetodipole radiation or by the strong magnetized flux with a prevailing electromagnetic energy flux, because the efficiency of the low-frequency wave transformation cannot be close to unity. Only if a considerable part of energy is connected with the relativistic particles will the star surface heating be rather effective. Therefore, the σ-problem—the problem of the efficient energy transfer from the electromagnetic field to particles in the pulsar wind—is one of the key problems of present-day astrophysics.

If in the 1970s particular attention was given to the relativistic particle motion in the intense electromagnetic wave of a rotating magnetic dipole (Max and Perkins, 1971; Asséo et al., 1978), starting from the 1980s when it was clear that the particles are of crucial importance in the pulsar wind, the MHD approach became the main line of investigation (Ardavan, 1976; Okamoto, 1978; Li et al., 1992; Begelman and Li, 1994; Bogovalov, 1997a). Recall that this approach was simultaneously discussed in the context of the problem of the formation of jets from the active galactic nuclei (Phinney, 1983; Blandford and Znajek, 1977; Macdonald and Thorne, 1982; Camenzind, 1986; Begelman et al., 1984) and the young stars (Mestel, 1968; Heyvaerts and Norman, 1989; Sakurai, 1990; Pelletier and Pudritz, 1992; Shu et al., 1994). In fact, we dealt with the possibility to construct a complete solution, i.e., extend the solutions obtained by the force-free approximation for the inner magnetosphere regions to the pulsar wind region.

The point is that, as was noted, the force-free approximation, by which the first results were obtained, encounters certain difficulties. First of all, within this approximation, it is impossible to determine the part of energy transported by relativistic particles. Besides, since in the force-free approximation the electric current $I(\Psi)$ is constant on the magnetic field lines, there is no hope to thoroughly study the current closure problem.

As to the MHD approach, both the transformation of the energy from the electromagnetic field to particles and the poloidal magnetic field structure can be described rather simply. Besides, as the electric current I must no longer be constant on the magnetic field lines, the current closure problem can be studied. Unfortunately, as was shown, it has nothing to do with the angular velocity of the plasma $\Omega_{\rm F}(\Psi)$, which remains constant on the magnetic surfaces in this approximation. Finally, what is very important, within the full MHD GS equation version, the value of the electric current circulating in the magnetosphere is no longer a free parameter but must be determined from the critical conditions on the singular surfaces (Bogovalov, 1992; Beskin, 1997). Otherwise, one of the main problems in theory—the construction of the current system and, as a result, the determination of the energy losses—can be posed mathematically rigorously.

It is interesting to note that the first results, within the MHD approach, were obtained already at the end of the 1960s (Michel, 1969; Goldreich and Julian, 1970). It was F. Michel who introduced the key relativistic parameter—the magnetization parameter σ (2.84)

$$\sigma = \frac{e\Omega\Psi_{\rm tot}}{4\lambda m_{\rm e}c^3}, \tag{5.1}$$

determining the ratio of the electromagnetic energy flux to the particle energy flux near the star surface. Here $\Psi_{\rm tot}$ is the total magnetic flux in the source and $\lambda = n/n_{\rm GJ}$ is the multiplicity parameter. We already used this parameter in Chap. 2 in the definition of the usability condition of the force-free limit $\sigma \to \infty$.

As was mentioned, for radio pulsars

$$\Psi_{\rm tot} = \pi B_0 R_0^2 \approx \pi B_0 R^2 \frac{\Omega R}{c}, \tag{5.2}$$

which corresponds to the magnetic flux only in the region of open magnetic field lines. As a result, we find

$$\sigma = \frac{1}{4\lambda}\left(\frac{\Omega R}{c}\right)^2\left(\frac{\omega_B R}{c}\right) \sim 3\times 10^3 \left(\frac{\lambda}{10^3}\right)^{-1}\left(\frac{P}{1\,{\rm s}}\right)^{-2}\left(\frac{B}{10^{12}\,{\rm G}}\right). \tag{5.3}$$

We see that for ordinary radio pulsars ($P \sim 1$ s, $B_0 \sim 10^{12}$ G, the multiplication parameter $\lambda = n/n_{\rm GJ} \sim 10^3$) we have $\sigma \sim 10^3$–10^4, and only for the fastest ones ($P \sim 0.1$ s, $B_0 \sim 10^{13}$ G) σ is $\sim 10^5$–10^6. The large value of σ just shows that the main contribution of the energy flux inside the light cylinder is made by the electromagnetic field.

Accordingly, the total particle ejection rate $\dot{N} = \lambda\pi R_0^2 n_{GJ} c$ is

$$\dot{N} \sim 10^{33}\,\mathrm{part/s}\left(\frac{\lambda}{10^3}\right)\left(\frac{P}{1\,\mathrm{s}}\right)^{-2}\left(\frac{B}{10^{12}\,\mathrm{G}}\right). \tag{5.4}$$

Finally, expression (4.178) for the critical period $P_{cr} = 2\pi/\Omega_{cr}$ has the form

$$P_{cr} = \pi\frac{R}{c}\left[\frac{2}{\lambda\gamma_{inj}^3}\left(\frac{\omega_B R}{c}\right)\right]^{1/2} \sim 10^{-3}\,\mathrm{s}\left(\frac{\lambda}{10^3}\right)^{-1/2}\left(\frac{\gamma}{10^3}\right)^{-3/2}\left(\frac{B}{10^{12}\,\mathrm{G}}\right)^{1/2}. \tag{5.5}$$

As we see, actually all radio pulsars should be referred to the slowly rotating sources. This implies that the particle energy remains constant up to the fast magnetosonic surface.

Unfortunately, none of the MHD models allowed one to construct an adequate pulsar wind model. All attempts to find the self-consistent solution containing the efficient particle acceleration failed. Recall that this conclusion is true exactly for the relativistic case; the acceleration efficiency for the nonrelativistic flows, on the contrary, must be high (see Sect. 4.4.4). Thus, there is an evident contradiction between the necessity of the efficient particle acceleration resulting from the observations and the absence of this acceleration in "smooth" MHD flows in which the electric current is determined from the critical conditions on the singular surfaces, the light surface located at infinity. It is not surprising, therefore, that various models are actively discussed, in which it is proposed, somehow, to go beyond the limits of the "classical" scheme.

First of all, the efficient particle acceleration can be produced by the above property of the relativistic flows—for weak longitudinal electric currents the light surface $|\mathbf{E}| = |\mathbf{B}|$ is located at a finite distance and for $i_0 \ll 1$ in the vicinity of the light cylinder. Therefore, if the interaction between the regions of closed and open field lines really leads to the limitation of the longitudinal current i_0 (or there are some other causes fixing the value $i_0 < 1$ in the plasma generation region), one should expect the occurrence of a light surface and the efficient particle acceleration (Beskin et al., 1993; Chiueh et al., 1998). The efficient acceleration can also be associated with the reconnection processes in the pulsar wind (Michel, 1982; Coroniti, 1990; Lyubarsky and Kirk, 2001; Kirk and Skjaeraasen, 2003). However, so far this model has not provided the explanation of the efficient particle acceleration at distances comparable with the light cylinder.

Thus, in spite of an understanding of the importance of the pulsar wind and particle acceleration problem and a great number of papers devoted to this theme, there is no adequate analytical model of the pulsar wind now. As was noted, one of the reasons is the impossibility to formulate sufficiently simple equations describing the behavior of the relativistic plasma when its energy density is comparable with that of the electromagnetic field. Therefore, there is nothing definite about either the energy spectrum of particles escaping from the magnetosphere or their radiation. It is only clear that already at small distances from the light cylinder the particles must transport a considerable part of energy as compared to the total energy flux.

To conclude, one should mention a number of papers devoted to the interaction between the pulsar wind and the supernova remnants (see, e.g., Kennel and Coroniti, 1984a,b; Gallant and Arons, 1994; Bogovalov and Khangoulyan, 2002, 2003) and to the problem of the formation of jets observed in the Crab and Vela pulsars (Komissarov and Lyubarsky, 2003). However, these processes occur at distances of order 10^{17} cm, i.e., many orders larger than the light cylinder radius. Therefore, these papers no longer completely deal with the pulsar wind theory.

5.1.2 Active Galactic Nuclei

The problems of the collimation and particle acceleration in the active galactic nuclei (AGN) are key to understanding the processes occurring in their "central engine." Unfortunately, the angular resolution of the current receivers does not allow one to directly observe the plasma flow on scales of order 10^{13}–10^{14} cm as compared to the black hole radius. Therefore, we have to judge the galactic nuclei activity only by indirect manifestations analyzing the processes on much larger scales.

Recall that the diffusive radio emission regions first discovered over 50 years ago were associated with the jets issuing from the nuclei of the active galaxies. It is the jets that supply these regions with matter and energy extracted by the jets from the nucleus (Begelman et al., 1984). The observations show that the jets can be accelerated and collimated very close to the central object. For example, in the case of our closest active galaxy M87, a jet is generated in the interior of the region of size $60\,r_{\rm g}$ (Junor et al., 1999).

The matter outflowing from AGN has high energies—the bulk Lorentz factor of the jet, as a whole, reaches, as a minimum, several units. For example, in the galaxy M87 this motion is observed directly, and the Lorentz factor of the outflowing matter is $\gamma \approx 6$ (Junor and Biretta, 1995). In many cases, the matter preserves the relativistic velocities up to enormous distances from the nucleus before it appreciably slows down due to the interaction with the intergalactic matter. The other remarkable property of the jets is a high degree of their collimation—the opening angle is several degrees only. The separate class is radio-quiet nuclei. Their jets are weakly collimated and move at subrelativistic velocities (up to $\sim 0.1c$). They are observed directly in the radio band and indirectly on the wide absorption lines visible in the optical and UV band of the spectrum in 10% sources.

Unfortunately, the observations do not allow us to find the reliable estimates of the particle energy flux and the mass in the jets, the value of the magnetic field both in the vicinity of the black hole and in the jet itself, and determine the jet composition. The spectrum of the jets (unlike, for example, the jets from young stars) does not show any spectral features of the moving matter, i.e., neither the atomic (ionic) lines nor the annihilation line of electron–positron pairs is observed.

There is indirect evidence for (Reynolds et al., 1996b; Hirotani et al., 1999) and against (Sikora and Madejski, 2000) the leading role of the electron–positron pairs. Therefore, we cannot say yet which mechanism of the energy transport to the jets is realized in reality. Possibly, the energy of the ultrarelativistic jet core, which is observed at high radiofrequencies and in the gamma range, is extracted from the rotational energy of the black hole, and the collimation is due to a moderately relativistic outflow accelerated by the inner parts of the disk (Sol et al., 1989). The external jet is collimated by the flows (wind) from the external parts of the disk.

The visible superluminal motion, flares in the gamma range, and the variability of the jet radio emission on the timescale of the order of a day (so-called intraday variability) point to not only the relativistic velocities but also the jet perturbations, the character and nature of which are not clear yet. Possibly, these perturbations were due to the jet propagation process. However, it is not improbable that they resulted from the perturbations generated by the central engine. It is also possible that these perturbations are shock waves on which the particles are accelerated. In any event, the representation of the jets with high resolution demonstrate that the jet is not homogeneous in length but consists of bright spots alternating with slightly radiating regions.

The study of very compact radio sources by ground interferometers with long bases (the intercontinental system VLBI and the American telescope VLBA), and, lately, by a space interferometer (VSOP, Japan), which allows one, on short waves, to attain the angular resolution of order 10^{-4} arcseconds, also shows that separate compact radio components, which may be "extinguished" in a few years, are ejected from the active nuclei with periodicity of a few months. These radio components move along a spiral trajectory and are withdrawn from the nucleus as a rather narrow cone (first with opening angle of several tens of degrees and then several degrees), with the observed velocities exceeding the velocity of light.

The AGN can be divided into classes depending on the accretion rate and the angular momentum of the black hole. The point is that the supercritical accretion gives rise to a strong wind and a high degree of ionization of the accretion disk. Therefore, these systems should not have spectral singularities in the X-ray radiation of the disk, because the superstrong radiation is to suppress the formation of a collimated jet. One can even assume that the type of galaxy depends on the parameters of the central black hole and the accretion flows. For example, if at the initial stage of evolution a very massive black hole is generated ($M \sim 10^9\,M_{\odot}$) and the supercritical accretion onto it persists, the produced noncollimated outflow can interrupt the disk formation. As a result, a radio-quiet elliptic galaxy is generated. Conversely, the jet is generated in the systems with accretion of the order of the critical one if only the black hole has a sufficient angular momentum. In the systems with low accretion rates, outflows and jets can be effectively generated though the luminosity of these sources due to the accretion is not high.

On the other hand, the strong anisotropy of the jets and their one-sided character lead to the conclusion of the considerable effect of the jet axis orientation to the observer. This idea given as an "unified scheme" was used to explain the diversity of the active nuclei types (Urry and Padovani, 1995). For example, at an angle of ejection $\vartheta > 45^\circ$, the observed radio source is supposed to be a radio galaxy, for $10^\circ < \vartheta < 45^\circ$, it is already a radio quasar, and for $\vartheta < 10^\circ$ an object of the BL Lacertae type that has no absorption lines. True, in the past years, there are increasingly more facts that the unified scheme fails to explain the differences in the properties of various AGN types. For example, it cannot explain the differences between the quasars and the lacertides which, seemingly, must not depend on the axis orientation. This scheme disregards the possibility of the evolutionary change in the active nuclei properties.

As to the physical nature of AGN, several particle acceleration and jet collimation mechanisms were proposed, but there is no confident answer yet which of them is realized. Possibly, in the various types of sources the different mechanisms are realized, or, on the contrary, all of them simultaneously.

Gas-Dynamic Acceleration. The jet acceleration and collimation may be associated with the existence of the environment with high pressure that drops with distance from the center (Blandford and Rees, 1974; Sauty and Tsinganos, 1994; Fabian and Rees, 1995; Tsinganos et al., 1996). It is necessary for the pressure in the jet to be lower than that in the hot environment, which, in principle, can be estimated from the observations in the X-ray range (Feretti et al., 1995). This mechanism, possibly, explains how weak jets are generated in the relativistic sources in our Galaxy or in some Seyfert galaxies. On the other hand, the observed pressure of the hot matter around the most powerful jets from the active nuclei is not high enough, and an alternative mechanism of plasma confinement should exist (Celotti and Blandford, 2001).

Radiation Acceleration. The matter acceleration mechanism in the jet due to the radiation pressure was also proposed, because the photon density in the vicinity of the central source can be very high (Proga et al., 2000; O'Dell, 1981; Cheng and O'Dell, 1981). Within this model, the inner parts of the disk are believed to work as a nozzle, which directs the matter fluxes accelerated by the photon pressure. However, there are some flaws in this mechanism. For example, there is no correlation between the jet power and the central source luminosity—many sources with very powerful jets have a weak luminosity of the compact object (Ghisellini et al., 1990). Another difficulty is that starting from rather low particle energies $\gamma \approx 3$ the radiation field slows down the particles much more effectively than accelerating them (Königl and Kartje, 1994). This contradicts the observations of the "superluminal" sources in which the particle energy is much higher. Besides, if the jet was generated in the system with a thin accretion disk whose radiation is more or less isotropic, one should use additional mechanisms to collimate the jet. The modification of this model, with a funnel formed in the thick disk of the accreting matter, can explain the initial jet collimation, but there are indications that this structure is not stable (Ghisellini et al., 1990, 1992).

Magnetohydrodynamic Mechanism. As was mentioned, most theorists prefer the MHD jet formation model (Blandford, 2002). Otherwise, it is believed that the electromagnetic energy flux—the Poynting vector flux—plays the main role in the energy transport from the central engine to the active regions as well as in the vicinity of radio pulsars. The MHD model was successfully used to describe a lot of processes in the active nuclei, in particular, in the context of the problem of the origin and stability of jets and the explanation of the energetics of the central black hole. The magnetic field is a natural connecting link between the disk and the jet. Therefore, the problem of the origin and structure of the magnetic field in the central engine is very important.

Indeed, as was demonstrated in Sect. 3.3.1, the BZ luminosity (3.68)

$$W_{\rm BZ} \approx 10^{45}\,{\rm erg/s}\left(\frac{a}{M}\right)^2\left(\frac{B_0}{10^4\,{\rm G}}\right)^2\left(\frac{M}{10^9\,M_\odot}\right)^2 \tag{5.6}$$

is in good agreement with the observed energy release. Besides, some remarks are necessary. First of all, as seen from Eq. (5.6), the necessary energy losses at the level 10^{45} erg/s characteristic of AGN can be attained only for the limiting values of the black hole mass of order $10^9\,M_\odot$, the maximal magnetic field in its neighborhood $B_0 \sim B_{\rm Edd}$ (3.4), and also for the maximum angular velocity $a \sim M$. Therefore, there have appeared papers, in which the efficiency of the BZ process for the real astrophysical objects was doubted (Ghosh and Abramowicz, 1997; Livio et al., 1999). However, this does not imply that the electromagnetic model studied encounters serious difficulties. The point is that the effectively working central engine can be connected with the inner regions of the accretion disk, where the fast rotation with the orbital velocity $v_\varphi \sim c$ obviously exists, and the field lines of the regular magnetic field can also pass through its surface. In particular, this process can be effective for the accretion onto the nonrotating (Schwarzschild) black hole.

On the other hand, for the electron–positron jets from AGN ($B_0 \sim 10^4$ G, $R \sim 10^{14}$ cm), the main uncertainty is in the value of the magnetization parameter σ. Indeed, this value depends on the efficiency of the pairs generation in the black hole magnetosphere, which, in turn, is specified by the hard γ-quanta density. As a result, if the density of the hard γ-quanta with energies $\mathcal{E}_\gamma > 1$ MeV in the vicinity of the black hole is sufficiently high, the particles are generated by the direct collision of photons $\gamma + \gamma \rightarrow e^+ + e^-$ (Svensson, 1984), which results in an abrupt increase in the multiplicity parameter $\lambda = n/n_{\rm GJ} \sim 10^{10}$–$10^{12}$. As a result, we find

$$\sigma \sim 10 - 10^3; \qquad \gamma_{\rm in} \sim 3 - 10. \tag{5.7}$$

It yields for the total particle ejection rate $\dot{N} = \lambda\pi r_{\rm g}^2 n_{\rm GJ} c$

$$\dot{N} \sim 10^{48}\,{\rm part/s}\left(\frac{a}{M}\right)\left(\frac{\lambda}{10^{10}}\right)\left(\frac{M}{10^9 M_\odot}\right)^{1/2}\left(\frac{B}{B_{\rm Edd}}\right). \tag{5.8}$$

For the low γ-quanta densities, when the generation of the electron–positron plasma, as shown above, can occur only in the regions with a nonzero longitudinal electric field, which are equivalent to an outer gap in the radio pulsar magnetosphere, the multiplicity of the particle generation appears insignificant, viz., $\lambda \sim 10$–100. In this case

$$\sigma \sim 10^{11} - 10^{13}; \qquad \gamma_{\text{in}} \sim 10, \tag{5.9}$$

and

$$\dot{N} \sim 10^{40}\,\text{part/s}\left(\frac{\lambda}{10^2}\right)\left(\frac{M}{10^9 M_\odot}\right)^{1/2}\left(\frac{B}{B_{\text{Edd}}}\right). \tag{5.10}$$

Thus, in its simplest form, the pattern is the following: the poloidal magnetic field generated in the disk links the rotating central engine (the disk and the black hole) and infinity. Thus, the plasma outflow and the energy flux occur along the magnetic field lines. Due to the differential rotation of the disk and the gas inertness, the field lines are twisted, the toroidal component of the field occurs, and the field pressure connected with this component can collimate the gas. The computations showed that this process can really lead to the collimation and the transferring of a certain part of the electromagnetic energy flux (Poynting flux) to the kinetic energy of particles. However, there are still a lot of unsolved problems in this model. In particular, as was demonstrated, the higher the outflowing plasma energy, the less the efficient collimation. Therefore, in the relativistic case, the collimation and acceleration efficiency does not appear high.

Thus, within the electrodynamic model, the jet collimation mechanism remains unclear. In particular, there is no answer yet to one of the crucial questions of whether the jets are strongly magnetized (and, hence, the electric current specifying the major energy release of the system really flows along the jet and is closed in the region of hot spots) or the electric current closure occurs on parsec scales and the observed jets already contain accelerated particles only. Below, we try to give the main model-independent results obtained by this approach.

5.1.3 Young Stellar Objects

Young stellar objects (YSO) were indirectly discovered in the early 1950s when G. Herbig and G. Haro (Herbig, 1950; Haro, 1950) detected a new class of extended diffusion objects generally existing in pairs and, as was apparent later, connected by thin jet flows from young fast-rotating stars (Lada, 1985). It was logical to connect the formation of these jets with the necessity to more effectively carry away the angular momentum of the star that hinders its formation. As we see, the situation here is quite analogous to AGN, when a number of different objects (quasars, Seyfert, and radio galaxies) were first discovered, and only later it was clear that the nature of the activity of all these objects is unique. Moreover, the similarity of the observed properties suggests that the collimation mechanism in the young stars can be similar to that of the jets in the active nuclei. Inspite of the physical conditions in the vicinity of the young stars (the mass of order 3–10 $M_\odot$, the total energy release of order 10^{31}–10^{36} erg/s) considerably differ from those in the AGN center. One of the main differences is the nonrelativistic character of the flow in the jets from the young stars.

Over 150 objects are already now known as Herbig–Haro objects. These are luminous condensations of dimension of several angular seconds (the linear size of order 500–1000 AU), which are generally surrounded by a luminous diffusion shell. Their spectra mainly consist of the emission lines of hydrogen and some other elements with low excitation energy. The main excitation source is, evidently, a shock wave propagating with velocity 40–200 km/s in a gas with density $\sim 10^2\ 1/\mathrm{cm}^3$ (Reipurth and Bally, 2001).

As in the case of radio galaxies, the activity of the Herbig–Haro objects is connected with jets well pronounced, for example, on the forbidden lines (Lada, 1985; Reipurth and Bally, 2001). In 60% of objects, both jets are visible; in other cases the distant jet is hidden from view by the accretion disk. The extension of the optical jets is 0.01–2 pc, the velocity of motion up to 600 km/s. The gas density in the jets is estimated as 10–100 $1/\mathrm{cm}^3$, and the ejection rate is 10^{-10}–$10^{-6}\ M_{\odot}$/years. The degree of the jet collimation (the observed length-to-width ratio) can be 30; the whole opening angle of jets is 5–10°. Besides the strongly extended jets, in the vicinity of the young stars the molecular flows are also observed; their degree of collimation is much lower. Their dimensions are 0.04–4 pc, and the gas velocity does not exceed 5–100 km/s. One should stress here that this velocity is much higher than the velocity of sound in their molecular matter as its temperature is 10–90° K only. The total mass of the ejected gas is estimated as 0.1–200 $M_{\odot}$, and the total kinetic energy of the molecular flows can amount to 10^{43}–10^{47} erg. In recent years, the most important discovery has been the direct observation of the jet rotation. The characteristic velocities at distances of 20–30 AU from the jet axis is 3–10 km/s (Bacciotti et al., 2002; Coffey et al., 2007). There is also a direct indication to the spiral structure of the magnetic field (Chrysostomou et al., 2007). All this is uniquely in favor of the MHD model.

Indeed, for the parameters characteristic of the young stars the general relation (4.204) yields

$$W_{\mathrm{tot}} \approx 10^{35}\,\mathrm{erg/s}\left(\frac{P}{10^6\,\mathrm{s}}\right)^{4/3}\left(\frac{B_{\mathrm{in}}}{10^3\,\mathrm{G}}\right)^{4/3}\left(\frac{R_{\mathrm{in}}}{10^{11}\,\mathrm{cm}}\right)^{8/3}\left(\frac{\dot{M}}{10^{-6}\,M_{\odot}/\mathrm{year}}\right)^{1/3}. \tag{5.11}$$

As we see, this value is really close to the energy release characteristic of YSO. Accordingly, the "mass loading" is

$$\mu_1 \approx 10^{-3}\left(\frac{P}{10^6\,\mathrm{s}}\right)^{-1}\left(\frac{B_{\mathrm{in}}}{10^3\,\mathrm{G}}\right)^{-2}\left(\frac{R_{\mathrm{in}}}{10^{11}\,\mathrm{cm}}\right)^{-1}\left(\frac{\dot{M}}{10^{-6}\,M_{\odot}/\mathrm{year}}\right). \tag{5.12}$$

On the other hand, for the ordinary young stars we have $\Omega_{\mathrm{crit}} \sim 10^{-6}\,1/\mathrm{s}$, i.e., $P_{\mathrm{crit}} = 2\pi/\Omega_{\mathrm{crit}}$ is about a few months. This implies that for young stars both the cases $\Omega > \Omega_{\mathrm{crit}}$ and $\Omega < \Omega_{\mathrm{crit}}$ can be realized. Hence, the central parts of a jet can be particle-dominated ones. On the other hand, the fast rotation of the inner region of the accretion disk ($\Omega_{\mathrm{K}} = (GM/r^3)^{1/2} \sim 10^{-3}$–$10^{-4}\,1/\mathrm{s}$) guarantees that the flow is of a magnetically dominated type in the vicinity of the young star. Finally, using relation (4.214) for the radius of the critical surfaces, we obtain

$$r_{\mathrm{f}} \approx 1\,\mathrm{AU}\left(\frac{P}{10^6\,\mathrm{s}}\right)^{1/3}\left(\frac{B_{\mathrm{in}}}{10^3\,\mathrm{G}}\right)^{2/3}\left(\frac{R_{\mathrm{in}}}{10^{11}\,\mathrm{cm}}\right)^{4/3}\left(\frac{\dot{M}}{10^{-6}\,M_{\odot}/\mathrm{year}}\right)^{-1/3}. \quad (5.13)$$

This implies that the fast magnetosonic surface radius is much smaller than the transverse dimension of jets $r_{\mathrm{jet}} \sim$ 10–100 AU. Therefore, it is not surprising that the nonrelativistic jets observed are supersonic ones.

As to the physical nature of the jet formation, this problem is far from completion. It is only clear that the central star energy is always enough to accelerate the outflows of matter; however, the energy transformation mechanism has not been established yet. We emphasize that, unlike the relativistic galactic objects (for example, micro-quasars), where the jet formation is associated with the supercritical accretion, the luminosity in the young stars never approaches the Eddington limit. On the other hand, there is no doubt that it is the accretion disks, the existence of which in this class of objects is beyond question, that play the key role in the jet formation. This is evident from the direct correlation between the energy of the gas outflow and the mass of the disk estimated by its luminosity.

The parameters of the disks can be rather diverse. Their masses, for example, are in the range from 0.1 to 60 $M_{\odot}$ and the radii from 10 AU to 1 pc. It is important that, unlike the disks surrounding the relativistic objects (neutron stars and black holes), the gas temperature in them is 20–100 K only. As a result, as in the AGN case, neither the radiation pressure nor the gas pressure can account for the high velocities observed in the jets. Therefore, the models, in which the key role was played by the magnetic field responsible for the effective interaction between the accretion disk and the jets, were again used to explain the jet formation and the particle acceleration. Since the real magnetic field structure in the vicinity of the young stars is not known now, both the models in which the magnetic field of the star itself is of vital importance (Shu et al., 1994) and the models in which the magnetic field of the disk is also of importance (Pudritz and Norman, 1986; Pelletier and Pudritz, 1992; Uchida and Shibata, 1984) were proposed here. As we see, one also has to deal with the same magnetic field structure problems as in the studies of the black hole magnetosphere.

Problem 5.1 Using the integrals of motion, show that, given from the observations the toroidal and poloidal velocities in the supersonic nonrelativistic jet are at a distance of r_0 from the rotation axis (and assuming that the observed jet region started from the accretion disk surface at a distance of r_{st} from the star), we can estimate both the "loading" μ_{l} and the distance r_{st}, while the value r_{st} is given by the simple formula (Anderson et al., 2003; Ferreira et al., 2006)

$$r_{\rm st} \approx 0.7\,{\rm AU} \left(\frac{r_0}{10\,{\rm AU}}\right)^{2/3} \left(\frac{v_\varphi}{10\,{\rm km/s}}\right)^{2/3} \left(\frac{v_{\rm p}}{100\,{\rm km/s}}\right)^{-4/3} \left(\frac{M}{M_\odot}\right)^{1/3}. \tag{5.14}$$

5.1.4 Microquasars, Cosmological Gamma-Bursters, etc.

As to the cosmic compact objects with jets, one cannot but mention microquasars and sources of cosmological gamma-bursters. *Microquasars* are galactic objects in which the jet formation is connected with the supercritical accretion onto a compact relativistic object (a neutron star or a black hole). It is a very small class of objects including about 10 sources (Fender, 2006) and only half of them have well-pronounced jets in which the particle velocity is a relativistic one ($v > 0.9c$). The characteristic longitudinal size of the jets, generally, amounts to 0.1 pc, and the opening angle does not exceed several degrees. The total energetics is estimated as 10^{37} erg/s (Mirabel and Rodriguez, 1994; Fender, 2006). In view of the relativistic velocities, the effect of the superluminal motion is observed in some sources, and the visible angular velocity, due to the relative closeness of these objects, is by several orders higher than that of the jets observed in AGN.

Historically, the first object of this class, which was discovered, was the famous source SS433 (Spencer, 1979) in which, however, the gas outflow velocity in the jets was $0.26\,c$ only. This velocity can be readily explained by the radiation pressure connected with the strongly heated inner regions of the accretion disk. As to the relativistic jets, the first microquasar was discovered only in 1994 (Mirabel and Rodriguez, 1994). Since the occurrence of near-light velocities due to the radiation or the gas pressure is problematic, it is not improbable that for their explanation one again has to use an electrodynamic model similar to the one used to explain the formation and collimation of the AGN jets (Blandford, 2002; Sauty et al., 2002). It counts in favor of this model that, except for object SS433, the emission lines are not observed in the microquasars, which indirectly points to the electron–positron composition of the jets (Fender, 2006).

As to the sources of *cosmological gamma-bursters*, there is only indirect, though rather reliable, evidence of the existence of the jets in them, which is just connected with the relativistic strongly magnetized flows considered in this chapter. The discovery of optical afterglow (Akerlof et al., 1999), as well as afterglow in other energy ranges, which permits one to determine by the red shift the distances to these objects, imposed very serious constraints on their energetics (Postnov, 1999). If the radiation in the gamma range is considered to be isotropic, at the distances of order 0.1–1 Gpc characteristic of these sources, we have to assume that their luminosity is 10^{54} erg/s. However, currently, the processes with this enormous energy release are not known. The short duration of the burst ($\sim$10 s) limits the size of the radiating region, which, in turn, does not allow one to explain the observed gamma radiation spectra, because the optical thickness of the source proves too high (Ruderman, 1975; Schmidt, 1978).

On the other hand, if we assume that the radiation is concentrated in a narrow cone $\vartheta \sim 1^{\circ}$, the liberated energy can be reduced to 10^{51} erg, which in order of magnitude is already close to the energy liberated during the supernova explosions. But this implies that the characteristic Lorentz factor of particles responsible for the observed radiation is to be of order ϑ^{-1}, i.e., $\gamma \sim 100$–300. As a result, it is possible to remove the compactness problem since the estimated size of the radiating region is 100–300 times larger. However, the ultrarelativistic character of the flow, in turn, imposes constraints on the particle composition in the jets, because the existence of protons with this energy would contradict the total energy release in the gamma-burst. Therefore, the contribution of protons should be only 10^{-5} of the total number of particles, so we can deal exactly with the electron–positron jets. The existence of the jet flow is supported by the characteristic break of the time dependence of the luminosity intensity when in about a few days after the burst the exponent $\alpha_{\rm t}$ in the dependence $W_{\rm tot} \propto t^{-\alpha_{\rm t}}$ varies from $\alpha_{\rm t} \approx 1.1$ to $\alpha_{\rm t} \approx 2.0$. This effect is due to the completion of the relativistic compression of the radiation cone in the particle motion rigorously to the observer.

As to the nature of the central engine responsible for the formation of the strongly magnetized jets, one usually discusses here either the collision of two neutron stars (Blinnikov et al., 1984; Eichler et al., 1989) or a neutron star and a black hole (Paczyński, 1991) or the collapse of the massive nucleus of an unusual supernova (Woosley, 1993; Paczyński, 1998). However, the fast-rotating black hole with mass of the order of the solar mass, which loses its rotational energy due to the BZ process, is thus generated in most models (Paczyński, 1991; Mésźaros and Rees, 1997; Katz, 1997; Lee et al., 2000; van Putten and Levinson, 2003); indeed, as we saw, it is this process that allows one to readily explain both the small quantity of baryons and the large Lorentz factors of particles in the jets. In other words, the model is again constructed by the same scheme as the model discussed in the context of activity of the galactic nuclei. In particular, the key processes here are those of the magnetic field generation in the plasma surrounding the black hole, the interaction between the black hole and the accretion disk connected by the magnetic field lines, as well as the particle generation process in the magnetosphere. For the observed energy release to be interpreted it is necessary to assume that the magnetic field in the vicinity of the black hole must reach 10^{14}–10^{15} G. It is believed that the generation of this field is possible for such nonstationary processes as the star collapse or the collision of neutron stars (Usov, 1992; Thompson and Duncan, 1995).

As a result, we can obtain for the energy release $W_{\rm tot} \approx W_{\rm BZ}$

$$W_{\rm tot} \approx 10^{51}\,{\rm erg/s} \left(\frac{P}{10^{-3}\,{\rm s}}\right)^2 \left(\frac{B}{10^{15}\,{\rm G}}\right)^2. \tag{5.15}$$

On the other hand, supposing that the jet is already particle dominated (i.e., $\sigma \approx \gamma \sim 100$–300), we can obtain for the multiplicity parameter

$$\lambda \approx \frac{1}{\sigma}\left(\frac{\Omega R}{c}\right)^2\left(\frac{\omega_B R}{c}\right) \sim 10^{15}\left(\frac{\sigma}{10^2}\right)^{-1}\left(\frac{P}{10^{-3}\,\mathrm{s}}\right)^{-2}\left(\frac{B}{10^{15}\,\mathrm{G}}\right). \tag{5.16}$$

This value corresponds to equipartition between particles and radiation field in the vicinity of the black hole.

5.2 Cylindrical Flows

5.2.1 Cylindrical Jets—The Force-Free Approximation

We now proceed to the discussion of the exact analytical solutions describing the magnetized plasma flow in the neighborhood of compact astrophysical objects. We emphasize once again that our goal here is not to construct self-consistent models but to clarify the general properties of the flows, which can be deduced from the analysis of the exact analytical solutions.

The problem of the jet formation mechanism is a key one when studying the magnetosphere structure of compact astrophysical objects. Indeed, the jets are observed in most compact sources, beginning with quasars and radio galaxies and ending with accreting neutron stars, solar mass black holes, and young stellar objects. Moreover, as was noted, the jets have recently been detected in young radio pulsars. This broad class of jets shows that their formation mechanism must be rather universal. In any case, it must be independent of the character of the flow, viz., whether it is a relativistic or a nonrelativistic one.

In many papers dealing with the MHD model of these objects, in which the jet formation was connected with the attraction of the longitudinal currents flowing in the magnetosphere, main attention was given to the proper collimation in the sense that the environment was supposed to be of no importance (Blandford and Payne, 1982; Heyvaerts and Norman, 1989; Pelletier and Pudritz, 1992; Sulkanen and Lovelace, 1990; Li et al., 1992; Sauty and Tsinganos, 1994). However, as shown below, this is possible only for the nonzero total current I flowing within a jet (Nitta, 1997) and the problem is to close it in the outer regions of the magnetosphere. On the other hand, as we will see, the longitudinal current must be limited by the regularity condition on by the fast magnetosonic surface, which does not always give rise to rather strong longitudinal currents needed for collimation.

Moreover, it is quite clear that the collimation problem cannot be solved ignoring the environment (see, e.g., Appl and Camenzind, 1992, 1993). In particular, this is already evident from the example of the magnetosphere of the compact object with the monopole magnetic field, since for any arbitrarily small external regular magnetic field the monopole solution (for which the poloidal magnetic field decreases as r^{-2}) cannot be extended to infinity. Moreover, as is well known from the example of moving space bodies such as Jupiter's satellites (Zheleznyakov, 1996) or the artificial Earth satellites (Alpert et al., 1965; Gurevich et al., 1975), and radio pulsars in binary systems (Tsygan, 1997), the external magnetic field can serve as an efficient transmission chain sometimes defining the total energy losses in the system.

Certainly, the external regular magnetic field existing in the neighborhood of compact objects is mainly open to question. In our Galaxy, the regular magnetic field, i.e., the field homogeneous on scales comparable with the galaxy sizes, is

$$B_{\rm ext} \sim 10^{-6}\,{\rm G} \tag{5.17}$$

and actually coincides with the chaotic component of the magnetic field, which changes on scales of several parsecs (Marochnik and Suchkov, 1996). However, if the collimation is assumed to be really due to the presence of the ambient pressure, it is possible to estimate the transverse jet size $r_{\rm jet}$. Indeed, assuming the poloidal magnetic field in the jet to be close to the external magnetic field (5.17), we obtain from the condition of the magnetic flux conservation

$$r_{\rm jet} \sim R_{\rm in}\left(\frac{B_{\rm in}}{B_{\rm ext}}\right)^{1/2}, \tag{5.18}$$

where again $R_{\rm in}$ and $B_{\rm in}$ are, respectively, the radius and the magnetic field of the compact object. Thus, for AGN ($B_{\rm in} \sim 10^4$ G, $R_{\rm in} \sim 10^{13}$ cm), we have

$$r_{\rm jet} \sim 1\,{\rm pc}, \tag{5.19}$$

which exactly corresponds to the observed transverse sizes of the jets (Begelman et al., 1984). Accordingly, for young stellar objects ($B_{\rm in} \sim 10^2$ G, $R_{\rm in} \sim 10^{10}$ cm) (Lada, 1985), we have $r_{\rm jet} \sim 10^{16}$ cm, which also corresponds to the observations. One can expect that the similar pattern retains for the environment with pressure $P_{\rm ext} \sim B_{\rm ext}^2/8\pi$. Therefore, the problem of the internal structure of the one-dimensional jet submerged in the external homogeneous magnetic field should also be included in our investigation.

In this section, we consider the one-dimensional relativistic cylindrical jets in which the poloidal magnetic field and the poloidal velocity of matter are directed along the z-axis. Otherwise, we do not discuss here the collimation itself but only the problem of the internal structure of the really observed cylindrical flows. In this case, the poloidal magnetic field should be written as

$$B_z(\varpi) = \frac{1}{2\pi\varpi}\frac{{\rm d}\Psi}{{\rm d}\varpi}, \tag{5.20}$$

where now the magnetic flux function Ψ depends on one coordinate ϖ only. Accordingly, it is convenient to write the toroidal component of the magnetic field and the electric field as

$$B_\varphi(\varpi) = -\frac{2I}{\varpi c}, \tag{5.21}$$

$$\mathbf{E} = -\frac{\Omega_{\mathrm{F}}}{2\pi c}\frac{\mathrm{d}\Psi}{\mathrm{d}\varpi}\mathbf{e}_{\varpi}, \tag{5.22}$$

where $I(\varpi_0)$ is the total current within the domain $\varpi < \varpi_0$.

We would like to emphasize several features that are used when investigating the structure of the relativistic jets characteristic of AGN and radio pulsars. For the active nuclei and pulsars the transverse jet size r_{jet} is much larger than the light cylinder radius $R_{\mathrm{L}} = c/\Omega$. Indeed, as seen from relation (3.68), the central engine efficiency can be high only for the sufficiently large rotation parameter $a/M \sim \Omega_{\mathrm{F}} r_{\mathrm{g}}/c$. For example, the light cylinder radius for the magnetic field lines passing through the black hole horizon (not to mention the field lines passing through the inner regions of the accretion disk) must be comparable with the black hole radius. This implies that when investigating the internal jet structure the corresponding equations must be written in full relativistic form. On the other hand, far from the compact object the gravitational forces can be disregarded. Finally, for simplicity, we consider the cold plasma ($\mu =$ const), which, as was shown, is valid at large distances from compact objects.

We first consider the relativistic jet structure in the force-free approximation. As was noted, the remarkable feature of the GS equation is that it can be integrated in the one-dimensional case. For example, the solution to the force-free equation (2.101) can be written as (see, e.g., Istomin and Pariev, 1994)

$$\frac{\Omega_{\mathrm{F}}^2(\Psi)\varpi^4}{c^2}B_z^2 = \varpi^2 B_{\varphi}^2 + \int_0^{\varpi} x^2\frac{\mathrm{d}}{\mathrm{d}x}(B_z)^2\mathrm{d}x. \tag{5.23}$$

In particular, it is easy to check that the homogeneous magnetic field $B_z =$ const is the solution to the nonlinear equation (2.101) for any values of the integrals of motion $\Omega_{\mathrm{F}}(\Psi)$ and $I(\Psi)$ satisfying the condition

$$4\pi I(\Psi) = 2\Omega_{\mathrm{F}}(\Psi)\Psi. \tag{5.24}$$

We emphasize that for small Ψ this relation fully coincides with (2.231) obtained for the conical flows.

Problem 5.2 Show that the relation similar to (5.24) can be obtained for the conical solutions $\Psi = \Psi(\theta)$, but only at large distances $r \gg R_{\mathrm{L}}$ from the compact object. It has the form (Ingraham, 1973; Michel, 1974)

$$4\pi I(\theta) = \Omega_{\mathrm{F}}(\theta)\sin\theta\,\frac{\mathrm{d}\Psi}{\mathrm{d}\theta}. \tag{5.25}$$

Problem 5.3 Show that for both the cylindrical and the conical flows at large distances from the rotation axis the condition $E_{\theta} = B_{\varphi}$ is satisfied.

Thus, in the one-dimensional case, the assignment of any two of three values $B_z(\varpi)$, $\Omega_F(\Psi)$, and $I(\Psi)$ fully defines the solution to the GS equation. However, one should remember that the direct problem (i.e., the determination of the poloidal magnetic field by the given values $B_\varphi(\varpi)$ and $\Omega_F(\varpi)$) does not always have the solution. Indeed for $B_\varphi = 0$ and $\Omega_F = \text{const}$, we find

$$B_z(\varpi) = \frac{B_z(0)}{1 - \dfrac{\Omega_F^2 \varpi^2}{c^2}}, \tag{5.26}$$

so that the solution cannot be extended beyond the light cylinder. On the other hand, given the magnetic field $B_z(\varpi)$, we can always find the values of $B_\varphi(\varpi)$ and $\Omega_F(\varpi)$, for which this field is the solution to the GS equation. In this case, however, the value of the longitudinal current is to be close to that of the GJ current I_{GJ}.

Analysis of the force-free equation can already lead to several important conclusions that remain valid in the general case, i.e., with account taken of the finite particle mass (see Fig. 5.1).

1. The key property of the relativistic flows is that the transverse size of the real jets r_{jet} is 3–5 orders larger than that of the light cylinder radius R_L. As a result, at large distances from the central engine the major part of the magnetic flux must be far beyond the light cylinder. However, according to the relation

$$\frac{B_\varphi}{B_p} = \frac{\Omega_F \varpi}{c}, \tag{5.27}$$

resulting from the definition of the magnetic field for $I \approx I_{GJ}$, this implies that the toroidal magnetic field must also be 3–5 orders larger than the poloidal magnetic field. Consequently, the magnetic field should have the strongly pronounced spiral structure. Accordingly, the electric field should also be 3–5 orders larger than the poloidal magnetic field.
2. Further, the energy flux in the jet must be concentrated in the peripheral part of the flow rather than in the central one. This is because both the electric and toroidal magnetic fields vanish on the rotation axis. Accordingly, the Poynting flux vanishes here. Therefore, the strongly magnetized flows should have the form of a "hollow cylinder" (see the cover photo, Kovalev (2008)).
3. For the same reason the particle energy turns out to be the largest in the peripheral regions. Indeed, as shown above, beyond the light cylinder the main motion of charged particles is a drift motion in the crossed electric and magnetic fields. As was demonstrated in Sect. 4.4.1, the drift velocity $U_{dr} = c|\mathbf{E}|/|\mathbf{B}|$ (4.147) together with the estimates (5.27) $E_\theta \approx B_\varphi$ results in the universal form of the Lorentz factor

$$\gamma \approx \frac{\Omega_F \varpi}{c}. \tag{5.28}$$

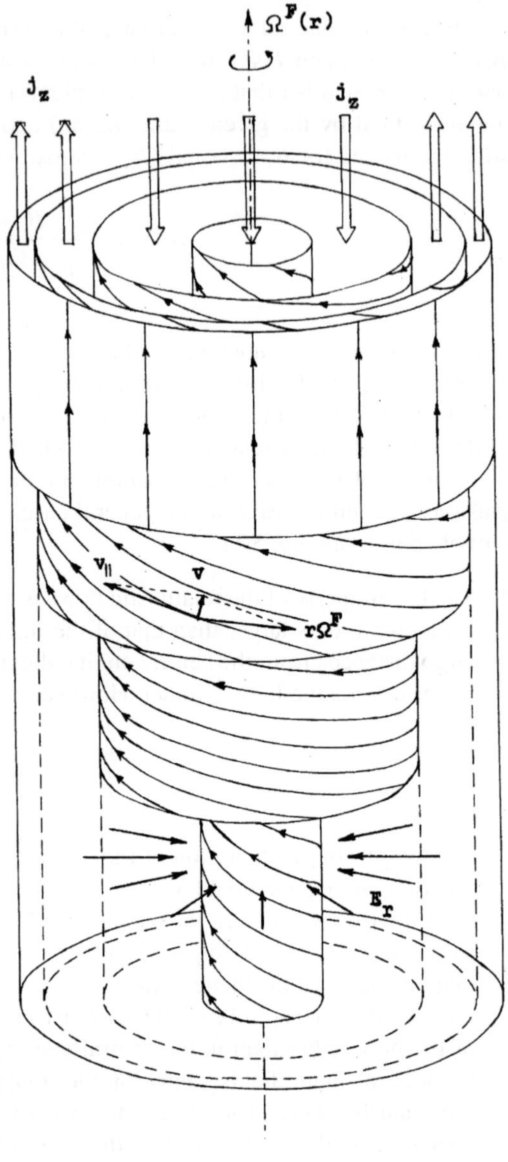

Fig. 5.1 The structure of the cylindrical strongly magnetized jet with zero total electric current (Istomin and Pariev, 1994). Longitudinal electric currents (*contour arrows*) and the toroidal magnetic field exist inside the jet only

Certainly, as shown below, the effects of the finite mass of particles limit their energy increase at large distances from the rotation axis. Nevertheless, the energy increase with distance from the rotation axis seems to be the general property of the strongly magnetized flows.

4. One should emphasize the nontrivial character of the particle motion, which is important for the jet radiation analysis. The point is that in the jet not only the toroidal magnetic field B_φ (5.27) but also the electric field **E** (directed from or

to the rotation axis) should be much greater than the poloidal magnetic field B_{p}. Therefore, the drift velocity of particles is actually directed along the poloidal magnetic field. But this implies that the particle motion may not be of a spiral character. Consequently, one should be careful using the standard synchrotron formulae to estimate the value of the magnetic field and the synchrotron lifetime of relativistic particles.

5. Finally, relation (5.24) shows that the homogeneous longitudinal magnetic field can be the solution to the GS equation for the total zero longitudinal electric current flowing along the jet. Indeed, according to (5.24), if the angular velocity $\Omega_{\mathrm{F}}(\Psi)$ vanishes on the jet boundary $\Psi = \Psi_0$, the total current $I(\Psi_0)$ must also be zero (Istomin and Pariev, 1994; Fendt, 1997b). As we saw, this was the case for the homogeneous magnetic field in the vicinity of the black hole horizon. But then the longitudinal magnetic field on the jet boundary does not vanish. Therefore, for the stability of these configurations it is necessary to assume that either the constant magnetic field or the finite gas pressure must exist beyond the jet. It is very important that in this case the relativistic jet is to be stable to helical kink (screw) modes (Istomin and Pariev, 1994; Tomimatsu et al., 2001).

As a result, we can formulate the latter assertion in the form of the important theorem.

Theorem 5.1 *The stationary cylindrical jet containing the finite magnetic flux Ψ_0 can occur either for the nonzero total longitudinal electric current $I(\Psi_0) \neq 0$ or in the presence of the environment with nonzero magnetic or gas pressure.*

We specially formulated these assertions here before we proceed to the discussion of the full MHD GS equation version as we would like to stress within the hypothesis of the strongly magnetized jets that these assertions are really universal and model independent.

5.2.2 Relativistic Jets

5.2.2.1 Basic Equations

We now proceed to a closer consideration of the structure of the one-dimensional jets in which the finite particle mass is taken into account. In this case, we should add to the definitions for the magnetic (5.20) and (5.21) and electric (5.22) fields the expression for the four-velocity (4.31)

$$\mathbf{u} = \frac{\eta}{n}\mathbf{B} + \gamma \frac{\Omega_{\mathrm{F}}\varpi}{c}\mathbf{e}_{\varphi}. \tag{5.29}$$

According to (1.64), to determine the flow structure for the cold plasma we must introduce four integrals of motion that should also be regarded as functions of the magnetic flux Ψ. These are, first of all, the values $\Omega_{\mathrm{F}}(\Psi)$ and $\eta(\Psi)$ available in definitions (5.22) and (5.29) and the fluxes of the angular momentum $L(\Psi) = I/2\pi + \mu\eta\varpi u_\varphi$ and the energy $E(\Psi) = \Omega_{\rm F} I/2\pi + \gamma\mu\eta$.

We emphasize that in the full statement of the problem the concrete form of these integrals of motion is to be determined from the boundary conditions on the compact object surface and also from the critical conditions on the singular surfaces. The possibility to use the integrals of motion derived from the analysis of the inner magnetosphere regions is not obvious. Indeed, as we will see, the flow beyond the fast magnetosonic surface for the cold plasma is fully defined by four boundary conditions on the rotating body surface. The one-dimensional flow can be generated by interacting with the environment, which produced disturbances or shock waves propagating from "acute angles" and other irregularities (see, e.g., Landau and Lifshits, 1987). Therefore, in the domain, where the usability conditions of ideal hydrodynamics are sure to be violated, the substantial redistribution of the energy E and the angular momentum L is to take place, let alone the part which can be lost as radiation. Nevertheless, below, for simplicity, we believe that the integrals of motion $E(\Psi)$ and $L(\Psi)$ remain exactly the same as in the inner magnetosphere.

As a result, far from gravitating bodies the GS equation (2.101) for the cold plasma ($\mu = $ const) is written as (Beskin and Malyshkin, 2000)

$$\frac{1}{\varpi}\frac{\mathrm{d}}{\mathrm{d}\varpi}\left(\frac{A}{\varpi}\frac{\mathrm{d}\Psi}{\mathrm{d}\varpi}\right) + \Omega_{\rm F}(\nabla\Psi)^2\frac{\mathrm{d}\Omega_{\rm F}}{\mathrm{d}\Psi} + \frac{64\pi^4}{\varpi^2}\frac{1}{2\mathcal{M}^2}\frac{\mathrm{d}}{\mathrm{d}\Psi}\left(\frac{G}{A}\right) - \frac{32\pi^4}{\mathcal{M}^2}\frac{\mathrm{d}(\mu^2\eta^2)}{\mathrm{d}\Psi} = 0. \tag{5.30}$$

Here again

$$G = \varpi^2(e')^2 + \mathcal{M}^2L^2 - \mathcal{M}^2\varpi^2E^2, \tag{5.31}$$

$$A = 1 - \Omega_{\rm F}^2\varpi^2 - \mathcal{M}^2, \tag{5.32}$$

$$e'(\Psi) = E(\Psi) - \Omega_{\rm F}(\Psi)L(\Psi), \tag{5.33}$$

and the derivative $\mathrm{d}/\mathrm{d}\Psi$ acts only on the integrals of motion. The other parameters of the jet are to be determined from algebraic relations (4.38), (4.39), and (4.40) which far from gravitating bodies are written as

$$\frac{I}{2\pi} = \frac{L - \Omega_{\rm F}\varpi^2E}{1 - \Omega_{\rm F}^2\varpi^2 - \mathcal{M}^2}, \tag{5.34}$$

$$\gamma = \frac{1}{\mu\eta}\frac{(E - \Omega_{\rm F}L) - \mathcal{M}^2E}{1 - \Omega_{\rm F}^2\varpi^2 - \mathcal{M}^2}, \tag{5.35}$$

$$u_{\hat\varphi} = \frac{1}{\varpi\mu\eta}\frac{(E - \Omega_{\rm F}L)\Omega_{\rm F}\varpi^2 - L\mathcal{M}^2}{1 - \Omega_{\rm F}^2\varpi^2 - \mathcal{M}^2}. \tag{5.36}$$

Hereafter we again take, for simplicity, $c = 1$. In the one-dimensional case studied, it is convenient to immediately reduce the second-order equation (5.30) to the system of two first-order equations for $\Psi(\varpi)$ and $\mathcal{M}^2(\varpi)$. Multiplying Eq. (5.30) by $2A(\mathrm{d}\Psi/\mathrm{d}\varpi)$, we get

$$\frac{\mathrm{d}}{\mathrm{d}\varpi}\left[\frac{A^2}{\varpi^2}\left(\frac{\mathrm{d}\Psi}{\mathrm{d}\varpi}\right)^2\right]+A\left(\frac{\mathrm{d}\Psi}{\mathrm{d}\varpi}\right)^2\frac{\mathrm{d}\Omega_{\mathrm{F}}^2}{\mathrm{d}\varpi}+\frac{64\pi^4 A}{\varpi^2\mathcal{M}^2}\frac{\mathrm{d}'}{\mathrm{d}\varpi}\left(\frac{G}{A}\right)-\frac{64\pi^4 A}{\mathcal{M}^2}\frac{\mathrm{d}}{\mathrm{d}\varpi}(\mu^2\eta^2)=0, \tag{5.37}$$

where the derivative $\mathrm{d}'/\mathrm{d}\varpi$ again acts only on the integrals of motion. Using now the explicit form of Bernoulli's equation (4.44) which, with account taken of the definition of the integrals of motion $E(\Psi)$ and $L(\Psi)$, can be written as

$$A^2\left(\frac{\mathrm{d}y}{\mathrm{d}x_r}\right)^2=\frac{(e')^2}{\mu^2\eta^2}\frac{x_r^2(A-\mathcal{M}^2)}{\mathcal{M}^4}+\frac{x_r^2E^2}{\mu^2\eta^2}-\frac{\Omega_{\mathrm{F}}^2(0)L^2}{\mu^2\eta^2}-\frac{x_r^2A^2}{\mathcal{M}^4}. \tag{5.38}$$

We introduced here the dimensionless variables

$$x_r=\Omega_{\mathrm{F}}(0)\varpi, \tag{5.39}$$

$$y=\sigma\frac{\Psi}{\Psi_0}, \tag{5.40}$$

where Ψ_0 is the total magnetic flux inside the jet and the magnetization parameter σ is again given by relation (5.3) for $\Omega_{\mathrm{F}}=\Omega_{\mathrm{F}}(0)$. As a result, substituting the right-hand side of Eq. (5.38) in the first term of Eq. (5.37) and performing the elementary differentiation, we obtain (Beskin, 1997)

$$\left[\frac{(e')^2}{\mu^2\eta^2}+\frac{\Omega_{\mathrm{F}}^2}{\Omega_{\mathrm{F}}^2(0)}x_r^2-1\right]\frac{\mathrm{d}\mathcal{M}^2}{\mathrm{d}x_r}=\frac{\mathcal{M}^6}{x_r^3A}\frac{\Omega_{\mathrm{F}}^2(0)L^2}{\mu^2\eta^2}-\frac{x_r\mathcal{M}^2}{A}\frac{\Omega_{\mathrm{F}}^2}{\Omega_{\mathrm{F}}^2(0)}\left[\frac{(e')^2}{\mu^2\eta^2}-2A\right]$$
$$+\frac{\mathcal{M}^2}{2}\frac{\mathrm{d}y}{\mathrm{d}x_r}\left[\frac{1}{\mu^2\eta^2}\frac{\mathrm{d}(e')^2}{\mathrm{d}y}+\frac{x_r^2}{\Omega_{\mathrm{F}}^2(0)}\frac{\mathrm{d}\Omega_{\mathrm{F}}^2}{\mathrm{d}y}-2\left(1-\frac{\Omega_{\mathrm{F}}^2}{\Omega_{\mathrm{F}}^2(0)}x_r^2\right)\frac{1}{\eta}\frac{\mathrm{d}\eta}{\mathrm{d}y}\right]. \tag{5.41}$$

The second equation is Bernoulli's equation (5.38) that should be regarded as an equation for the derivative $\mathrm{d}y/\mathrm{d}x_r$. The system of equations (5.38) and (5.41) allows one to construct the general solution for the cylindrical relativistic jet with zero temperature. In the general case of the finite temperature (and with account taken of the dimension), Eq. (5.41) has the form (Beskin and Nokhrina, 2009)

$$\left[\frac{(e')^2}{\mu^2\eta^2}-1+\frac{\Omega_{\mathrm{F}}^2\varpi^2}{c^2}-A\frac{c_s^2}{c^2}\right]\frac{\mathrm{d}\mathcal{M}^2}{\mathrm{d}\varpi}=\frac{\mathcal{M}^6L^2}{A\varpi^3\mu^2\eta^2}$$
$$+\frac{\Omega_{\mathrm{F}}^2\varpi\mathcal{M}^2}{c^2}\left[2-\frac{(e')^2}{A\mu^2\eta^2}\right]+\mathcal{M}^2\frac{e'}{\mu^2\eta^2}\frac{\mathrm{d}\Psi}{\mathrm{d}\varpi}\frac{\mathrm{d}e'}{\mathrm{d}\Psi}$$
$$+\mathcal{M}^2\frac{\varpi^2}{c^2}\Omega_{\mathrm{F}}\frac{\mathrm{d}\Psi}{\mathrm{d}\varpi}\frac{\mathrm{d}\Omega_{\mathrm{F}}}{\mathrm{d}\Psi}-\mathcal{M}^2\left(1-\frac{\Omega_{\mathrm{F}}^2\varpi^2}{c^2}+2A\frac{c_s^2}{c^2}\right)\frac{\mathrm{d}\Psi}{\mathrm{d}\varpi}\frac{1}{\eta}\frac{\mathrm{d}\eta}{\mathrm{d}\Psi}$$
$$-\left[\frac{A}{n}\left(\frac{\partial P}{\partial s}\right)_n+\left(1-\frac{\Omega_{\mathrm{F}}^2\varpi^2}{c^2}\right)T\right]\frac{\mathcal{M}^2}{\mu}\frac{\mathrm{d}\Psi}{\mathrm{d}\varpi}\frac{\mathrm{d}s}{\mathrm{d}\Psi}. \tag{5.42}$$

We would like to point to one important advantage of the system of first-order equations (5.38) and (5.41) as compared to the initial second-order equation (5.30).

As was mentioned, the relativistic equation (5.30), which is actually a force balance equation, contains the electromagnetic force $\mathbf{F}_{em} = \rho_e \mathbf{E} + \mathbf{j} \times \mathbf{B}$, in which at large distances from the rotation axis $\varpi \gg R_L$ the electric and magnetic contributions practically compensate one another. Therefore, in the analysis of Eq. (5.30) we have to retain all terms of a higher-order infinitesimal $\sim \gamma^{-2}$. In Eq. (5.41) the zero-order values $\rho_e \mathbf{E}$ and $\mathbf{j} \times \mathbf{B}$ are analytically eliminated by Bernoulli's equation, so all terms of this equation are of the same order of magnitude. Finally, it is important here that the exact equation (5.41) has no singularity in the vicinity of the rotation axis. Otherwise, its solution does not comprise the δ-shaped current $I \propto \delta(\varpi)$ flowing along the jet axis; the necessity to introduce it was pointed out in a number of papers (Heyvaerts and Norman, 1989; Lyubarskii, 1997).

Some comments of the general character, which deal with the peculiarity of the formulation of the problem for the cylindrical flows, are appropriate here. In the one-dimensional case studied, the plasma poloidal velocity is always parallel to the singular surfaces. Therefore, the absence of a singularity in this case shows the absence of a tangential discontinuity rather than a shock wave. Further, for the cylindrical flows the curvature of magnetic surfaces R_c^{-1} is always zero. Consequently, one of the leading terms in expressions (4.228) and (4.229) is identically equal to zero. Therefore, as we will see, the balance of forces for the cylindrical flows is only possible when the force contribution $j_\varphi B_p$ is taken into account.

Finally, the cylindrical flow is a one-dimensional one. This implies that we initially impose an additional self-similarity condition that substantially confines the class of possible solutions. In particular, as was already noted, we confine the class of possible disturbances, which immediately changes the entire structure of the singular surfaces. As a result, the one-dimensional equation (5.30) has no singularity on the fast magnetosonic surface at all. Indeed, as is readily checked, if the stream function Ψ depends only on one coordinate x_1, in the GS equation, the coefficient of the higher derivative $\mathrm{d}^2/\mathrm{d}x_1^2$ is $(D+1)$. Therefore, the singularity occurs on the cusp surface rather than on the fast magnetosonic one. Consequently, the cusp surface is absent for the cold plasma; the GS equation can have a singularity on the Alfvén surface only. As a result, for the super-Alfvén flows ($i = 4$, $s' = 0$), Eq. (5.30) requires six boundary conditions.

For example, for a jet submerged in the environment, we can first choose, as boundary conditions, the external homogeneous magnetic field

$$B_z(r_{\mathrm{jet}}) = B_{\mathrm{ext}}, \tag{5.43}$$

and the regularity condition on the rotation axis $\varpi \to 0$

$$\Psi(\varpi) \to C\varpi^2. \tag{5.44}$$

Besides, all the four integrals Ω_F, E, L, and η are to be given. As to the other values characterizing the flow such as the transverse jet size r_{jet} and the outflowing plasma energy, they must be found as the solution of the above-posed problem. Analogously, it is the solution of the problem that should give the answer to the question whether the jet flow is a supersonic one. On the other hand, if we consider an isolated jet instead of the boundary condition (5.43), we should choose the vanishing condition of all the fields for $\varpi \to \infty$.

Clearly, the flow structure is specified by the choice of the integrals of motion. Moreover, it turned out that most of the key properties can be grasped from the analysis of the solution behavior in the vicinity of the rotation axis, where the integrals of motion are written in a rather universal form. Therefore, we begin our analysis by considering the solution in the inner jet regions ($\Psi \ll \Psi_0$).

We first emphasize that this analysis cannot be made by the force-free approximation. As was noted, both the electric and toroidal magnetic fields are exactly zero on the rotation axis. As a result, the Poynting vector is zero too and, therefore, all energy losses here are connected with the particle flow. The particle energy, as is immediately evident from relation (4.39), far from gravitating bodies remains constant near the axis. Therefore, the Bernoulli integral should be written as

$$E(\Psi) = \gamma_{\text{in}}\mu\eta(\Psi) + \Omega_{\text{F}}(\Psi)L(\Psi), \tag{5.45}$$

which was already used above when defining the solution behavior (4.240) at large distances. We take into account the fact that $E(0) = \gamma_{\text{in}}\mu\eta \neq 0$, as we will see, just defined the basic properties of the flows in the vicinity of the rotation axis. In the following, for simplicity, we assume that in the vicinity of the axis

$$\gamma_{\text{in}} = \text{const.} \tag{5.46}$$

Note that the value γ_{in} available in expression (5.45) has the meaning of the ejection Lorentz factor near the origin and does not coincide with the particle Lorentz factor in the whole jet. Thus, we see that the electromagnetic field contribution is crucial only for $\Psi > \Psi_{\text{in}}$, where

$$\Psi_{\text{in}} = \frac{\gamma_{\text{in}}}{\sigma}\Psi_0. \tag{5.47}$$

In the central part of the jet $\Psi < \Psi_{\text{in}}$, the energy is transported by relativistic particles and their Lorentz factor remains constant and equal to their initial value of γ_{in}.

Further, it is logical to normalize the angular momentum L associated with the longitudinal current $I \approx 2\pi L$ in the vicinity of the compact object to the GJ current needed for the smooth passage of the critical surfaces in the neighborhood of the compact object. As to $\Omega_{\text{F}}(\Psi)$ and $\eta(\Psi)$, in the vicinity of the rotation axis they are naturally taken to be constant. As a result, the integrals of motion for $\Psi \ll \Psi_0$, with adequate accuracy, can be written as

$$L(\Psi) = \frac{\Omega_{\rm F}}{4\pi^2}\Psi, \tag{5.48}$$

$$\Omega_{\rm F}(\Psi) = \Omega_0 = \text{const}, \tag{5.49}$$

$$\eta(\Psi) = \eta_0 = \text{const}. \tag{5.50}$$

Using definition (5.40), we readily find that in the inner jet regions $\Psi \ll \Psi_0$, relation $\Omega_{\rm F}L/\mu\eta_0 = 2y$ holds true. Therefore, the Bernoulli integral can be written as

$$E(y) = \mu\eta_0(\gamma_{\rm in} + 2y). \tag{5.51}$$

Thus, Eqs. (5.38) and (5.41) can be rewritten as

$$(1 - x_r^2 - \mathcal{M}^2)^2\left(\frac{{\rm d}y}{{\rm d}x_r}\right)^2 = \frac{\gamma_{\rm in}^2 x_r^2}{\mathcal{M}^4}(1 - x_r^2 - 2\mathcal{M}^2) + x_r^2(\gamma_{\rm in} + 2y)^2 - 4y^2 - \frac{x_r^2}{\mathcal{M}^4}(1 - x_r^2 - \mathcal{M}^2)^2, \tag{5.52}$$

$$(\gamma_{\rm in}^2 + x_r^2 - 1)\frac{{\rm d}\mathcal{M}^2}{{\rm d}x_r} = 2x_r\mathcal{M}^2 - \frac{\gamma_{\rm in}^2 x_r \mathcal{M}^2}{(1 - x_r^2 - \mathcal{M}^2)} + \frac{4y^2\mathcal{M}^6}{x_r^3(1 - x_r^2 - \mathcal{M}^2)}. \tag{5.53}$$

The system of equations (5.52) and (5.53) describing the structure of the central jet region admits an analytical solution. We can verify by direct substitution that for $x_r \ll \gamma_{\rm in}$ the asymptotic behavior is

$$\mathcal{M}^2(x_r) = \mathcal{M}_0^2 = \text{const}, \tag{5.54}$$

$$y(x_r) = \frac{\gamma_{\rm in}}{2\mathcal{M}_0^2}x_r^2, \tag{5.55}$$

which corresponds to the constant magnetic field

$$B_z = B_z(0) = \frac{4\pi\,\gamma_{\rm in}\mu\eta_0}{\mathcal{M}_0^2} = \frac{\gamma_{\rm in}}{\sigma\mathcal{M}_0^2}B(R_{\rm L}). \tag{5.56}$$

Here $B(R_{\rm L}) = \Psi_0/2\pi R_{\rm L}^2$. We, of course, suppose that $\gamma_{\rm in} \gg 1$, which is characteristic of the jets from AGN and radio pulsars. The current density $j_{\parallel}$, by definition (5.48), is also constant and equal to the GJ current here

$$j_{\parallel} = \frac{\Omega_{\rm F}B_z}{2\pi}. \tag{5.57}$$

Thus, Eqs. (5.52) and (5.53) really have no singularity in the vicinity of the rotation axis.

As to the flow structure beyond the internal core $\varpi > \varpi_{\rm c}$, where

$$\varpi_{\rm c} = \gamma_{\rm in} R_{\rm L}. \tag{5.58}$$

Here the solution to Eqs. (5.52) and (5.53) substantially depends on the relation between $\gamma_{\rm in}$ and $\mathcal{M}_0 = \mathcal{M}(0)$.

5.2.2.2 Core Solution

For $\mathcal{M}_0^2 > \gamma_{\rm in}^2$, when, according to (5.56), the magnetic field on the axis is rather small, the total magnetic flux $\Psi_{\rm core} = \Psi(\varpi_{\rm c})$ within the central core $\varpi_{\rm c} = \gamma_{\rm in} R_{\rm L}$

$$\Psi_{\rm core} \approx \pi \gamma_{\rm in}^2 R_{\rm L}^2 B_z(0) \tag{5.59}$$

can be written as

$$\Psi_{\rm core} \approx \frac{\gamma_{\rm in}^2}{\mathcal{M}_0^2} \Psi_{\rm in}. \tag{5.60}$$

Here the terminating magnetic flux $\Psi_{\rm in}$ is again defined by relation (5.47). Recall that on the rotation axis the Alfvén condition $A = 0$ coincides with the fast magnetosonic surface condition $D = 0$ and, therefore, the condition $\mathcal{M}^2 > 1$ implies that the flow here is a supersonic one.

We see that under the condition $\mathcal{M}_0^2 > \gamma_{\rm in}^2$ the magnetic flux $\Psi_{\rm core}$ within the central core $\varpi_{\rm c}$ is less than $\Psi_{\rm in}$, so that beyond this region the main contribution to $E(\Psi)$ is still due to the particles, and the electromagnetic field contribution can be disregarded. As a result, at large distances as compared to the internal scale $\varpi_{\rm c}$ (i.e., for $\varpi \gg \gamma_{\rm in} R_{\rm L}$) the solution to Eqs. (5.52) and (5.53) yields a quadratic increase in $\mathcal{M}^2$ and a power decrease in the magnetic field (Chiueh et al., 1991; Eichler, 1993; Bogovalov, 1995)

$$\mathcal{M}^2(x_r) = \mathcal{M}_0^2 \frac{x_r^2}{\gamma_{\rm in}^2} \gg x_r^2, \tag{5.61}$$

$$B_z(x_r) = B_z(0) \frac{\gamma_{\rm in}^2}{x_r^2}. \tag{5.62}$$

It is interesting to note here that dependence (5.61) exactly corresponds to the conservation of the invariant

$$H = \frac{\mu \eta \Omega_{\rm F} x_r^2}{\mathcal{M}^2}, \tag{5.63}$$

following from relation (4.238) obtained under the assumption $\varpi \gg R_{\rm L}$. Thus, analysis of the cylindrical flows shows that the condition $H = {\rm const}$ is violated on the scale $\gamma_{\rm in} R_{\rm L}$ coinciding with the internal scale $\varpi_{\rm c}$. The value $\mathcal{M}^2$ itself remains finite in the vicinity of the rotation axis.

The above-constructed flow structure externally looks like a jet of the characteristic transverse size $r_{\rm jet} = \gamma_{\rm in} R_{\rm L}$. However, since it contains only a small part of the total magnetic flux, this solution does not require the description of the real jets (not to mention the fact that the scale $\varpi_{\rm c}$ is much smaller than the transverse sizes of the observed jets). Besides, according to (5.62), within this model, the magnetic flux very slowly (logarithmically) increases with distance from the rotation axis:

$$\Psi(x_r) \propto \ln(x_r/\gamma_{\rm in}). \tag{5.64}$$

This behavior of the magnetic field, in turn, shows that the transition value of the flow $\Psi = \Psi_{\rm in}$ is attained exponentially far from the rotation axis, i.e., for

$$r_{\rm jet} \sim R_{\rm L} \exp(\mathcal{M}_0^2/\gamma_{\rm in}^2), \tag{5.65}$$

where the poloidal magnetic field must be exponentially small. Accordingly, the energy density of the magnetic field must also be small for $\varpi \sim r_{\rm jet}$. But this configuration cannot exist in the presence of the finite external gas or magnetic pressure. On the other hand, as shown below, it may be of interest in the analysis of the magnetized flow in vacuum.

Problem 5.4 Show that for the core flow for $\varpi \gg \varpi_{\rm c}$

- the electric and toroidal magnetic fields have the form

$$|\mathbf{E}| \approx B_z(0)\,\frac{\gamma_{\rm in}^2}{x_r}, \tag{5.66}$$

$$B_\varphi \approx B_z(0)\,\frac{\gamma_{\rm in}^2}{x_r}, \tag{5.67}$$

- the particle Lorentz factor is independent of the distance from the rotation axis, viz., $\gamma(x_r) \approx \gamma_{\rm in}$.

5.2.2.3 Homogeneous Solution

Thus, if the jet is submerged in the finite pressure medium (for example, with the constant external magnetic field), the value of the Mach number on the rotation axis is to be bounded from above

$$\mathcal{M}^2(0) < \mathcal{M}_{\rm max}^2 = \gamma_{\rm in}^2. \tag{5.68}$$

According to (5.56), in this case, the magnetic field $B(0)$ on the rotation axis cannot be much less than $B_{\min}$, where

$$B_{\min} = \frac{1}{\sigma \gamma_{\rm in}} B(R_{\rm L}). \tag{5.69}$$

At first, let us consider the case

$$B_{\rm ext} > B_{\min}. \tag{5.70}$$

Then, it is readily checked that the solution to Eqs. (5.52) and (5.53) outside the central core $\varpi_{\rm c} \ll \varpi \ll r_{\rm jet}$ yields the constant magnetic field (5.56)

$$B_{\rm p} = B_{\rm ext}. \tag{5.71}$$

It corresponds to the solution

$$y(x_r) = \frac{\gamma_{\rm in}}{2\mathcal{M}_0^2} x_r^2, \tag{5.72}$$

and to the linear increase in the square of the Mach number

$$\mathcal{M}^2(x_r) = \mathcal{M}_0^2 \frac{x_r}{\gamma_{\rm in}} \ll x_r^2. \tag{5.73}$$

Problem 5.5 Show that for the homogeneous flow for $\varpi \gg \varpi_{\rm c}$

- the electric and toroidal magnetic fields have the form

$$|\mathbf{E}| = B_z(0) x_r, \tag{5.74}$$

$$B_\varphi = B_z(0) x_r, \tag{5.75}$$

- the particle Lorentz factor linearly increases with decreasing distance from the rotation axis: $\gamma(x_r) \approx x_r$.

Note that the unavailability of the core solution $B_z \propto \varpi^{-2}$ is associated with the first term on the right-hand side of Eq. (5.41) proportional to L^2. This term substantially changing the character of the solution was, evidently, omitted in most papers dealing with the jets. As was noted, it is not surprising, since, when analyzing the second-order equation (5.30), the corresponding term in it is small. On the other hand, at large distances from the rotation axis $\varpi \gg \varpi_{\rm c}$ Eq. (5.41) can be rewritten as

$$\frac{\mathrm{d}}{\mathrm{d}\varpi}\left(\frac{\mu\eta\Omega_{\mathrm{F}}\varpi^2}{\mathcal{M}^2}\right) - \frac{\mathcal{M}^2}{\mu\eta\Omega_{\mathrm{F}}\varpi^3(\Omega_{\mathrm{F}}^2\varpi^2 + \mathcal{M}^2)}L^2 = 0, \tag{5.76}$$

in which all terms are of the same order of magnitude. Disregarding the term proportional to L^2, we again have the conservation of H (5.63). In particular, $\mathcal{M}^2 \propto x_r^2$ for $\Omega_{\mathrm{F}} = \mathrm{const}$ and $\eta = \mathrm{const}$. However, as we see, in cylindrical geometry for the relativistic jets the H conservation does not occur. To be exact, the second term in Eq. (5.76) proves to be substantial for all models with the near-constant density of the longitudinal electric current in the central part of the jet when the invariant $L(\Psi)$ for $\Psi < \Psi_0$ linearly increases with increasing the magnetic flux Ψ.

Problem 5.6 Show that the first term in Eq. (5.76) corresponds to Ampére's force $\rho_{\mathrm{e}}|\mathbf{E}| + j_{\parallel}B_{\varphi}$ and the second one to Ampére's force $j_{\varphi}B_{\mathrm{p}}$.

Thus, the homogeneous solution can be matched to the environment by the boundary condition (5.43). Naturally, the flow structure greatly depends on the choice of the integrals of motion. Below, we consider one simple example of this flow and here we point to two general properties typical of this class of solutions. First, as was noted above, the jet can have a zero longitudinal electric current. This is the case if on the jet boundary, where there is no longitudinal motion of matter, the integrals of motion vanish:

$$\Omega_{\mathrm{F}}(\Psi_0) = 0, \quad \eta(\Psi_0) = 0. \tag{5.77}$$

Here again Ψ_0 is the finite total magnetic flux inside the jet. This case corresponds to the absence of tangential discontinuities on the jet boundary. The total electric current within the jet, according to (5.34), automatically proves equal to zero.

Second, as is readily verified, in the relativistic case studied, we can take $\gamma = u_z$ for the estimate with good accuracy. Then at large distances from the rotation axis $x_r \gg \gamma_{\mathrm{in}}$ ($\varpi \gg \varpi_{\mathrm{c}}$), Eq. (5.38) under the condition $\mathcal{M}^2 \ll x_r^2$ can be rewritten as

$$\frac{\mathrm{d}\Psi}{\mathrm{d}\varpi} = \frac{8\pi^2 E(\Psi)}{\varpi\,\Omega_{\mathrm{F}}^2(\Psi)}, \tag{5.78}$$

or, what is the same,

$$B_z(\varpi) = \frac{4\pi\,E(\Psi)}{\varpi^2\Omega_{\mathrm{F}}^2(\Psi)}. \tag{5.79}$$

As we see, Eq. (5.78) does not comprise $\mathcal{M}^2$ and can thus be integrated independently. This must be the case, since Eq. (5.78) coincides with the asymptotic behavior of the force-free equation obtained from (5.38) in the limit $\mathcal{M}^2 \to 0$. Supposing in (5.79) $B_z(r_{\mathrm{jet}}) = B_{\mathrm{ext}}$, we find, in particular, for the transverse size of the jet

$$r_{\rm jet}^2 = \lim_{\Psi\to\Psi_0} \frac{4\pi E(\Psi)}{\Omega_{\rm F}^2(\Psi)B_{\rm ext}}. \tag{5.80}$$

Hence, the jet size is determined by the limit of the ratio $E(\Psi)/\Omega_{\rm F}^2(\Psi)$ for $\Psi \to \Psi_0$. However, since for the strongly magnetized flows $E(\Psi) \propto \Omega_{\rm F}^2$, the estimate of the value $r_{\rm jet}$, in fact, coincides with expression (5.18).

We consider, as an example, the internal structure of the relativistic jet submerged in the external homogeneous magnetic field $B_{\rm ext} > B_{\rm min}$ parallel to the jet axis (Beskin and Malyshkin, 2000). We use the integrals of motion $L(\Psi) = I(\Psi)/2\pi$, $E(\Psi) = \Omega_{\rm F}(\Psi)L(\Psi)$ (3.107) and (3.108), and $\Omega_{\rm F}(\Psi)$ deduced for the force-free black hole magnetosphere. Note at once that these integrals of motion satisfy conditions (5.77) and, hence, can be directly used to study the jet structure that has the zero total electric current. As a result, with allowance for the particle contribution to be taken into account in the vicinity of the rotation axis, we have up to the terms $\sim \sigma^{-1}$

$$\Omega_{\rm F}(\Psi) = \frac{2\sqrt{1-\Psi/\Psi_0}}{1+\sqrt{1-\Psi/\Psi_0}}\,\Omega_{\rm F}(0), \tag{5.81}$$

$$L(\Psi) = \frac{1}{2\pi^2}\frac{\sqrt{1-\Psi/\Psi_0}}{1+\sqrt{1-\Psi/\Psi_0}}\,\Omega_{\rm F}(0)\Psi, \tag{5.82}$$

$$E(\Psi) = \gamma_{\rm in}\mu\eta + \Omega_{\rm F}(\Psi)L(\Psi). \tag{5.83}$$

In particular, for the values of $E(\Psi)$ and $\Omega_{\rm F}(\Psi)$ given by formulae (5.81), (5.82), and (5.83), we have

$$\lim_{\Psi\to\Psi_0} \frac{E(\Psi)}{\Omega_{\rm F}^2(\Psi)} = \frac{1}{4\pi^2}\Psi_0, \tag{5.84}$$

so that the limit (5.80) really exists. Thus, we get

$$r_{\rm jet} = \left(\frac{\Psi_0}{\pi B_{\rm ext}}\right)^{1/2}, \tag{5.85}$$

which naturally coincides with (5.18). According to (5.40) and (5.72), the transverse jet size can be written as

$$r_{\rm jet} = \left(\frac{\sigma\mathcal{M}_0^2}{\gamma_{\rm in}}\right)^{1/2} R_{\rm L}, \tag{5.86}$$

which is equivalent to (5.85). Moreover, it is easy to verify that for the invariants (5.81), (5.82), and (5.83) the constant magnetic field $B_z = B(0)$ turns out to be an exact solution to Eq. (5.78) up to the jet boundary $\varpi = r_{\rm jet}$. Therefore, we can take $B(0) = B_{\rm ext}$. According to (5.56), we obtain

$$\mathcal{M}_0^2 = \frac{\gamma_{\rm in}}{\sigma}\frac{B(R_{\rm L})}{B_{\rm ext}}. \qquad (5.87)$$

Using relation (5.87), we can express all the other parameters of the jet in terms of the external magnetic field. Recall once again that the external magnetic field is to be larger than $B_{\rm min}$, which is given by relation (5.69).

In order for the energy distribution in the jet and the particle Lorentz factor to be found it is convenient to return to the value

$$q(x_r) = \frac{\mathcal{M}^2}{x_r^2}, \qquad (5.88)$$

which, as we saw, at large distances from the rotation axis is simply the particle-to-electromagnetic energy flux ratio. As a result, from relation (5.73) for $x_r \gg \gamma_{\rm in}$ we have

$$\frac{W_{\rm part}}{W_{\rm tot}} \approx \frac{\mathcal{M}_0^2}{\gamma_{\rm in}} x_r^{-1} \ll 1. \qquad (5.89)$$

Accordingly, from formulae (4.137) and (5.88) for the particle Lorentz factor for $x_r \gg \gamma_{\rm in}$ ($\varpi > \varpi_{\rm c}$) we get

$$\gamma(x_r) = x_r, \qquad (5.90)$$

which exactly corresponds to estimate (5.28). Finally, expression (5.36) for the four-velocity $u_{\hat{\varphi}}$ yields for $\varpi > \varpi_{\rm c}$

$$u_{\hat{\varphi}} = 1. \qquad (5.91)$$

Accordingly, $v_{\hat{\varphi}}(x_r) = u_{\hat{\varphi}}/\gamma = 1/x_r$.

On the other hand, according to expression (5.88), we arrive at the most important conclusion that far from the light cylinder

$$q \ll 1. \qquad (5.92)$$

Hence, according to (4.138), for $B_{\rm ext} > B_{\rm min}$, the particle contribution to the total energy flux is insignificant. For example, for $B_{\rm ext} \sim B_{\rm min}$ and $\varpi \sim r_{\rm jet}$, we have

$$\frac{W_{\rm part}}{W_{\rm tot}} \sim \left(\frac{\gamma_{\rm in}}{\sigma}\right)^{1/2}, \qquad (5.93)$$

and, accordingly, $\gamma \sim (\gamma_{\rm in}\sigma)^{1/2}$. In the general case, we get (Beskin and Malyshkin, 2000)

$$\frac{W_{\rm part}}{W_{\rm tot}} \sim \frac{1}{\sigma}\left[\frac{B(R_{\rm L})}{B_{\rm ext}}\right]^{1/2}. \qquad (5.94)$$

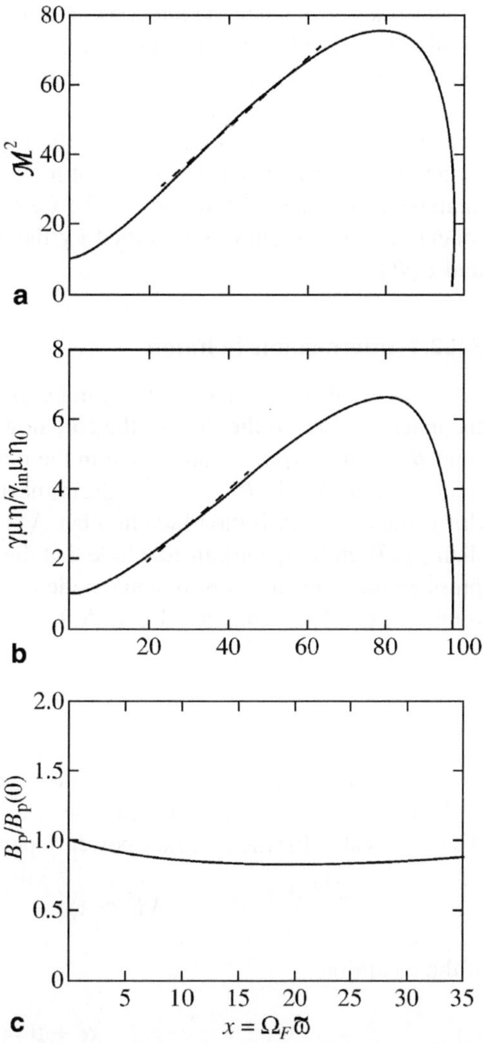

Fig. 5.2 Internal structure of the relativistic cylindrical jet for $B_{\rm ext} > B_{\rm min}$ obtained by the numerical integration of Eqs. (5.38) and (5.41) for the parameters $\mathcal{M}_0^2 = 16$, $\gamma_{\rm in} = 8$, and $\sigma = 1000$. The radial dependence is shown for (**a**) the Mach number $\mathcal{M}^2$, (**b**) the particle energy flux $\gamma\mu\eta$, and (**c**) the poloidal magnetic field B_z. The *dashed lines* indicate the behavior of these values following from the analytic asymptotic behavior (5.73) and (5.90) (Beskin and Malyshkin, 2000)

We arrive at the sufficiently nontrivial conclusion that in the one-dimensional jet the particle acceleration efficiency depends on the environmental parameters. However, this must be the case, because the external magnetic field $B_{\rm ext}$ (ambient pressure $P_{\rm ext}$) determines the transverse jet size that prescribes, as in the nonrelativistic case, the "sling" radius fixing the maximum particle energy.

As an example, Fig. 5.2a, b shows the radial dependence of the Mach number $\mathcal{M}^2$ and the particle energy flux $\gamma\mu\eta$ for the parameters $\mathcal{M}_0^2 = 16$, $\gamma_{\rm in} = 8$, and $\sigma = 1000$ resulting from the numerical integration of Eqs. (5.38) and (5.41) for the integrals of motion (5.81), (5.82), and (5.83) corresponding to the condition $B_{\rm ext} > B_{\rm min}$ (Beskin and Malyshkin, 2000). The dashed lines indicate the analytic behavior of these values that follow from (5.73) and (5.90). As we see, for rather small values of x_r when the integrals of motion (5.81), (5.82), and (5.83) are close to (5.48), (5.49), and (5.50), the analytic expressions coincide with their exact values. On the other hand, for $\Psi = \Psi_0$, i.e., on the jet edge, as was expected, the values of $\gamma\mu\eta$ and $\mathcal{M}^2$ are zero. Figure 5.2c shows the dependence of the poloidal magnetic field $x_r^{-1}\mathrm{d}y/\mathrm{d}x_r$ for the inner parts of the jet $\Psi < \Psi_0$. As seen from the graph, the magnetic field remains nearly constant for $x_r > \gamma_{\mathrm{in}}$, which is in agreement with the analytical estimates (5.56) and (5.72). Certainly, in the general case, the poloidal magnetic field structure is specified by the concrete choice of the integrals $E(\Psi)$ and $L(\Psi)$.

5.2.2.4 Intermediate Solution

As to the weaker external magnetic fields $B_{\mathrm{ext}} < B_{\mathrm{min}}$, in this case the structure of the inner regions has the form of the core flow. However, since the external magnetic field B_{ext} is not exponentially small in the real conditions, the quadratic decrease in the longitudinal field $B_z \propto \varpi^{-2}$ (5.62) is never really realized. But, as shown above, this is the case only if the Mach number $\mathcal{M}_0$ on the rotation axis is not much larger than γ_{in}. Therefore, one can conclude that the relativistic jets submerged in the finite pressure medium cannot have a magnetic field much smaller than B_{min} (5.69) on the rotation axis. Analyzing now Eq. (5.53), which can now be written as

$$x_r^2\frac{\mathrm{d}\mathcal{M}^2}{\mathrm{d}x_r} = 2x_r\mathcal{M}^2 - \frac{4y^2\mathcal{M}^6}{x_r^3(x_r^2+\mathcal{M}^2)}, \tag{5.95}$$

it is easy to show that the solution has no structure $\mathcal{M}^2 \propto \varpi^2$ (i.e., $B_z \propto \varpi^{-2}$) only if the latter term in order of magnitude is close to the other terms in this equation. This is possible for the arbitrary power dependencies

$$\mathcal{M}^2 \propto x_r^{\alpha}, \qquad y \propto x_r^{\beta}, \tag{5.96}$$

if the condition

$$\alpha + \beta = 3 \tag{5.97}$$

is satisfied.

The results of the numerical calculation confirm that, as the external magnetic field decreases, the value α gradually increases, and β, on the contrary, decreases, the condition $\alpha + \beta = 3$ remaining true (Beskin and Nokhrina, 2006). As a result, if for $B_{\mathrm{ext}} \sim B_{\mathrm{min}}$ we have $\alpha \approx 1$ and $\beta \approx 2$, for the weaker external magnetic fields $B_{\mathrm{ext}} \sim B_{\mathrm{eq}}$, where

$$B_{\mathrm{eq}} = \frac{1}{\sigma^2}\,B(R_{\mathrm{L}}), \tag{5.98}$$

we have $\alpha \approx 2$ and $\beta \approx 1$. Thus, we proved the following theorem:

Theorem 5.2 *In the relativistic case, in the presence of the environment with rather high pressure ($B_{\rm ext} > B_{\rm min}$) the poloidal magnetic field inside the jet remains practically constant: $B_{\rm p} \approx B_{\rm ext}$. For small external pressure ($B_{\rm ext} < B_{\rm min}$) in the vicinity of the rotation axis there must form a core region $\varpi < \varpi_{\rm c} = \gamma_{\rm in} R_{\rm L}$ with the magnetic field $B_{\rm p} \approx B_{\rm min}$ (5.69) containing only a small part of the total magnetic flux Ψ_0:*

$$\frac{\Psi_{\rm core}}{\Psi_0} \approx \frac{\gamma_{\rm in}}{\sigma}. \tag{5.99}$$

For $\varpi > \varpi_{\rm c}$, the poloidal magnetic field $B_{\rm p}$ decreases as

$$B_{\rm p} \propto \varpi^{\,\beta-2}, \tag{5.100}$$

where $\beta < 2$.

The existence of the central core with $\varpi_{\rm c} \approx \gamma_{\rm in} R_{\rm L}$ was predicted in a lot of papers (see, e.g., Heyvaerts and Norman, 1989; Bogovalov, 1996). However, the magnetic flux

$$\Psi_{\rm core} = \pi \varpi_{\rm c}^2 B_{\rm min} \tag{5.101}$$

within the central core was not specified.

On the other hand, by definition (4.137), for $B_{\rm ext} > B_{\rm eq}$ (i.e., $q \ll 1$) the particle Lorentz factor

$$\gamma \approx q\,\frac{E}{\mu\eta} \approx x_r^{\alpha+\beta-2} \tag{5.102}$$

is still described by the universal asymptotic solution $\gamma = x_r$, so relation (5.94) appears valid for these external magnetic fields. As a result, for $B_{\rm ext} < B_{\rm eq}$ (i.e., $\alpha \approx 2$), the particle contribution to the energy flux becomes crucial in the whole jet volume. Consequently, one can conclude that the full transformation from the electromagnetic energy flux to the particle energy flux can occur only for rather weak external magnetic fields $B_{\rm ext} < B_{\rm eq}$.

Problem 5.7 Show that for $B_{\rm ext} \sim B_{\rm eq}$ the poloidal magnetic field in the jet behaves as $B_z \propto x_r^{-1}$.

Problem 5.8 Find expression (5.98) for the equipartition magnetic field $B_{\rm eq}$.

Problem 5.9 Show that the transverse size of the jet for the external magnetic fields $B_{\rm eq} < B_{\rm ext} < B_{\rm min}$ is, as before, given by the expression

$$r_{\rm jet} \sim \left(\frac{\Psi_0}{\pi B_{\rm ext}}\right)^{1/2}. \qquad (5.103)$$

In conclusion, it is interesting to compare the particle energy in the jet with the ultimate energy gained by the particles in the outflow from the compact object magnetosphere with the monopole magnetic field. As is shown in Sect. 5.3.4, for the quasimonopole relativistic plasma outflow the particle Lorentz factor beyond the fast magnetosonic surface $r > \sigma^{1/3} R_{\rm L}$ in the absence of the environment can be written as

$$\gamma(y) = y^{1/3}, \qquad y > \gamma_{\rm in}^3, \qquad (5.104)$$

$$\gamma(y) = \gamma_{\rm in}, \qquad y < \gamma_{\rm in}^3, \qquad (5.105)$$

where y is defined by formula (5.40). On the other hand, relations (5.72) and (5.90) for the jet yield

$$\gamma(y) = \left(\frac{\mathcal{M}_0^2}{\gamma_{\rm in}}\right)^{1/2} y^{1/2}, \qquad y > \frac{\gamma_{\rm in}^3}{\mathcal{M}_0^2}, \qquad (5.106)$$

$$\gamma(y) = \gamma_{\rm in}, \qquad y < \frac{\gamma_{\rm in}^3}{\mathcal{M}_0^2}. \qquad (5.107)$$

As shown in Fig. 5.3a, for $\mathcal{M}_0^2 > 1$, i.e., for $B_{\rm min} < B_{\rm ext} < B_{\rm cr}$, where

$$B_{\rm cr} = \frac{\gamma_{\rm in}}{\sigma} B(R_{\rm L}), \qquad (5.108)$$

the particle Lorentz factor in the jet (5.106) is always larger than the Lorentz factor gained by the particles in the outflow from the magnetosphere with the monopole magnetic field, but, of course, always less than the limiting Lorentz factor

$$\gamma_{\rm in} = 2y, \qquad (5.109)$$

corresponding to the full transformation of the electromagnetic energy to the particle energy. This implies that for $B_{\rm ext} < B_{\rm cr}$ in the collimation process connected with the interaction between the outflowing plasma and the environment there must be additional particle acceleration. If $B_{\rm ext} > B_{\rm cr}$, in the internal jet regions for $\varpi < \varpi_{\rm f}$, where

$$\varpi_{\rm f} = \frac{\gamma_{\rm in}}{\mathcal{M}_0^2} R_{\rm L}, \qquad (5.110)$$

the particle energy on the given magnetic field line proves even smaller than the values obtained for the monopole magnetic field, which is shown in Fig. 5.3b.

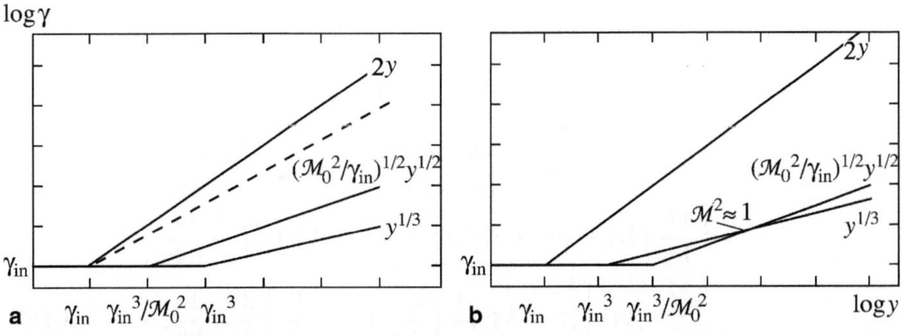

Fig. 5.3 The particle Lorentz factor γ versus the magnetic flux $y = \sigma\Psi/\Psi_0$ for the monopole field $\gamma \sim y^{1/3}$, for the one-dimensional jet $\gamma \sim \left(\mathcal{M}_0^2/\gamma_{\rm in}\right)^{1/2} y^{1/2}$ and in the case of the full transformation of the electromagnetic energy to the particle energy $\gamma = 2y$. (**a**) $\mathcal{M}_0^2 > 1$ ($B_{\rm ext} < B_{\rm cr}$). The *dashed line* indicates the behavior of the Lorentz factor for $B_{\rm eq} < B_{\rm ext} < B_{\rm min}$. (**b**) $\mathcal{M}_0^2 < 1$ ($B_{\rm ext} > B_{\rm cr}$). The intersection of the lines ($\mathcal{M}^2 \approx 1$) defines the fast magnetosonic surface location (Beskin and Malyshkin, 2000)

This result can be readily explained. Indeed, for the studied integrals of motion (5.48), (5.49), and (5.50) the factor D, whose equality to zero specifies the fast magnetosonic surface location, for the cold plasma can be rewritten as

$$\mathcal{M}^2 D \equiv A + \frac{B_\varphi^2}{B_{\rm p}^2} = A + \frac{4x_r^2 y^2 \mathcal{M}^4}{4y^2\mathcal{M}^4 - 2x_r^2\mathcal{M}^2 - x_r^4}. \tag{5.111}$$

It is easy to show that for the values of y and $\mathcal{M}^2$ given by formulae (5.72) and (5.73), expression (5.111) is negative for $\mathcal{M}^2 > 1$, i.e., just for the values of x_r corresponding to (5.110). Hence, one can conclude that for sufficiently large external magnetic fields $B_{\rm ext} > B_{\rm cr}$ (5.108), when $\mathcal{M}_0^2 < 1$, in the inner regions $\varpi < \varpi_{\rm f}$, where

$$\varpi_{\rm f} \approx \sigma \left[\frac{B_{\rm ext}}{B(R_{\rm L})}\right] R_{\rm L}, \tag{5.112}$$

a subsonic flow inevitably occurs. The occurrence of the subsonic flow region far from the compact object is possible either in the presence of the shock wave or due to the strong distortion of the magnetic field lines within the fast magnetosonic surface located in the neighborhood of the compact object. In both cases, the magnetic field disturbance results in a decrease in the particle energy.

5.2.3 Nonrelativistic Jets

We now discuss the structure of the nonrelativistic jets from young stars. In the general case of nonzero temperature, Eq. (5.41) has the form (Beskin and Nokhrina, 2009)

$$
\begin{aligned}
\left[2e_{\rm n} - 2w + \Omega_{\rm F}^2\varpi^2 - (1-\mathcal{M}^2)c_{\rm s}^2\right]\frac{{\rm d}\mathcal{M}^2}{{\rm d}\varpi} = \\
\frac{\mathcal{M}^6}{1-\mathcal{M}^2}\frac{L_{\rm n}^2}{\varpi^3} - \frac{\Omega_{\rm F}^2\varpi}{1-\mathcal{M}^2}\mathcal{M}^2(2\mathcal{M}^2-1) \\
+\mathcal{M}^2\frac{{\rm d}\Psi}{{\rm d}\varpi}\frac{{\rm d}e_{\rm n}}{{\rm d}\Psi} + \mathcal{M}^2\varpi^2\Omega_{\rm F}\frac{{\rm d}\Psi}{{\rm d}\varpi}\frac{{\rm d}\Omega_{\rm F}}{{\rm d}\Psi} \\
+\mathcal{M}^2\frac{{\rm d}\Psi}{{\rm d}\varpi}\left[2e_{\rm n} - 2w + \Omega_{\rm F}^2\varpi^2 - 2(1-\mathcal{M}^2)c_s^2\right]\frac{1}{\eta_{\rm n}}\frac{{\rm d}\eta_{\rm n}}{{\rm d}\Psi} \\
-\mathcal{M}^2\left[(1-\mathcal{M}^2)\frac{1}{\rho}\left(\frac{\partial P}{\partial s}\right)_\rho + \frac{T}{m_{\rm p}}\right]\frac{{\rm d}\Psi}{{\rm d}\varpi}\frac{{\rm d}s}{{\rm d}\Psi},
\end{aligned}
\tag{5.113}
$$

where $e_{\rm n} = E_{\rm n} - \Omega_{\rm F}L_{\rm n}$. In the central part of the jet ($\Psi \ll \Psi_0$) where it is logical to assume that $j_\parallel = i_0 j_{\rm GJ} = {\rm const}$, the integrals of motion can be written as

$$
E_{\rm n}(\Psi) \approx \frac{v_{\rm in}^2}{2} + i_0\frac{\Omega_0^2}{4\pi^2 c\eta_0}\Psi, \tag{5.114}
$$

$$
L(\Psi) \approx i_0\frac{\Omega_0}{4\pi^2 c\eta_0}\Psi, \tag{5.115}
$$

where $v_{\rm in}$ has the same meaning as $\gamma_{\rm in}$ for the relativistic flows. Accordingly, we can take $\Omega_0 = \Omega_{\rm F}(0) = {\rm const}$ and $\eta_0 = \eta_{\rm n}(0) = {\rm const}$. As a result, the terminating magnetic flux $\Psi_{\rm in}$ containing a particle-dominated flow is given by

$$
\Psi_{\rm in} = 2\pi^2\frac{v_{\rm in}^2 c\eta_0}{i_0\Omega_0^2}. \tag{5.116}
$$

We emphasize once again that the value $v_{\rm in}$, as for relativistic flows, has the meaning of the characteristic constant rather than the real ejection velocity in the source, and i_0 fixes the angular momentum L rather than the real current in the jet. If the angular momentum is consistent with the critical condition on the fast magnetosonic surface, the value of i_0 is to be determined by relation (4.207).

Problem 5.10 Show that Eq. (5.113) has a singularity on the cusp surface, i.e., for $D = -1$.

As a result, Eq. (5.113) together with nonrelativistic Bernoulli's equation (4.88) for $\Psi < \Psi_0$ is written as

$$\left(1+x_r^2\right)\frac{\mathrm{d}\mathcal{M}^2}{\mathrm{d}x_r} = 2x_r\mathcal{M}^2 + 4i_0^2\frac{v_{\rm in}^2}{c^2}\frac{y^2\mathcal{M}^6}{x_r^3(1-\mathcal{M}^2)} - \frac{x_r\mathcal{M}^2}{(1-\mathcal{M}^2)}, \tag{5.117}$$

$$\mathcal{M}^4\left(\frac{\mathrm{d}y}{\mathrm{d}x_r}\right)^2 = x_r^2 + 4i_0\frac{v_{\rm in}}{c}x_r^2 y - \frac{[x_r^2 - 2i_0(v_{\rm in}/c)y\mathcal{M}^2]^2}{(1-\mathcal{M}^2)^2} - 2x_r^2\frac{2i_0(v_{\rm in}/c)y - x_r^2}{1-\mathcal{M}^2}. \tag{5.118}$$

Here $x_r = \Omega_0\varpi/v_{\rm in}$, $y = \sigma_{\rm n}\Psi/\Psi_0$, the "nonrelativistic" magnetization parameter $\sigma_{\rm n}$ is

$$\sigma_{\rm n} = \frac{\Omega_0^2\Psi_0}{8\pi^2 v_{\rm in}^3\eta_0}, \tag{5.119}$$

and we restrict ourselves to the cold plasma flow for simplicity. As we see, Eqs. (5.117) and (5.118) contain the inner scale

$$\varpi_{\rm c} = \frac{v_{\rm in}}{\Omega_0}. \tag{5.120}$$

Problem 5.11 Show that the condition $\sigma_{\rm n} \gg 1$ coincides with the condition of the fast rotation $\Omega_{\rm F} \gg \Omega_{\rm crit}$.

At first, we consider the super-Alfvén flow $\mathcal{M}^2(0) > 1$ in more detail. Since, as was noted, the Alfvén surface in the vicinity of the rotation axis must coincide with the fast magnetosonic one, the flow in the vicinity of the rotation axis is to be supersonic. As a result, Eqs. (5.117) and (5.118) can be rewritten as

$$\left(1+x_r^2\right)\frac{\mathrm{d}\mathcal{M}^2}{\mathrm{d}x_r} = 2x_r\mathcal{M}^2 - 4i_0^2\frac{v_{\rm in}^2}{c^2}\frac{y^2\mathcal{M}^4}{x_r^3}, \tag{5.121}$$

$$\mathcal{M}^4\left(\frac{\mathrm{d}y}{\mathrm{d}x_r}\right)^2 = x_r^2 + 4i_0\frac{v_{\rm in}}{c}x_r^2 y - 4i_0^2\frac{v_{\rm in}^2}{c^2}y^2 - 2\frac{x_r^4}{\mathcal{M}^2}. \tag{5.122}$$

It is easy to verify that under the condition $\mathcal{M}_0^2 \gg 1$, the last term in (5.121) can be dropped. As a result, we have

$$\mathcal{M}^2(\varpi) = \mathcal{M}_0^2\left(1+\frac{\varpi^2}{\varpi_c^2}\right). \tag{5.123}$$

Accordingly, under the condition $\mathcal{M}_0^2 \gg 1$, we can disregard the last three terms in Eq. (5.122). Thus, we have

$$\frac{1}{x_r}\frac{\mathrm{d}y}{\mathrm{d}x_r} = \frac{1}{\mathcal{M}^2(x_r)}, \tag{5.124}$$

which again corresponds to the core solution (5.62) (Eichler, 1993; Bogovalov, 1995)

$$B_z(\varpi) = \frac{B_z(0)}{1 + \dfrac{\varpi^2}{\varpi_c^2}}. \tag{5.125}$$

As a result, at large distances from the rotation axis $\varpi \gg \varpi_c$, we again have the power dependence $\mathcal{M}^2 \propto \varpi^2$, which corresponds to the conservation of the invariant H (5.63). On the other hand, the solution studied has no singularity on the rotation axis. In particular, $I(\varpi) \to 0$ for $\varpi \to 0$.

Further, using definitions (5.21) and (5.22), we find for $\varpi \gg \varpi_c$

$$B_\varphi = -\frac{v_{in}}{\Omega_0 \varpi} B_0, \tag{5.126}$$

$$E_\varpi = -\frac{v_{in}^2}{c\Omega_0 \varpi} B_0, \tag{5.127}$$

so that the electric field is always smaller than the magnetic one. Finally, the velocity components, according to (5.29), can be written as

$$v_p = v_{in}, \tag{5.128}$$

$$v_\varphi \ll v_{in}. \tag{5.129}$$

As a result, the energy flux is fully determined by the particle flux:

$$\frac{W_{part}}{W_{tot}} \sim 1. \tag{5.130}$$

But, as was already noted, in this case the magnetic flux is to increase logarithmically slowly with distance from the rotational axis

$$\Psi(\varpi) \propto \ln \varpi. \tag{5.131}$$

It is clear that in the presence of the environment with finite pressure this solution can exist only if the magnetic flux within the core $\Psi_{core} = \Psi(\varpi_c)$ is comparable with the total magnetic flux of the jet, i.e., $\Psi_{core} \approx \Psi_0$. As

$$\frac{\Psi_{core}}{\Psi_0} \approx \frac{1}{\mathcal{M}_0^2} \frac{4\pi \rho_{in} v_{in}^2}{B_{in}^2} \left(\frac{v_{in}}{\Omega_0 R_{in}}\right)^2 \approx \frac{1}{\mathcal{M}_0^2} \left(\frac{\Omega_F}{\Omega_{crit}}\right)^{-2}, \tag{5.132}$$

where Ω_{crit} (4.205) terminates the particle and magnetically dominated flows near the origin, we proved the following theorem:

Theorem 5.3 *In the presence of the environment the cold nonrelativistic supersonic cylindrical flow can exist only if the flow is particle dominated in the vicinity of the central object* ($\Omega_{\rm F} < \Omega_{\rm crit}$).

In this case, the magnetic field on the axis cannot be much smaller than $B_{\rm min}$, where

$$B_{\rm min} = \frac{\Psi_0}{\pi \varpi_{\rm c}^2}. \tag{5.133}$$

Integrating now Eqs. (5.117) and (5.118), one can find that

$$B_0 \approx \frac{B_{\rm min}}{\ln(1 + B_{\rm min}/B_{\rm ext})}. \tag{5.134}$$

Accordingly,

$$\Psi_{\rm core} \approx \frac{\Psi_0}{\ln(1 + B_{\rm min}/B_{\rm ext})}. \tag{5.135}$$

This structure was reproduced numerically as well (see Sect. 5.5.3).

On the other hand, for the fast rotation $\Omega_{\rm F} \gg \Omega_{\rm crit}$ the adequate solution cannot be realized as the core magnetic flux $\Psi_{\rm core}$ is much smaller than even the terminating flux $\Psi_{\rm in}$ (5.116) within the central part of the flow. Indeed, according to definitions (4.77) and (4.87), one can write for $\mathcal{M}_0^2 > 1$

$$\frac{\Psi_{\rm core}}{\Psi_{\rm in}} \approx \frac{i_0 v_{\rm in}}{2c\mathcal{M}_0^2} < \left(\frac{\Omega_{\rm F}}{\Omega_{\rm cr}}\right)^{-2/3} \ll 1. \tag{5.136}$$

This implies that the jet boundary $\Psi = \Psi_0$ ($y = \sigma_{\rm n}$) is to locate exponentially far from the axis:

$$r_{\rm jet} \sim \varpi_{\rm c} \exp(\mathcal{M}_0^2 \sigma_{\rm n}), \tag{5.137}$$

where the energy density of the magnetic field $B(r_{\rm jet})^2/8\pi$ becomes small. This is in contradiction with the finite energy density of the environment. This implies that the cold nonrelativistic cylindrical flow resulting from the interaction between the supersonic, fast-rotating wind and the environment cannot be realized.

To resolve this contradiction, one can suppose that the finite temperature may play an important role in the force balance of the observed nonrelativistic supersonic jets. For example, the additional heating can be connected with the shock wave formation at the jet foot (Bogovalov and Tsinganos, 2005; Bromberg and Levinson, 2007). This additional heating is really needed to explain the emission lines observed in some YSO (Schwartz, 1983). This situation is to be analogous to the case of the interaction between the supersonic flow and a solid wall. This gas-dynamical analogy is all the more appropriate because the nonrelativistic supersonic MHD flow must be weakly magnetized outside the fast magnetosonic surface.

Finally, contrary to the relativistic jets, we can give the following theorem:

Theorem 5.4 *The cylindrical trans-Alfvénic flow cannot be realized in the center part of the nonrelativistic flow.*

To prove this theorem we should make two assumptions. We assume that the derivative of the Mach number remains finite on the Alfvén surface: $L_{\rm n} - \Omega_{\rm F}\varpi^2\big|_{\rm A} = 0$. We also assume that the total current I is not closing on the Alfvén surface.

Let us suppose that the flow in the center of the cylindrical jet is sub-Alfvénic and is about to cross the Alfvén surface: $\mathcal{M}^2 = 1 - \delta_1$, $L_{\rm n}^2 = \Omega_{\rm F}^2\varpi^4 - \delta_2$, where $\delta_1 > 0$, $\delta_2 > 0$. In this case, we can write the leading terms of Eq. (5.113) as

$$\left(v_{\rm in}^2 + \Omega_{\rm F}^2\varpi^2\right)\frac{{\rm d}\mathcal{M}^2}{{\rm d}\varpi} = \frac{1-\delta_1}{\delta_1}\left[\Omega_{\rm F}^2\varpi\delta_1^2 - \frac{\delta_2}{\varpi^3} + \right.$$
$$\left. \delta_1\frac{{\rm d}\Psi}{{\rm d}\varpi}\left(\frac{{\rm d}e_{\rm n}}{{\rm d}\Psi} + \frac{\varpi^2}{2}\frac{{\rm d}\Omega_{\rm F}^2}{{\rm d}\Psi} + \left(v_{\rm in}^2 + \Omega_{\rm F}^2\varpi^2\right)\frac{1}{\eta}\frac{{\rm d}\eta}{{\rm d}\Psi}\right)\right]. \quad (5.138)$$

The term on the right-hand side of the equation is zero for the inner part of the flow. As the total current does not close on the Alfvén surface,

$$\frac{L_{\rm n} - \Omega_{\rm F}\varpi^2}{1 - \mathcal{M}^2} = \left.\frac{\delta_2}{\delta_1(L_{\rm n} + \Omega_{\rm F}\varpi^2)}\right|_{\rm A} \rightarrow {\rm const} \neq 0, \quad (5.139)$$

i.e., $\delta_2 = O(\delta_1)$; we can disregard the first term in the right-hand side bracket in (5.138). Thus, we can conclude that the derivative of the Mach number in the vicinity of the Alfvén surface is negative. However, if we assume that $\mathcal{M}^2$ should be unity, there must be at least one point in the vicinity of the Alfvén surface at which the derivative is positive. We see a contradiction, so the transition of the Alfvén surface is impossible.

Problem 5.12 Show that for the subsonic nonrelativistic flows in the vicinity of the rotation axis, i.e., for $\mathcal{M}^2 \ll 1$,

- the square of the Mach number $\mathcal{M}^2$ linearly increases with distance from the axis and the poloidal magnetic field B_z remains constant regardless of the choice of the integrals of motion

$$\mathcal{M}^2(x_r) \approx \mathcal{M}_0^2\left(1 + x_r\right), \quad (5.140)$$
$$y(x_r) \approx \frac{1}{2}\frac{1}{\mathcal{M}_0^2}x_r^2, \quad (5.141)$$

(Hint: the energy density of the poloidal magnetic field is larger than that of all the other fields and the particle energy density.)

- the matter velocity increases with distance from the rotation axis

$$v_{\rm p} = \sqrt{v_{\rm in}^2 + \Omega_{\rm F}^2 \varpi^2}, \tag{5.142}$$

$$v_{\varphi} = \Omega_{\rm F} \varpi, \tag{5.143}$$

- the minimum external magnetic field for which the subsonic regime can be realized is

$$B_1 = \left(\frac{4\pi \rho_{\rm in} v_{\rm in}^2}{B_{\rm in}^2}\right)^{2/3} \left(\frac{\Omega_0 R_{\rm in}}{v_{\rm in}}\right)^{2/3} B_{\rm in}. \tag{5.144}$$

5.2.4 General Properties and Application

We can now make the main conclusions based on the above analysis of the exact solutions to the GS equation for the one-dimensional cylindrical flows.

1. In the vicinity of the rotation axis ($\varpi < \varpi_{\rm c}$) of the nonrelativistic supersonic cylindrical flows there always occurs a core region with a higher value of the longitudinal magnetic field $B \sim B_{\rm min}$ (5.133) containing a considerable part of the total magnetic flux. If the flow is magnetically dominated near the origin, the gas pressure connected, say, with the additional heating in the oblique shock near the jet base is to play an important role in the force balance.
2. The structure of the relativistic cylindrical flows substantially depends on the external pressure. For the jets submerged in the medium with rather large pressure ($B_{\rm ext} > B_{\rm min}$ or $P_{\rm ext} > B_{\rm min}^2/8\pi$), the longitudinal magnetic field weakly depends on the distance from the rotation axis, so the flow has no central core. However, when the external magnetic field is smaller than $B_{\rm min}$, the relativistic flow must also have the core structure. The central core, in this case, contains only a small part of the total magnetic flux.
3. The fraction of the energy transported by the particles $W_{\rm part}/W_{\rm tot}$ is defined by the environmental parameters. Thus, for example, in the relativistic case for the values of $\sigma \gg \sigma_{\rm cr}$ (i.e., for $B_{\rm ext} \ll B_{\rm eq}$), where

$$\sigma_{\rm cr} = \left[\frac{B(R_{\rm L})}{B_{\rm ext}}\right]^{1/2}, \tag{5.145}$$

the particle energy flux $W_{\rm part}$ is only a small part of the energy flux $W_{\rm em}$ transported by the electromagnetic field. Consequently, the jet remains strongly magnetized ($W_{\rm part} \ll W_{\rm tot}$) only at rather large values of the parameter σ. If the magnetization parameter is not larger than $\sigma_{\rm cr}$, the jet is to be particle dominated. This implies that in the jet collimation process a considerable part of the energy is to be transferred from the electromagnetic field to the plasma. It is interesting to note that for both AGN and the fast radio pulsars the value of $\sigma_{\rm cr}$ is about identical:

$$\sigma_{\rm cr} \approx 10^5 - 10^6. \tag{5.146}$$

4. For sufficiently large external magnetic fields $B_{\rm ext} > B_{\rm cr}$ (5.108), the central part of the relativistic cylindrical flow must be subsonic.
5. The important result is the assertion that, with account taken of the external regular magnetic field, the magnetohydrodynamics equations make it possible to construct the self-consistent model of the jet in which the total longitudinal electric current $I(\Psi_0)$ is zero.

Thus, the general relations allow us to make direct predictions whose validity can be checked by the observations. As we saw, they actually depend on three value only, viz., the magnetization parameter σ, the Lorentz factor in the generation region $\gamma_{\rm in}$ (velocity $v_{\rm in}$ for the nonrelativistic flows), and the external magnetic field $B_{\rm ext}$ (ambient pressure $P_{\rm ext}$). Note that the above results can be applied to both the electron–positron (Wardle et al., 1998) and electron–proton (Sikora and Madejski, 2000) jets.

As a result, for AGN with $\sigma \sim 10$–10^3, $\gamma_{\rm in} \sim 3$–10, the characteristic magnetic fields $B_{\rm min} \sim 1$ G and $B_{\rm eq} \sim 10^{-3}$ G are much larger than the magnetic fields in the interstellar medium. On the other hand, for $\sigma \sim 10^{11}$–10^{13}, $\gamma_{\rm in} \sim 10$, the value $B_{\rm min} \sim 10^{-8}$ G appears too small as compared to the external magnetic field. Finally, for the fast radio pulsars of the Crab and Vela type ($B_{\rm in} \sim 10^{13}$ G, the polar cap radius $R_{\rm in} \sim 10^5$ cm, and $\lambda = n/n_{\rm GJ} \sim 10^4$), in which the jets are observed, we have (Bogovalov, 1998)

$$B_{\rm min} \sim 10^{-1}\ {\rm G}, \qquad B_{\rm cr} \sim 10^3\ {\rm G}. \tag{5.147}$$

As a result, for AGN with large values of $\sigma \sim 10^{12}$, the particles transport only a small fraction of energy as compared to the electromagnetic flux, so the flow within the jet slightly differs from the force-free one. Besides, the external magnetic field $B_{\rm ext} \sim 10^{-6}$ G is larger than the critical magnetic field $B_{\rm cr} \sim 10^{-7}$ G. According to (5.56) and (5.108), this implies that a subsonic region must exist in the inner regions of these jets. On the other hand, in the sources with the magnetization parameter $\sigma \sim 100$ (e.g., gamma-bursters) a considerable part of energy in the jet collimation process is to be transferred to the plasma particles, and the subsonic region in the vicinity of the rotation axis is not generated. As to the fast radio pulsars, we have $B_{\rm ext} \ll B_{\rm cr}$, so that the subsonic flow in the central regions of the jet is not realized either.

Finally, as we saw, the homogeneous magnetic field coinciding with the external magnetic field $B_{\rm ext}$ can be the solution for the inner regions of relativistic jets. Consequently, the transverse size of the AGN jets can be explained in a natural way. Simultaneously, the small efficiency of the transformation of the electromagnetic energy into the particle energy can account for the energy transport from the compact object to the energy release region. Moreover, the extension of the MHD solution to the jet region requires extremely high particle energies ($\sim 10^4$ MeV) that are not detected now. However, for the discussion of the outflowing plasma energy problem it is necessary to correctly take into account the interaction with the environment (for example, with the background radiation), which can greatly affect the particle energy (Li et al., 1992; Beskin et al; 2004).

5.3 Cold Quasimonopole Outflows

We consider now another characteristic case in which the plasma outflows from the surface of the rotating body with the monopole magnetic field. As was mentioned, this model can be regarded as the first approximation to the outer regions in the magnetosphere of the black hole located in the AGN center. It can also be of interest for the description of the pulsar wind, since beyond the light cylinder the structure of the magnetic field lines can also be close to the monopole one. It is not improbable that this geometry of the magnetic field occurs in the vicinity of the young stars. Below, we discuss in detail the basic properties of the exact solutions obtained in the analysis of this set of problems.

5.3.1 Relativistic Slowly Rotating Outflows

At first, let us consider the radial cold outflows of the relativistic plasma, with the gravitational forces neglected (Bogovalov, 1992). In the absence of rotation (i.e., for $\Omega_{\rm F} = 0$, $L = 0$) the monopole magnetic field $\Psi = \Psi_0(1 - \cos\theta)$ for $\theta < \pi/2$ is again the solution to the GS equation, because the radially moving plasma (which for $\Omega_{\rm F} = 0$ should be electrically neutral) does not disturb the radial magnetic field. Therefore, the monopole magnetic field can really be chosen as an exact zero approximation.

According to the general relation $b = 2 + i - s'$ (1.64), the GS equation for the cold plasma requires four boundary conditions on the surface $r = R$ (four integrals of motion, two critical surfaces). These four boundary conditions can be the angular velocity $\Omega_{\rm F}$, the Lorentz factor γ, the plasma concentration n, and the magnetic flux function $\Psi(R, \theta)$. For the nonrotating sphere, they are naturally chosen as

$$\Omega_{\rm F}(R,\theta) = 0,$$

$$\gamma(R,\theta) = \gamma_{\rm in} = {\rm const}, \qquad (5.148)$$

$$n(R,\theta) = n_{\rm in} = {\rm const},$$

$$\Psi(R,\theta) = \Psi_0(1 - \cos\theta). \qquad (5.149)$$

Recall that the value n is the concentration in the comoving reference frame. It is connected with the concentration in the laboratory system by the relation $n^{(\mathrm{lab})} = n/\gamma_{\mathrm{in}}$. At a distance of

$$r_a = R\frac{B_{\mathrm{p}}}{\sqrt{4\pi n_{\mathrm{in}} u_{\mathrm{in}}^2 m_{\mathrm{e}} c^2}}, \tag{5.150}$$

where $u_{\mathrm{in}} = (\gamma_{\mathrm{in}}^2 - 1)^{1/2}$ and $B_{\mathrm{p}} = \Psi_0/2\pi R^2$ is the magnetic field on the surface $r = R$, the plasma flow smoothly crosses the Alfvén surface and the fast magnetosonic surface which is coincident with it for $I = 0$. This expression directly follows from Bernoulli's equation (4.44) in which we must take $\Omega_{\mathrm{F}} = 0$ and $L = 0$ (which can be deduced from the relation $\mathcal{M}^2 = 1$ as well).

Suppose now that a body begins to rotate with small angular velocity Ω, and the parameter

$$\varepsilon_{\mathrm{b}} = \frac{\Omega r_a}{v_{\mathrm{in}}} \tag{5.151}$$

($v_{\mathrm{in}} = c(1 - \gamma_{\mathrm{in}}^{-2})^{1/2}$) thus becomes much less than unity. Then due to the high conductivity of the central body (another additional assumption), we can take $\Omega_{\mathrm{F}} = \Omega$.

As to the determination of the longitudinal current I, one should stress that, with account taken of the finite particle mass (when, besides the Alfvén surface, the plasma must cross the fast magnetosonic surface), the longitudinal current I is no longer a free function but must be determined from the critical condition on the singular surfaces. This is quite similar to the hydrodynamical flows when the critical conditions on the sonic surface fix the accretion rate.

Indeed, for the nonzero rotation, when at once two parameters, $\Omega_{\mathrm{F}}(\Psi)$ and $L(\Psi)$, appear to be different from zero, their ratio, generally speaking, may be arbitrary. Accordingly, the location of the Alfvén surface (4.154)

$$r_{\mathrm{A}}^2 = \frac{L}{\Omega_{\mathrm{F}} E \sin^2\theta} \tag{5.152}$$

proves arbitrary too. Here now $E = \gamma_{\mathrm{in}}\mu\eta = \mathrm{const}$.

However, as shown in Fig. 5.4, for small values of ε_{b} the smooth passage from a subsonic to a supersonic branch is possible only if the location of the Alfvén surface (5.152) in zero order with respect to the small value of ε_{b} coincides with that of the fast magnetosonic surface (5.150). As a result, we have

$$L(\theta) = \frac{\Omega_{\mathrm{F}}}{8\pi^2}\Psi_0 \sin^2\theta, \tag{5.153}$$

so that the longitudinal current $I = 2\pi L$ must actually coincide here with the Michel current I_M (2.225). It is not surprising, therefore, that, as shown below, the

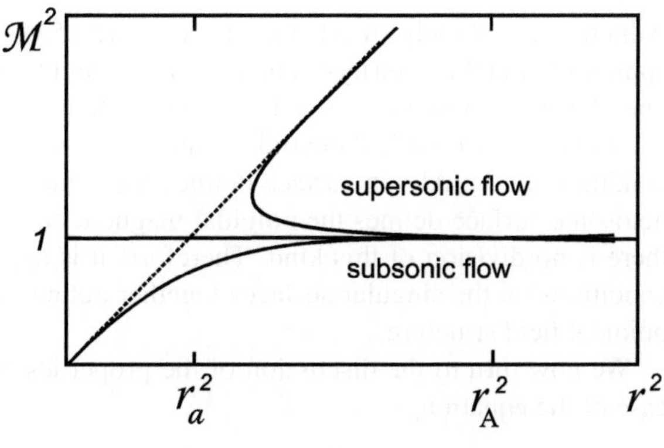

Fig. 5.4 The motion of the roots of relativistic Bernoulli's equation (4.44) in the vicinity of the singular surfaces. The *dotted line* indicates the Mach number $\mathcal{M}^2(r)$ in the absence of rotation, the *solid lines*—$\Omega_F \neq 0$. The transonic flow becomes possible when the Alfvén radius r_A is close to r_a

collimation of the magnetic surfaces along the rotation axis is actually absent for this current.

As a result, the solution to the GS equation can still be sought as a small perturbation of the monopole outflow $\Psi(r,\theta) = \Psi_0[1 - \cos\theta + \varepsilon_b^2 f(x,\theta)]$, and the linearized equation has the form (Bogovalov, 1992)

$$\varepsilon_b^2(x^2-1)\frac{\partial^2 f}{\partial x^2} + 2\varepsilon_b^2 x\frac{\partial f}{\partial x} - \frac{\varepsilon_b^2}{x^2}\sin\theta\frac{\partial}{\partial\theta}\left(\frac{1}{\sin\theta}\frac{\partial f}{\partial\theta}\right) = \frac{2}{\gamma_{in}^2}\left(\frac{\Omega r_a}{v_{in}}\right)^2\sin^2\theta\cos\theta. \tag{5.154}$$

Here $x = r/r_a$. Note that when deriving Eq. (5.154) we use relation (5.153). Otherwise, the right-hand side of this equation would have an additional singularity for $x = 1$. Besides, we again used here the fact that the disturbance of the monopole magnetic field remains insignificant and, therefore, the value $L(\Psi)$ considered here as a small perturbation can be rewritten as $L(\theta)$.

Thus, Eq. (5.154) has the same properties as the equations considered for the hydrodynamical flows.

- It is linear.
- The angular operator coincides with the operator $\hat{\mathcal{L}}_\theta$ (1.120).
- The solution to Eq. (5.154) can be expanded in terms of the eigenfunctions of the GS operator $\hat{\mathcal{L}}_\theta$.

Therefore, its solution can be written as

$$\Psi(r,\theta) = \Psi_0[1 - \cos\theta + \varepsilon_b^2 g_2(x)\sin^2\theta\cos\theta], \tag{5.155}$$

where the radial function $g_2(x)$ must be determined from the equation

$$(x^2-1)\frac{d^2 g_2}{dx^2} + 2x\frac{dg_2}{dx} + \frac{6}{x^2}g_2 = \frac{2}{\gamma_{in}^2}. \tag{5.156}$$

We emphasize that the singularity for $x = 1$ in Eq. (5.156) corresponds to the singularity on the fast magnetosonic surface rather than to the Alfvén singularity (which was already used for deriving (5.153)). Thus, though in the zero approximation these surfaces coincide, they yield two critical conditions rather than one. We should remember that the example studied is degenerate and, therefore, we can use these two critical conditions independently of each other. Thus, the critical condition on the Alfvén surface defines the value $L(\Psi)$ and that on the fast magnetosonic surface defines the poloidal magnetic field structure. In the general case, there is no division of this kind. Therefore, it is appropriate to say that all critical conditions on the singular surfaces together define the longitudinal current and the poloidal field structure.

We now turn to the discussion of the properties of Eq. (5.156). It is the special case of the equation

$$(x^2-1)\frac{\mathrm{d}^2 g_m}{\mathrm{d}x^2} + 2x\frac{\mathrm{d}g_m}{\mathrm{d}x} + \frac{m(m+1)}{x^2}g_m = a_m \tag{5.157}$$

for $m = 2$. The boundary conditions for this equation, as before, are

1. the boundary condition on the body surface $g_m(R/r_a) = 0$;
2. the absence of the singularity for $x = 1$.

As a result, the general solution to Eq. (5.157) is written as (Bogovalov, 1992)

$$g_m(t) = -a_m C_1(t)\mathcal{P}_m(t) + a_m C_2(t)\mathcal{Q}_m(t). \tag{5.158}$$

Here $t = 1/x$,

$$C_1(t) = \frac{\mathcal{Q}_m(r_a/R)}{\mathcal{P}_m(r_a/R)}\int_1^{r_a/R}\frac{\mathcal{P}_m(y)}{y^2}\mathrm{d}y + \int_{r_a/R}^{t}\frac{\mathcal{Q}_m(y)}{y^2}\mathrm{d}y, \tag{5.159}$$

$$C_2(t) = \int_1^{t}\frac{\mathcal{P}_m(y)}{y^2}\mathrm{d}y, \tag{5.160}$$

and $\mathcal{P}_m(t)$ and $\mathcal{Q}_m(t)$ are the Legendre polynomials of the first and second kinds, respectively (see Appendix D). These expressions solve the problem posed.

The asymptotic behavior of the solution for $r \to \infty$, as in most hydrodynamical problems studied in Chap. 1, is fully defined by the inhomogeneous solution to Eq. (5.154), i.e., it is independent of the boundary conditions. The remarkable fact is that the particular solutions to the homogeneous equations

$$(x^2-1)\frac{\mathrm{d}^2 g_m}{\mathrm{d}x^2} + 2x\frac{\mathrm{d}g_m}{\mathrm{d}x} + \frac{m(m+1)}{x^2}g_m = 0, \tag{5.161}$$

$g_m^{(1)}(x) = c_1$ and $g_m^{(2)}(x) = c_2 x^{-1}$, are limited for $x \to \infty$, whereas the specific solution of the inhomogeneous equation increases with increasing x. As a result, we have at distances $r \gg r_a$

$$\Psi(r,\theta) = \Psi_0 \left[1 - \cos\theta + 2\varepsilon_{\rm b}^2 \frac{1}{\gamma_{\rm in}^2} \ln\left(\frac{r}{r_a}\right) \sin^2\theta \cos\theta + \cdots\right]. \tag{5.162}$$

As we see, though the magnetic field lines have a tendency for collimation along the rotation axis (perturbation $\delta\Psi > 0$ for $\theta < \pi/2$), this process is extremely weak, so that in the vicinity of the light cylinder $R_{\rm L} \approx \varepsilon_{\rm b}^{-1} r_a$ that, in the problem studied, is much farther than the singular surfaces (5.150), the perturbation of the monopole magnetic field $\sim \varepsilon_{\rm b}^2 \ln(1/\varepsilon_{\rm b})$ is much less than unity. As to the particle energy, in the problem studied, it does not actually change as compared to the initial energy $\gamma_{\rm in} m_{\rm p} c^2$.

Problem 5.13 Show that in the first approximation to $\varepsilon_{\rm b}^2$ the motion is radial: $v_\varphi = 0$.

5.3.2 Relativistic Outflows with Differential Rotation

The above results can be readily generalized to the inhomogeneous outflows with differential rotation (Beskin and Okamoto, 2000). Indeed, at zero temperature, as for the accretion of dust described in Sect. 1.4.3, the radial motion can be realized for the arbitrary plasma concentration. It is convenient here to give the boundary conditions as

$$u(R,\theta) = u_{\rm in} q_1(\theta), \tag{5.163}$$
$$n(R,\theta) = n_{\rm in} q_2(\theta), \tag{5.164}$$
$$\Omega_{\rm F}(R,\theta) = \Omega\,\omega(\theta), \tag{5.165}$$
$$\Psi(R,\theta) = \Psi_0(1 - \cos\theta), \tag{5.166}$$

where the functions $q_1(\theta) \sim 1$ and $q_2(\theta) \sim 1$ describe the inhomogeneities in the initial four-velocity and concentration distribution, and the function $\omega(\theta)$ describes the differential rotation that, for example, can be connected with the differential rotation of the disk (see Fig. 5.5).

As a result, for the zero rotation the location of the Alfvén and fast magnetosonic surfaces is given by the relation

$$r_{\rm A}^2(\theta) = \frac{r_a^2}{q_1^2(\theta) q_2(\theta)}, \tag{5.167}$$

where r_a is again given by relation (5.150). The regularity condition on the Alfvén surface yields the expression for the angular momentum

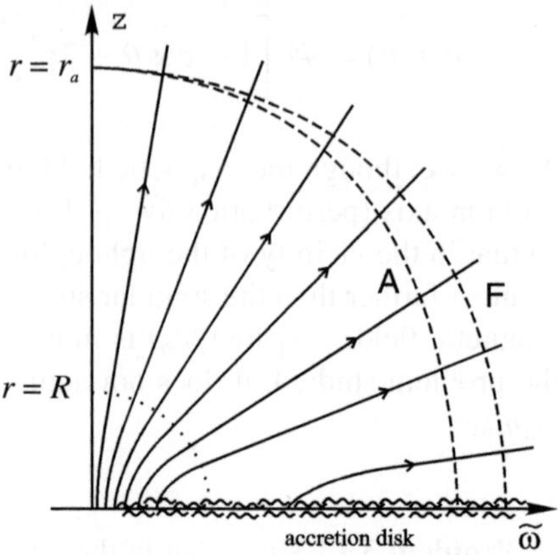

Fig. 5.5 The differential rotation of the disk in the vicinity of the compact object results in the angular dependence of the angular velocity $\Omega_{\rm F}(\theta)$. The boundary conditions can be given on the surface $r = R$, where the poloidal magnetic field is already close to the monopole one (Beskin and Okamoto, 2000)

$$L(\theta) = \frac{1}{c}\Omega_{\rm F}(\theta)E_0(\theta)r_{\rm A}^2(\theta)\sin^2\theta, \tag{5.168}$$

where $E_0(\theta) = \gamma(\theta)\mu\eta(\theta)$ can also be readily determined from the boundary conditions (5.163), (5.164), (5.165), and (5.166). Thus, the longitudinal current I is written as

$$I(r,\theta) = \frac{2\pi}{c}\Omega_{\rm F}(\theta)E_0(\theta)r_{\rm A}^2(\theta)\sin^2\theta. \tag{5.169}$$

Recall that due to the small perturbation of the monopole magnetic field these expressions hold for any radius r.

The latter relation shows that, with account taken of the differential rotation, as in the cylindrical jet problem, we can model any longitudinal current profile varying the angular velocity $\Omega_{\rm F}(\theta)$. Thus, if $\Omega_{\rm F}$ vanishes on a magnetic surface $\theta = \theta_0$, according to (5.169), the total electric current flowing in the domain $\theta < \theta_0$ is zero.

As to the linearized equation for the function $f(r,\theta)$, it now has the form

$$q_1^2(\theta)q_2(\theta)\frac{\partial}{\partial x}\left(x^2\frac{\partial f}{\partial x}\right) - \frac{\partial^2 f}{\partial x^2} - \frac{\sin\theta}{x^2}\frac{\partial}{\partial\theta}\left(\frac{1}{\sin\theta}\frac{\partial f}{\partial\theta}\right) = \frac{1}{\gamma_{\rm in}^2}\frac{\omega(\theta)\sin\theta}{q_1(\theta)}\frac{\rm d}{{\rm d}\theta}\left[\frac{\omega(\theta)\sin^2\theta}{q_1(\theta)}\right], \tag{5.170}$$

where again $x = r/r_a$ and $\gamma_{\rm in}^2 = 1 + u_{\rm in}^2$. As we see, the separation of the variables is possible only if $q_1^2(\theta)q_2(\theta) = {\rm const}$. In particular, for $q_1 = q_2 = 1$, the exact solution can again be written as

$$f(r,\theta) = \sum_m g_m(r)Q_m(\theta), \tag{5.171}$$

where the radial functions $g_m(r)$ are to be specified from Eq. (5.157) in which the coefficients a_m are defined as

$$\sum_m a_m Q_m(\theta) = \frac{1}{2\sin\theta}\frac{\mathrm{d}}{\mathrm{d}\theta}(\omega^2 \sin^4\theta). \tag{5.172}$$

On the other hand, the asymptotic behavior of the solution to Eq. (5.170) is again independent of the boundary conditions and can be obtained directly from the analysis of the leading terms in the limit $r \to \infty$. It has the form

$$f(x,\theta) = \frac{\ln x}{2\sin\theta q_1^2(\theta) q_2(\theta)}\frac{\mathrm{d}}{\mathrm{d}\theta}\left[\frac{\omega^2 \sin^4\theta}{u_{\mathrm{in}}^2(\theta)}\right], \tag{5.173}$$

where $u_{\mathrm{in}}(\theta) = u_{\mathrm{in}}\, q_1(\theta)$.

Problem 5.14 Using relation (5.169), show that expression (5.173) can be rewritten as

$$f(x,\theta) = \frac{\ln x}{2I_{\mathrm{A}}^2 \sin\theta q_1^2(\theta) q_2(\theta)}\frac{\mathrm{d}}{\mathrm{d}\theta}\left[\frac{I^2}{\gamma_{\mathrm{in}}^2(\theta)}\right], \tag{5.174}$$

where $I_{\mathrm{A}} = 2\pi c \eta_{\mathrm{in}} r_a^2 \Omega$.

Problem 5.15 Show that in the case of the small perturbation of monopole magnetic field the curvature radius of the magnetic surfaces can be written as (see, e.g., Begelman and Li, 1994)

$$\frac{1}{R_{\mathrm{c}}} = \frac{\varepsilon^2}{r\sin\theta}\frac{\partial}{\partial r}\left(r^2\frac{\partial f}{\partial r}\right). \tag{5.175}$$

Therefore, the asymptotic solution of Eq. (5.170) for $r \to \infty$ can be rewritten as

$$\frac{\mu n u^2}{R_{\mathrm{c}}} \sim \frac{1}{2\pi r^3 c^2 \sin^2\theta}\frac{\mathrm{d}}{\mathrm{d}\theta}\left[\frac{I^2}{\gamma_{\mathrm{in}}^2(\theta)}\right]. \tag{5.176}$$

In particular, for $\gamma_{\mathrm{in}} = \mathrm{const}$, we return to relation (4.229).

5.3.3 Nonrelativistic Slowly Rotating Outflows

Here we give the basic relations for the problem of the nonrelativistic magnetized plasma outflow from the surface of the slowly rotating body (Beskin and Okamoto, 2000). In this case, the four boundary conditions look like

$$v_{\rm p}(R,\theta) = v_{\rm in}q_1(\theta), \tag{5.177}$$

$$\rho(R,\theta) = \rho_{\rm in}q_2(\theta), \tag{5.178}$$

$$\Omega_{\rm F}(R,\theta) = \Omega\omega(\theta), \tag{5.179}$$

$$\Psi(R,\theta) = \Psi_0(1-\cos\theta). \tag{5.180}$$

As a result, the radius of the Alfvén and fast magnetosonic surfaces for the nonrotating body is written as

$$r_{\rm A}^2(\theta) = \frac{r_a^2}{q_1^2(\theta)q_2(\theta)}, \tag{5.181}$$

where

$$r_a^2 = R^2\frac{B_0^2}{4\pi\rho_{\rm in}v_{\rm in}^2}. \tag{5.182}$$

The critical condition on the Alfvén surface yields the expression for the angular momentum $L_{\rm n}$ and the current I

$$L_{\rm n}(r,\theta) = \frac{\Omega_{\rm F}(\theta)r_a^2\sin^2\theta}{q_1^2(\theta)q_2(\theta)}, \tag{5.183}$$

$$I(r,\theta) = 2\pi c\frac{v_{\rm in}\rho_{\rm in}r_a^2}{B_0}\frac{\Omega_{\rm F}(\theta)\sin^2\theta}{q_1(\theta)}. \tag{5.184}$$

Together with (5.182), for $q_1(\theta) = q_2(\theta) = 1$ it yields

$$I(r,\theta) = \frac{c}{v_{\rm in}}I_{\rm M}^{(\rm A)}\sin^2\theta, \tag{5.185}$$

i.e., $i_0 = c/v_{\rm in}$. We have already obtained this relation in Sect. 4.4.4. Accordingly, the linearized equation for the function $f(r,\theta)$ looks like

$$q_1^2(\theta)q_2(\theta)\frac{\partial}{\partial x}\left(x^2\frac{\partial f}{\partial x}\right) - \frac{\partial^2 f}{\partial x^2} - \frac{\sin\theta}{x^2}\frac{\partial}{\partial\theta}\left(\frac{1}{\sin\theta}\frac{\partial f}{\partial\theta}\right) = \frac{\omega(\theta)\sin\theta}{q_1(\theta)}\frac{\mathrm{d}}{\mathrm{d}\theta}\left[\frac{\omega(\theta)\sin^2\theta}{q_1(\theta)}\right], \tag{5.186}$$

where again $x = r/r_a$. As a result, for $q_1 = q_2 = 1$, the solution to this equation can again be found by the above procedure; we should only take in Eq. (5.170) $\gamma_{\rm in} = 1$. In the asymptotically far region $r \gg r_a$, this solution has the form

$$f(x,\theta) = \frac{\ln x}{2I_{\rm A}^2 \sin\theta q_1^2(\theta) q_2(\theta)} \frac{{\rm d}I^2}{{\rm d}\theta}, \tag{5.187}$$

where $I_{\rm A} = (c/v_{\rm in})\Omega\Psi_0/4\pi$. Therefore, the nonrelativistic GS equation, as was expected, can be rewritten in physically explicit form as $\rho v^2/R_{\rm c} = j_{\parallel} B_{\varphi}/c$ (4.228). Since $B_{\varphi} < 0$ for $\Psi > 0$, one can conclude that in the nonrelativistic limit in the vicinity of the rotation axis there must be the collimation of the magnetic surfaces ($R_{\rm c} > 0$ for $j_{\parallel} < 0$), whereas one should expect the decollimation of the magnetic field lines ($R_{\rm c} < 0$ for $j_{\parallel} > 0$) in the bulk closing current region (see Fig. 5.6). The longitudinal current profile itself can be completely arbitrary. On the other hand, in the relativistic case, the sign of the curvature radius $R_{\rm c}$ no longer specifies the current I but I/γ. We emphasize again that this is not the case at mathematical infinity, where the disturbance of the monopole magnetic field $\varepsilon_{\rm b}^2 f$ is of the order of unity.

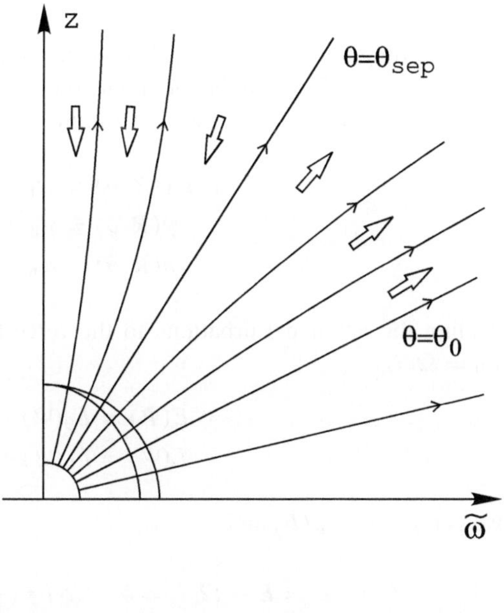

Fig. 5.6 The structure of magnetic surfaces in the presence of the bulk closing current outside the critical surfaces for the nonrelativistic flow. In the vicinity of the rotation axis $\theta < \theta_{\rm sep}$ there is the collimation, whereas in the bulk closing current region $\theta_{\rm sep} < \theta < \theta_0$ there is the decollimation of the magnetic surfaces (Beskin and Okamoto, 2000). The profile of the longitudinal current itself (*contour arrows*) can be arbitrary

Problem 5.16 Show that for the homogeneous outflow ($q_1 = q_2 = 1$) the toroidal magnetic and electric fields have the form (Parker, 1958)

$$B_\varphi = -\frac{\Omega_F R}{v_{in}} \frac{R}{r} B_0 \sin\theta, \tag{5.188}$$

$$E_\theta = -\frac{\Omega_F R}{c} \frac{R}{r} B_0 \sin\theta, \tag{5.189}$$

so in the nonrelativistic case the electric field is always much smaller than the magnetic one.

5.3.4 Relativistic Fast-Rotating Outflow

Now we consider the cold relativistic plasma outflow from the fast-rotating body surface (Beskin et al., 1998). The analytic analysis is possible because, as a zero approximation, we can use the Michel monopole solution (Michel, 1973a) (see Sect. 2.6.1.4) that is the exact solution to the force-free GS equation for the special choice of the longitudinal current $I_M(\theta)$ (2.225) and the angular velocity Ω_F. In other words, we consider here a small perturbation associated with the finite particle mass, while the magnetization parameter is supposed to be much larger than unity: $\sigma \gg 1$.

According to (1.64), the problem again requires four boundary conditions on the surface $r = R$. They can be the monopole condition $\Psi(R, \theta) = \Psi_0(1 - \cos\theta)$ and three other physical quantities. For simplicity, we again consider the case

$$\Omega_F(R, \theta) = \Omega_F = \text{const}, \tag{5.190}$$

$$\gamma(R, \theta) = \gamma_{in} = \text{const}, \tag{5.191}$$

$$n(R, \theta) = n_{in} = \text{const}. \tag{5.192}$$

Adding the small perturbations to the force-free integrals $L_0 = I_M(\theta)/2\pi$ and $E_0 = \Omega_F L_0$

$$E(\Psi) = E_0(\Psi) + b(\Psi), \tag{5.193}$$

$$L(\Psi) = L_0(\Psi) + l(\Psi), \tag{5.194}$$

we get $\eta = n_{in} u_{in} / B_p$ and

$$e' = E - \Omega_F L = b - \Omega_F l = \frac{B_p}{4\pi} \mathcal{M}^2(R) = \gamma_{in} \mu \eta, \tag{5.195}$$

the integrals of motion η, Ω_F, and e' being constant over the entire space. In relation (5.195), $\mathcal{M}(R)$ is the Mach number and $B_p = \Psi_0 / 2\pi R^2$ is the radial magnetic field on the surface $r = R$. Recall that according to (2.225), the value E_0 can be represented as $E_0 = E_A \sin^2\theta$, where $E_A = \sigma \mu \eta$. Under the condition $\sigma \gg 1$, we also have $e'/E \ll 1$. Finally, according to (5.195), we can write

$$b(\Psi) = e' + \Omega_{\rm F} l(\Psi). \tag{5.196}$$

As to the value $l(\Psi)$ (which for small perturbations of the monopole field can again be regarded as the function of the angle θ, i.e., $l = l(\theta)$), it must be specified from the regularity condition on the fast magnetosonic surface.

We again seek the solution to Eq. (4.66) as a small perturbation of the monopole magnetic field $\Psi(r,\theta) = \Psi_0[1 - \cos\theta + \varepsilon f(r,\theta)]$, where $\varepsilon \sim \gamma_{\rm in}\sigma^{-1} \ll 1$. Substituting this expansion in (4.66), we find (Beskin et al., 1998) (hereafter $c = 1$)

$$\begin{aligned}
&\varepsilon A\frac{\partial^2 f}{\partial r^2} + \varepsilon A\frac{(D+1)}{Dr^2}\sin\theta\frac{\partial}{\partial\theta}\left(\frac{1}{\sin\theta}\frac{\partial f}{\partial\theta}\right) - 2\varepsilon\Omega_{\rm F}^2 r\sin^2\theta\frac{\partial f}{\partial r}\\
&-2\varepsilon\Omega_{\rm F}^2\sin\theta\cos\theta\frac{\partial f}{\partial\theta} + 2\varepsilon\Omega_{\rm F}^2(3\cos^2\theta - 1)f\\
&-2\varepsilon\frac{A}{Dr^2}\frac{\cos\theta}{\sin\theta}\frac{\partial f}{\partial\theta} + 2\varepsilon\frac{A}{Dr^2}\frac{(1+\cos^2\theta)}{\sin^2\theta}f =\\
&-\frac{8\pi^2\Omega_{\rm F}}{\Psi_0}\frac{1}{\sin\theta}\frac{\mathrm{d}}{\mathrm{d}\theta}(l\sin^2\theta) + \frac{64\pi^4 A}{D\mathcal{M}^4}\frac{1}{\Omega_{\rm F}^2}\left(\frac{e'}{\Psi_0}\right)^2\frac{\cos\theta}{\sin^2\theta}\\
&-\frac{2A(1-\mathcal{M}^2)}{D}\frac{1}{\Omega_{\rm F}^2 r^4}\frac{\cos\theta}{\sin^2\theta} - \frac{16\pi^2 A}{D}\frac{e'}{\Psi_0}\frac{1}{\Omega_{\rm F}^2 r^2}\frac{\cos\theta}{\sin^2\theta}\\
&+\frac{8\pi^2 A}{D}\frac{1}{\Omega_{\rm F} r^2}\frac{\sin\theta}{\Psi_0}\frac{\mathrm{d}}{\mathrm{d}\theta}\left(\frac{l}{\sin^2\theta}\right) + \frac{A}{D}\frac{1}{\Omega_{\rm F}^2 r^4}\frac{\cos\theta}{\sin^2\theta}\\
&-\frac{2\mathcal{M}^2}{A}\Omega_{\rm F}^2\sin^2\theta\cos\theta + \frac{16\pi^2}{A}\frac{e'}{\Psi_0}\Omega_{\rm F}^2 r^2\sin^2\theta\cos\theta.
\end{aligned} \tag{5.197}$$

It is clear that in the limit $e' \to 0$, $\mathcal{M}^2 \to 0$, and $D \to \infty$, it becomes the force-free equation (2.243). Unfortunately, unlike the above-studied flows, this problem cannot be solved analytically, because we fail to separate the variables. In particular, we can impose only the constraints on the function $l(\theta)$.

For the same reason, we can only roughly define the location of the fast magnetosonic surface. Using the general relations (4.172) and (4.174) and the explicit dependence E_0 on θ, we obtain for $\gamma_{\rm in} \ll \sigma^{1/3}$

$$r_{\rm f}(\theta) \approx R_{\rm L}\sigma^{1/3}\sin^{-1/3}\theta \tag{5.198}$$

for $\theta > \sigma^{-1/2}$, and

$$r_{\rm f} \approx R_{\rm L}(\sigma/\gamma_{\rm in})^{1/2} \tag{5.199}$$

in the vicinity of the rotation axis. Accordingly, from (4.175), we obtain for the particle Lorentz factor on the fast magnetosonic surface

$$\gamma(r_{\rm f}, \theta) = \left(\frac{E}{\mu\eta}\right)^{1/3} = \sigma^{1/3} \sin^{2/3}\theta. \tag{5.200}$$

The locations of the Alfvén and fast magnetosonic surfaces are shown in Fig. 5.7.

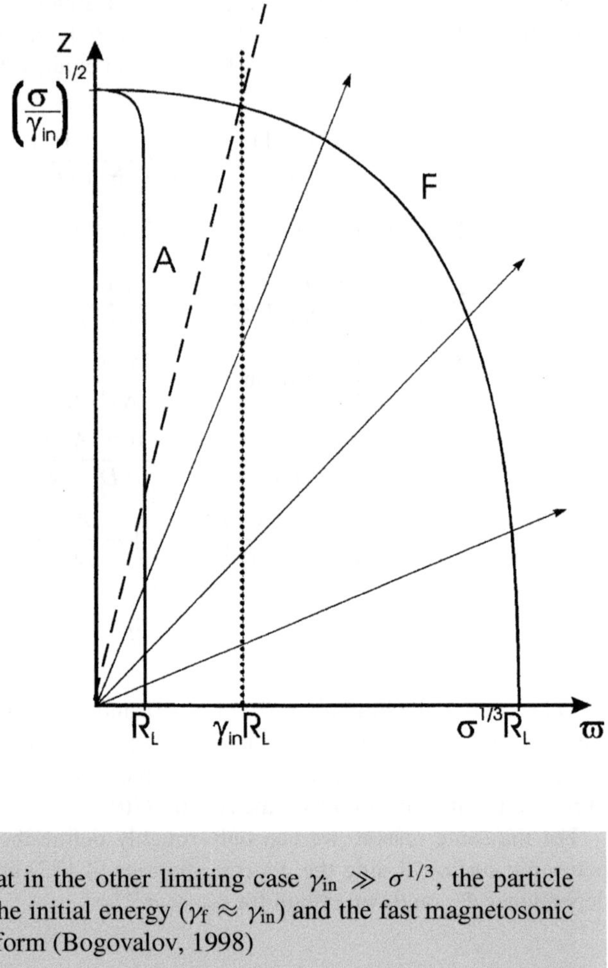

Fig. 5.7 Locations of the Alfvén (A) and fast magnetosonic (F) surfaces in the case $\gamma_{\rm in} \ll \sigma^{1/3}$ (Beskin et al., 1998). The *dashed line* indicates the magnetic field line $\Psi = \Psi_{\rm in}$ dividing the strongly ($W_{\rm part} \ll W_{\rm em}$) and weakly ($W_{\rm part} \gg W_{\rm em}$) magnetized flow regions. There is no particle acceleration inside the cylinder $\varpi = \gamma_{\rm in} R_{\rm L}$ (*dotted line*)

Problem 5.17 Show that in the other limiting case $\gamma_{\rm in} \gg \sigma^{1/3}$, the particle energy coincides with the initial energy ($\gamma_{\rm f} \approx \gamma_{\rm in}$) and the fast magnetosonic surface has a spherical form (Bogovalov, 1998)

$$r_{\rm f} \approx \left(\frac{\sigma}{\gamma_{\rm in}}\right)^{1/2} R_{\rm L}. \tag{5.201}$$

We now proceed to the determination of the magnetic field structure. It is necessary to determine the physical root $\mathcal{M}^2(r)$ of the third-order equation (4.161). It is again convenient to use the value $q = \mathcal{M}^2/\Omega_{\rm F}^2\varpi^2$ (4.135) and also introduce the new variable $\xi = 1 - \Sigma_{\rm r}^2$. In our case, we have

$$\xi = 2\frac{e' + \Omega_{\rm F} l}{E} - \frac{2\varepsilon}{\sin\theta}\frac{\partial f}{\partial\theta} + 4\varepsilon\frac{\cos\theta}{\sin^2\theta} f. \tag{5.202}$$

As is readily verified, $\xi = 0$ for the Michel force-free monopole solution, and ξ is much smaller than unity for $\sigma^{-1} \ll 1$. We emphasize that exactly the dependence of ξ on εf allows us to analyze our problem self-consistently. As a result, Eq. (4.161) can be rewritten as

$$q^3 - \frac{1}{2}\left(\xi + \frac{1}{\Omega_{\rm F}^2 r^2 \sin^2\theta}\right) q^2 + \frac{\mu^2\eta^2}{2E^2} + \frac{(e')^2}{2E^2}\frac{1}{\Omega_{\rm F}^2 r^2 \sin^2\theta} = 0. \tag{5.203}$$

We first consider the domain in the vicinity of the rotation axis $r\sin\theta < \gamma_{\rm in} R_{\rm L}$, $r < r_{\rm f}$ (see Fig. 5.7). In this domain, the physical root of Eq. (5.203) (i.e., the solution corresponding to the subsonic flow $D > 0$) is

$$q = \frac{e'}{E}, \tag{5.204}$$

i.e., it is independent of the radius r. Hence, there is no particle acceleration in this domain (see Fig. 5.8)

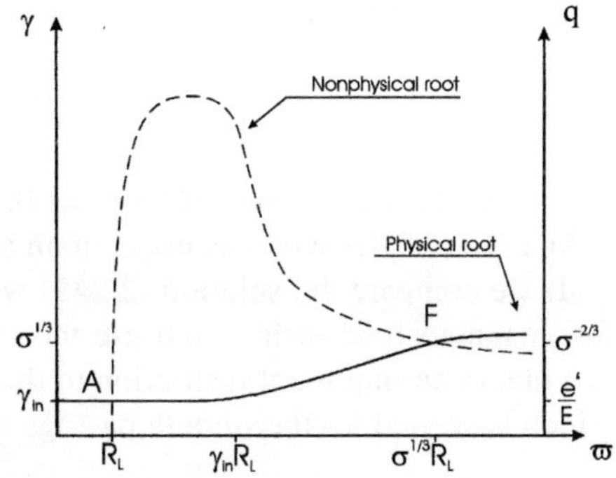

Fig. 5.8 The change in the particle energy along magnetic field lines in the quasimonopole magnetic field for the case $\gamma_{\rm in} \ll \sigma^{1/3}$ for $\theta \approx \pi/2$. The *dashed line* indicates the nonphysical root (Beskin et al., 1998)

$$\gamma(r,\theta) = \gamma_{\rm in}. \tag{5.205}$$

In particular, as we saw, for $\sigma < \gamma_{\rm in}^3$, the particle energy remains constant up to the fast magnetosonic surface.

In the intermediate domain $\gamma_{\rm in} R_{\rm L} < r\sin\theta$, $r < r_{\rm f}$, which exists for $\sigma > \gamma_{\rm in}^3$, the physical root $D > 0$ of the algebraic equation (5.203) has the form

$$q = \left(\frac{\mu\eta}{E}\right)\Omega_{\rm F} r\sin\theta. \tag{5.206}$$

It yields

$$\gamma(r,\theta) = \Omega_{\rm F} r \sin\theta, \tag{5.207}$$

$$\mathcal{M}^2(r,\theta) = \sigma^{-1}\Omega_{\rm F}^3 r^3 \sin\theta, \tag{5.208}$$

and $D = \mathcal{M}^{-2}$. As we see, the linear increase in the particle energy $\gamma m_{\rm p}c^2$ is to take place here. We emphasize that all physical characteristics of the flow in this domain do not depend on ξ at all and, hence, on the field disturbance $\varepsilon f(r,\theta)$. In particular, the universal dependence (5.207) is again defined by the drift particle motion in the crossed fields. Therefore, to determine the basic parameters of the flow inside the fast magnetosonic surface we do not need to know the exact solution to the GS equation (5.197).

Further, the conditions $Q = 0$ and $\partial Q/\partial r = 0$ on the fast magnetosonic surface, which correspond to the conditions $D = 0$ and $N_r = 0$, are rewritten as

$$\xi(r_{\rm f},\theta) + \frac{1}{\Omega_{\rm F}^2 r_{\rm f}^2 \sin^2\theta} = 3\left(\frac{\mu\eta}{E}\right)^{2/3}, \tag{5.209}$$

$$r_{\rm f}\left.\frac{\partial\xi}{\partial r}\right|_{r=r_{\rm f}} - \frac{2}{\Omega_{\rm F}^2 r_{\rm f}^2 \sin^2\theta} = 0. \tag{5.210}$$

Thus, supposing that $r_{\rm f}\,\partial\xi/\partial r|_{r=r_{\rm f}} \approx \xi$, we conclude that relations (5.209) and (5.210), besides the above expression for the sonic surface radius $r_{\rm f}$, yield one more important condition

$$\xi(r_{\rm f}) \approx \sigma^{-2/3}. \tag{5.211}$$

As to the third condition $\partial Q/\partial\theta = 0$, in the approximation studied, it, in fact, reduces to (5.210) and does not contain any new information.

If we compare the solution (2.245) with the regularity condition (5.211) on the fast magnetosonic surface and use the explicit form of the function ξ (5.202), we can obtain an important restriction to the perturbation of the longitudinal current l, which is needed for the smooth passage through the fast magnetosonic surface

$$l/L_0 \sim \sigma^{-4/3}. \tag{5.212}$$

Hence, for the transonic flow to exist the longitudinal current must actually coincide with the force-free current $I_{\rm M}$. Thus, we justified our choice of the electric current $I = 2\pi L_0(\Psi)$ as a zero approximation. Finally, comparing relations (5.211) and (5.212) with expression (5.202), we find

$$\varepsilon f(r_{\rm f}) \sim \sigma^{-2/3}. \tag{5.213}$$

This implies that the perturbation of the monopole magnetic field remains insignificant at least up to the fast magnetosonic surface $r = r_{\rm f}$. Therefore, our assumption of the possibility to investigate this problem by the perturbation theory seems justified.

As to the asymptotic domain $r \gg r_{\rm f}$, the physical root of equation (5.203), which corresponds to the supersonic flow $D < 0$, is

$$q = \frac{\xi}{2}\left(1 - 4\frac{\mu^2\eta^2}{\xi^3 E^2}\right). \tag{5.214}$$

We easily check that the corresponding particle energy $m_{\rm p}c^2\gamma = m_{\rm p}c^2 q(E/\mu\eta)$ here also corresponds to the Lorentz factor $\gamma = (1 - U_{\rm dr}^2)^{-1/2}$ defined by the drift velocity $U_{\rm dr} = |\mathbf{E}|/|\mathbf{B}| \approx |\mathbf{E}|/B_{\varphi}$. On the other hand, the GS equation (5.197) can be rewritten in simple form as

$$\varepsilon r^2\frac{\partial^2 f}{\partial r^2} + 2\varepsilon r\frac{\partial f}{\partial r} = \sin\theta\frac{D+1}{D}\frac{\partial q}{\partial\theta}, \tag{5.215}$$

where q is given by formula (5.214) and

$$D + 1 = 8\frac{\mu^2\eta^2}{E^2\xi^3} \ll 1. \tag{5.216}$$

We emphasize that if the flow is supposed to be exactly radial ($\partial/\partial r = 0$), Eq. (5.215) would have the form $\mathrm{d}q/\mathrm{d}\theta = 0$, which just corresponds to the earlier found integral (4.238) (Heyvaerts and Norman, 1989). However, this relation, as was shown above, can occur only at mathematical infinity, where the radial derivatives can be really disregarded. In the physical infinity domain, allowance for the radial derivatives proves absolutely necessary. As a result, according to (5.215), in the asymptotically far domain $r \gg r_{\rm f}$ the perturbation of the magnetic field εf can be written as (Tomimatsu, 1994)

$$\varepsilon f(r,\theta) = \sigma^{-2/3}a(\theta)\ln^{1/3}\left(\frac{r}{r_{\rm f}}\right), \tag{5.217}$$

where $a(\theta) \sim 1$. Otherwise, for the strongly magnetized flow the collimation appears weaker than the case $W_{\rm part} \gg W_{\rm em}$, where, as we saw, $\varepsilon f \propto \ln(r/r_{\rm f})$. Accordingly, because of (4.137) and (5.214), beyond the fast magnetosonic surface the particle energy actually ceases to increase

$$\gamma(r \gg r_{\rm f}) \approx \sigma^{1/3}\ln^{1/3}\left(\frac{r}{r_{\rm f}}\right)\sin^{2/3}\theta. \tag{5.218}$$

Problem 5.18 Show that Eq. (5.215) exactly corresponds to the condition $\mathcal{F}_{\rm c} = \mathcal{F}_{\rm em}$, where the electromagnetic force $\mathcal{F}_{\rm em}$ is given by the right-hand side of relation (5.176), and in the expression for the centrifugal force the Poynting vector flux S is of major importance now: $\mathcal{F}_{\rm c} = S/(cR_{\rm c})$.

Problem 5.19 Show that the magnetic field line passing through the point of intersection of the fast magnetosonic surface with the cylinder $\varpi = \gamma_{\rm in} R_{\rm L}$ corresponds to the magnetic field line $\Psi = \Psi_{\rm in}$ (5.47) separating the strongly ($W_{\rm part} \ll W_{\rm em}$) and weakly ($W_{\rm part} \gg W_{\rm em}$) magnetized flow domains (see Fig. 5.7).

Problem 5.20 Show that in the vicinity of the axis ($\Psi < \Psi_{\rm in}$ for $r \gg r_{\rm f}$) an increase in the disturbance of the monopole magnetic field corresponds to the weakly magnetized flow $\varepsilon f \propto \ln(r/r_{\rm f})$ (Lyubarsky and Eichler, 2001).

Thus, we constructed one more example of the solution in which the regularity conditions on the singular surfaces limit the longitudinal current, so beyond the fast magnetosonic surface both the collimation and the particle acceleration become inefficient. On the other hand, it is shown that the Michel expression (Michel, 1969) for the particle energy $\gamma_{\rm f} \sim \sigma^{1/3}$ (within the model studied) remains valid.

5.3.5 Relativistic Outflow in the Parabolic Magnetic Field

The above example was often used as an argument against the very possibility of the efficient particle acceleration in the magnetized relativistic wind. However, as we saw, in the cylindrical jets the particle energy flux can approach the energy flux of the electromagnetic field (see Sect. 4.4.5 as well). Therefore, it is interesting to analyze the strongly magnetized flow structure using the parabolic solution of the force-free GS equation (2.237) (Beskin and Nokhrina, 2006). In this case, already in the zero approximation the magnetic surfaces are collimated to the rotation axis, and the main problem thus becomes the particle acceleration problem. On the other hand, unlike the previous example, the total magnetic flux extends to infinity here. At large distances, as we will see, the flow, in fact, becomes one-dimensional, i.e., close to the cylindrical jets submerged into the homogeneous magnetic field, which were studied in Sect. 5.2.2.

Recall that in the parabolic magnetic field the angular velocity $\Omega_{\rm F}$ cannot be constant on all magnetic surfaces, since, as shown in Fig. 2.19, the value $\Omega_{\rm F}(\Psi)$ is determined by the angular velocity of the disk. Below, for simplicity, we discuss only the domain in the vicinity of the rotation axis, where $\Omega_{\rm F}$ can be taken to be constant. In this case, the magnetization parameter σ is naturally determined by the magnetic flux Ψ_0 enclosed at the equator within the light cylinder $R_{\rm L} = 1/\Omega_{\rm F}$, which yields the expression ($c = 1$)

$$\sigma = \frac{C\Omega_{\rm F}}{4\pi \mu \eta}. \tag{5.219}$$

Otherwise, the construction of the solution is fully equivalent to the problem of the relativistic plasma outflow in the monopole magnetic field. Therefore, we do not discuss the computation in detail and give only the main results for the case $\gamma_{\rm in} \ll \sigma^{1/3}$.

We first consider the subsonic domain $r < r_{\rm f}$. We seek the magnetic flux function, with the perturbation taking into account the finite particle mass in the form

$$\Psi(X, Y) = \Psi^{(0)}(X) + \varepsilon f(X, Y). \tag{5.220}$$

Here $\Psi^{(0)}(X) \approx \pi C X$ is the force-free solution (2.237), $\varepsilon f(X, Y)$ is the small perturbation, and we introduce again the orthogonal coordinates

$$X = r(1 - \cos\theta), \tag{5.221}$$

$$Y = r(1 + \cos\theta). \tag{5.222}$$

In this case, the "working volume" $\Psi < \Psi_0$ (where, as we suppose, $\Omega_{\rm F} = {\rm const}$) is written as $\Omega_{\rm F} X \approx \Omega_{\rm F} r\theta^2/2 < 1$. Now linearizing the GS equation, we obtain the equation for the function $\varepsilon f(X, Y)$ (Beskin and Nokhrina, 2006)

$$\varepsilon X \frac{\partial^2 f}{\partial X^2} + \varepsilon \frac{\partial f}{\partial X} - \varepsilon \frac{f}{X} + \varepsilon \frac{\partial}{\partial Y}\left(Y \frac{\partial f}{\partial Y}\right) = -\pi C \left(X \frac{\partial q}{\partial X} + 2q\right), \tag{5.223}$$

where q is again given by relation (4.135). The most important consequence of this equation is that in the studied domain $\Omega_{\rm F} X < 1$ ($\Omega_{\rm F} = {\rm const}$) and $Y \gg X$ ($\theta \ll 1$) the term $\varepsilon \partial/\partial Y (Y \partial f/\partial Y)$ associated with the curvature of the magnetic surfaces becomes inessential and can be disregarded. This implies that the parabolic jet structure should be close to the one-dimensional cylindrical jets. Only the "external" homogeneous magnetic field changes. If we define the value q from Bernoulli's equation (5.203)

$$q = \frac{1}{\sigma}\left(\frac{Y}{X}\right)^{1/2}, \tag{5.224}$$

and substitute it in Eq. (5.223), we obtain for the perturbation to the magnetic flux function for $r < r_{\rm f}$

$$\varepsilon f(X, Y) = \frac{2\pi C}{\sigma} X^{1/2} Y^{1/2}. \tag{5.225}$$

We now specify the location of the fast magnetosonic surface $r = r_{\rm f}$ and the particle Lorentz factor $\gamma_{\rm f} = \gamma(r_{\rm f})$. It is necessary again to use Bernoulli's equation (4.161). As a result, in the vicinity of the rotation axis $\theta < \gamma_{\rm in}^2/\sigma$, where the particle energy flux is always larger than the energy flux of the electromagnetic field, the location of the fast magnetosonic surface and the particle Lorentz factor $\gamma_{\rm f} = \gamma(r_{\rm f})$ are

$$r_{\rm f} \approx \frac{\sigma}{\gamma_{\rm in}} R_{\rm L}, \tag{5.226}$$

$$\gamma_{\rm f} \approx \gamma_{\rm in}. \tag{5.227}$$

In the opposite case $\theta \gg \gamma_{\rm in}^2/\sigma$, as shown in Fig. 5.9, we have

$$r_{\rm f} \approx \left(\frac{\sigma}{\theta}\right)^{1/2} R_{\rm L}, \tag{5.228}$$

$$\gamma_{\rm f} \approx \sigma^{1/2}\theta^{1/2}. \tag{5.229}$$

Note that on the outer boundary of the studied domain $\Omega_{\rm F}X = 1$ (i.e., $r \approx \sigma^{2/3} R_{\rm L}$ and $\theta \approx \sigma^{-1/3}$) the particle Lorentz factor on the sonic surface is exactly equal to $\sigma^{1/3}$.

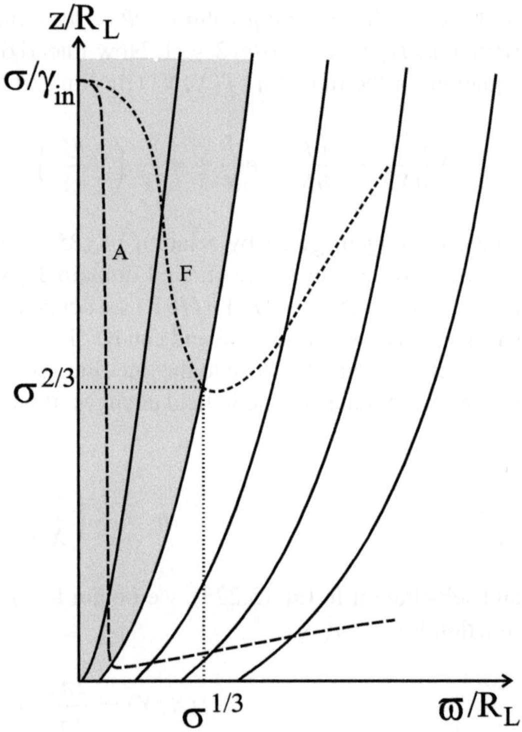

Fig. 5.9 The location of the Alfvén (*long dashed line*) and fast magnetosonic (*short dashed line*) surfaces for the parabolic magnetic field. The shaded region corresponds to the "working volume" $\Omega_{\rm F}X < 1$ ($\Omega_{\rm F} = \text{const}$). The outer region can be described by the self-similar equations (Beskin and Nokhrina, 2006)

If we estimate now the relative role of the perturbation $\varepsilon f(X, Y)$ to the magnetic flux function $\Psi^{(0)} \approx \pi C X$ on the fast magnetosonic surface, we find that

$$\frac{\varepsilon f}{\Psi^{(0)}} \approx \frac{1}{\sigma\theta} \ll 1 \tag{5.230}$$

for $\theta > \gamma_{\rm in}^2/\sigma$. Hence, our assumption of the smallness of the perturbation to the magnetic flux function inside the fast magnetosonic surface is confirmed. Consequently, at least up to the distances $r < (\sigma/\gamma_{\rm in})R_{\rm L}$ the poloidal magnetic field is close to the force-free one. Moreover, if we take, as the external magnetic field $B_{\rm ext}$, the poloidal field of the force-free parabolic solution $B = \mathcal{C}/2r$, we readily see that expression (5.112) for the location of the fast magnetosonic surface for the cylindrical jet exactly coincides with (5.228). In the case of the smallness of the angle θ in the studied domain $\Omega_{\rm F}X < 1$, the poloidal magnetic field is actually constant: $B_z \approx$ const.

Problem 5.21 Show that, as for the monopole magnetic field

- for $r < r_{\rm f}$, the particle Lorentz factor is defined by the universal dependence $\gamma \approx x_r$;
- on the fast magnetosonic surface we have $\gamma_{\rm f} \approx y^{1/3}$ for $y > \gamma_{\rm in}^3$;
- expression (5.230) exactly matches the asymptotic expression (5.225) for $r < r_{\rm f}$.

Clearly, the above results suggest that the flow remains one-dimensional beyond the fast magnetosonic surface, and we can use the results obtained for the cylindrical flow for the supersonic region. Otherwise, up to the distance $r \sim r_1$, where

$$r_1 = \gamma_{\rm in}\sigma\, R_{\rm L}, \tag{5.231}$$

the poloidal field in the domain $\Omega_{\rm F}X < 1$ for $r =$ const can be regarded as constant and for $r_1 < r < r_{\rm eq}$, where

$$r_{\rm eq} = \sigma^2\, R_{\rm L}, \tag{5.232}$$

the poloidal magnetic field $B_z(r, \theta)$ has the core structure. In this case, the magnetic field on the rotation axis for $r > r_1$ slightly differs from the value $B_{\rm min}$ (5.69). The recent numerical simulation (Komissarov et al., 2009) fully confirmed this structure (see Sect. 5.5.3).

Further, the contribution of the particle energy flux to the entire domain $r > r_{\rm f}$ increases with distance from the equatorial plane

$$\frac{W_{\rm part}}{W_{\rm tot}} \approx \left(\frac{r}{r_{\rm eq}}\right)^{1/2}. \tag{5.233}$$

At large distances $z \gg r_{\rm f}$, the energy of particles moving along the magnetic field line $\Omega_{\rm F}X \approx 1$ increases as

$$\gamma \approx \left(\frac{z}{R_{\rm L}}\right)^{1/2}, \tag{5.234}$$

in agreement with the general relation (4.234). Accordingly, $u_{\hat{\varphi}} \approx 1$. Finally, for $r > r_{\rm eq}$, the particle energy flux becomes comparable with the energy flux of the electromagnetic field. Thus, in the parabolic magnetic field, as was demonstrated in Sect. 4.4.5, the full transformation of the electromagnetic energy flux into the particle energy flux becomes possible. Note that on the fast magnetosonic surface the flow structure can appreciably change, so for $r \gg r_{\rm f}$ the disturbance of the force-free parabolic field is no longer small though the perturbation εf is not much larger than the magnetic flux $\Psi^{(0)}$ in the zero approximation (see Fig. 5.2c).

5.3.6 General Properties

To sum up, we again turn to the most important conclusions that can be made by analyzing the above exact solutions.

1. Within the full MHD GS equation version, i.e., with account taken of the finite particle mass, the longitudinal current I (to be exact, the angular momentum $L(\Psi)$) is no longer a free parameter and must be found from the condition of the smooth passage through all critical surfaces. In the relativistic case, the longitudinal current in the vicinity of the fast magnetosonic surface proves close to the GJ current. For the nonrelativistic particle-dominated flow $i_0 \approx c/v_{\rm in} \gg 1$.
2. The closeness of the longitudinal current I to the characteristic current $I_{\rm GJ}$ implies that the current I is to be close to the minimum possible current $I_{\rm min}$ for which, within the force-free approximation, the fast magnetosonic surface extends to infinity. Consequently, in the analysis of the strongly magnetized flows we can use a much simpler force-free GS equation version and determine the current from the condition
$$I = I_{\rm min} \tag{5.235}$$
(certainly, it is impossible to describe the regions located beyond the fast magnetosonic surface). Condition (5.235) can be rewritten as the "boundary condition at infinity" $E_\theta(r \to \infty) = B_\varphi(r \to \infty)$, i.e., as (Okamoto, 2001)
$$4\pi I = \Omega_{\rm F} \sin\theta \, \frac{{\rm d}\Psi}{{\rm d}\theta}, \tag{5.236}$$
which is equivalent to the "boundary condition on the horizon."
3. In the relativistic case, if the curvature of the magnetic surfaces can be disregarded beyond the fast magnetosonic surface $r \gg r_{\rm f}$, the efficient acceleration regime is realized here, so that $\gamma \approx \Omega\varpi/c$ (5.28). If the curvature $R_{\rm c}$ appears to be substantial, the acceleration is inefficient and $\gamma \approx (R_{\rm c}/\varpi)^{1/2}$ (4.231).
4. The possibility of the particle acceleration is closely connected with the magnetic field structure. The efficient acceleration is possible only for the strongly colli-

mated flows. The particle energy flux for the quasimonopole outflow is always much smaller than the energy flux of the electromagnetic field.

5. As in the nonrelativistic case, the magnetic field works as a sling increasing the particle energy with distance from the rotation axis. However, for the quasi-monopole flows, the curvature of the magnetic field lines, which increases with distance, suppresses the acceleration efficiency. In the parabolic field the curvature of the field lines, though present initially, decreases with distance and at large distances does not hinder the efficient particle acceleration.

5.4 Black Hole Magnetosphere

We now discuss the main results that can be obtained in the analysis of the full GS equation version, which describes the rotating black hole magnetosphere. Here we are, first of all, interested in the general properties of the flows rather than in the self-consistent model of real rotating black holes (a brief survey of the astrophysical subject was given in Sect. 3.1). Therefore, it is necessary to formulate the main questions we will try to answer in the analysis of the exact solutions.

First of all, the determination of the energy losses of the rotating black hole is not connected so much with the value of the current I, as for the radio pulsar magnetosphere, as with the value of the angular velocity $\Omega_{\rm F}(\Psi)$. Indeed, in the case of the black hole magnetosphere, the angular velocity $\Omega_{\rm F}(\Psi)$ is in no way connected with the angular velocity of the black hole $\Omega_{\rm H}$ and is to be determined from the solution of the complete problem. It is shown below that, with account taken of the finite particle mass, the number of the critical surfaces is enough to determine not only the longitudinal current I but also the angular velocity $\Omega_{\rm F}$.

Further, the exact solutions allow us to determine the behavior of the flow in the vicinity of the black hole and thus confirm our conclusion of the absence of a singularity on the event horizon. Finally, having established where the boundary conditions must be given for the flow structure to be fully determined, one can make conclusions about the slowing-down mechanism of the rotating black hole.

At present, there are only two cases in which the problem of the two-dimensional magnetosphere structure of the black hole has been solved analytically. Below, we discuss in detail the basic properties of these solutions, which, though the problem is greatly simplified, make it possible to clear up most key features in the real flows.

5.4.1 Slowly Rotating Black Hole Surrounded by a Thin Disk

Consider again the black hole with the (split) monopole magnetic field generated in the thin accretion disk (see Fig. 5.10) (Beskin and Kuznetsova, 2000a). Clearly, in the absence of rotation the monopole field $\Psi = \Psi_0(1 \pm \cos\theta)$ is an exact solution to the full GS equation version, because the radial motion of the plasma (which is again uncharged for $\Omega_{\rm F} = 0$) can, in no way, disturb the magnetic structure. We

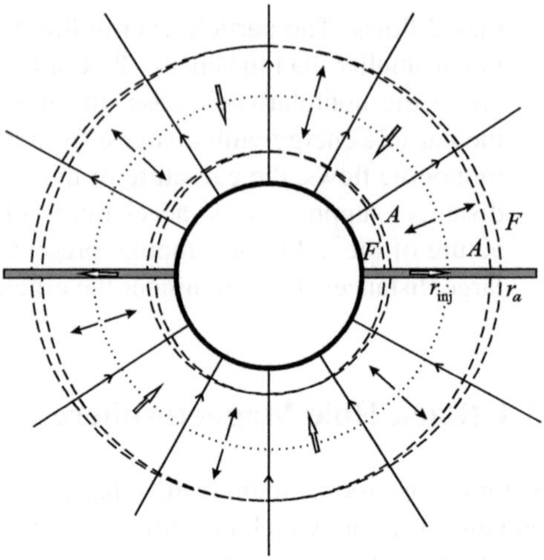

Fig. 5.10 The structure of the singular surfaces (*dashed lines*) and the plasma generation region $r = r_{\mathrm{inj}}$ (*dotted line*) for the black hole submerged in the monopole magnetic field. In the absence of rotation the monopole magnetic field is the exact solution to the GS equation. *Contour arrows* indicate the currents flowing in the magnetosphere, ordinary arrows – motion of particles

consider here, for simplicity, the flow in which the energy density of the magnetic field is much higher than that of the plasma

$$\frac{\epsilon_{\mathrm{part}}}{\epsilon_{\mathrm{em}}} \ll 1, \tag{5.237}$$

but, on the other hand, the black hole rotational velocity is so small that the main energy flux is determined by particles rather than the Poynting flux

$$\frac{W_{\mathrm{part}}}{W_{\mathrm{em}}} \gg 1. \tag{5.238}$$

Otherwise, we do not discuss the generalization of the force-free BZ solution to the case of the finite particle mass but the generalization of the solution discussed in Sect. 5.3.1 to the accretion onto the black hole. Besides, we again consider the plasma to be cold ($T = 0$, i.e., $\mu = m_{\mathrm{p}} = \mathrm{const}$).

Further, we assume that even in the absence of the black hole rotation, plasma is generated in the magnetosphere, and two particle flows are thus generated in it, one of them extends to infinity and the other moves to the event horizon. Undoubtedly, in this statement of the problem there is an obvious inconsistency, because the solution considered below has no physically meaningful limit $\Omega_{\mathrm{H}} \to 0$, since the plasma generation mechanism studied in Sect. 3.2.5 can effectively work only in the presence of the longitudinal electric field that, in turn, can be produced by the black hole rotation. Nevertheless, as we saw, a relatively slow rotation is enough for the plasma generation to be initiated in the vicinity of the surface $\rho_{\mathrm{GJ}} = 0$. Therefore, the particles are generated for the parameter $\varepsilon_3 = a/M \ll 1$. This surface location (for example, according to (3.60), $r_{\mathrm{inj}} = 2^{1/3} r_{\mathrm{g}}$ for $\Omega_{\mathrm{F}} = \Omega_{\mathrm{H}}/2$) is independent of the black hole rotation. This justifies the possibility to consider the solution of our problem again as a small perturbation of the monopole magnetic field $\Psi = \Psi_0(1 \pm \cos\theta)$. As to the plasma generation region $\rho_{GJ} \approx 0$ assumed to be an infinitely thin surface, its location itself is to be defined from the solution of the complete problem. The value of $r_{inj} = 2^{1/3} r_g$ can be taken as a zero approximation.

We now consider in detail the number of boundary conditions which are to be specified in order to fully determine all flow parameters. One should remember that the problem studied contains, in fact, two domains in which the integrals of motion differ from one another. This is evident, at least, from the analysis of the value η available in the definition $\alpha n \mathbf{u}_p = \eta \mathbf{B}_p$, which is to have different signs for the outflowing and accreting plasma. Therefore, according to the general formula (1.64) $b = 2 + i - s' = 4$ (four integrals of motion, two singular surfaces) valid for the cold plasma, for each of the domains on the particle generation surface $\rho_{GJ} = 0$, four functions should be given. Therefore, the complete problem requires eight boundary conditions.

Clearly, four of them, viz., the concentration n_{inj} and the Lorentz factors γ_{inj} for the outflowing and accreting plasma, can be chosen in the same way as in the problem of the cold plasma outflow from the surface of the rotating sphere. We consider here, for simplicity, the constant values of the concentrations and the particle Lorentz factor

$$n^{\pm}_{inj} = n_{inj} = \text{const}, \tag{5.239}$$

$$\gamma^{\pm}_{inj} = \gamma_{inj} = \text{const}. \tag{5.240}$$

These four values define the four integrals of motion, two each for the outflowing and accreting plasma

$$E^{(out)} = \alpha_{inj}\, \mu\, \eta_{inj}\, \gamma_{inj}, \tag{5.241}$$

$$E^{(in)} = -\alpha_{inj}\, \mu\, \eta_{inj}\, \gamma_{inj}, \tag{5.242}$$

$$\eta^{(out)} = \eta_{inj} = \frac{\alpha_{inj}\, n_{inj}\sqrt{\gamma^2_{inj} - 1}}{B_{inj}}, \tag{5.243}$$

$$\eta^{(in)} = -\eta_{inj} = -\frac{\alpha_{inj}\, n_{inj}\sqrt{\gamma^2_{inj} - 1}}{B_{inj}}. \tag{5.244}$$

Here $B_{inj} = \Psi_0/(2\pi r^2_{inj})$, $\mu = m_p = \text{const}$ for the cold plasma. As we see, when the energy flux is connected with the particles (5.238), the energy integrals $E^{(in)}$ and $E^{(out)}$ have different signs for the outer and inner regions in the magnetosphere. Note that in this section our formulation of the problem differs from that in Hirotani et al. (1992), in which the particle velocity was supposed to be close to zero in the plasma generation region.

As to two other pairs of the functions, we cannot take, as boundary conditions, the angular velocity $\Omega_F(r_{inj})$ and the magnetic flux $\Psi(r_{inj})$, because they are to be determined by the solution of the problem. On the other hand, we can give, as boundary conditions, the following relations and values:

1. The continuity condition of the magnetic flux in the plasma generation region

$$\Psi|_{r=r_{\rm inj}-0} = \Psi|_{r=r_{\rm inj}+0}\,. \tag{5.245}$$

2. Two components of the surface electric current $\mathbf{J}_{\rm s}$ flowing along the particle generation region

$$\mathbf{J}|_{r=r_{\rm inj}} = \mathbf{J}_{\rm s}. \tag{5.246}$$

This two-dimensional vector yields two other boundary conditions.

3. The drop of the electric potential $V_{\rm g}$ along the magnetic field lines in the particle generation region (which must be defined by the concrete plasma generation mechanism)

$$V_{\rm g} = V|_{r=r_{\rm inj}+0} - V|_{r=r_{\rm inj}-0}\,. \tag{5.247}$$

As we will see, relations (5.239), (5.240), (5.245), (5.246), and (5.247) fully define the solution of the problem posed.

As a result, in the absence of rotation when $E = \text{const}$, $\eta = \text{const}$, $\Omega_{\rm F} = 0$ $L = 0$, the Alfvén and fast magnetosonic surfaces (having, obviously, a spherical form) coincide with one another. In the following we denote the values corresponding to the spherically symmetric solution, by "0". Using now Bernoulli's equations

$$\frac{1}{64\pi^4}\frac{\mathcal{M}^4(\nabla\Psi)^2}{\varpi^2} = E_0^2 - \alpha_0^2\mu^2\eta_0^2, \tag{5.248}$$

we can obtain the expressions for the location of the outer ($\mathcal{M}^2 = 1$)

$$r_a^{\rm (out)} = r_{\rm f}^{\rm (out)} = \left(\frac{\Psi_0}{8\pi^2}\frac{1}{\sqrt{E_0^2 - \mu^2\eta_0^2}}\right)^{1/2} \tag{5.249}$$

and inner ($\mathcal{M}^2 = \alpha^2$)

$$\alpha_a^{\rm (in)} = \alpha_{\rm f}^{\rm (in)} = \left(\frac{8\pi^2 r_{\rm g}^2}{\Psi_0}|E_0|\right)^{1/2} \tag{5.250}$$

surfaces. Here we used the additional assumption $\gamma_{\rm inj} \gg 1$ and relation (5.237) by which the outer singular surfaces are at large distances from the black hole and the inner ones, on the contrary, are in the vicinity of the event horizon

$$r_{\rm f}^{\rm (out)} \gg r_{\rm g}, \tag{5.251}$$

$$\alpha_{\rm f}^{\rm (in)} \ll 1. \tag{5.252}$$

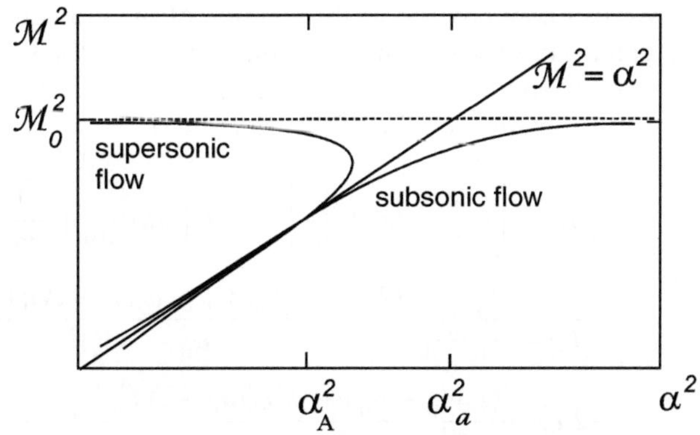

Fig. 5.11 The motion of the roots to Bernoulli's equation (4.44) in the vicinity of inner Alfvén surface. The *dotted line* indicates the dependence $\mathcal{M}^2(\alpha)$ for the zero rotation and *solid lines* for $\Omega_F \neq 0$. The transonic accretion regime occurs when the Alfvén surface location α_A is close to α_a

Note that in relations (5.250) we have the modulus of the Bernoulli integral E_0, because, as was noted, the value η is negative for the accreting flow.

Further, we use the above fact that for the slowly rotating body with the monopole magnetic field the problem of determining the integral of motion can be solved independent of the magnetic field structure. Indeed, as in the above example, for nonzero rotation when simultaneously two parameters, $\Omega_F(\Psi)$ and $L(\Psi)$, differ from zero, their ratio can again be arbitrary. Accordingly, the location of the Alfvén surface α_A (4.156) proves arbitrary too. The motion of the roots of Bernoulli's equation for the inner Alfvén surface is fully equivalent to the outflow case shown in Fig. 5.4, where for the inner surface the corresponding curves have the form shown in Fig. 5.11. For the transonic flow to exist we must again suppose that the Alfvén surface location defined by the numerators in relation (4.38) must coincide with expressions (5.249) and (5.250).

As a result, comparing relations (4.154) and (4.156) with expressions (5.249) and (5.250), we obtain the values of the invariants $L^{(\mathrm{in})}$ and $L^{(\mathrm{out})}$

$$L^{(\mathrm{out})} = \frac{\Omega_F^{(\mathrm{out})} - \omega_A^{(\mathrm{out})}}{8\pi^2} \frac{E_0}{\sqrt{E_0^2 - \mu^2\eta_0^2}} \Psi_0 \sin^2\theta, \tag{5.253}$$

$$L^{(\mathrm{in})} = -\frac{\Omega_F^{(\mathrm{in})} - \omega_A^{(\mathrm{in})}}{8\pi^2} \Psi_0 \sin^2\theta. \tag{5.254}$$

In particular, when the outer surfaces are far from the black hole (i.e., for $\omega_A^{(\mathrm{out})} \approx 0$) and the inner ones in the vicinity of the event horizon (i.e., for $\omega_A^{(\mathrm{in})} \approx \Omega_H$), we simply have

$$L^{(\mathrm{out})} = \frac{\Omega_F^{(\mathrm{out})}}{8\pi^2} \frac{E_0}{\sqrt{E_0^2 - \mu^2\eta_0^2}} \Psi_0 \sin^2\theta, \tag{5.255}$$

$$L^{(\mathrm{in})} = \frac{\Omega_H - \Omega_F^{(\mathrm{in})}}{8\pi^2} \Psi_0 \sin^2\theta. \tag{5.256}$$

We can now proceed to the solution of the GS equation. For the slow rotation we can again seek its solution in the form $\Psi = \Psi_0[1 - \cos\theta + \varepsilon_3^2 f(r,\theta)]$, where $\varepsilon_3 = a/M$ and Ψ_0 is the total magnetic flux passing through the black hole. As a result, we obtain upon linearizing Eq. (4.64)

$$\varepsilon_3^2\alpha_0^2\frac{\partial}{\partial r}\left[\left(\alpha_0^2 - \mathcal{M}_0^2\right)\frac{\partial f}{\partial r}\right] + \frac{\varepsilon_3^2}{r^2}\alpha_0^2\sin\theta\frac{\partial}{\partial\theta}\left(\frac{1}{\sin\theta}\frac{\partial f}{\partial\theta}\right) =$$

$$-\frac{\alpha_0^2E_0^2}{E_0^2 - \alpha_0^2\mu^2\eta_0^2}\,\frac{(2\alpha_0^2 - \mathcal{M}_0^2)(\Omega_{\rm F} - \omega_{\rm A})^2 - \mathcal{M}_0^2(\Omega_{\rm F} - \omega)^2}{(\alpha_0^2 - \mathcal{M}_0^2)^2}\sin^2\theta\cos\theta$$

$$-2\alpha_0^2\frac{E_0^2(2\mathcal{M}_0^2 - \alpha_0^2) + \mu^2\eta_0^2(\alpha_0^2 - \mathcal{M}_0^2)^2}{(E_0^2 - \alpha_0^2\mu^2\eta_0^2)(\alpha_0^2 - \mathcal{M}_0^2)^2}(\Omega_{\rm F} - \omega)^2\sin^2\theta\cos\theta$$

$$+2{\rm sign}\eta_0\frac{\alpha_0^2\mathcal{M}_0^2E_0^2}{(E_0^2 - \alpha_0^2\mu^2\eta_0^2)(\alpha_0^2 - \mathcal{M}_0^2)^2}(\Omega_{\rm F} - \omega)(\Omega_{\rm F} - \omega_{\rm A})\sin^2\theta\cos\theta$$

$$+\alpha_0^2\frac{a^2r_{\rm g}}{r^5}\left(\frac{\mu^2\eta_0^2}{E_0^2 - \alpha_0^2\mu^2\eta_0^2} + 2\right)\sin^2\theta\cos\theta - \alpha_0^2\frac{a^2}{r^4}\mathcal{M}_0^2\sin^2\theta\cos\theta, \quad (5.257)$$

where $\alpha_0(r)$, $\mathcal{M}_0(r)$, and E_0 refer to the zero rotation. As for ω, with adequate accuracy, it can be considered to be dependent on the radial coordinate r only. Besides, note that, as we will see, the condition $\Omega_{\rm F} \approx$ const is satisfied. Therefore, in Eq. (5.257) we disregarded the terms containing $d\Omega_{\rm F}/d\theta$.

First, we emphasize that when deriving Eq. (5.257) we essentially used Eqs. (5.253) and (5.254) for the angular momentum $L(\Psi)$. Otherwise, this equation would have a singularity on the Alfvén surfaces. Second, we should remember that the equations describing the outflowing and accreting plasma differ from one another. They have different values of the Lense–Thirring angular velocity on the Alfvén surface $\omega_{\rm A}$, different values of the integrals η_0, and, what is especially important, different values of the integrals $\Omega_{\rm F}$. If the values of $\omega_{\rm A}$ and η_0 are defined directly from the boundary conditions (5.239) and (5.240) for defining the integrals $\Omega_{\rm F}$ we should already use the implicit conditions (5.245), (5.246), and (5.247). This procedure is performed below.

Otherwise, Eq. (5.257) is fully equivalent to the equations studied in Chap. 1.

- It is linear.
- The angular operator coincides with the operator $\hat{\mathcal{L}}_\theta$ (1.120).
- Since all terms in this equation contain the small parameter $\varepsilon_3^2 \sim \Omega_{\rm H}^2r_{\rm g}^2$, the functions $\alpha_0(r)$, $\mathcal{M}_0(r)$, etc., can be taken from the zero approximation and the Lense–Thirring angular velocity in the form $\omega = \Omega_{\rm H}(r_{\rm g}/r)^3$.
- Since for the spherically symmetric flow the functions $\alpha_0(r)$, $\mathcal{M}_0(r)$, etc., are independent of θ, the solution to Eq. (5.257) can be expanded in terms of the eigenfunctions of the operator $\hat{\mathcal{L}}_\theta$.

In particular, for this reason we can disregard the disturbance of the location of the critical surfaces. We emphasize here that the singularities $\alpha_0^2 = \mathcal{M}_0^2$ correspond to the critical conditions on the fast magnetosonic surfaces which were not yet used for the construction of the solution. Therefore, as before, though in the zero approximation the locations of the two singular surfaces coincide, they yield two critical conditions rather than one.

Finally, one should especially stress that Eq. (5.257) has no singularity on the event horizon $\alpha^2 = 0$. Indeed, all coefficients in this equation are proportional to α_0^2, so when reduced by this factor, Eq. (5.257) proves regular in the vicinity of the surface $r = r_g$. This important property present in the analysis of the hydrodynamical flows results from the above general behavior of the GS equation (4.66) in the vicinity of the event horizon.

Substituting now the disturbance function in the form

$$f(r,\theta) = g_2(r)\sin^2\theta\cos\theta, \tag{5.258}$$

we obtain the ordinary differential equation for the radial function $g_2(r)$

$$\begin{aligned}
&r_g^2\frac{d}{dr}\left[\left(\alpha_0^2 - \mathcal{M}_0^2\right)\frac{dg_2(r)}{dr}\right] - 6\left(\frac{r_g}{r}\right)^2 g_2(r) = \\
&-\frac{1}{4}\frac{1}{(\alpha_0^2-\mathcal{M}_0^2)^2}\frac{E_0^2}{E_0^2-\alpha_0^2\mu^2\eta_0^2}\left[(2\alpha_0^2-\mathcal{M}_0^2)\frac{(\Omega_F-\omega_A)^2}{\Omega_H^2} - \mathcal{M}_0^2\frac{(\Omega_F-\omega)^2}{\Omega_H^2}\right] \\
&-\frac{1}{2}\frac{(\alpha_0^2-\mathcal{M}_0^2)^2\mu^2\eta_0^2 + E_0^2(2\mathcal{M}^2-\alpha_0^2)}{(\alpha_0^2-\mathcal{M}_0^2)^2(E_0^2-\alpha_0^2\mu^2\eta_0^2)}\frac{(\Omega_F-\omega)^2}{\Omega_H^2} \\
&+\frac{1}{2}\mathrm{sign}\eta_0\frac{\mathcal{M}_0^2}{(\alpha_0^2-\mathcal{M}_0^2)^2}\frac{E_0^2}{E_0^2-\alpha_0^2\mu^2\eta_0^2}\frac{(\Omega_F-\omega)(\Omega_F-\omega_A)}{\Omega_H^2} \\
&+\frac{1}{4}\left(\frac{r_g}{r}\right)^5\left(\frac{\mu^2\eta_0^2}{E_0^2-\alpha_0^2\mu^2\eta_0^2}+2\right) - \frac{1}{4}\left(\frac{r_g}{r}\right)^4\mathcal{M}_0^2.
\end{aligned} \tag{5.259}$$

The boundary conditions (5.245), (5.246), and (5.247) in the plasma generation region $r = r_{inj}$ now have the form

$$g_2|_{r=r_{inj}-0} = g_2|_{r=r_{inj}+0}, \tag{5.260}$$

$$\left.\frac{dg_2}{dr}\right|_{r=r_{inj}-0} = \left.\frac{dg_2}{dr}\right|_{r=r_{inj}+0} + \Delta j, \tag{5.261}$$

$$I|_{r=r_{inj}-0} = I|_{r=r_{inj}+0} + \Delta I, \tag{5.262}$$

$$\Omega_F^{(out)} = \Omega_F^{(in)} + \Delta\Omega_F. \tag{5.263}$$

Here the values of Δj and ΔI are two components of the surface current $\mathbf{J}_S$ (5.246) flowing in the plasma generation region. As was noted, they are to be defined by the concrete particle generation mechanism. The value $\Delta\Omega_F$ is connected with the potential drop V_g in the plasma generation region by the relation

$$\Delta\Omega_{\rm F} \approx \Omega_{\rm H} \frac{V_{\rm g}}{V_{\rm max}}, \tag{5.264}$$

where $V_{\rm max} \approx \psi_{\rm max}$ is the maximum possible potential drop in the vicinity of the horizon. Finally, as in all examples studied, to define the solution we should use the regularity conditions on the fast magnetosonic surfaces $\alpha_0^2 = \mathcal{M}_0^2$.

Thus, the solution to Eq. (5.259) can be readily constructed by analogy with the solutions to the ordinary differential equations that were earlier analyzed in detail. But since the main characteristics of the flow such as the electric current structure and the energy losses in the studied quasimonopole geometry are defined independently of the poloidal field structure, we only enumerate the basic properties of the solution to Eq. (5.259).

First of all, the absence of a singularity for $\alpha^2 = 0$ implies that for $\varepsilon_3 \ll 1$ the disturbance of the monopole magnetic field remains small up to the event horizon

$$\varepsilon_3^2 f(r_{\rm g}, \theta) \sim \varepsilon_3^2 \ll 1. \tag{5.265}$$

On the other hand, note that at large distances $r \gg r_{\rm a}^{\rm (out)}$, Eq. (5.257) exactly coincides with Eq. (5.154) describing the cold plasma outflow from the rotating sphere surface. Therefore, we can use the asymptotic solution (5.162) independent of the boundary conditions

$$\varepsilon_3^2 f(r, \theta) = 2\left(\frac{\Omega_{\rm F} r_a}{v_{\rm inj}}\right)^2 \frac{1}{\gamma_{\rm inj}^2} \ln\left(\frac{r}{r_a}\right) \sin^2\theta \cos\theta. \tag{5.266}$$

Finally, the most important result is that in the case studied, not only the longitudinal current I (to be exact, the angular momentum $L(\Psi)$) but also the angular velocity $\Omega_{\rm F}$ is not a free parameter and must be determined from the solution of the problem. Thus, the total energy release is fixed.

Indeed, using the expressions for the angular momentum (5.253) and (5.254) and the boundary conditions (5.262) and (5.263), we get

$$\Omega_{\rm F} = \frac{\omega_{\rm A}^{\rm (in)} + \omega_{\rm A}^{\rm (out)} + \Delta\Omega_{\rm F} - \varepsilon^2(2\omega_{\rm inj} + \Delta\Omega_{\rm F})}{2(1-\varepsilon^2)} + \frac{2\pi(\alpha_{\rm inj}^2 - \mathcal{M}_{\rm inj}^2)\Delta I}{\alpha_{\rm inj}^2 \Psi_0 (1-\varepsilon^2)\sin^2\theta}. \tag{5.267}$$

Here

$$\varepsilon^2 = \frac{8\pi^2 r_{\rm inj}^2 E_0}{\alpha_{\rm inj}^2 \Psi_0}, \tag{5.268}$$

so that $\varepsilon \sim \epsilon_{\rm part}/\epsilon_B$. Thus, under the condition $\varepsilon \ll 1$ we have $\omega_{\rm A}^{\rm (out)} \ll \Omega_{\rm H}$ and $\omega_{\rm A}^{\rm (in)} \approx \Omega_{\rm H}$, so that

$$\Omega_{\mathrm{F}}^{(\mathrm{out})} = \frac{1}{2}\left[\Omega_{\mathrm{H}} + \Delta\Omega_{\mathrm{F}} + \frac{4\pi(\alpha_{\mathrm{inj}}^2 - \mathcal{M}_{\mathrm{inj}}^2)\Delta I}{\alpha_{\mathrm{inj}}^2 \Psi_0 \sin^2\theta}\right], \tag{5.269}$$

$$L = \frac{\Omega_{\mathrm{F}}}{8\pi^2}\Psi_0 \sin^2\theta. \tag{5.270}$$

In particular, for $\Delta I \ll I_{\mathrm{GJ}}$ and $\Delta\Omega_{\mathrm{F}} \ll \Omega_{\mathrm{H}}$ we simply have

$$\Omega_{\mathrm{F}} = \frac{\Omega_{\mathrm{H}}}{2}. \tag{5.271}$$

Hence, according to (5.44),

$$W_{\mathrm{em}} = \frac{1}{24}\left(\frac{a}{M}\right)^2 B_{\mathrm{n}}^2 r_{\mathrm{g}}^2 c. \tag{5.272}$$

As we see, the self-consistent analysis of the GS equation really shows that its solution is fully determined by the physical boundary conditions in the plasma generation region and is, in no way, connected with the properties of the flow in the vicinity of the event horizon. Besides, as we saw, the presence of additional critical surfaces fixes not only the longitudinal current but also the angular velocity Ω_{F} that, under reasonable assumptions, $\Delta I \ll I_{\mathrm{GJ}}$ and $\Delta\Omega_{\mathrm{F}} \ll \Omega_{\mathrm{H}}$ proves close to $\Omega_{\mathrm{H}}/2$. Accordingly, the current I (5.270) is close to the Michel current I_{M}. Recall once again that in the force-free BZ solution, in which the fast magnetosonic surfaces were absent, the solution $\Omega_{\mathrm{F}} = \Omega_{\mathrm{H}}/2$ depends on the additional hypothesis for the monopole magnetic field at large distances from the black hole.

5.4.2 Slowly Rotating Black Hole Surrounded by a Rotating Shell

We now consider the case in which the magnetic field lines do not extend to infinity but are frozen in a spherical shell rotating with angular velocity Ω_{d}, which, due to its high conductivity, defines the angular velocity Ω_{F} (see Fig. 5.12). Thus, we try to model the magnetosphere in which the magnetic field lines cross both the event horizon and the accretion disk surface. For simplicity, we consider the case in which the differential rotation of the shell is absent, viz., $\Omega_{\mathrm{d}} = \mathrm{const}$. Obviously, in the presence of the spherical shell the outer singular surfaces are absent.

Clearly, for the slow rotation of the black hole and the external shell $\Omega_{\mathrm{H}} r_{\mathrm{g}}/c \ll 1$ and $\Omega_{\mathrm{d}} r_{\mathrm{d}}/c \ll 1$, the problem must be solved in the same way as in the above example (the values of Ω_{H} and $\Omega_{\mathrm{F}} = \Omega_{\mathrm{d}}$ are two independent parameters now). In particular, the linearized equation, in fact, coincides with Eq. (5.257). As a result, in this example the GS equation has no singularity on the event horizon either. Moreover, here we can define the longitudinal current I (angular momentum L) separately from the analysis of the poloidal structure of the magnetic field. Therefore, we again consider only the latter problem and do not discuss the solution to the GS equation.

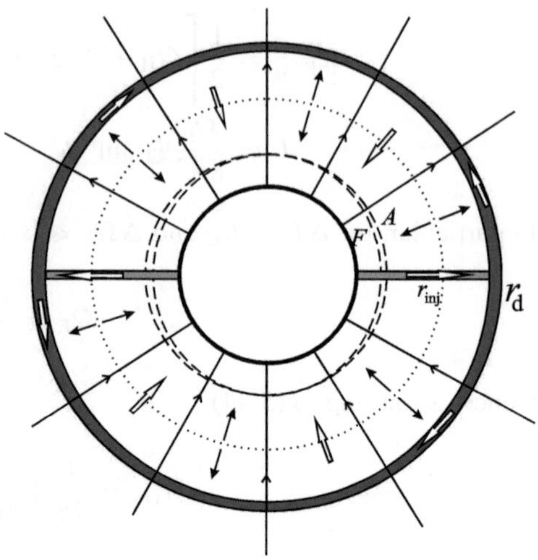

Fig. 5.12 The structure of the singular surfaces (*dashed lines*) and the plasma generation region (*dotted line*) for the fast-rotating ($\Omega_H > \Omega_d$) black hole submerged in the monopole magnetic field in the presence of the distant accretion disk modeled by the spherical shell with radius $r = r_d$. The plasma generated in the vicinity of the surface $r = r_{inj}$ flows both to the black hole and to the external shell

We first consider the flow in which the angular velocity of the black hole is less than that of the shell

$$\Omega_H < \Omega_d. \tag{5.273}$$

This case is fully equivalent to that described in Sect. 5.3.1 in which, however, the outflow from the surface $r = r_d$, on which the boundary conditions are to be given, occurs in the inner magnetosphere regions $r < r_d$ rather than in the outer ones. Therefore, the energy is transferred from the shell to the black hole.

Problem 5.22 Show that under the condition $\Omega_H < \Omega_d$ the electric charge density does not change the sign in the entire region $r_g < r < r_d$, so that it is not necessary to consider two domains with different values of η.

As a result, at zero temperature the problem again requires four boundary conditions. As such boundary conditions, we can take the concentration and the particle Lorentz factor, as well as the angular velocity of the shell and the magnetic flux on its surface

$$n(r_d, \theta) = n_{inj}, \tag{5.274}$$
$$\gamma(r_d, \theta) = \gamma_{inj}, \tag{5.275}$$
$$\Omega_F = \Omega_d, \tag{5.276}$$
$$\Psi(r_d, \theta) = \Psi_0(1 - \cos\theta). \tag{5.277}$$

These relations immediately define the integrals of motion E and η and the location of the inner singular surfaces for the slow rotation

$$E_0 = -\alpha_{\mathrm{inj}}\, \mu\, \eta_{\mathrm{inj}}\, \gamma_{\mathrm{inj}}\,, \tag{5.278}$$

$$\eta_{(\mathrm{inj})} = -\frac{\alpha_{\mathrm{inj}}\, n_{\mathrm{inj}} \sqrt{\gamma_{\mathrm{inj}}^2 - 1}}{B_{\mathrm{inj}}}, \tag{5.279}$$

$$\alpha_a^{(\mathrm{in})} = \alpha_{\mathrm{f}}^{(\mathrm{in})} = \left(\frac{8\pi^2 r_{\mathrm{g}}^2}{\Psi_0} |E_0| \right)^{1/2}. \tag{5.280}$$

Here $\alpha_{\mathrm{inj}} = \alpha(r_{\mathrm{d}})$ and $B_{\mathrm{inj}} = B(r_{\mathrm{d}})$. Thus, comparing again the locations of the Alfvén and fast magnetosonic surfaces, we can define the angular momentum L needed for the smooth crossing of these surfaces:

$$L^{(\mathrm{in})} = -\frac{\Omega_{\mathrm{F}} - \Omega_{\mathrm{H}}}{8\pi^2} \Psi_0 \sin^2\theta. \tag{5.281}$$

We see that for $\Omega_{\mathrm{H}} \ll \Omega_{\mathrm{F}}$, the electric current flowing in the magnetosphere is equal to the GJ current. The negative value of L shows that in the accretion process the angular momentum of the shell is transferred to the black hole. Therefore, the angular velocity of the rotating black hole increases with time and that of the shell decreases. We stress that in the foregoing we implicitly supposed that the flow in the shell region is subsonic ($\alpha_{\mathrm{inj}} > \alpha_{\mathrm{f}}$). If the boundary conditions on the shell surface correspond to the supersonic flow when the singular surfaces are absent, the value of the angular momentum can be arbitrary.

Problem 5.23 Having specified the direction of the electric currents flowing in the spherical shell, show that a slowing down of the shell is again connected with Ampére's forces.

Problem 5.24 Consider an example in which the shell and the black hole rotate in different directions.

If, on the contrary, the angular velocity of the shell Ω_{d} satisfies the condition $0 < \Omega_{\mathrm{d}} < \Omega_{\mathrm{H}}$ (3.69), the situation appears more complicated. Recall first that when the condition (3.69) is satisfied, the energy flux is directed from the black hole to the shell. This implies that, besides the acceleration of its rotation, the shell must absorb the energy and, consequently, must be heated. This situation is unlikely to take place near the stationary sources, because the strong heating can lead to the destruction of the shell (which models the inner regions of the accretion disk). However, it may be of interest for transient objects (microquasars, gamma-bursters).

Further, it is obvious that the problem statement depends on the existence or nonexistence of the zero GJ charge density surface between the shell and the black hole horizon $\rho_{GJ} = 0$, where plasma is to be generated. Indeed, if the radius of the shell r_d is close to that of the black hole, in the entire domain $r_g < r < r_d$ there is only a plasma flow to the black hole. For this reason it is enough to have four boundary conditions on the surface $r = r_d$. In this case, the problem is again equivalent to the problem described in Sect. 5.3.1. As a result, analysis of the critical conditions (which is carried out in the same way as in the previous example) shows that the angular momentum L is again defined by (5.281). However, now L is positive and the energy and angular momentum fluxes are directed from the black hole to the external shell.

If the shell is far from the black hole, we have two flow regions (see Fig. 5.12). The singular surfaces (Alfvén and fast magnetosonic ones) exist on the accretion region only, because the shell cannot rotate with the velocity larger than that of light. Therefore, in this region we must give $b = 2 + 4 - 2 = 4$ boundary conditions, whereas in the outflow region $b = 2 + 4 - 0 = 6$. Thus, the complete problem requires 10 boundary conditions. Eight of them must again be given in the particle generation region. These may be concentrations $n^{\pm}_{\rm inj}$ (5.239) and Lorentz factors $\gamma^{\pm}_{\rm inj}$ (5.240) of the generated plasma and the four conditions (5.245), (5.246), and (5.247) specifying the magnetic flux continuity, the voltage drop, and the surface current in the particle generation region. The other two conditions must be the values of the angular velocity Ω_d and the magnetic flux $\Psi(r_d, \theta)$ on the shell surface. As we see, it is not necessary to define here the parameters of the plasma since it does not outflow but inflows into the shell surface.

Thus, for the small values of the potential drop in the particle generation region ($\Delta\Omega_F \ll \Omega_H$) and in the absence of the surface current ($\Delta I = 0$), the angular velocity Ω_F is defined by the rotational velocity of the shell

$$\Omega_F = \Omega_d. \tag{5.282}$$

The angular momentum is again defined by (5.281). This implies that for $\Omega_d \ll \Omega_H$ the longitudinal current $I = 2\pi L$ appears much stronger than the GJ current defined by the value of the angular velocity $\Omega_F = \Omega_d$

$$I = \frac{\Omega_H}{\Omega_F} I_{GJ}. \tag{5.283}$$

The above example is, of course, too simplified, because, in reality, the angular velocity of the external parts of the disk is much smaller than that of the internal ones, and, besides, the disk itself is in the equatorial plane. On the other hand, relation (5.283) is undoubtedly model independent, because to derive it we need only the condition $\Omega_F = \Omega_d$ along the magnetic field line and the "boundary condition on the horizon." But if this is the case, it is easy to show that the outer regions of the disk cannot be connected by the magnetic field lines with the event horizon (Uzdensky,

2005). Indeed, as was shown in Chap. 2, under the condition $I \gg I_{\rm GJ}$, the magnetic surfaces must be collimated within the light surface (to be exact, within the distance $\varpi = (I_{\rm GJ}/I)R_{\rm L}$ from the rotation axis). Otherwise, under condition (5.283), the magnetic field lines are bended to the rotation axis rather than to the accretion disk. Therefore, our initial assumption that the black hole horizon is connected with the accretion disk by the magnetic field lines (and its angular velocity $\Omega_{\rm F}$ is thus defined by the rotational velocity of the disk) is not valid. Therefore, one can conclude that at least part of the magnetic surfaces crossing the event horizon in the vicinity of the rotation axis must diverge and extend to infinity.

Problem 5.25 Show that for the constant angular velocity of the shell the plasma generation region is a sphere of radius

$$r_{\rm inj} = \left(\frac{\Omega_{\rm H}}{\Omega_{\rm d}}\right)^{1/3} r_{\rm g}. \tag{5.284}$$

Problem 5.26 Show that the above 10 boundary conditions are quite enough to determine not only the value of the longitudinal current but also the disturbance of the magnetic surfaces $\varepsilon f(r,\theta)$, which is determined by the GS equation (5.257).

To sum up, we repeat some conclusions of a general character, which are model independent.

1. The full MHD GS equation version has no singularity on the event horizon. Because of the general form (4.255), it is regular for $\alpha^2 \to 0$. Therein lies the main difference of the horizon from the infinitely distant region that is a singular point of the GS equation.
2. The horizon is in the hyperbolic domain of the flow. Hence, the GS equation does not require any additional boundary conditions that could affect the solution structure in the outer space. As a result, the solution of the problem is fully defined by the boundary conditions in the plasma generation region which is causally connected with the outer space.
3. The slow-down mechanism of the rotating black hole is connected with the long-range gravitomagnetic forces which act on the plasma generation region forming the longitudinal electric currents circulating in the magnetosphere. Otherwise, the BZ mechanism, in fact, is the electrodynamic realization of the Penrose mechanism.

5.5 Other Methods

5.5.1 Analysis of the Algebraic Relations

As was noted, the remarkable property of the GS equation method is that, given the poloidal structure of the magnetic field and the integrals of motion, all the other characteristics (the particle energy, the toroidal magnetic field, thermodynamic functions) can be found from the analysis of the system of implicit algebraic relations. It is due to this simplicity that so many papers were devoted to this direction from the nonrelativistic stellar (solar) wind (Weber and Davis, 1967; Cao and Spruit, 1994; Paatz and Camenzind, 1996; Breitmoser and Camenzind, 2000) and the relativistic pulsar wind (Okamoto, 1978; Kennel et al., 1983) to the purely hydrodynamical (Abramowicz and Zurek, 1981; Abramowicz and Kato, 1989; Chakrabarti, 1990) and MHD accretion (Takahashi et al., 1990; Daigne and Drenkhahn, 2002) onto black holes. In a great number of cases, the poloidal magnetic field structure (the hydrodynamical flow) was assumed to be radial, though some other cases were studied. Using some examples below, we try to briefly survey the main results obtained by this approach.

5.5.1.1 Hydrodynamical Accretion onto the Black Hole

We consider, as the first example, the hydrodynamical accretion with nonzero angular momentum (Abramowicz and Zurek, 1981; Chakrabarti, 1990). We again consider the angular momentum to be so small that it cannot hinder the matter accretion onto the black hole. The corresponding solution can be obtained from the analysis of hydrodynamical Bernoulli's equation (1.255). As was noted, in a great number of papers the motion of nonrelativistic particles was analyzed in Paczyński–Wiita's model potential (1.7) $\varphi_{\rm PW} = -GM/(r - r_{\rm g})$, the radial motion of the accreting matter being supposed, viz., $\Phi = \Phi(\theta)$. Since we are now interested in the qualitative behavior of the solution, we restrict ourselves to the discussion of only these model solutions, as they, with good accuracy, reproduce all the main characteristics of the accretion with angular momentum onto the black hole.

As shown in Fig. 5.13, the occurrence of the nonzero angular momentum in the accreting matter substantially changes the whole topology of the phase portrait. Instead of one saddle point in Fig. 1.1, for nonzero angular momentum the phase portrait can contain two saddle points and also one center. The analytical solution studied in Chap. 1 proves equivalent to the upper curve passing through the right saddle point.

At first sight, the upper curve must correspond to the true solution of the problem. However, it does not always appear stable (Kovalenko and Eremin, 1998; Das et al., 2003). Therefore, even in the ideal (dissipation-free) case, the flow under certain conditions passes to the lower curve. Clearly, this passage can occur only in the presence of the shock wave. Therefore, the entropy of the accreting matter $s(\Phi)$ on the lower curve differs from the matter entropy at large distances. This example

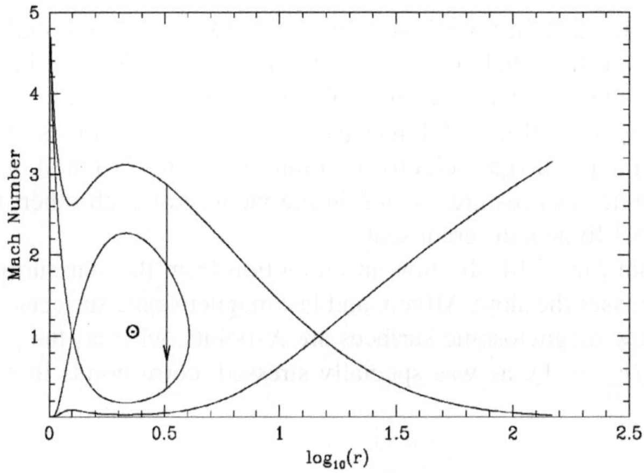

Fig. 5.13 The structure of hydrodynamical accretion with nonzero angular momentum onto the black hole (Das et al., 2003). The presence of the angular momentum results in a substantial change in the phase portrait (hydrodynamical Mach number $\mathcal{M} = v/c_\mathrm{s}$ versus the distance r) as compared to the spherically symmetric accretion (see Fig. 1.1). Under certain conditions, the flow with the shock wave (*vertical arrow*) crossing two sonic surfaces is stable (Reproduced by permission of the AAS, Fig. 3 from Das, T.K., Pendharkar, J.K., Mitra, S.: Multitransonic black hole accretion disks with isothermal standing shocks. ApJ **592**, 1078–1088 (2003))

shows once again how important is the stability checking of the solution obtained (recall that this can be successfully done only beyond the approach studied here).

5.5.1.2 Nonrelativistic Stellar (Solar) Wind

As was noted, E. Weber and L. Davis' paper (Weber and Davis, 1967) was the first one in which algebraic Bernoulli's equation was thoroughly analyzed for the

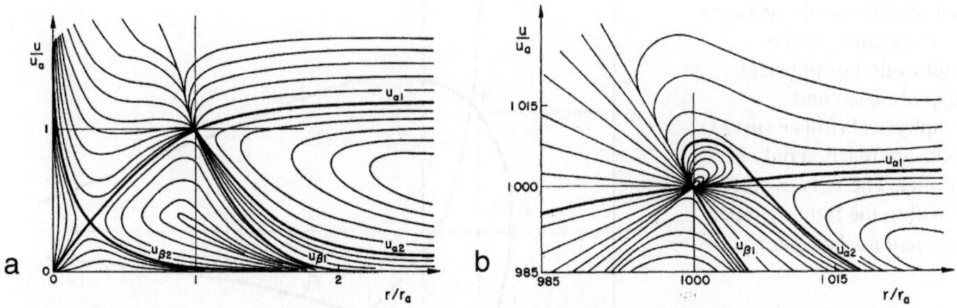

Fig. 5.14 Nonrelativistic magnetized wind structure in the equatorial plane (Weber and Davis, 1967). *Solid lines* indicate the dependence of the poloidal Mach number u/u_a as a function of r/r_a for different Bernoulli integrals E_n. *Bold solid lines* correspond to transonic flows, X-points to the slow and fast magnetosonic surfaces, whereas the Alfvén surface is a higher-order singularity (Reproduced by permission of the AAS, Fig. 1, 2 from Weber, E.J., Davis, L. Jr.: The angular momentum of the solar wind. ApJ **148**, 217–227 (1967))

strongly magnetized stellar wind (see Fig. 5.14). Only the equatorial region under the assumption of the radial character of the flow was studied. Solid lines indicate the dependence of the poloidal Mach number u/u_a as a function of r/r_a for different Bernoulli integrals $E_{\rm n}$. Bold solid lines indicate the transonic flows. As for parameters studied in the paper (gas velocity at infinity is about 425 km/s), the Alfvén and fast magnetosonic surfaces are located in the vicinity of each other; this region is shown in Fig. 5.14b on a different scale.

As seen from Fig. 5.14, the flow in its motion from the star surface to infinity successively crosses the slow, Alfvén, and fast magnetosonic surfaces. It is seen that the fast and slow magnetosonic surfaces are X-points, whereas the Alfvén surface ($u/u_a = 1$, $r/r_a = 1$), as was specially stressed, corresponds to a higher-order singularity.

5.5.1.3 Cold Relativistic Plasma Outflow from the Radio Pulsar Magnetosphere

The cold relativistic plasma outflow corresponds to the particle acceleration problem in the pulsar wind (Okamoto, 1978). Figure 5.15 demonstrates the dependence of the poloidal four-velocity $\gamma v_{\rm p}/c$ on the distance from the rotation axis $\varpi/R_{\rm L}$. Both the physical (lower curve) and nonphysical (upper curve) roots are shown. As the poloidal magnetic field was assumed to be exactly a monopole one, the plasma reaches the fast magnetosonic surface only at infinity, the limiting Lorentz factor γ_∞ being equal to the standard value $\sigma^{1/3}$. It is also seen that there is no nonphysical root at small distances from the neutron star and the Alfvén surface A is within the light cylinder (vertical dashed line).

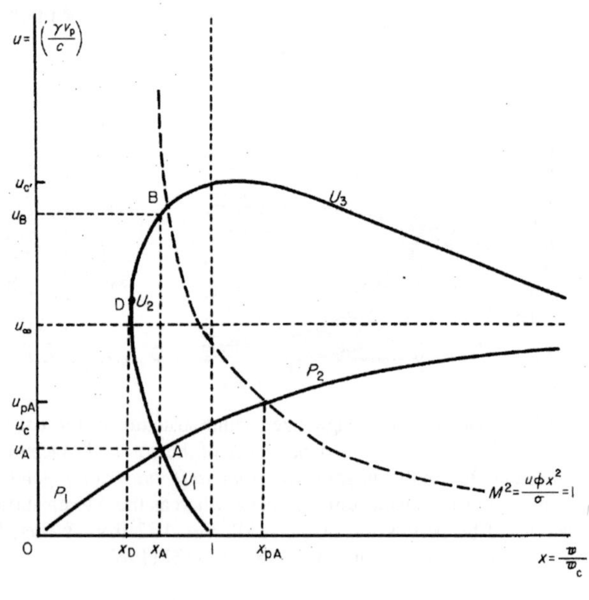

Fig. 5.15 Cold relativistic wind structure in the monopole magnetic field (Okamoto, 1978). The fast magnetosonic surface corresponding to the intersection of physical (*upper curve*) and nonphysical (*lower curve*) roots are reached only at infinity. The Alfvén surface A is within the light cylinder (*vertical dashed line*)

5.5.1.4 MHD Accretion onto the Black Hole

The strongly magnetized radial cold accretion onto the black hole was first studied in Takahashi et al. (1990) (see Fig. 5.16). Here we should first mention that the flow from infinity to the black hole horizon is impossible. Indeed, as was demonstrated, the plasma can cross the outer Alfvén surface at a positive radial velocity only, whereas the inner Alfvén surface can be crossed only at a negative radial velocity. Therefore, it is necessary to generate plasma in some source region S in the magnetosphere. The algebraic relations on the Alfvén surface do not impose any constraints on the flow parameters, because all physically reasonable trajectories with $E^2 > 0$ pass through it. Further, it is seen in Fig. 5.16 that, as in the hydrodynamical case, besides the physically pronounced transonic accretion, there is the infinitely large number of subsonic flows with the zero radial velocity on the event horizon. Later, in Takahashi (2002), the accretion with nonzero temperature was studied as well.

Thus, analysis of the algebraic equations makes it possible to determine the main characteristics of the axisymmetric stationary flows. On the other hand, it is necessary to emphasize once again that the algebraic approach does not give answers to all questions. Moreover, it can lead to incorrect results in some cases.

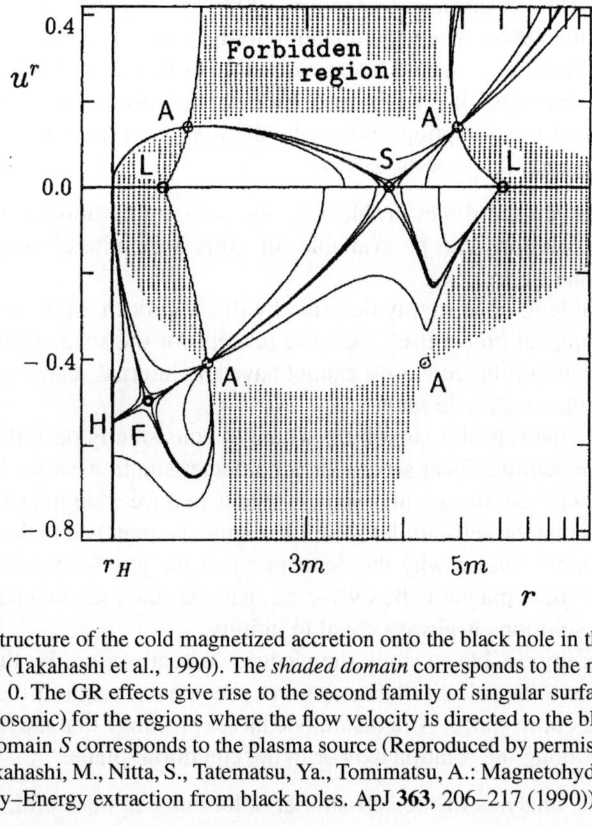

Fig. 5.16 The structure of the cold magnetized accretion onto the black hole in the vicinity of the equatorial plane (Takahashi et al., 1990). The *shaded domain* corresponds to the nonphysical solutions with $E^2 < 0$. The GR effects give rise to the second family of singular surfaces (A—Alfvén, F—fast magnetosonic) for the regions where the flow velocity is directed to the black hole horizon ($u^r < 0$). The domain S corresponds to the plasma source (Reproduced by permission of the AAS, Fig. 7a from Takahashi, M., Nitta, S., Tatematsu, Ya., Tomimatsu, A.: Magnetohydrodynamic flows in Kerr geometry–Energy extraction from black holes. ApJ **363**, 206–217 (1990))

- As was noted, Bernoulli's algebraic equation has no singularity on the Alfvén surface, whereas the Alfvén surface for the GS equation is a singular one. The critical condition on the Alfvén surface is generally used as an additional boundary condition for the GS equation determining the magnetic flux function $\Psi(r, \theta)$.
- The unsuccessful choice of the poloidal magnetic field can greatly change the key results of the flow structure. For example, the choice of the monopole poloidal field leads to the incorrect conclusion that the fast magnetosonic surface for the cold plasma outflow must be at infinity.

5.5.2 Self-Similar Solutions

As was mentioned, the remarkable property of the GS equation is that it has a rather broad class of self-similar solutions. In fact, we dealt with them in the analysis of the conical and cylindrical flows. It is not surprising, therefore, that a great number of papers were devoted to the study of the properties of the self-similar solutions, which was reduced to the analysis of the ordinary differential equation (Bisnovatyi-Kogan et al., 1979; Blandford and Payne, 1982; Tsinganos and Sauty, 1992; Contopoulos and Lovelace, 1994; Sauty and Tsinganos, 1994; Tsinganos et al., 1996; Ostriker, 1997; Sauty et al., 1999). The generalization to the relativistic case was made in Li et al. (1992). Therefore, it would be unreasonable to disregard this important trend of research. Nevertheless, it is advisable to point at once to the main restrictions that cannot be avoided in attempting to describe the real two-dimensional flows by the self-similar approach (Heyvaerts, 1996).

- We cannot study the direct problem by the self-similar approach, because the self-similar solutions can be available for quite a definite class of self-similar boundary conditions.
- It is impossible to consistently describe by this approach the flows in the vicinity of the compact object itself, because (except for the special case considered below) the self-similar equations cannot have any internal scales such as the star radius R or the black hole radius r_{g}.
- Accordingly, the self-similar solutions cannot consistently describe the jets that also have the definite linear scale—their characteristic transverse size r_{jet}. Moreover, as we will see, the angular velocity Ω_{F} is to have a singularity for $\Psi \to 0$. This implies that the self-similar solutions cannot be matched to the rotation axis.
- There is another reason why the description of the jets is impossible—all real jets have the finite magnetic flux $\Psi_0 < \infty$, whereas the total magnetic flux in the self-similar solutions is always equal to infinity.
- Finally, it is impossible to study the electric current closure by the self-similar approach, since the self-similarity requires the constant sign of the current density $j_{\parallel}$ in the entire space. As a result, one has to postulate that the inverse current flows either along the rotation axis or in the equatorial plane.

Nevertheless, analysis of the self-similar solutions can be useful. For example, this is the case with the solutions in which, as in the conical flows, one must introduce the cylindrical flow in the vicinity of the rotation axis, with which the main longitudinal electrical current is contained. Then, the flow structure at the jet periphery can have a self-similar form. In other words, of major interest are the self-similar solutions in which all values diverge exactly in the vicinity of the rotation axis.

We consider, as an example, the classical solution obtained by Blandford and Payne (1982) for the cold nonrelativistic plasma outflow from the surface of the thin Keplerian disk. The basis for the construction of the self-similar solution is the power dependence of the Keplerian angular velocity $\Omega_{\rm K}$ on the radius r

$$\Omega_{\rm K} \propto r^{-3/2}, \tag{5.285}$$

which is defined by the power dependence of the gravitational potential $\varphi_{\rm g} \propto r^{-1}$. Therefore, for the definite choice of the power $1/\beta$ in the self-similar substitution

$$\Psi(r, \theta) = r^{1/\beta}\Theta(\theta) \tag{5.286}$$

it may happen that all terms in the GS equation and Bernoulli's equation depend on the radius r only in terms of power with identical exponents $\beta_{\rm GS}$ and $\beta_{\rm B}$. Clearly, this can be the case only if

- the square of the Mach number $\mathcal{M}^2$, which is commonly available in the combination $(1 - \mathcal{M}^2)$, is independent of the radius r;
- all four invariants $E_{\rm n}(\Psi)$, $L_{\rm n}(\Psi)$, $\Omega_{\rm F}(\Psi)$, and $\eta_{\rm n}(\Psi)$ also depend on the magnetic flux Ψ in terms of power.

As a result, upon canceling by the factors r^{β_i}, the GS and Bernoulli's equations depend only on the function $\Theta(\theta)$ and its first and second derivatives. Therefore, the solution can be constructed in the same way as in the case of the cylindrical flows.

Analyzing Bernoulli's equation (4.88), we can at once conclude that, in addition to relation (5.285), the other three integrals must have the following dependence on the coordinate r in the equatorial plane:

$$E_{\rm n} \propto r^{-1}, \tag{5.287}$$

$$L_{\rm n} \propto r^{1/2}, \tag{5.288}$$

$$\eta_{\rm n} \propto r^{1/\beta-3/2}. \tag{5.289}$$

These relations show that the density $\rho = 4\pi\eta_{\rm n}^2/\mathcal{M}^2$ must depend on the radius as

$$\rho \propto r^{2/\beta-3}. \tag{5.290}$$

Thus, the GS equation (4.102) also has a self-similar form. Since the angular velocity $\Omega_{\rm F}$

- must be an integral of motion (i.e., depends only on the magnetic flux Ψ),
- must coincide with the Keplerian angular velocity $\Omega_{\rm K}$ on the disk surface,

it can be defined as

$$\Omega_F(\Psi) = \Omega_0(\Psi/\Psi_b)^{-3\beta/2}, \tag{5.291}$$

where Ω_0 is a constant. Accordingly,

$$E_n(\Psi) = E_0(\Psi/\Psi_b)^{-\beta}, \tag{5.292}$$
$$L_n(\Psi) = L_0(\Psi/\Psi_b)^{\beta/2}, \tag{5.293}$$
$$\eta_n(\Psi) = \eta_0(\Psi/\Psi_b)^{1-3\beta/2}. \tag{5.294}$$

As we see, all freedom in the self-similar simulation reduces to the choice of the constants $\Omega_0, \ldots, \eta_0$, as well as Ψ_b and β. In particular, R. Blandford and D. Payne studied the case $\beta = 4/3$, so that

$$\Psi(r, \theta) = r^{3/4}\Theta(\theta), \tag{5.295}$$

and $B \propto r^{-5/4}$. As was noted, the magnetic flux $\Psi(r)$ diverges for $r \to \infty$.

As a result, Bernoulli's equation can be used to define the (generally speaking, implicit) dependence $\mathcal{M}^2$ on θ, while the value $\mathcal{M}^2$ appears to be a function of the first derivative $d\Theta(\theta)/d\theta$, the function $\Theta(\theta)$ itself, and the angle θ. If we substitute the obtained expression in the compact form of the GS equation comprising both the second derivatives Θ with respect to θ and the first derivatives $\mathcal{M}^2$ with respect to θ, we obtain the desired second-order ordinary differential equation for the function $\Theta(\theta)$. All singular surfaces, as shown in Fig. 5.17, have the conical form $\theta = \text{const}$. We emphasize again that this solution is available only for $\Psi > \Psi_b$. In the cen-

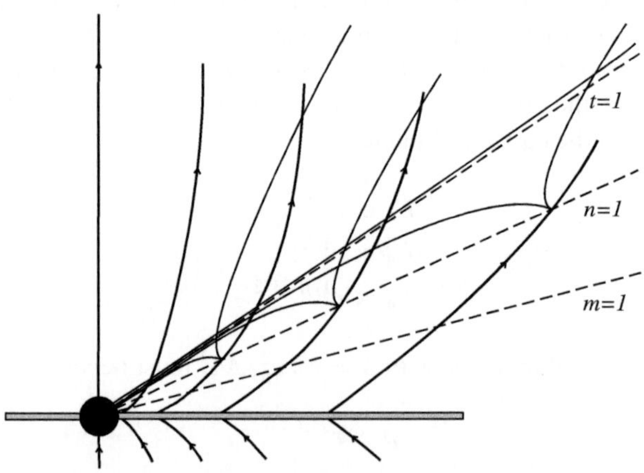

Fig. 5.17 The self-similar solution to the GS equation in which the magnetic surfaces (*bold lines*) are repeated according to the substitution (5.296) and (5.297). In this case the field lines are inclined to the surfaces $\theta = \text{const}$ at a constant angle. In particular, all singular surfaces (*dashed lines*) including the fast magnetosonic surface $n = 1$ and the singular surface $t = 1$ also have the conical form $\theta = \text{const}$. The *fine lines* indicate the behavior of the characteristic surfaces

tral domain $\Psi < \Psi_b$, the self-similar approximation is obviously inapplicable (see Fig. 5.9).

If now, as was done in Blandford and Payne (1982), instead of the independent variable θ and the unknown function $\Theta(\theta)$, we take the coordinate $z = r_0\chi$ and the function $\xi(\chi)$ prescribing the form of the magnetic field line by the self-similar substitution

$$z = r_0\chi, \tag{5.296}$$
$$\varpi = r_0\xi(\chi) \tag{5.297}$$

(here the parameter r_0 has the meaning of the radius on which the given field line crosses the equator so that $\xi(0) = 1$), the GS equation is rewritten as

$$\xi f^2 T(m-1)^2(t-1)J^{-1}S^{-2}\xi'' + H(\chi, \xi, \xi') = 0, \tag{5.298}$$

where

$$\begin{aligned} H(\chi, \xi, \xi') = {} & (m-1)^2[\xi T + (n-m-1)f^2 J]T \\ & + m(m-1)[(t-1)\xi T S \\ & - \xi f^2(\chi + \xi\xi')(\xi\xi' - \xi\xi' S^3 - \chi S^3)] \\ & + (m-1)[m\xi^2(\xi T - m f^2 J) - 5/4\,(n-1)\xi T^2] \\ & + 2m^2(\xi^2 - \lambda)(m f^2 J - \xi T). \end{aligned} \tag{5.299}$$

Here $\xi' = \mathrm{d}\xi/\mathrm{d}\chi$, $\xi'' = \mathrm{d}\xi^2/\mathrm{d}\chi^2$, and we designate

$$S = (\chi^2 + \xi^2)^{-1/2}, \tag{5.300}$$
$$T = \xi^2 + 2S - 3, \tag{5.301}$$
$$J = \xi - \chi\xi', \tag{5.302}$$

and

$$m = \kappa\xi f J, \tag{5.303}$$
$$n = \kappa\xi J[1 + (\xi')^2]/T, \tag{5.304}$$
$$t = \kappa\xi f^3 J^3 S^2/T. \tag{5.305}$$

The values κ and λ are the dimensionless constants of the problem

$$\kappa = 8\pi^2\eta_0(GM)^{1/2}\Theta_0^{-1}\frac{[1 + (\xi_0')^2]^{1/2}}{[9/16\,\Theta_0^2 + (\Theta_0')^2]^{1/2}}, \tag{5.306}$$

$$\lambda = \frac{L_0\Theta_0^{2/3}}{(GM)^{1/2}}, \tag{5.307}$$

where $\Theta_0 = \Theta(\pi/2)$ and $\Theta_0' = \mathrm{d}\Theta/\mathrm{d}\theta$ for $\theta = \pi/2$. Finally, the function $f = f(\chi)$ specifying the poloidal velocity of matter by the relation

$$(v_{\varpi}, v_z) = [\xi' f(\chi), f(\chi)] \left(\frac{GM}{r_0}\right)^{1/2} \tag{5.308}$$

can be found from the algebraic equation

$$T - f^2 U = \left[\frac{(\lambda - \xi^2)m}{(1-m)\xi}\right]^2. \tag{5.309}$$

Problem 5.27 Show that in the self-similar solution studied the value $f(\chi)$ in (5.308) must really be a function of the coordinate χ only.

Problem 5.28 Show that the value $m = \kappa \xi f J$ is nothing but the square of the Mach number relative to the Alfvén velocity ($m = \mathcal{M}^2$), the value $n = \kappa \xi J[1 + (\xi')^2]/T$ is the square of the Mach number relative to the velocity of the fast magnetosonic wave $V_{(2)}$ (4.16), and the value $t = \kappa \xi f^3 J^3 S^2/T$ is defined by the ratio of the θ-component of the velocity to the velocity of the fast magnetosonic wave:

$$m = \frac{4\pi\rho v_{\mathrm{p}}^2}{B_{\mathrm{p}}^2}, \tag{5.310}$$

$$n = \frac{4\pi\rho v_{\mathrm{p}}^2}{B^2}, \tag{5.311}$$

$$t = \frac{4\pi\rho v_{\theta}^2}{B^2}. \tag{5.312}$$

Problem 5.29 Show that the value in square brackets in expression (5.309) is always positive and the critical condition ($\lambda = \xi^2$ for $m = 1$) exactly corresponds to condition (4.155) on the Alfvén surface.

As we see, even in the simplest case the self-similar equation proves extremely cumbersome. Therefore, we do not analyze its solutions. Nevertheless, some general properties of Eq. (5.298) merit discussion in more detail.

1. The self-similar equation (5.298) has a singularity on the Alfvén surface $m = 1$. This fact is not unusual, because all GS equation versions, for the cylindrical geometry as well, have a singularity on this surface. We call attention to the coefficient in (5.298) that comprises the factor $(1 - m)^2$, which just shows that the GS equation remains elliptic when crossing the Alfvén surface.
2. It is unexpected at first sight that there is no singularity on the fast magnetosonic surface $n = 1$ and there occurs a singularity on the surface $t = 1$ (it is sometimes called the modified fast magnetosonic surface). However, we readily understand that this must be exactly the case for the self-similar flows. Indeed, as was noted, postulating the definite symmetry (axisymmetry, one-dimensionality) of the problem, we thus restrict the direction of the disturbances. Since, as we saw, the self-similar equations actually depend on the angle θ only, it is not surprising that a singularity occurs on the surface, where exactly the θ-component of the velocity of matter becomes equal to that of the fast magnetosonic wave (Blandford and Payne, 1982). We emphasize that this displacement of the singularity is the known property of the self-similar solutions, which is often discussed (von Mises, 1958; Bogovalov, 1997a; Vlahakis et al., 2000).
3. The displacement of the singular surface can be interpreted differently. Since, in the nonrelativistic case, the denominator D (4.101) at zero temperature can be written as

$$D = -1 + \frac{B^2}{\mathcal{M}^2 B_{\rm p}^2}, \tag{5.313}$$

we readily show that the coefficient $\mathcal{C}$ (4.71) in (4.68) can be represented as

$$\mathcal{C} = \frac{1}{\mathcal{M}^2 B_{\rm p}^2}\left(B^2 - \mathcal{M}^2 B_\theta^2\right). \tag{5.314}$$

Since

$$\frac{v_\theta}{v_{\rm p}} = \frac{B_\theta}{B_{\rm p}}, \tag{5.315}$$

the condition $t = 1$ exactly coincides with the condition $\mathcal{C} = 0$. As a result, according to the general equation (1.129) for the characteristic surfaces, we have $\mathrm{d}\theta/\mathrm{d}r = 0$ for $t = 1$. This implies that, as shown in Fig. 5.17, the singular surface $t = 1$ coincides with one of the characteristic surfaces, which appears to be perpendicular to the direction of the change of the coordinate θ. It is not surprising, therefore, that the singularity in the self-similar equation occurs exactly on this surface.
4. As the singular surface $t = 1$ actually coincides with the separatrix characteristic, this gives occasion to consider this example as the confirmation that exactly the separatrix characteristic rather than the sonic surface is the true singularity in the GS equation (Bogovalov, 1996; Tsinganos et al., 1996). However, it is not obvious that this assertion unconditionally valid for the self-similar solutions can be extended to the general case of the axisymmetric flows.

Thus, the above example really shows that the GS equation has a rather broad class of self-similar solutions. It is not surprising, therefore, that the above approach in the following was studied in many papers, including the relativistic case (Contopoulos and Lovelace, 1994; Vlahakis et al., 2000; Vlahakis and Königl, 2003). Since the gravitational field was disregarded in these papers, the self-similar dependence of the magnetic flux Ψ on the radius r was due to the fact that the Alfvén factor $A = 1 - \Omega_{\rm F}^2\varpi^2 - \mathcal{M}^2$ comprises the term $\Omega_{\rm F}^2\varpi^2$ that, as before, must be independent of the radius r. Therefore, within the relativistic self-similar approach, it was possible to describe only the flows in which the angular velocity in the equatorial plane is inversely proportional to the radius

$$\Omega_{\rm F} \propto r^{-1} \tag{5.316}$$

(this example, in our opinion, very clearly shows the limitation of the self-similar approach). Thus, it is easy to show that in the relativistic case the self-similar solution is possible only for the following dependencies of the integrals of motion on the magnetic flux Ψ:

$$\Omega_{\rm F}(\Psi) = \Omega_0(\Psi/\Psi_{\rm b})^{-\beta'}, \tag{5.317}$$
$$E(\Psi) = E_0(\Psi/\Psi_{\rm b})^{1-2\beta'}, \tag{5.318}$$
$$L(\Psi) = L_0(\Psi/\Psi_{\rm b})^{1-\beta'}, \tag{5.319}$$
$$\eta(\Psi) = \eta_0(\Psi/\Psi_{\rm b})^{1-2\beta'}. \tag{5.320}$$

Then, the flux can be written as

$$\Psi(r,\theta) = r^{1/\beta'}\Theta(\theta). \tag{5.321}$$

The GS equation again has the form (5.298) and the coefficient of ξ'' vanishes both on the Alfvén and on the modified fast magnetosonic surfaces (Li et al., 1992).

Finally, one cannot but mention the other family of self-similar solutions, which was first introduced in Low and Tsinganos (1986). It is based on the substitution

$$\Psi(r,\theta) = R(r)\sin^2\theta \tag{5.322}$$

and on the hypothesis that there are solutions in which the density ρ depends only on the radius r

$$\rho = \rho(r). \tag{5.323}$$

It is also necessary to suppose that E and L are linearly dependent on Ψ

$$E(\Psi) = E^{(0)} + E_0\Psi, \tag{5.324}$$
$$L(\Psi) = L_0\Psi, \tag{5.325}$$

and the integrals $\Omega_{\rm F}$, η are constant (this choice is natural when considering the flows in the vicinity of spherical objects rather than disk ones). Then the Mach number $\mathcal{M}^2 = 4\pi\eta^2/\rho$ depends only on the coordinate r and, therefore, the form of the Alfvén surface is a spherical one.

As a result, it is easy to verify that the GS equation (4.102) does not actually comprise the coordinate θ. As to Bernoulli's equation (4.88), it, unfortunately, takes the form

$$G_1(R, R', E_0, L_0, \Omega_{\rm F}, \eta, \mathcal{M}^2) + G_2(R, R', E_0, L_0, \Omega_{\rm F}, \eta, \mathcal{M}^2)\sin^2\theta = 0 \tag{5.326}$$

($R' = {\rm d}R/{\rm d}r$) and, therefore, splits into two equations

$$G_1(R, R', E_0, L_0, \Omega_{\rm F}, \eta, \mathcal{M}^2) = 0, \tag{5.327}$$

$$G_2(R, R', E_0, L_0, \Omega_{\rm F}, \eta, \mathcal{M}^2) = 0. \tag{5.328}$$

Thus, in the general case, the standard procedure for defining $\mathcal{M}^2$ by the integrals of motion and the function $R(r)$ is impossible. There may be agreement only if there is energy release or generation in the volume, which, besides, has a special self-similar form. This class of solutions was also studied in a great number of papers (Tsinganos and Sauty, 1992; Sauty and Tsinganos, 1994; Tsinganos et al., 1996; Sauty et al., 1999).

5.5.3 Computational Results

At present, there are a lot of papers devoted to the numerical analysis of the relativistic and nonrelativistic accretion onto the compact objects and the jet outflows. However, it is hardly possible to encompass all of them. For example, there are a lot of papers devoted to the generation process, the internal structure, and stability of the jets (Ouyed and Pudritz, 1997; Lucek and Bell, 1997; Hardee et al., 1998; Nishikawa et al., 1998; Hardee, 2003), which actually cannot be compared with the analytical theory predictions discussed here. Accordingly, in Kudoh et al. (1998), Koide et al. (1999, 2000), and Semenov et al. (2002), the nonstationary regime of the MHD accretion onto black holes was studied. The familiar "magnetic tower" (Lynden-Bell, 2003; Sherwin and Lynden-Bell, 2007) is, in fact, a nonstationary configuration as well. Clearly, these flows cannot be described by the stationary equations either. Further, in Toropin et al. (1999), Toropina et al. (2001, 2003), and Romanova et al. (2003), the accretion onto the neutron star with a strong dipole magnetic field was studied, which was also far from a stationary one. Finally, a lot of three-dimensional MHD calculations have been carried out recently, in which the flow was supposed to be strongly turbulent and the location of the singular surfaces could not be actually specified with sufficient accuracy (Balbus and Hawley, 1998; Igumenshchev et al., 2000; de Villiers and Hawley, 2002; Krolik and Hawley, 2002). Therefore, we consider here only the papers directly related to the analytical results obtained by the GS equation method. As was noted, we are interested in exactly the general properties of the flows rather than in the possibility to interpret the real observations.

Let us first recall the results of the hydrodynamical flows. As was shown, the analytical solution (1.154) corresponding to the Bondi–Hoyle accretion is in excellent agreement with the results of the numerical calculations (Hunt, 1979; Petrich et al., 1989). Further, the conclusion about the disk formation in the accretion of matter with angular momentum and in the transonic ejection from the surface of rotating stars is not new either. Here the analytical solutions only confirm the known results obtained by the numerical methods earlier (Lammers and Cassinelli, 1999). As for the structure of the thin accretion disk in the vicinity of the black hole horizon, the comparison, unfortunately, is impossible here, since all papers on numerical simulation (including the simulation of three-dimensional flows done in recent years (Papaloizou and Szuszkiewicz, 1994; Igumenshchev and Beloborodov, 1997; Krolik and Hawley, 2002; Daigne and Font, 2004)) deal with thick disks only, for which the above effects (strong vertical flow compression, nozzle formation) must not occur. Indeed, as was demonstrated in Sect. 1.4.7, the visible space oscillations of the disk thickness inside the marginally stable orbit can be realized for $H/r < 10^{-2}$ only.

Some results of the numerical simulation within the force-free approximation were already mentioned in the previous chapters. For example, as was noted at the end of Chap. 3, the existence of the equatorial current sheet inside the ergosphere of the fast-rotating black hole was confirmed (Komissarov, 2005) (see Fig. 3.14). As in the case of the radio pulsar magnetosphere, the magnetic field lines do not cross the equatorial plane inside the inner Alfvén surface. As a result, the electromagnetic energy propagates from the black hole to infinity. The above simple analytic dependencies $\gamma \approx (z/R_{\rm L})^{1/2}$ ($u_{\hat{\varphi}} \approx 1$) (5.234) for the energy of particles accelerated in the parabolic magnetic field were also reproduced. As shown in McKinney (2006b), these dependencies are seen in a wide range of distances from the central source.

We now proceed to the discussion of the results obtained in the numerical experiments for the MHD flows. Here the "hydrogen atom" is again the problem of the magnetosphere structure of the rotating body with the monopole magnetic field. First of all, an important result was obtained clarifying the limitation of the stationary solutions when studying the current closure process (Bogovalov and Tsinganos, 1999) (see also Komissarov, 2004b). The authors pose the problem in which there is a nonrotating sphere with the monopole magnetic field, which at time $t = 0$ begins to rotate with angular velocity Ω. As a result, as shown in Fig. 5.18, a switching-on wave begins to propagate outward with velocity c, and the magnetic field beyond it thus remains monopole and the electric currents are absent, whereas within the switching-on wave (it is a very important result) the solution rapidly approaches the stationary transonic regime fully consistent with the solution (5.173). Thus, the assumption of the stationary solution is confirmed, in which the longitudinal currents practically flow along the magnetic surfaces.

As to the current closure, it occurs in the switching-on wave, where the flow is essentially nonstationary, so the time-dependent term $\partial\rho_{\rm e}/\partial t$ plays the leading role.

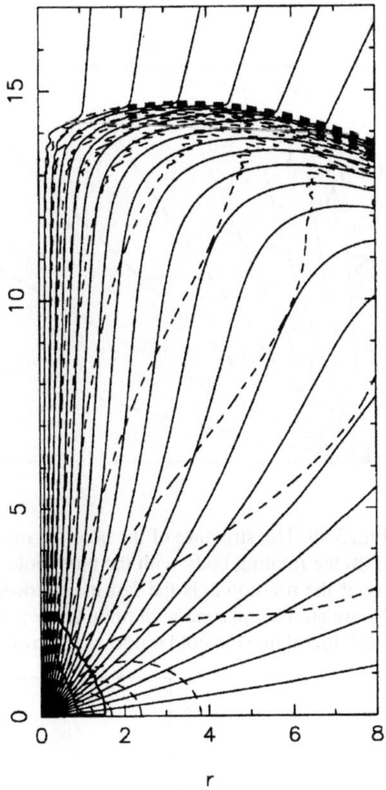

Fig. 5.18 The electric current "closure" in the switching-on wave propagating with velocity c from the compact object (Bogovalov and Tsinganos, 1999). Inside the switching-on wave, the time-dependent term $\partial\rho_e/\partial t$ plays the leading role. Within the switching-on wave, the flow structure rapidly becomes stationary, which is consistent with the analytical estimate (5.173)

With the environment disregarded, this switching-on wave propagates to infinity. This example shows that, within the stationary approach, one can really suppose that the electric current closure occurs at infinity as was generally recognized. But, in reality, for any finite external pressure, the current closure takes place in a shock wave to be generated at a finite distance from the compact object. It is important that this region cannot affect the flow structure at smaller distances. We emphasize that in this point the transonic flows substantially differ from the subsonic ones when the electric current circulating in the magnetosphere is determined by the conductivity of the boundary of the plasma-filled region (see, e.g., Rafikov et al., 1999).

We can mention, as the next result, the confirmation of the pronounced collimation for the nonrelativistic case (Sakurai, 1985) (Fig. 5.19a) and, conversely, the confirmation of the absence of collimation for ultrarelativistic flows (Bogovalov, 2001) (Fig. 5.19b). One can also see that in the nonrelativistic case the Alfvén and fast magnetosonic surfaces are at about the same distance from the origin (see Fig. 5.19a). Moreover, much subtler predictions of the theory were verified. For example, for the relativistic flows the additional "linear" acceleration $\gamma = \varpi/R_{\rm L}$ up to the values of $\gamma = \sigma^{1/3}$ within the fast magnetosonic surface and the very slow one close to the law $\gamma \propto (\ln r)^{1/3}$ at larger distances $r \gg r_{\rm f}$ was confirmed (see Fig. 5.20). In particular, the condition $\gamma_{\rm f} = \sigma^{1/3}$ is satisfied with high accuracy.

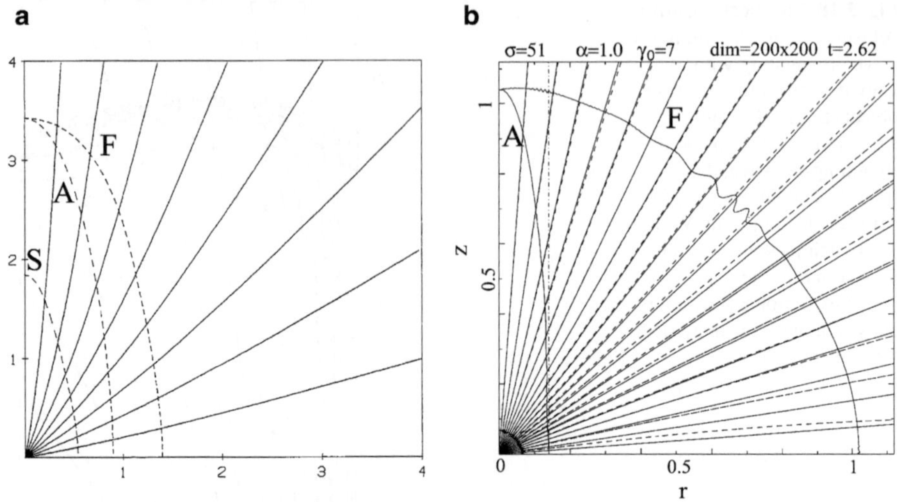

Fig. 5.19 The structure of the poloidal magnetic field and singular surfaces for plasma outflowing from the rotating body with the monopole magnetic field. It is seen that in both cases in the vicinity of the rotation axis the fast magnetosonic surface (F) coincides with the Alfvén one (A). (**a**) Nonrelativistic plasma with finite temperature (Sakurai, 1985). There is a pronounced collimation. (**b**) Ultrarelativistic cold outflow (Bogovalov, 2001). There is not any appreciable collimation

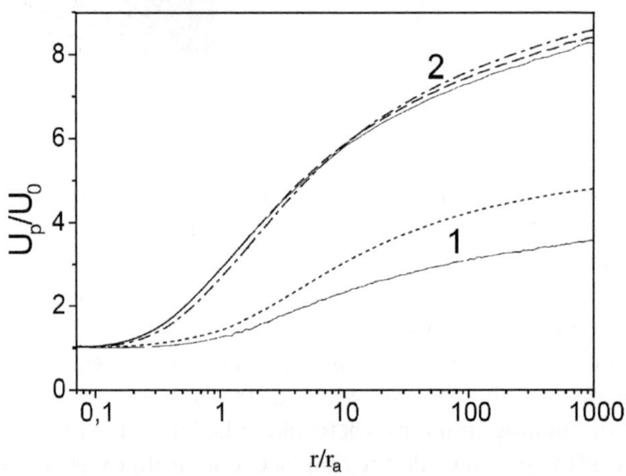

Fig. 5.20 The increase in the particle energy γ with distance r for the ultrarelativistic flow along the quasimonopole magnetic field (Bogovalov, 2001). Various curves correspond to various magnetic surfaces. Attention should be given to the difference in the scales (a logarithmic one along the horizontal axis and a linear one along the vertical one), so the increase in the particle energy for $r \gg r_{\rm f}$ thus appears extremely slow

We emphasize that this figure shows the results corresponding to the rather small magnetization parameter $\sigma = 10$, when there is no, in fact, large parameter $\sigma^{1/3}/\gamma_{\rm in}$ in the problem.

Further, for the nonrelativistic flows both the decollimation in the bulk closing current region and the asymptotic behavior $j_{\parallel} = 0$ at large distances from the

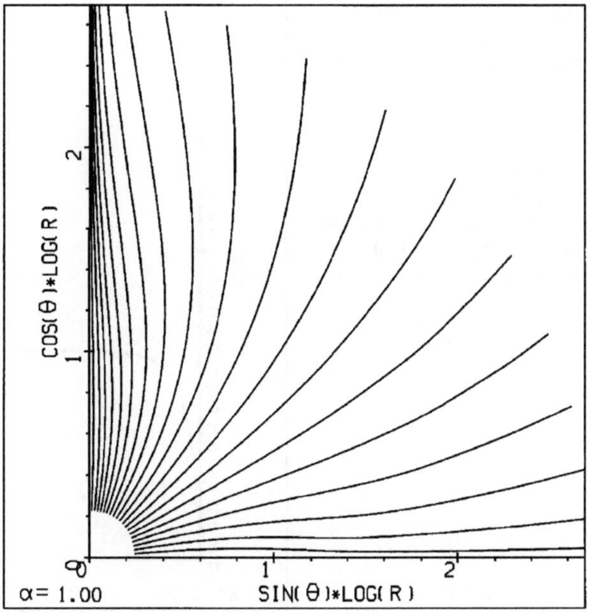

Fig. 5.21 The structure of magnetic surfaces for the nonrelativistic plasma flow from the body with the monopole magnetic field (Bogovalov and Tsinganos, 1999). The differential rotation of the central body results in bulk currents closing the longitudinal currents flowing in the magnetosphere and, as a consequence, to decollimation in the vicinity of the equatorial plane. However, resulting from the redistribution of the longitudinal currents, at larger distances the collimation is present in the entire region of the open field lines

compact object were demonstrated (Bogovalov and Tsinganos, 1999). The bulk closing current region was simulated in the same way as in the analytical example described in Sect. 5.3.2, viz., by the dependence of the angular velocity $\Omega_{\rm F}$ on the magnetic flux Ψ. As shown in Fig. 5.21, at small distances from the fast magnetosonic surface both the collimation and decollimation of the magnetic surfaces really occur. However, "nature abhors a vacuum" and, therefore, at large distances the divergence of the magnetic field lines is replaced by collimation even for the magnetic surfaces which were initially deflected from the rotation axis. This is the case only if, with account taken of the finite particle mass, the electric current I is no longer an integral of motion, which makes it possible to redistribute the longitudinal current over the region of the open magnetic field lines.

Thus, at large distances (to be exact, at mathematical infinity), in good agreement with the theoretical predictions, the outflowing current is concentrated in the vicinity of the rotation axis and the closing one in the vicinity of the equatorial plane. It is seen in Fig. 5.22 that the longitudinal current $I(\Psi) \propto x_r B_\varphi$ in the collimation region attains a constant value, at least, on a larger part of the open field lines (a further increase in the current I, according to the authors of the paper, is associated with boundedness of the computational domain, which hinders the obtaining of asymptotic values). As shown in Fig. 5.23 also borrowed from Bogovalov and Tsinganos (1999), in the vicinity of the rotation axis the dependencies $B_{\rm p} \propto \varpi^{-2}$ (5.62) and $B_\varphi \propto \varpi^{-1}$ (5.67) for the poloidal and toroidal magnetic fields and also the logarithmic increase $\Psi(\varpi) \propto \ln \varpi$ (5.64) for the magnetic flux hold with good accuracy.

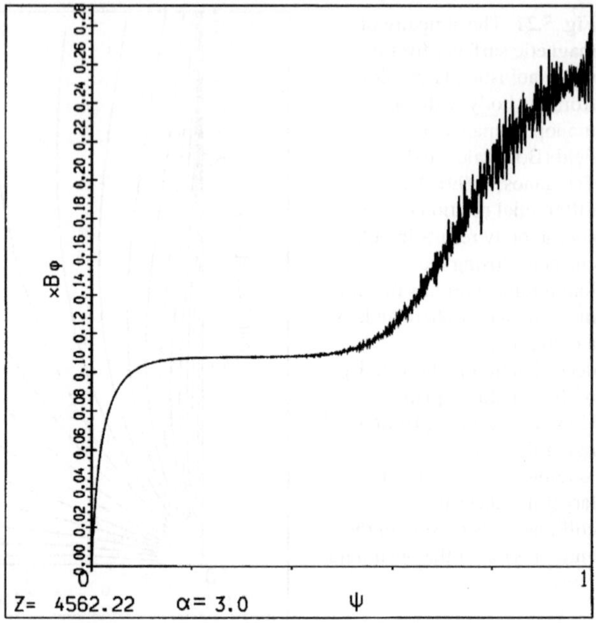

Fig. 5.22 The longitudinal current $I \propto x_r B_\varphi$ at large distances from the compact object for the nonrelativistic plasma outflow as a function of the magnetic flux Ψ (Bogovalov and Tsinganos, 1999). The plateau region corresponds to the zero longitudinal current density $j_\parallel$

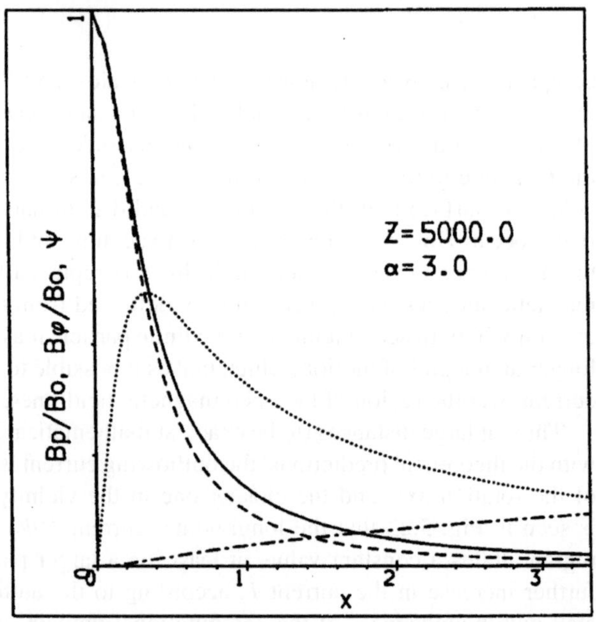

Fig. 5.23 The core structure of the poloidal magnetic field B_p (*solid line*), the toroidal magnetic field B_φ (*dotted line*), and the magnetic flux Ψ (*dash-dotted line*) in the vicinity of the rotation axis at large distances from the compact object (Bogovalov and Tsinganos, 1999). All these dependencies are in good agreement with analytical estimates (5.62) for B_p (*dashed line*), (5.67) and (5.64)

The presence of the central core was independently confirmed in other papers as well. For example, Fig. 5.24a demonstrates the poloidal magnetic field B_z inside the nonrelativistic jets as a function of the distance from the rotational axis ϖ (Lery et al., 1999). Different curves correspond to the different angular velocities Ω_F

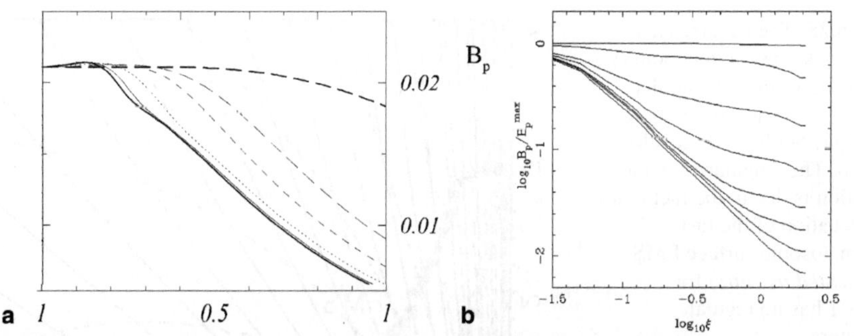

Fig. 5.24 Central core in the cylindrical jets. (**a**) Nonrelativistic flow (Lery et al., 1999). Different curves correspond to different angular velocities $\Omega_{\rm F}$, which changes the radius of the central core $\varpi_{\rm c} = v_{\rm in}/\Omega_{\rm F}$. (**b**) Relativistic flow (Komissarov et al., 2009). Different curves correspond to the different external magnetic field. For sufficiently large external pressure the poloidal magnetic field remains practically constant, while, as the external pressure diminishes, the magnetic field structure follows the intermediate regime $B_z \propto \varpi^{\beta-2}$ (5.96), the power β indeed changing from $\beta = 2$ for the homogeneous flow to $\beta \approx 0$ for the core one

which change the radius of the central core $\varpi_{\rm c} = v_{\rm in}/\Omega_{\rm F}$. As we see, the poloidal magnetic field outside the central core really decreases as $B_z \propto \varpi^{-2}$. Figure 5.24b shows the internal structure of relativistic flows in the parabolic magnetic field, and different curves correspond to the different distance from the equatorial plane (or, what is the same, to the different external magnetic field) (Komissarov et al., 2009). As is seen, for sufficiently large external pressure the poloidal magnetic field remains practically constant, while, as the external pressure diminishes, the magnetic field structure follows the intermediate regime $B_z \propto \varpi^{\beta-2}$ ($\Psi \propto \varpi^{\beta}$) (5.96), the power β indeed changing from $\beta = 2$ for the homogeneous flow to $\beta \approx 0$ for the core one.

Finally, in the numerical experiment in Bogovalov (1996), the effect of analyticity loss in the vicinity of the nonstandard singular point was, evidently, observed. The problem of the cold relativistic outflow from the surface of the body with the monopole magnetic field was again studied. It turned out that for the slow increase of the rotational velocity Ω when the parameter $\varepsilon_a = \Omega r_{\rm A}/c$ becomes of the order of unity, in the region where the fast magnetosonic surface crosses the equator (the nonstandard singular point is to be located here), a shock wave abruptly occurs (see Fig. 5.25). Extensive study shows that for small values of $\varepsilon_a \ll 1$ the expression for the logarithmic derivative $\eta_1 = (r_*/n_*)({\rm d}n/{\rm d}r)_*$ has the form (Beskin and Kuznetsova, 1998)

$$\eta_1 = -\frac{4}{3} - \frac{2}{3}\varepsilon_a^2 - \sqrt{\frac{4}{9} - \frac{2}{9}\varepsilon_a^2}. \tag{5.329}$$

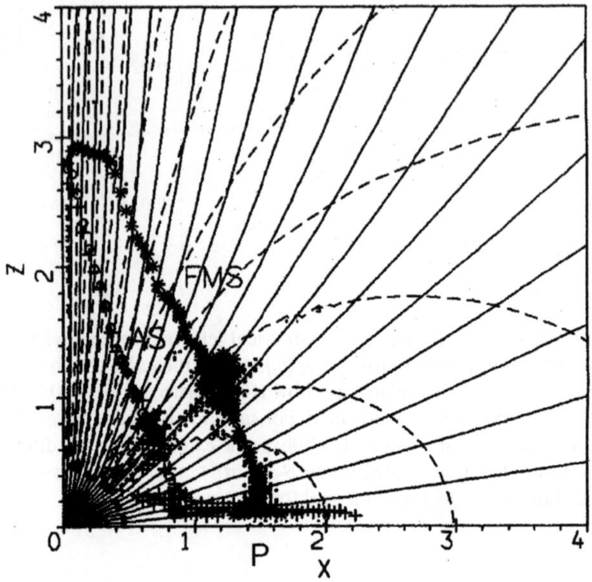

Fig. 5.25 The occurrence of the shock wave in the vicinity of the nonstandard singular point for the quasimonopole cold gas outflow (Bogovalov, 1996). The singularity in the solution is due to the fact that the location of the fast magnetosonic surface FMS (*indicated by stars*) for $\varepsilon_a > 1$ has no regular structure

In particular, for $\varepsilon_a = 0$, we simply obtain $\eta_1 = -2$, which just corresponds to the dependence $n \propto r^{-2}$ valid for the undisturbed monopole outflow. The corresponding expression for $D_1 = r_*(\mathrm{d}D/\mathrm{d}r)_*$ looks like

$$D_1^2 = \frac{4}{9} - \frac{2}{9}\varepsilon_a^2. \tag{5.330}$$

Recall that it is the condition $D_1^2 < 0$ that corresponds to the analyticity loss described in Sect. 1.3.3. Unfortunately, for $\varepsilon_a \sim 1$, the values η_1 and D_1 become dependent on the other parameters that cannot be analytically related to ε_a. Nevertheless, the general trend for vanishing the value D_1^2 as ε_a tends to unity makes one hope that the above interpretation is realistic.

To conclude, note that various groups have recently carried out the new investigations of the central engine structure. Generally, the initial regular magnetic field in the computations was supposed to be given, for example, a homogeneous or close to a monopole one. In most papers, the nonrelativistic approximation was studied and the problem was to show that the rotation of the disk, in which the magnetic field lines are frozen-in, actually results in the generation of a jet with a strong toroidal magnetic field transferring the energy and the angular momentum by the electromagnetic energy flux (Ustyugova et al., 1995, 2000).

On the other hand, in some papers (Koide et al., 2000; Koide, 2003) the GR effects were exactly taken into account, because the Kerr metric was used. Accordingly, the plasma motion was described in a self-consistent way. Therefore, it was possible to model both the accretion disk and the magnetically dominated magnetosphere. An important property in the models studied was that the environment was also taken into account in a rather consistent way, which is, certainly, of importance in the jet generation process. As a result, it was demonstrated how the accreting matter rotation results in the generation of a jet transporting the energy away from the compact object. In particular, the structure of the longitudinal currents flowing in the magnetosphere was specified. Recall once again that in the latter case the nonstationary problem was studied.

5.6 Conclusion

Thus, the existence of the exact solutions to the GS equation made it possible to get important information concerning the properties of the strongly magnetized flows in the neighborhood of compact objects. In particular, it was possible to show for the simplest geometry, though tentatively, the possibility of the efficient energy release from the rotating black hole. The main element, as was noted, is the existence of the regular poloidal magnetic field along which the energy is transported and the angular momentum is removed.

The above examples are, certainly, too simplified. Unfortunately, they are, presently, the only exact analytical solutions that could, though incompletely, describe the magnetosphere of the real compact objects. In particular, the problem of the interaction between the accretion disk and the rotating black hole remains unsolved. Nevertheless, the solutions obtained allow one to pay attention to several key points that can help one judge the basic properties of the central engine.

First of all, when it was possible to rather reliably determine the angular velocity $\Omega_{\rm F}$ (provided that the magnetic field lines extend to infinity), it turned out to be close to half the angular velocity of the black hole $\Omega_{\rm F}$. This implies that the efficiency of the energy release of the rotating black hole is really close to the maximum one. Since, according to the papers on the secular evolution of the accreting black holes (Moderski and Sikora, 1996; Moderski et al., 1998), their angular velocity can really be large enough ($a \sim M$), one can conclude that the BZ process can be of importance for the energy release mechanism of the central engine.

Further, the above-formulated assertions of the absence of the strong self-collimation for the ultrarelativistic flows and the appreciable collimation for the nonrelativistic ones were confirmed. Finally, one should stress that, while there is search of the mechanisms providing the stationary outflow regime, there are indications that in some objects this process is not stationary.

As for the observations, as was already mentioned, the present-day telescopes and the large distances to AGN are not sensitive enough to resolve the inner structure of the central engine. Therefore, now we really know nothing about the flow of matter in the vicinity of the central engine or the magnetic fields in the jet generation region. Not long ago there was information (Junor et al., 1999) according to which the opening angle of the flow of matter in the immediate vicinity of the central engine is much larger than that of the jet. If this is the case, the quasimonopole magnetic field model has been observationally confirmed. Accordingly, in Lobanov et al. (2003) and Kovalev (2008), where the internal structure of the jet was specified, it was shown that it actually has the form of a hollow cylinder, and in Gabuzda et al.

(1992) and Zhang et al. (2004) it was asserted that the jets have a strong toroidal magnetic field. However, all these papers can be regarded as preliminary ones and much has to be done to prove that the really existing jets are just the magnetized flows studied in this chapter.

Chapter 6
Conclusion

Thus, the GS equation method was shown to be a rather powerful instrument that helps, in some cases, formulate and solve the direct problems of magnetohydrodynamics that belong to a very broad class of problems. The potentialities of the GS equation method were demonstrated when investigating the concrete astrophysical objects.

On the other hand, it was shown that the GS equation method is, to a great degree, a dead-end trend in astrophysics. First of all, there is no possibility to generalize this equation to the case in which the dissipative processes (viscosity, heat conductivity, plasma radiation and its interaction with radiation, kinetic effects, etc.) are of importance. This is because the existence of the integrals of motion is intrinsic in the approach itself. The violation of this assumption makes it impossible, in the general case, to reduce the full system of equations to a single second-order equation. Therefore, only in the exceptional cases was it possible to take advantage of the dissipative processes (Li et al., 1992; Beskin and Rafikov, 2000; Soldatkin and Chugunov, 2003; Beskin et al., 2004) (for example, for the spherically symmetric hydrodynamical accretion onto the black hole describing the motion of matter, with account taken of the interaction with its radiation (Thorne et al., 1981; Nobili et al., 1991)). For the same reason, it was essentially impossible to generalize this method to the case of the nonaxisymmetric and nonstationary flows and, hence, study the obtained solutions in terms of stability. Therefore, we can mention only several papers in which there was some progress made in this direction (Park and Vishniac, 1989, 1990; Lynden-Bell, 1996; Istomin and Pariev, 1996; Timokhin et al., 1999; Bogovalov, 2001).

As we saw, the possibility to solve the direct problems was associated with the availability of the exact analytical solution to the GS equation for spherically symmetric flows. In this sense, state of the art in astrophysics proved simpler than in the other applied problems, because the simplest spherically symmetric flow, in most cases, describes the real situation quite well. In the general case in which the exact solution is not known, the direct problem cannot be posed and solved. Therefore, in the numerical calculations the relaxation method was used, which, even its statement, differs from the GS equation method.

V.S. Beskin, *MHD Flows in Compact Astrophysical Objects,* Astronomy and Astrophysics Library, DOI 10.1007/978-3-642-01290-7_7,
© Springer-Verlag Berlin Heidelberg 2010

Nevertheless, one can hope that the above results are of considerable interest. First of all, within the sufficiently simple model problems, one can describe all the main features of the transonic flows in the vicinity of the real compact objects. This makes it possible to derive rather simple analytical expressions, in fact, for all key quantities characterizing these flows. Further, there is hope that most of the results obtained by the model solutions (the absence of the inherent collimation in the magnetized relativistic wind outflow, the connection between the current and the potential in the radio pulsar magnetosphere, the secondary plasma generation in the black hole magnetosphere) qualitatively correctly describe the real processes, especially, as many of the other solutions, for example, the disk formation in the presence of the transonic accretion and ejection, confirmed the earlier obtained results. Finally, the exact analytical solutions can be used as an adequate verification in various numerical codes.

To conclude, we point to a number of important problems that, in our opinion, have not yet been comprehensively studied by the GS equation method.

1. One should first mention the separatrix characteristic problem. Undoubtedly, it is this surface that separates the causally unconnected space regions. Moreover, as we saw, the self-similar flows have singularities on this surface. Therefore, one could assume that the true singularity in the GS equation is exactly in the separatrix characteristic (Bogovalov, 1994; Tsinganos et al., 1996) and, hence, the whole solution procedure must be reconsidered. Nevertheless, some arguments presented above (e.g., the necessity to use the critical condition $N_\theta = 0$ on the sonic surface) demonstrate that the question is open. In any case, one should stress that, in the general case, no matter whether the true singularity in the GS equation is on the sonic or separatrix surfaces, its location is not known beforehand. Therefore, the corresponding critical condition does not allow one to constructively establish the connection between the physical quantities on the arbitrarily chosen boundary. As to the model problems, for which the exact analytical solutions were obtained, they, unfortunately, cannot be used as evidence in this case, since in the studied approximation $\varepsilon \ll 1$ the singular and separatrix surfaces coincide with one another.
2. Further, the cusp surface problem is not solved yet. The full GS equation does not have the same singularity as on the fast and slow magnetosonic surfaces. However, the type of the equation also changes from an elliptic to a hyperbolic one here. Moreover, as shown above, in the one-dimensional cylindrical case, a singularity occurs exactly on the cusp surface rather than on the sonic surface. Evidently, this singularity is associated with the supposed one-dimensionality of the problem (in the same way as in the self-similar solutions the singularity displaces from the sonic surface to the separatrix characteristic). However, there is no clarity in this problem yet.
3. The other direction that has not been adequately studied yet, where the above method could be successfully used, is the problem of the two-dimensional structure of the flows with shock waves. Indeed, the well-known relations can be readily rewritten for the shock waves in the language of the main terms in the GS

equation, because the conservation of the flow of matter, energy, and transverse momentum is simply written as

$$\{\Psi\} = 0, \qquad \{E\} = 0, \qquad \{L\} = 0. \tag{6.1}$$

As to the entropy jump (and also the surface location), they must be found as the solution of the problem. The investigations in this area started not long ago, but the first important results have already been obtained both for the case of accretion onto the black hole (Takahashi et al., 2002) and for the interaction between the pulsar wind and the supernova remnant (Bogovalov and Khangoulyan, 2002, 2003).

4. Finally, the flows with anisotropic pressure call for further investigation as there are only preliminary results here (Marsch et al., 1982; Hu et al., 1997; Krasheninnikov and Catto, 2000).

We are hopeful that the thorough discussion of the GS equation method in the monograph will contribute to the solution of both these and other problems in present-day astrophysics.

Appendix A
From Euler to Grad–Shafranov—The Simplest Way

Here we show how the subsonic GS equation version (1.345) can be directly derived from the Euler equation. We consider, for simplicity, only the nonrelativistic flow and the small angles $\Theta = \pi/2 - \theta$ in the vicinity of the equatorial plane.

The θ-component of the Euler equation is

$$v_r \frac{\partial v_\theta}{\partial r} + v_\theta \frac{\partial v_\theta}{r\partial\theta} + \frac{v_r v_\theta}{r} - \frac{v_\varphi^2}{r}\cot\theta = -\frac{\nabla_\theta P}{\varrho n} - \nabla_\theta \varphi_{\rm g}. \tag{A.1}$$

If we add $v_\varphi \partial v_\varphi / r\partial\theta$ to both sides and add and subtract $v_r \partial v_r / r\partial\theta$ on the left-hand side, we find

$$v_r \frac{\partial v_\theta}{\partial r} - v_r \frac{\partial v_r}{r\partial\theta} + \nabla_\theta\left(\frac{v^2}{2}\right) + \frac{v_r v_\theta}{r} = \frac{v_\varphi^2}{r}\cot\theta + v_\varphi \frac{\partial v_\varphi}{r\partial\theta} - \frac{\nabla_\theta P}{m_{\rm p} n} - \nabla_\theta \varphi_{\rm g}. \tag{A.2}$$

By the definition of the Bernoulli integral $E_{\rm n} = v^2/2 + w + \varphi_{\rm g}$ and according to the thermodynamic relationship ${\rm d}P = \varrho {\rm d}w - nT{\rm d}s$, we obtain for $E_{\rm n} = {\rm const}$

$$v_r \frac{\partial v_\theta}{\partial r} - v_r \frac{\partial v_r}{r\partial\theta} + \frac{v_r v_\theta}{r} = \frac{r v_\varphi \sin\theta}{\varpi^2} \frac{\partial}{r\partial\theta}(r v_\varphi \sin\theta) + \frac{T}{m_{\rm p}} \frac{\partial s}{r\partial\theta}. \tag{A.3}$$

The definition (1.90) yields

$$v_r = \frac{1}{2\pi n r^2 \sin\theta} \frac{\partial \Phi}{\partial\theta}, \qquad v_\theta = -\frac{1}{2\pi n r \sin\theta} \frac{\partial\Phi}{\partial r}. \tag{A.4}$$

Assuming now $n \approx {\rm const}$ (this is the case for the subsonic flow), we get

$$-\frac{1}{4\pi^2 n^2}\left(\frac{\partial\Phi}{\partial\theta}\right)\left[\frac{\partial^2\Phi}{\partial r^2} + \frac{\sin\theta}{r^2}\frac{\partial}{\partial\theta}\left(\frac{1}{\sin\theta}\frac{\partial\Phi}{\partial\theta}\right)\right]$$
$$= r v_\varphi \sin\theta \frac{\partial}{\partial\theta}(r v_\varphi \sin\theta) + \varpi^2 \frac{T}{m_{\rm p}} \frac{\partial s}{\partial\theta}. \tag{A.5}$$

Finally, if we divide both sides by $-(\partial\Phi/\partial\theta)$, we obtain (1.345).

V.S. Beskin, *MHD Flows in Compact Astrophysical Objects,* Astronomy and Astrophysics Library, DOI 10.1007/978-3-642-01290-7,
© Springer-Verlag Berlin Heidelberg 2010

Therefore, if the first term on the left-hand side of (1.345) really corresponds to the component $v_r \partial v_\theta/\partial r$ and the last term on the right-hand side ($c_s = \text{const}$) to the pressure gradient, the role of the term $\propto L_n \partial L_n/\partial\Phi$ proves less trivial. It comprises both the effective potential gradient and, actually, the component $v_\theta \partial v_\theta/\partial\theta$. The former is the leading one near the marginally stable orbit $r \approx 3\, r_g$; the latter becomes important only when approaching the sonic surface.

Appendix B
Nonrelativistic Force-Free Grad–Shafranov Equation

As was noted, the nonrelativistic version of the force-free equation (2.101) reduces to zero of Ampere's force $[\nabla \times \mathbf{B}] \times \mathbf{B}$, which implies that the contribution of the electric field is not taken into account. Thus, the GS equation has the form

$$-\nabla^2\Psi + \frac{2}{\varpi}\frac{\partial\Psi}{\partial\varpi} - \frac{16\pi^2}{c^2} I \frac{\mathrm{d}I}{\mathrm{d}\Psi} = 0. \tag{B.1}$$

As we see, the main difference from the relativistic version is that there is no integral of motion Ω_{F} in Eq. (B.1). Hence,

- the nonrelativistic GS equation version has no critical surface;
- according to the general formula $b = 2 + i - s'$, for a number of boundary conditions we have $b = 3$, i.e., the problem requires three boundary conditions.

The nonrelativistic GS equation can be substantially simplified if we consider a one-dimensional cylindrical configuration; the current $I(\Psi)$ is the linear function of the magnetic flux Ψ. It is convenient to write this relation as

$$I(\Psi) = \frac{c}{4\pi\varpi_0}\Psi, \tag{B.2}$$

where ϖ_0 is the constant of length dimension. In this case, Eq. (B.1) becomes linear:

$$\frac{\mathrm{d}^2\Psi}{\mathrm{d}x_r^2} - \frac{1}{x_r}\frac{\mathrm{d}\Psi}{\mathrm{d}x_r} + \Psi = 0. \tag{B.3}$$

Here $x_r = \varpi/\varpi_0$. The solution to Eq. (B.3) is the known fields

$$B_z = B_0 J_0(x_r), \tag{B.4}$$

$$B_\varphi = B_0 J_1(x_r), \tag{B.5}$$

where $J_0(x_r)$ and $J_1(x_r)$ are the Bessel functions. As we see, in a stable cylindrical discharge at some distance from the axis, the longitudinal magnetic field must change its direction, which is observed experimentally (see, e.g., Kadomtsev, 1988).

Appendix C
Part-Time Job Pulsars

Recently, Kramer et al. (2006) have reported on the thorough study of radio pulsar $B1931{+}24$. The difference between this pulsar and ordinary radio pulsars is that the former has 5–10 d active phases. Therefore, the radio emission is switched off in less than 10 s and is actually undetectable for the next 25–35 d. It is very important that the neutron star spin-down is different in the on and off states:

$$\dot{\Omega}_{\rm on} = 1.02 \times 10^{-14}\,1/{\rm s}^2, \tag{C.1}$$

$$\dot{\Omega}_{\rm off} = 0.68 \times 10^{-14}\,1/{\rm s}^2. \tag{C.2}$$

Hence

$$\frac{\dot{\Omega}_{\rm on}}{\dot{\Omega}_{\rm off}} = 1.5. \tag{C.3}$$

Later, the same ratio was detected for PSR $J1832{+}0031$ ($t_{\rm on} \sim 300$ d, $t_{\rm off} \sim 700$ d).

This discovery offers a unique opportunity to observe both energy-loss mechanisms for the same radio pulsar (Beskin and Nokhrina, 2007; Gurevich and Istomin, 2007). Besides, we can clarify the pulsar braking mechanism. Thus, it is logical to suppose that in the on state the energy release is due to the current losses only and in the off state to the magnetodipole radiation (in this case, it is not the plasma-filled magnetosphere). Using (2.5) and (2.178), we find

$$\frac{\dot{\Omega}_{\rm on}}{\dot{\Omega}_{\rm off}} = \frac{3 f_*^2}{2} \cot^2\chi, \tag{C.4}$$

which yields the reasonable value of the inclination angle $\chi \approx 60^\circ$.

On the other hand, if we take relation (2.260) for the on-state energy losses (Spitkovsky, 2006)

$$W_{\rm tot} = \frac{1}{4}\frac{B_0^2 \Omega^4 R^6}{c^3}(1 + \sin^2\chi), \tag{C.5}$$

we get

$$\frac{\dot{\Omega}_{\rm on}}{\dot{\Omega}_{\rm off}} = \frac{3}{2}\,\frac{(1+\sin^2\chi)}{\sin^2\chi}. \tag{C.6}$$

It is obvious that this ratio cannot be equal to 1.5 for any inclination angle.

Comparing the theory with the observations of radio pulsar $B1931{+}24$, we can make the following conclusions (Gurevich and Istomin, 2007):

1. It is for the first time that in the PSR $B1931{+}24$ off state the spin-down of the magnetized neutron star rotation due to the magnetodipole radiation energy losses was observed.
2. There are really current losses which are fully different from the vacuum ones.
3. As the switching time between the on and off states is very short, the plasma source is to be located in the open part of the magnetosphere.

Appendix D
Special Functions

D.1 Legendre Polynomials

The Legendre polynomials of the first and second kind $\mathcal{P}_m(x)$ and $\mathcal{Q}_m(x)$ are the solutions to the equation

$$\frac{\mathrm{d}}{\mathrm{d}x}\left[(1-x^2)\frac{\mathrm{d}}{\mathrm{d}x}\right] f(x) = q f(x). \tag{D.1}$$

They have the eigenfunctions

$$\begin{aligned}
\mathcal{P}_0(x) &= 1, && \text{(D.2)}\\
\mathcal{P}_1(x) &= x, && \text{(D.3)}\\
\mathcal{P}_2(x) &= \frac{3x^2-1}{2}, && \text{(D.4)}\\
\mathcal{P}_3(x) &= \frac{5x^3-3x}{2}, && \text{(D.5)}\\
&\cdots
\end{aligned}$$

and

$$\begin{aligned}
\mathcal{Q}_0(x) &= \frac{1}{2}\ln\left(\frac{1+x}{1-x}\right), && \text{(D.6)}\\
\mathcal{Q}_1(x) &= -1+\frac{x}{2}\ln\left(\frac{1+x}{1-x}\right), && \text{(D.7)}\\
\mathcal{Q}_2(x) &= -\frac{3x}{2}+\frac{3x^2-1}{4}\ln\left(\frac{1+x}{1-x}\right), && \text{(D.8)}\\
\mathcal{Q}_3(x) &= \frac{2}{3}-\frac{5x^2}{2}+\frac{5x^3-3x}{4}\ln\left(\frac{1+x}{1-x}\right), && \text{(D.9)}\\
&\cdots
\end{aligned}$$

and the eigenvalues

$$q_m = -m(m+1). \tag{D.10}$$

The general expression for $\mathcal{P}_m(x)$ is

$$\mathcal{P}_m(x) = \frac{1}{2^m m!}\frac{\mathrm{d}^m}{\mathrm{d}x^m}(x^2-1). \tag{D.11}$$

Since the Legendre polynomials are a complete orthogonal system in the interval $-1 < x < 1$, any function $f(x)$ can be given as

$$f(x) = \sum_{m=0}^{\infty}(f)_m \mathcal{P}_m(x), \tag{D.12}$$

where

$$(f)_m = \frac{2m+1}{2}\int_{-1}^{1}\mathcal{P}_m(x)f(x)\mathrm{d}x. \tag{D.13}$$

D.2 Bessel Functions

The Bessel functions of the first kind $J_m(x)$ are the solutions to the equation

$$x^2\frac{\mathrm{d}^2 f(x)}{\mathrm{d}x^2} + x\frac{\mathrm{d}f(x)}{\mathrm{d}x} + (x^2-m^2)f(x) = 0, \tag{D.14}$$

having no singularity at $x = 0$:

$$J_m(x) = \left(\frac{x}{2}\right)^m \sum_{k=0}^{\infty}\frac{1}{k!\Gamma(m+k+1)}\left(\frac{-x^2}{4}\right)^k. \tag{D.15}$$

In particular,

$$J_0(0) = 1 - \frac{x^2}{4} + \frac{x^4}{64} - \frac{x^6}{2304} + \cdots, \tag{D.16}$$

$$J_1(x) = \frac{x}{2} - \frac{x^3}{16} + \frac{x^5}{384} - \frac{x^7}{18432} + \cdots, \tag{D.17}$$

as $x \to 0$, and

$$J_m(x) = \sqrt{\frac{\pi}{2x}}\cos\left(x - \frac{m\pi}{2} - \frac{\pi}{4}\right), \tag{D.18}$$

as $x \to \infty$.

The Macdonald (modified Bessel) functions $K_m(x)$ are the solutions to equation

$$x^2 \frac{\mathrm{d}^2 f(x)}{\mathrm{d}x^2} + x\frac{\mathrm{d}f(x)}{\mathrm{d}x} - (x^2 + m^2)f(x) = 0. \tag{D.19}$$

They have a singularity at $x = 0$, but vanish at infinity. In particular,

$$K_0(x) = -\ln x + \ln 2 - \gamma, \tag{D.20}$$

$$K_1(x) = \frac{1}{x}, \tag{D.21}$$

as $x \to 0$, and

$$K_m(x) = \sqrt{\frac{\pi}{2x}}\, e^{-x}, \tag{D.22}$$

as $x \to \infty$. Here $\gamma \approx 0.577$ is the Euler constant.

D.3 Hypergeometric Function

The hypergeometric function $F(a, b, c, x)$ is the solutions to the equation

$$x(1-x)\frac{\mathrm{d}^2 f(x)}{\mathrm{d}x^2} + [c - (a+b+1)x]\frac{\mathrm{d}f(x)}{\mathrm{d}x} - abf(x) = 0, \tag{D.23}$$

having no singularity at $x = 0$:

$$\begin{aligned} F(a,b,c,x) = 1 &+ \frac{ab}{c}\frac{x}{1!} + \frac{a(a+1)b(b+1)}{c(c+1)}\frac{x^2}{2!} \\ &+ \frac{a(a+1)(a+2)b(b+1)(b+2)}{c(c+1)(c+2)}\frac{x^3}{3!} + \cdots . \end{aligned} \tag{D.24}$$

Thus, if a or b is an integer less than or equal to zero, the function $F(a, b, c, x)$ reduces to a polynomial.

Appendix E
List of Symbols

Table E.1 List of symbols

Variable	Meaning
a	specific angular momentum
a	characteristics of the critical point
a_n	expansion coefficients
A_{ik}	matrix of the coefficients
$\mathbf{A}$	vector potential
$\mathcal{A}, \mathcal{B}, \mathcal{C}$	coefficients in the second-order partial differential equation
b	internal radius of a disk
b	number of boundary conditions
b_n	expansion coefficients
$b(\theta)$	dimensionless perturbation of the sonic velocity
$\mathbf{b}$	disturbance of the magnetic field
$\mathbf{B}$; $\mathbf{B}_{\rm n}$; $\mathbf{B}_{\rm p}$; $\mathbf{B}_{\rm T}$	magnetic field; normal component; poloidal; additional
B_0	magnetic field on the neutron star surface
$B_\hbar$	critical quantum magnetic field
$B_{\rm cr}$; $B_{\rm Edd}$; $B_{\rm ext}$; $B_{\rm in}$; $B_{\rm inj}$	critical magnetic field; Eddington magnetic field; external; inside the star; injection
c	speed of light
c_n	expansion coefficients
$c_\parallel$, $c_\perp$	coefficients
$c_{\rm s}$; c_*; c_R; c_∞	sound velocity; at the sonic surface; at the star surface; at infinity
$\mathcal{C}$	constant
d_i	vector of variables
$d(\theta)$	dimensionless perturbation of the sonic radius
d; D; $\mathcal{D}$	denominators
D_1	the derivatives of the denominators
$\mathcal{D}(\lambda)$, $\mathcal{D}_0(\lambda)$, $\mathcal{D}_1(\lambda)$	normalized functions
e'; $e_{\rm n}$	hydrodynamical integral; nonrelativistic
$\mathbf{e}_r$, $\mathbf{e}_\theta$, $\mathbf{e}_\varphi$; $\mathbf{e}_m$	unit vectors; along the magnetic axis
E; $E_{\rm n}$; E_0	energy (Bernoulli integral); nonrelativistic; zero approximation
$\mathbf{E}$; $\mathbf{E}_{\rm H}$; $\mathbf{E}_{\rm in}$; $\mathbf{E}^{Q}$	electric field; on the horizon; inside the neutron star; electric quadrupole
$E_\parallel$	electric field parallel to magnetic one
$\mathcal{E}_{\rm e}$; $\mathcal{E}_{\rm ph}$	electron energy; photon energy
f	perturbation of hydrodynamic and magnetic fluxes

Table E.2

f; f_*	dimensionless area of the polar cap
F; F_n	functions
$F(a,b,c,x)$, $\mathcal{F}(r)$	hypergeometric function
$F^{\alpha\beta}$	electromagnetic tensor
F^+	viscous heating
F^-	thermal radiation of a disk
$\mathbf{F}_\mathrm{A}$; $\mathbf{F}_\mathrm{em}$	Ampére force; electromagnetic force
$\mathcal{F}_\mathrm{c}$; $\mathcal{F}_\mathrm{em}$	bulk centrifugal force; bulk electromagnetic force
$\mathbf{g}$	gravitational field
g_H	surface gravity
$g_m(r)$	radial eigenfunctions
g_{ik}	metrics
G	gravitational constant
G, G_1, G_2	self-similar functions
h	additional potential
h_1, h_2	components of the additional potential
h_E	Endean nonphysical potential
h_*	coefficient
$\hbar$	Planck constant
$h(\theta)$	dimensionless perturbation of the radial velocity
H	disk half-thickness
H	inner gap height
H	Heyvaerts–Norman invariant
H_cr	critical height
H_RS	Ruderman–Sutherland height
$\mathbf{H}$	gravitomagnetic vector
H_{ik}	gravitomagnetic tensor
i	number of invariants
i_0	dimensionless longitudinal current
i_A	asymmetric dimensionless longitudinal current
i_S	symmetric dimensionless longitudinal current
$i_\parallel$	current coefficient
I; I_bulk; I_GJ; I_M; I_sep	electric current; bulk; GJ; Michel current; current along the separatrix
$\mathcal{I}_\mathrm{cur}$	invariant for current losses
$\mathcal{I}_\mathrm{md}$	invariant for magnetodipole losses
I_r	moment of inertia of a star
$\mathbf{j}$; $\mathbf{j}_\mathrm{p}$	current density; poloidal current density
$j_\parallel$	velocity coefficient
$j_\parallel$	current density parallel to the magnetic field
j_back	back current density
j_GJ	Goldreich–Julian current density
$\mathbf{J}$	angular moment
J	self-similar function
$J_0(x)$, $J_1(x)$	Bessel functions
$\mathbf{J}_\mathrm{H}$	horizon surface current
$\mathbf{J}_\mathrm{s}$	surface current
k	constants
K	angular momentum losses
K; K_n	functions
K_tot	total angular momentum losses

Table E.3

K_1, K_2	components of the $K_\perp$
$K_0(x)$, $K_1(x)$	Macdonald functions
$\mathbf{K}$	torque
$K_\parallel$, $K_\perp$, $K_\dagger$	components of the torque
l	dimensionless sound velocity
$l(\theta)$	perturbation of the electric current
l_γ	photon free path
$l_{\rm acc}$	acceleration path
L; $L_{\rm cr}$; $L_{\rm n}$	specific angular momentum; critical; nonrelativistic
L	luminosity
L	characteristic length
$\mathcal{L}_{\rm ph}$	photon angular momentum
$\hat{\mathcal{L}}$	GS operator
$\hat{\mathcal{L}}_2$	vacuum operator
$\hat{\mathcal{L}}_{\rm psr}$	pulsar operator
$\hat{\mathcal{L}}_{\boldsymbol{\beta}}$	Lie derivative
m	self-similar parameter (Alfvén)
$m(\theta)$	asymptotic function
$m_{\rm e}$	electron mass
$m_{\rm p}$	particle mass
$\mathbf{m}$	magnetic moment of a neutron star
M, M_*	star mass
$M_{\rm bulge}$	bulge mass
$M_\odot$	solar mass
$\dot{M}$	mass accretion/ejection rate
$\mathcal{M}$; $\mathcal{M}_0$; $\mathcal{M}_{\rm f}$	Alfvénic Mach number; zero approximation; on the fast magnetosonic surface
$\hat{\mathcal{M}}$; $\hat{\mathcal{M}}_\infty$; $\hat{\mathcal{M}}_*$	hydrodynamic "Mach number"; at infinity; at the sonic surface
n; $n_{\rm GJ}$; $n_{\rm in}$; $n_{\rm inj}$; n_R; n_*; n_∞	concentration; GJ; initial; injection; on the star surface; at the sonic surface; at infinity
n	self-similar parameter (fast magnetosonic)
$\mathbf{n}_1$, $\mathbf{n}_2$, $\mathbf{n}_\perp$	unit vectors
$n_{\rm br}$	braking index
N	numerators
N^β	particle flux
$p(\theta)$	dimensionless perturbation of the sonic enthalpy
P	pressure
P; $\dot{P}$; $P_{\rm cr}$; $P_{\rm crit}$	pulsar period; period derivative; critical period; nonrelativistic
P	parameter
$P_{\rm n}$; $P_{\rm s}$	anisotropic pressure
$\mathcal{P}$	Legendre polynomials
$\mathcal{P}_t$	dissipation-free energy exchange
q; $q_{\rm f}$	particle-to-electromagnetic flux ratio; on the Alfvén surface
$q(\theta)$	dimensionless perturbation of the sonic concentration
Q	pulsar parameter
Q_*	electric charge of a neutron star
Q_{ik}	quadrupole electric moment
$Q_m(\theta)$	angular eigenfunction
r	radial coordinate
r_*; $r_{\rm inj}$; $r_{\rm d}$	sonic radius; injection radius; shell radius
r_0	radius of marginally stable orbit
$r_{\rm a}$; $r_{\rm A}$	Alfvén radius
r_B	Larmor radius

Table E.4

r_e	the classical electron radius
r_{eq}	equipartition radius
r_f	radius of the fast magnetosonic surface
r_g	black hole radius
r_{in}	inner radius of the hollow cone
r_{jet}; r_{rad}	jet radius; radiating radius
R, R_*	star radius
R	dimensionless distance from the sonic surface
R_0	polar cap radius
R_c	curvature radius
R_{cap}	capture radius
R_{in}	radius of the compact object
R_L	light cylinder radius
$R_\lambda(\varpi)$	radial function
R_t, R_t'	radial scale
$\mathcal{R}$	gas constant
$\mathcal{R}$	horizon resistance
s; s_∞	specific entropy; at infinity
s_1, s_2, S	anisotropic integrals
s'	number of critical surfaces
S	self-similar function
S, S_{cap}	area, polar cap area
S_H	black hole area
$\mathbf{S}$	energy flux (Poynting vector)
t	time
t	self-similar parameter (shifted fast magnetosonic)
T; T_{br}; T_e; T_i	temperature; brightness temperature; electron temperature; ion temperature
T	self-similar function
$t_{r\varphi}$	stress tensor
T_H	black hole temperature
$T^{\alpha\beta}_{part}$	particle energy–momentum tensor
$T^{\alpha\beta}_{em}$	electromagnetic energy–momentum tensor
u	dimensionless concentration
$\mathbf{u}$	four-velocity
u_A; $u_{A,p}$	Alfvén four-velocity; poloidal component
u_p	poloidal four-velocity
u_φ; $u_{\hat{\varphi}}$	toroidal four-velocity; physical component
u_θ; $u_{\hat{\theta}}$	azimuthal four-velocity; physical component
u_*	four-velocity at the sonic surface
v; v_p; v_r; v_φ; v_θ	velocity; poloidal; radial; toroidal azimuthal;
v_R; v_*; v_K; v_∞; v_{inj}	velocity on the star surface; at the sonic surface; Keplerian; at infinity; injection
$\mathbf{v}$	group velocity
V; V_{crit}; V_g	voltage drop; critical; in the gap
V; $V_{1,...,4}$	phase velocity
V_A	Alfvén velocity
V_{cusp}	cusp velocity
w; w_∞	specific enthalpy; at infinity
w	probability of the one-photon conversion
W_d	radio window width

Table E.5

$w, w_{1,2}$	parameters of the separatrix characteristics
$W; W_{\rm em}; W_{\rm part}; W_{\rm tot}$	energy losses; electromagnetic energy losses; particle energy losses; total
$W_{\rm BZ}$	Blandford–Znajek energy losses
x, y, z	Cartesian coordinates
$x; x_{\rm f}; x_{\rm n}$	dimensionless radius; on the Alfvén surface; nonrelativistic
x_0	dimensionless polar cap radius
$x_{\rm L}$	dimensionless radius of the light cylinder
$x_{\rm m} = \sin\theta_{\rm m}$	dimensionless magnetic pole coordinate
x_r	dimensionless cylindrical coordinate
x_*	sonic surface in planar geometry
X, Y	parabolic coordinates
z	cylindrical coordinate
z_*	position of the zero point
Z	atomic number
$\alpha; \alpha_0; \alpha_{\rm inj}; \alpha_*$	lapse function (gravitational red shift); zero approximation; in the injection region; on the sonic surface
α_1	dimensionless current
$\alpha_{\rm t}$	power
$\alpha_{\rm A}$	lapse function on the internal Alfvén surface
$\alpha_{\rm dyn}$	α-ω dynamo parameter
$\alpha_{\rm fin}$	fine structure constant
$\alpha_{\rm SS}$	Shakura parameter
β, β'	self-similar exponents
β_0	dimensionless potential drop
$\beta_{\rm a}$	anisotropy parameter
$\boldsymbol{\beta}, \beta^i$	Lense–Thirring vector
$\boldsymbol{\beta}_{\rm R}$	corotation vector
$\gamma; \gamma_{\rm in}; \gamma_{\rm inj}; \gamma_\infty; \gamma_{\rm f}$	particle Lorentz factor; initial; injection; at infinity; on the fast magnetosonic surface
Γ	polytropic index
Γ^i_{jk}	Christoffel symbols
$\delta(x); \delta_{ik}$	δ-function; Kronecker delta
Δ	metric coefficient
$\epsilon_{\rm part}; \epsilon_{\rm em}$	energy density of the particles; of the electromagnetic field
ε	energy density
$\varepsilon_i; \varepsilon_{\rm A}; \varepsilon_{\rm g}$	small parameters; small geometrical parameter; small gravitational parameter
η	efficiency
$\eta; \eta_{\rm n}; \eta_0; \eta_{\rm inj}$	particle-to-magnetic flux ratio; nonrelativistic; zero approximation; in the injection region
η_1	logarithmic derivative of the concentration
$\eta(\theta)$	dimensionless perturbation of the concentration at the star surface
θ	spherical coordinate
$\theta_{\rm b}$	an angle between the magnetic field and the rotation axis
Θ	latitude
$\Theta(\theta)$	self-similar function
Θ_m	initial latitude
ϑ	angular distance from the critical point
κ	self-similar constant
κ_m	expansion coefficient
λ	multiplication parameter
λ	expansion function
λ	self-similar constant
$\lambda_{\rm C}$	Compton wavelength
λ_m	expansion coefficient

Table E.6

Λ	logarithmic factor
μ; μ_*; μ_∞	relativistic enthalpy; at the sonic surface; at infinity
$\mu_\parallel$, $\mu_\perp$	coefficients
ν	radiofrequency
ν_m	expansion coefficient
ξ	dimensionless surface electric potential
ξ	self-similar coordinate
ξ'	surface electric potential
ϖ; ϖ_g; ϖ_f	cylindrical coordinate; on the horizon; on the fast magnetosonic surface
ϖ_A; ϖ_c; ϖ_{core}	cylindrical radius of the Alfvén surface; of the light surface; of the central core
ϱ	mass density
ρ_e; ρ_{GJ}	electric charge density; Goldreich–Julian charge density
ϱ_K	metric coefficient
$\varrho(\theta)$	dimensionless perturbation of the star radius
ϱ_m	relativistic energy density
σ	Michel magnetization parameter
σ_H	horizon surface charge density
σ_m	expansion coefficient
σ_s	surface charge density
Σ	metric coefficient
$\Sigma_\perp$; $\Sigma_{\parallel}$	Hall conductivity; Pedersen conductivity
Σ_r; Σ_f; Σ_n	outflow parameter; on the Alfvén surface; nonrelativistic
τ	proper time
$\tau(\theta)$	dimensionless perturbation of the star temperature
τ_D	dynamical age
τ_s	synchrotron lifetime
τ_χ	characteristic age
φ	spherical coordinate
φ_g; φ_{eff}	gravitational potential; effective potential
ϕ	velocity potential for a flat flow
Φ; Φ_{cr}	particle flux; critical
Φ_e; Φ_e^Q	electric potential; of an electric quadrupole
φ	spherical coordinate
χ	inclination angle
χ	self-similar coordinate
ψ	planar hydrodynamical potential
ψ; ψ_{max}; ψ_{in}	potential drop; maximal; inside the star
Ψ	stream function
Ψ_0, Ψ_{tot}	total magnetic flux
Ψ_{erg}; Ψ_*	magnetic flux through the ergosphere; through the black hole horizon
Ψ_{core}; Ψ_{in}	magnetic flux within the central core; within the particle dominated core
ω; ω_A	Lense–Thirring angular velocity; at the Alfvén surface
ω_{cur}	characteristic frequency of the curvature radiation
ω_n	dimensionless nonrelativistic parameter
ω_{syn}	characteristic frequency of the synchrotron radiation
$\omega(\theta)$	dimensionless differential rotation
ω_B	cyclotron frequency
Ω; Ω_{cr}; Ω_{crit}; Ω_d; Ω_K	angular velocity; critical; nonrelativistic critical; of a shell; Keplerian
$\dot{\Omega}$	angular velocity derivative
Ω_F	"field" angular velocity
Ω_H	black hole angular velocity

References

Abramowicz, M.A., Chen, X., Kato, S., Lasota, J.-P., Regev, O.: Thermal equilibria of accretion disks. ApJ **438**, L37–L39 (1995)

Abramowicz, M.A., Czerny, B., Lasota, J.-P., Szuszkiewicz, E.: Slim accretion disks. ApJ **332**, 646–658 (1988)

Abramowicz, M.A., Kato, S.: Constraints for transonic black hole accretion. ApJ **336**, 304–312 (1988)

Abramowicz, M.A., Kluźniak, W., Lasota, J.-P.: No observational proof of the black-hole event-horizon. A&A **396**, L31–L34 (2002)

Abramowicz, M.A., Lanza, A., Percival, M.J.: Accretion disks around Kerr black holes: vertical equilibrium revisited. ApJ **479**, 179–183 (1997)

Abramowicz, M.A., Lasota, J.-P., Igumenshchev, I.V.: On the absence of winds in advection-dominated accretion flows. MNRAS **314**, 775–781 (2000)

Abramowicz, M.A., Zurek, W.: Rotation-induced bistability of transonic accretion onto a black hole. ApJ **246**, 314–320 (1981)

Abrassart, A., Czerny, B.: Toy model of obscurational variability in active galactic nuclei. A&A **356**, 475–489 (2000)

Adler, S.L.: Photon splitting and photon dispersion in a strong magnetic field. Ann. Phys. **67**, 599–647 (1971)

Agapitou, V., Papaloizou, C.B.: Accretion disc-stellar magnetosphere interaction: field line inflation and the effect on the spin-down torque. MNRAS **317**, 273–288 (2000)

Akerlof, C., Balsano, R., Barthelemy, S. et al.: Observation of contemporaneous optical radiation from a gamma-ray burst. Nature **398**, 400–402 (1999)

Akhiezer, A.I., Akhiezer, I.A., Polovin, R.V., Sitenko, A.G., Stepanov K.N.: Plasma Electrodynamics. Pergamon Press, Oxford (1975)

Al'ber, Ya.I., Krotova, Z.N., Eidman, V.Ya.: Cascade process in strong magnetic and electric fields under astrophysical conditions. Astrophysics **11**, 189–195 (1975)

Alfven, H., Fälthammar, C.-G.: Cosmic electrodynamics. Clarendon press, Oxford, (1963)

Allen, M.P., Horvath, J.E.: Implications of a constant observed braking index for young pulsars' spin-down. ApJ **488**, 409–412 (1997)

Alpert, Ya.L, Gurevich, A.V. Pitaevsky, L.P.: Space Physics with Artificial Satellites. Consultants Bureau, New York (1965)

Anderson, M.: Sonic surfaces for adiabatic flow in a stationary axisymmetric spacetime. MNRAS **239**, 19–54 (1989)

Anderson, J.M., Li, Z.-Y., Krasnopolsky, R., Blandford, R.: Locating the launching region of T Tauri winds: the case of DG Tauri. ApJ **590**, L107–109 (2003)

Anderson, J.M., Li, Z.-Y., Krasnopolsky, R., Blandford, R.: The structure of magnetocentrifugal winds. I. Steady mass loading. ApJ, **630**, 945–957 (2005)

Appl, S., Camenzind, M.: The stability of current-carrying jets. A&A **256**, 354–370 (1992)

Appl, S., Camenzind, M.: The structure of relativistic MHD jets: a solution to the nonlinear Grad-Shafranov equation. A&A **274**, 699–706 (1993)

Ardavan, H.: The pulsar equation including the inertial term—its first integrals and its Alfvenic singularity. ApJ **204**, 889–896 (1976)
Ardavan, H.: The stellar-wind model of pulsar magnetospheres. MNRAS **189**, 397–412 (1979)
Arons, J.: Pair creation above pulsar polar caps—steady flow in the surface acceleration zone and polar CAP X-ray emission. ApJ **248**, 1099–1116 (1981)
Arons, J.: General relativistic polar particle acceleration and pulsar death. In: Shibazaki, N., Kawai, N., Shibata, S., Kifune, T. (eds.) Neutron Stars and Pulsars : Thirty Years after the Discovery, pp. 339–351. Universal Academy Press, Tokio (1998)
Arons, J., Scharlemann, E.T.: Pair formation above pulsar polar caps—structure of the low altitude acceleration zone. ApJ **231**, 854–879 (1979)
Arlt, R., Rüdiger, G.: Global accretion disk simulations of magneto-rotational instability. A&A **374**, 1035–1048 (2001)
Artemova, I.V., Bisnovatyi-Kogan, G.S., Igumenshchev, I.V., Novikov, I.D.: On the structure of advective accretion disks at high luminosity. ApJ **549**, 1050–1061 (2001)
Asséo, E., Beaufils, D.: Role of the pressure anisotropy in the relativistic pulsar wind. Ap&SS **89**, 133–141 (1983)
Asséo, E., Kennel, F.C., Pellat, R.: Synchro-Compton radiation damping of relativistically strong linearly polarized plasma waves. A&A **65**, 401–408 (1983)
Asséo, E., Khechinashvili, D.: The role of multipolar magnetic fields in pulsar magnetospheres. MNRAS **334**, 743–759 (2002)
Baade, W., Zwicky, F.: On super-novae. Proc. Nat. Acad. Sci. **20**, 254–258 (1934)
Bacciotti, F., Ray, T.P., Mundt, R. , Eislöffel, J., Solf, J.: Space Telescope/STIS spectroscopy of the optical outflow from DG Tauri: Indications for rotation in the initial jet channel. ApJ **576**, 222–231 (2002)
Balbus, S.A., Hawley, J.F.: A powerful local shear instability in weakly magnetized disks. ApJ **376**, 214–233 (1991)
Balbus, S.A., Hawley, J.F.: Instability, turbulence, and enhanced transport in accretion disks. Rev. Mod. Phys. **70**, 1–53 (1998)
Bardeen, J.M., Petterson, J.A.: The Lense-Thirring effect and accretion disks around Kerr black hole. ApJ **195**, L65–68 (1975)
Bardou, A., Heyvaerts, J.: Interaction of a stellar magnetic field with a turbulent accretion disk. A&A **307**, 1009–1022 (1996)
Baring, M.G., Harding, A.K.: Magnetic photon splitting: computations of proper-time rates and spectra. ApJ **482**, 372–376 (1997)
Baring, M.G., Harding, A.K.: Radio-quiet pulsars with ultrastrong magnetic fields. ApJ **507**, L55–58 (1998)
Barkov, M.V., Komissarov, S.S.: Magnetic acceleration of ultra-relativistic GRB and AGN jets. Intern. J. Mod. Phys. D **17**, 1669–1675 (2008)
Baykal, A., Alpar, A.M., Boynton, P.E., Deeter, J.E.: The timing noise of PSR 0823+26,1706-16, 1749-28, 2021+51 and the anomalous brakingindices. MNRAS **306**, 207–212 (1999)
Baym, G., Pethick, C.: Physics of neutron stars. ARA&A **17**, 415–443 (1979)
Becker, P.A., Subramanian, P., Kazanas, D.: Relativistic outflows from advection-dominated accretion disks around black holes. ApJ **552**, 209–220 (2001)
Beckert, T.: Advection-dominated accretion with infall and outflows. ApJ **539**, 223–234 (2000)
Begelman, M.C., Blandford, R.D., Rees, M.J.: Theory of extragalactic radio sources. Rev. Mod. Phys. **56**, 255–351 (1984)
Begelman, M., Chiueh, T.: Thermal coupling of ions and electrons by collective effects in two-temperature accretion flows. ApJ **332**, 872–890 (1988)
Begelman, M.C., Li, Zh.-Yu.: Asymptotic domination of cold relativistic MHD winds by kinetic energy flux. ApJ **426**, 269–278 (1994)
Begelman, M., Rees, M.J.: The fate of dense stellar systems. MNRAS **185**, 847–860 (1978)
Beloborodov, A.M.: Super-Eddington accretion discs around Kerr black holes. MNRAS **297**, 739–746 (1998)

Berestetsky, V.B., Lifshits, E.M., Pitaevsky L.P.: Relativistic Quantum Theory. Pergamon, Oxford (1982)

Berezinsky, V.S., Dokuchaev, V.I.: Hidden source of high-energy neutrinos in collapsing galactic nucleus. Astropart. Phys. **15**, 87–96 (2001)

Beskin, V.S.: Pair creation in a strong magnetic field. Astrophysics **8**, 266–272 (1982)

Beskin, V.S.: General relativity effects on electrodynamic processes in radio pulsars. Sov. Astron. **16**, 286–293 (1990)

Beskin, V.S.: Axisymmetric stationary flows in compact astrophysical objects. Phys. Uspekhi **40**, 659–688 (1997)

Beskin, V.S.: Radio pulsars. Phys. Uspekhi **42**, 1071–1095 (1999)

Beskin, V.S.: Axisymmetric steady flows in astrophysics. Phys. Uspekhi **46**, 1209–1211 (2003)

Beskin, V.S., Gurevich, A.V., Istomin, Ya.N.: The electrodynamics of a pulsar magnetosphere. Sov.Phys. JETP **58**, 235–253 (1983)

Beskin, V.S., Gurevich, A.V., Istomin, Ya.N.: Spin-down of pulsars by the current — comparison of theory with observations. Ap&SS **102**, 301–326 (1984)

Beskin, V.S., Gurevich, A.V., Istomin, Ya.N.: Theory of the radio emission of pulsars. Ap&SS **146**, 205–281 (1988)

Beskin, V.S., Gureich, A.V., Istomin, Ya.N.: Physics of the Pulsar Magnetosphere. Cambridge University Press, Cambridge (1993)

Beskin, V.S., Istomin, Ya.N., Pariev, V.I.: On pair creation in a black hole magnetosphere. In: Roland, J., Sol, H., Pelletier, G. (eds.) Extragalactic Radio Sources—From Beams to Jets. pp. 45-51. Cambridge University Press, Cambridge (1992a)

Beskin, V.S., Istomin, Ya.N., Pariev, V.I.: Filling the magnetosphere of a supermassive black hole with plasma. Sov. Astron. **36**, 642–649 (1992b)

Beskin, V.S.; Kompaneetz, R.Yu.; Tchekhovskoy, A.D.: The Two-dimensional structure of thin accretion disks. Astron. Lett. **28**, 543–552 (2002)

Beskin, V.S.; Kuznetsova, I.V.: Continuity of magnetohydrodynamic flows near singular points. JETP **86**, 421–428 (1998)

Beskin, V.S., Kuznetsova, I.V.: On the Blandford-Znajek mechanism of the energy loss of a rotating black hole. Nuovo Cimento B, **115**, 795–814 (2000a)

Beskin, V.S., Kuznetsova, I.V.: Grad-Shafranov equation with anisotropic pressure. ApJ **541**, 257–260 (2000b)

Beskin, V.S., Kuznetsova, I.V., Rafikov, R.R.: On the MHD effects on the force-free monopole outflow. MNRAS **299**, 341–348 (1998)

Beskin, V.S., Malyshkin, L.M.: Hydrodynamic accretion of gas with angular momentum onto a black hole. Astron. Lett. **22**, 475-481 (1996)

Beskin, V.S., Malyshkin, L.M.: On the self-consistent model of an axisymmetric radio pulsar magnetosphere. MNRAS **298** 847–853 (1998)

Beskin, V.S., Malyshkin, L.M.: On the internal structure of relativistic jets. Astron. Lett. **26** 208–218 (2000)

Beskin, V.S., Nokhrina, E.E.: Weakly radiating radio pulsars. Astron. Lett. **30**, 685-693 (2004)

Beskin, V.S., Nokhrina, E.E.: The effective acceleration of plasma outflow in the paraboloidal magnetic field. MNRAS **367**, 375–386 (2006)

Beskin, V.S., Nokhrina, E.E.: On the role of the current loss in radio pulsar evolution. Ap&SS **308**, 569–573 (2007)

Beskin, V.S., Nokhrina, E.E.: On the central core in MHD winds and jets. MNRAS **397**, 1486–1497 (2009)

Beskin V.S., Okamoto I.: On the magnetohydrodynamic decollimation in compact objects. MNRAS **313**, 445–453 (2000)

Beskin, V.S., Pariev, V.I.: Axially symmetric steady-state flows in the vicinity of a Kerr black hole and the nature of the activity of galactic nuclei. Phys. Uspekhi **36**, 529–539 (1993)

Beskin, V.S., Pidoprygora, Yu.N.: Hydrodynamic accretion onto black holes. JETP **80** 575–585 (1995)

Beskin, V.S., Pidoprygora, Yu.N.: The direct problem for two-dimensional transonic outflow of gas from the surface of a slowly-rotating star. Astron. Rep. **42**, 71–80 (1998)

Beskin V.S., Rafikov R.R.: On the particle acceleration near the light surface of radio pulsars. MNRAS **313**, 433–444 (2000)

Beskin, V.S., Tchekhovskoy, A.D.: Two-dimensional structure of thin transonic discs: Theory and observational manifestations. A&A **433**, 619–628 (2005)

Beskin, V.S., Zakamska, N.L., Sol, H.: Radiation drag effects on magnetically dominated outflows around compact objects. MNRAS **347**, 587–600 (2004)

Bialynicka-Birula, Z., Bialynicka-Birula, I.: Nonlinear effects in quantum electrodynamics. Photon propagation and photon splitting in an external field. Phys. Rev. D **2**, 2341–2345 (1970)

Bičák, J., Dvořák, L.: Stationary electromagnetic fields around black holes. II. General solutions and the fields of some special sources near a Kerr black hole. Gen. Rel. Grav. **7**, 959–983 (1976)

Bisnovatyi-Kogan, G.S.: Evolution of dense star clusters. Sov. Astron. Lett. **4**, 69–71 (1978)

Bisnovatyi-Kogan, G. S.: Stellar Physics. Springer, Berlin (2001)

Bisnovatyi-Kogan, G.S., Blinnikov, S.I.: Disk accretion onto a black hole at subcritical luminosity. A&A **59**, 111–125 (1977)

Bisnovatyi-Kogan, G.S., Blinnikov, S.I.: Spherical accretion on to compact X-ray sources with preheating—no thermal limit for the luminosity. MNRAS **191**, 711–719 (1980)

Bisnovatyi-Kogan, G.S., Kazhdan, Ya. M., Klypin, A.A., Lutskii, A.E., Shakura, N.I.: Accretion onto a rapidly moving gravitating center. Sov. Astron. **23**, 201–209 (1979)

Bisnovatyi-Kogan, G.S., Lovelace, R.V.E.: Influence of ohmic heating on advection-dominated accretion flows. ApJ **486**, L43–47 (1997)

Bisnovatyi-Kogan, G.S., Lovelace, R.V.E.: Advective accretion disks and related problems including magnetic fields. New Astron. Rev. **45**, 663–742 (2001)

Bisnovatyi-Kogan, G.S., Ruzmaikin, A.A.: The accretion of matter by a collapsing star in the presence of a magnetic field. Ap&SS **28**, 45–59 (1973)

Bjorkman, J.E., Cassinelli, J.P.: Equatorial disk formation around rotating stars due to Ram pressure confinement by the stellar wind. ApJ **409**, 429–449 (1993)

Blackman, E.G.: On particle energization in accretion flows. MNRAS **302**, 723–730 (1999)

Blandford, R.D.: Amplification of radiation by relativistic particles in a strong magnetic field. MNRAS **170**, 551–557 (1975)

Blandford, R.D.: Accretion disc electrodynamics—a model for double radio sources. MNRAS. **176**, 468–481 (1976)

Blandford, R.D.: To the lighthouse. In: Gilfanov, M., Sunyaev, R., Churazov, E. (eds.) Lighthouses of the universe: the most luminous celestial objects and their use for cosmology, pp. 381. Springer, Berlin (2002)

Blandford, R.D., Begelman, M.C.: On the fate of gas accreting at a low rate on to a black hole. MNRAS **303**, L1–L5 (1999)

Blandford, R.D., Payne, D.G.: Hydromagnetic flows from accretion discs and the production of radio jets. MNRAS **199**, 883–903 (1982)

Blandford, R.D., Rees, M.J.: A 'twin-exhaust' model for double radio sources. MNRAS **169**, 395–415 (1974)

Blandford, R.D., Romani, R.W.: On the interpretation of pulsar braking indices. MNRAS **234**, 57P–60P (1988)

Blandford, R.D., Znajek, R.L.: Electromagnetic extraction of energy from Kerr black holes. MNRAS **179**, 433–456 (1977)

Blinnikov, S.I., Novikov, I.D., Perevodchikova, T.V., Polnarev, A.G.: Exploding neutron stars in close binaries. Sov.Astron. Lett. **10**, 177–179 (1984)

Bogovalov, S.V.: Particle acceleration and gamma-ray production near the light cylinder of an axially symmetric rotator. Sov. Astron. Lett. **16**, 362–367 (1990)

Bogovalov, S.V.: Poloidal field of a slowly rotating axisymmetric rotator. Sov. Astron. Lett. **18**, 337–340 (1992)

Bogovalov, S.V.: On the theory of magnetohydrodynamic winds from a magnetosphere of axisymmetric rotators. MNRAS **270**, 721–733 (1994)

Bogovalov, S.V.: Formation of jets during the ejection of plasma by an axisymmetric rotator. Astron. Lett. **21**, 565–571 (1995)

Bogovalov, S.V.: Plasma flow in the magnetosphere of an axisymmetric rotator. MNRAS **280**, 39–52 (1996)

Bogovalov, S.V.: Boundary conditions and critical surfaces in astrophysical MHD winds. A&A **323**, 634–643 (1997a)

Bogovalov, S.V.: Acceleration of relativistic plasma in the magnetosphere of an axisymmetric rotator. A&A. **327**, 662–670 (1997b)

Bogovalov, S.V.: Magnetic collimation of astrophysical winds. Astron. Lett. **24**, 321–331 (1998)

Bogovalov, S.V.: On the physics of cold MHD winds from oblique rotators. A&A **349**, 1017–1026 (1999)

Bogovalov, S.V.: Acceleration and collimation of relativistic plasmas ejected by fast rotators. A&A **371**, 1155–1168 (2001)

Bogovalov, S.V., Khangoulyan, D.V.: The Crab Nebula: interpretation of Chandra observations. Astron. Lett. **28**, 373–385 (2002)

Bogovalov, S.V., Khangoulyan, D.V.: The role of a magnetic field in the formation of jet-like features in the Crab Nebula. Astron. Lett. **29**, 495–501 (2003)

Bogovalov, S.V., Tsinganos K.: On the magnetic acceleration and collimation of astrophysical outflows. MNRAS **305**, 211–224 (1999)

Bogovalov, S.V., Tsinganos, K. Shock formation at the magnetic collimation of relativistic jets. MNRAS **357**, 918–928 (2005)

Bondi, H.: On spherically symmetrical accretion. MNRAS **112**, 195–204 (1952)

Bondi, H., Hoyle, F.: On the mechanism of accretion by stars. MNRAS **104**, 272–282 (1944)

Brandenburg, A., Sokoloff, D.D.: Local and nonlocal magnetic diffusion and alpha-effect tensors in shear flow turbulence. Geophys. Ap. Fluid Dyn. **96**, 319–344 (2002)

Brandenburg, A.: Testing Cowling's antidynamo theorem near a rotating black hole. ApJ **465**, L115–L119 (1996)

Brandenburg, A., Nordlund, A., Stein, R.F., Torkelsson, U.: Dynamo-generated turbulence and large-scale magnetic fields in a Keplerian shear flow. ApJ **446**, 741–754 (1995)

Braginsky, S.I.: Transport processes in a plasma. In: Leontovich, M.A. (ed.) Reviews of Plasma Physics, **1**, pp. 201–292. Consultants Bureau, New York (1965)

Breitmoser, E., Camenzind M.: Collimated outflows of rapidly rotating young stellar objects. Wind equation, GSS equation and collimation. A&A **361**, 207–225 (2000)

Brown, J.C., Taylor, A.R.: The structure of the magnetic field in the outer Galaxy from rotation measure observations through the disk. ApJ **563**, L31–L34 (2001)

Bromberg, O., Levinson, A.: Hydrodynamic collimation of relativistic outflows: Semianalytic solutions and application to gamma-ray bursts. ApJ **671**, 678–688 (2007)

Cao, X., Spruit, H.C.: Magnetically driven wind from an accretion disk with low-inclination field lines. A&A **287**, 80–86 (1994)

Camenzind M.: Hydromagnetic flows from rapidly rotating compact objects. I - Cold relativistic flows from rapid rotators. A&A **162**, 32–44 (1986)

Camenind, M.: Compact Objects in Astrophysics. Springer, Heidelberg, (2007)

Campbell, C.G.: Launching of accretion disc winds along dynamo-generated magnetic fields. MNRAS **310**, 1175–1184 (1999)

Campbell, C.G., Papaloizou, J.C.B., Agapitou, V.: Magnetic field bending in accretion discs with dynamos. MNRAS **300**, 315–320 (1998)

Celotti, A., Blandford, R.D.: On the formation of jets. In: Kaper, L., van den Heuvel, E.P.J., Woudt, P.A. (eds.) Black Holes in Binaries and Galactic Nuclei: Diagnostics, Demography and Formation, p. 206. Springer, Berlin (2001)

Celotti, A., Fabian, A.C., Rees, M.J.: Dense thin clouds in the central regions of active galactic nuclei. MNRAS **255**, 419–422 (1992)

Celotti, A., Rees, M.J.: Reprocessing of radiation by multi-phase gas in low-luminosity accretion flows. MNRAS **305**, L41–L45 (1999)

Chakrabarti, S.K.: Theory of Transonic Astrophysical Flows. World Scientific, Singapore (1990)

Chen H.H., Ruderman M.A., Sutherland P.G.:Structure of solid iron in superstrong neutron-star magnetic fields. ApJ **191**, 473–478 (1974)

Chen, K., Ruderman, M.A., Zhu, T.: Millisecond pulsar alignment: PSR J0437-4715. ApJ **493**, 397–403 (1998)

Chen, X., Abramowicz, M.A., Lasota, J.-P.: Advection-dominated accretion: global transonic solutions. ApJ **476**, 61–69 (1997)

Cheng, A.Y.S., O'Dell, S.L.: Compton losses, Compton rockets. ApJ **251**, L49–L51 (1981)

Cheng, K.S., Ho, C., Ruderman, M.A.: Energetic radiation from rapidly spinning pulsars. I – Outer magnetosphere gaps. II – VELA and Crab. ApJ **300**, 500–539 (1986)

Cheng, K.S., Ruderman, M.A., Zhang, L.: A three-dimensional outer magnetospheric gap model for gamma-ray pulsars: geometry, pair production, emission morphologies, and phase-resolved spectra. ApJ **537**, 964–976 (2000)

Chew, G.F., Goldberger, M.L., Low, F.E.: The Boltzmann equation and the one-fluid hydrodynamic equations in the absence of particles collisions. Proc. R. Soc. **236**, 112–118 (1956)

Chiang, J., Romani, R.W.: An outer gap model of high-energy emission from rotation-powered pulsars. ApJ **436**, 754–761 (1994)

Chistyakov, M.V., Kuznetsov, A.V., Mikheev, N.V.: Photon splitting above the pair creation threshold in a strong magnetic field. Phys. Lett. B **434**, 67–73 (1998)

Chitre, D.M., Vishveshwara, C.V.: Electromagnetic field of a current loop around a Kerr black hole. Phys. Rev. D **12**, 1538–1543 (1975)

Chiueh, T., Li, Zh.-Yu., Begelman, M.C.: Asymptotic structure of hydromagnetically driven relativistic winds. ApJ **377**, 462–466 (1991)

Chiueh, T., Li, Zh.-Yu., Begelman, M.C.: A critical analysis of ideal magnetohydrodynamic models for Crab-like pulsar winds. ApJ **505**, 835–843 (1998)

Chrysostomou, A., Lucas, P.W., Hough, J.H.: Circular polarimetry reveals helical magnetic fields in the young stellar object HH135-136. Nature **450**, 71–73 (2007)

Clemmow, P.C., Dougherty, J.P.: Electrodynamics of Particles and Plasmas. Addison-Wesley, London (1969)

Coffey, D., Bacciotti, F., Ray, T.P., Eislöffel, J., Woitas, J.: Further indications of jet rotation in new ultraviolet and optical Hubble Space Telescope STIS spectra. ApJ **663**, 350–364 (2007)

Colgate, S.: Stellar coalescence and the multiple supernova interpretation of Quasi-Stellar Sources. ApJ **150**, 163–192 (1967)

Contopoulos, I.: The coughing pulsar magnetosphere. A&A **442**, 579–586 (2005)

Contopoulos, I., Kazanas, D., Fendt, C.: The axisymmetric pulsar magnetosphere. ApJ **511**, 351–558 (1999)

Contopoulos, J., Lovelace, R.V.E.: Magnetically driven jets and winds: Exact solutions. ApJ **429**, 139–152 (1994)

Coroniti, F.V.: Magnetically striped relativistic magnetohydrodynamic winds – The Crab Nebula revisited. ApJ **349**, 538–545 (1990)

Daigne, F., Drenkhahn, G.: Stationary equatorial MHD flows in general relativity. A&A **381**, 1066–1079 (2002)

Daigne, F., Font, J.A.: The runaway instability of thick discs around black holes - II. Non-constant angular momentum discs. MNRAS **349**, 841–868 (2004)

Das, T.K., Pendharkar, J.K., Mitra, S.: Multitransonic black hole accretion disks with isothermal standing shocks. ApJ **592**, 1078–1088 (2003)

Daugherty, J.K., Harding, A.K.: Electromagnetic cascades in pulsars. ApJ **252**, 337–347 (1982)

Daugherty, J.K., Harding, A.K.: Pair production in superstrong magnetic fields. ApJ **273**, 761–773 (1983)

Davis, L., Goldstein, M.: Magnetic-dipole alignment in pulsars. ApJ **159**, L81–L86 (1970)

Denton, R.E., Anderson, B.J., Gary, S.P., Fuselier, S.A.: Bounded anisotropy fluid model for ion temperatures. J. Geophys. Res. **99**, 11225–11242 (1994)

Deutsch, A.J.: The electromagnetic field of an idealized star in rigid rotation in vacuo. Ann. d'Astrophys. **18**, 1–10 (1955)

de Villiers, J.-P., Hawley, J.F.: Three-dimensional hydrodynamic simulations of accretion tori in Kerr spacetimes. ApJ **577**, 866–879 (2002)

DiMatteo, T., Fabian, A.: Advectively dominated flows in the cores of giant elliptical galaxies: application to M60 (NGC 4649). MNRAS **286**, L50–L53 (1997)

di Matteo, T., Fabian, A., Rees, M.J., Carilli, C. L., Ivison, R.J.: Strong observational constraints on advection-dominated accretion in the cores of elliptical galaxies. MNRAS **305**, 492–504 (1999)

Djorgovski, S., Evans, C.R.: Photometry and the light curve of the optical counterpart of the eclipsing millisecond pulsar 1957+20. ApJ **335**, L61–L64 (1988)

Dokuchaev, V.I.: Joint evolution of a galactic nucleus and central massive black hole. MNRAS **251**, 564–574 (1991a)

Dokuchaev, V.I.: Birth and life of massive black holes. Sov. Phys. Uspekhi **34**, 447–470 (1991b)

Eichler, D.: Magnetic confinement of jets. ApJ **419**, 111–116 (1993)

Eichler, D., Livio, M., Piran, T., Schramm, D.N.: Nucleosynthesis, neutrino bursts and gamma-rays from coalescing neutron stars. Nature **340**, 126–128 (1989)

Endean, V.G.:The pulsar oblique rotator – Numerical solution of an illustrative problem. MNRAS **204**, 1067–1079 (1983)

Evans, C.R., Kochanek, C.S.: The tidal disruption of a star by a massive black hole. ApJ **346**, L13–L16 (1989)

Faber, S.M., Tremaine, S., Ajhar, E.A. et al.: The tidal disruption of a star by a massive black hole. AJ **114**, 1771–1796 (1997)

Fabian, A.C., Rees, M.J.: The accretion luminosity of a massive black hole in an elliptical galaxy. MNRAS **277**, L55–L57 (1995)

Fawley, W.M., Arons, J., Scharlemann, E.T.: Potential drops above pulsar polar caps – Acceleration of nonneutral beams from the stellar surface. ApJ **217**, 227–243 (1977)

Fender, R.P.: Jets from X-ray binaries. In: Lewin,W.H.G., van den Klis, M.(eds.) Compact Stellar X-ray Sources, pp. 381–419. Cambridge University Press, Cambridge (2006)

Fendt, C.: Collimated jet magnetospheres around rotating black holes. General relativistic force-free 2D equilibrium. A&A **319**, 1025–1035 (1997a)

Fendt, C.: Differentially rotating relativistic magnetic jets. Asymptotic trans-field force-balance including differential rotation. A&A **323**, 999–1010 (1997b)

Fendt, C., Camenzind, M., Appl, S.: On the collimation of stellar magnetospheres to jets. I. Relativistic force-free 2D equilibrium. A&A **300**, 791–907 (1995)

Fendt, C., Elstner, D.: Long-term evolution of a dipole type magnetosphere interacting with an accretion disk. II. Transition into a quasi-stationary spherically radial outflow. A&A **363**, 208–222 (2000)

Fendt, C., Geiner, J.: Magnetically driven superluminal motion from rotating black holes. Solution of the magnetic wind equation in Kerr metric. A&A **369**, 308–322 (2001)

Feretti, L., Fanti, R., Parma, P., Massaglia, S., Trussoni, E., Brinkmann, W.: ROSAT observations of the B2 radio galaxies 1615+35 and 1621+38: implications for the radio source confinement. A&A **298**, 699–710 (1995)

Ferraro, V.C.A.L.: The non-uniform rotation of the Sun and its magnetic field. MNRAS **97**, 458–472 (1937)

Ferreira, J., Dougados, K., Cabrit, S.: Which jet launching mechanism(s) in T Tauri stars? A&A **453**, 785–796 (2006)

Ferreira, J., Pelletier, G.: Magnetized accretion-ejection structures. 1. General statements. A&A **276**, 625–636 (1993a)

Ferreira, J., Pelletier, G.: Magnetized accretion-ejection structures. II. Magnetic channeling around compact objects. A&A **276**, 637–647 (1993b)

Ferreira, J., Pelletier, G.: Magnetized accretion-ejection structures. III. Stellar and extragalactic jets as weakly dissipative disk outflows. A&A **295**, 807–832 (1995)

Fitzpatrick, R., Mestel, L.: Pulsar electrodynamics. MNRAS **232**, 277–321 (1988)

Flowers, E., Lee, J.-F., Ruderman, M.A., Sutherland, P.G., Hillebrandt, W., Müller, E.: Variational calculation of ground-state energy of iron atoms and condensed matter in strong magnetic fields. ApJ **215**, 291–301 (1977)

Frankl', F. I.: On the problems of Chaplygin for mixed sub- and supersonic flows. Izv. Acad. Nauk. USSR Ser. Math. **9**, 121–143 (1945)

Frolov, V.P., Novikov, I.D.: Black Hole Physics. Kluwer Academic Publishers, Dordrecht (1998)

Frolov, V.P., Khokhlov, A.M., Novikov, I.D., Pethick, C.J.: Relativistic tidal interaction of a white dwarf with a massive black hole. ApJ **432**, 680–689 (1994)

Gabuzda, D., Cawthorne, T., Roberts, D., Wardle, J.: A survey of the milliarcsecond polarization properties of BL Lacertae objects at 5 GHz. ApJ **338**, 40–54 (1992)

Galeev, A.A., Rosner, R., Vaiana, G.S.: Structured coronae of accretion disks. ApJ **229**, 318–326 (1979)

Gallant, Y.A., Arons, J.: Structure of relativistic shocks in pulsar winds: A model of the wisps in the Crab Nebula. ApJ **435**, 230–260 (1994)

Gammie, C.F., Popham, R.: Advection-dominated accretion flows in the Kerr metric. I. Basic equations. ApJ **498**, 313–326 (1998)

Garcia, M.R., McClintock, J.E., Narayan, R. et al.: New evidence for black hole event horizons from Chandra. ApJ **553**, L47–L50 (2001)

Ghisellini, G., Bodo, G., Trussoni, E.: Radiative acceleration by synchrotron sources. ApJ **401**, 87–92 (1992)

Ghisellini, G., Bodo, G., Trussoni, E., Rees, M.J.: Acceleration by synchrotron absorption and superluminal sources. ApJ **362**, L1–L4 (1990)

Ghosh, P.: The structure of black hole magnetospheres – I. Schwarzschild black holes. MNRAS **315**, 89–97 (2000)

Ghosh, P., Abramowicz, M.A.: Electromagnetic extraction of rotational energy from disc-fed black holes – The strength of the Blandford-Znajek process. MNRAS **292**, 887–895 (1997)

Ghosh, P., Lamb, F.: Accretion by rotating magnetic neutron stars. II – Radial and vertical structure of the transition zone in disk accretion. ApJ **232**, 259–276 (1979)

Giacconi, R., Gursky, H., Kellogg, E., Schreier, E., Tananbaum, H.: Discovery of Periodic X-Ray Pulsations in Centaurus X-3 from UHURU. ApJ **167**, L67–L70 (1971)

Gil, Ya., Melikidze, G.: On the formation of inner vacuum gaps in radio pulsars. ApJ **577**, 909–916 (2002)

Ginzburg, V.L.: The magnetic fields of collapsing masses and the nature of superstars. Sov. Phys. Doklady **9**, 329–334 (1964)

Ginzburg, V.L.: Pulsars (theoretical concepts). Sov. Phys. Uspekhi **14**, 83–103 (1971)

Ginzburg, V.L., Zheleznyakov, V.V., Zaitsev, V.V.: Coherent radio-emission mechanisms and magnetic pulsar models. Sov. Phys. Uspekhi **12**, 378–398 (1969)

Ginzburg, V.L., Kirzhniz, D.A.: Superconductivity in white dwarfs and pulsars. Nature **220**, 148–149 (1968)

Ginzburg, V.L., Usov, V.V.: Concerning the atmosphere of magnetic neutron stars (pulsars). JETP Lett. **15**, 196–198 (1972)

Gold, T.: Rotating neutron stars as the origin of the pulsating radio sources. Nature **218**, 731–732 (1968)

Goldreich, P., Julian, W.H.: Pulsar electrodynamics. ApJ **157**, 869–880 (1969)

Goldreich, P., Julian, W.H.: Stellar winds. ApJ **160**, 971–977 (1970)

Goodwin, S.P., Mestel, J., Mestel, L., Wright, G.A.E.: An idealized pulsar magnetosphere: the relativistic force-free approximation. MNRAS **349**, 213–224 (2004)

Grad, H.: Reducible problems in magneto-fluid dynamic steady flows. Rev. Mod. Phys. **32**, 830–847 (1960)

Gruzinov, A.: Radiative efficiency of collisionless accretion. ApJ **501**, 787–791 (1998)

Gruzinov, A.: Power of an axisymmetric pulsar. Phys. Rev. Lett. **94**, 021101 (2005)

Guderley, K.G.: Theorie Schallnaher Stromungen. Springer, Berlin (1957)

Guilbert, P.W., Rees, M.J.: 'Cold' material in non-thermal sources. MNRAS **233**, 475–484 (1988)
Gurevich, A.V., Dimant, Ya.S., Zybin, K.P.: Accretion onto a neutron star in the presence of intense radiation. JETP **77**, 5–13 (1993)
Gurevich, A.V., Istomin, Ya.N.: Electron-positron plasma generation in a pulsar magnetosphere. Sov. Phys. JETP **62**, 1–11 (1985)
Gurevich, A.V., Istomin, Ya.N.: The energy loss of a rotating magnetized neutron star. MNRAS **377**, 1663–1667 (2007)
Gurevich, A.V., Krylov, A.L., Fedorov, E.N.: Inductive interaction of conducting bodies with a magnetized plasma. Sov. Phys. JETP **48**, 1074–1078 (1975)
Gurevich, A.V., Zybin, K.P.: Large-scale structure of the Universe. Analytic theory. Phys. Uspekhi **38**, 687–722 (1995)
Gurevich, A.V., Zybin, K.P., Siorota V.A.: Small-scale structure of dark matter and microlensing. Phys. Uspekhi **40**, 869–898 (1997)
Haardt, F., Marashi, L.: A two-phase model for the X-ray emission from Seyfert galaxies. ApJ **380**, L51–L54 (1991)
Haardt, F., Maraschi, L.: X-ray spectra from two-phase accretion disks. ApJ **413**, 507–517 (1993)
Haardt, F., Maraschi, L., Ghisellini, G.: A model for the X-ray and ultraviolet emission from Seyfert galaxies and galactic black holes. ApJ **432**, L95–L99 (1994)
Haehnelt, M., Kauffmann, G.: A unified model for the evolution of galaxies and quasars. MNRAS **311**, 576–588 (2000)
Haehnelt, M., Rees, M.J.: The formation of nuclei in newly formed galaxies and the evolution of the quasar population. MNRAS **263**, 168–198 (1993)
Haensel, P., Potekhin, A.Y., Yakovlev, D.G.: Neutron Stars. 1: Equation of State and Structure. Springer, New-York (2007)
Halpern, J.P., Marshall, H.L.: A long EUVE observation of the Seyfert galaxy RX J0437.4-4711. ApJ **464**, 760–764 (1996)
Hardee, P.E.: Modeling helical structures in relativistic jets. ApJ **597**, 798–808 (2003)
Hardee, P.E., Rosen, A., Hughes, P.A., Duncan, G.: Time-dependent structure of perturbed relativistic jets. ApJ **500**, 599–609 (1998)
Harding, A.K., Muslimov, A.G.: Particle acceleration zones above pulsar polar caps: electron and positron pair formation fronts. ApJ **508**, 328–346 (1998)
Harding, A.K., Muslimov, A.G.: Pulsar polar cap heating and surface thermal X-ray emission. II. Inverse Compton radiation pair fronts. ApJ **568**, 862-877 (2002)
Haro, G.: Faint stars with strong emission in and around the Orion nebula. AJ **55**, 72–73 (1950)
Hawking, S.W.: Gravitationally collapsed objects of very low mass. MNRAS **152**, 75–78 (1971)
Hawking, S.W.: Black holes from cosmic strings. Phys. Lett. B **231**, 237–239 (1989)
Hawking, S.W., Moss, I.G., Stewart, J.M.: Bubble collisions in the very early universe. Phys. Rev. D **26**, 2681–2693 (1982)
Hawley, J.F., Gammie, C.F., Balbus, S.A.: Local three-dimensional magnetohydrodynamic simulations of accretion disks. ApJ **440**, 742–763 (1995)
Hawley, J.F., Smarr, L.L., Wilson, J.R.: Numerical study of nonspherical black hole accretion. II - Finite differencing and code calibration. ApJ Suppl. **277**, 296–246 (1984)
Heinemann, M., Olbert, S.: Axisymmetric ideal MHD stellar wind flow. J. Geophys. Res. **83**, 2457–2460 (1978)
Helfand, D.J., Gotthelf, E.V., Halpern, J.P.: Vela pulsar and its synchrotron nebula. ApJ **556**, 380–391 (2001)
Henri, G., Pelletier, G.: Relativistic electron-positron beam formation in the framework of the two-flow model for active galactic nuclei. ApJ **383**, L7–L10 (1991)
Henriksen, R.N., Norton, J.A.: Oblique rotating pulsar magnetospheres with wave zones. ApJ **201**, 719–728 (1975)
Herbig, G.H.: The spectrum of the nebulosity surrounding T Tauri. ApJ **111**, 11–14 (1950)
Herrnstein, J., Greenhill, L., Moran, J., Diamond, P.J., Inoue, M., Nakai, N., Miyoshi, M.: VLBA continuum observations of NGC 4258: Constraints on an advection-dominated accretion flow. ApJ **497**, L69–L72 (1998)

Hewish, A., Bell, S.J., Pilkington, J.D. et al.: Observation of a Rapidly Pulsating Radio Source. Nature **217**, 709–713 (1968)

Heyvaerts, J.: Rotating MHD Winds. In: Chiuderi, C., Einaudi, G. (eds.) Plasma Astrophysics, pp. 31–54. Springer, Berlin (1996)

Heyvaerts, J., Norman, J.: The collimation of magnetized winds. ApJ **347**, 1055–1081 (1989)

Heyvaerts, J., Norman, J.: Global asymptotic solutions for relativistic magnetohydrodynamic jets and winds. ApJ **596**, 1240–1255 (2003a)

Heyvaerts, J., Norman, J.: Kinetic energy flux versus poynting flux in magnetohydrodynamic winds and jets: The intermediate regime. ApJ **596**, 1256–1269 (2003b)

Heyvaerts, J., Norman, J.: Global asymptotic solutions for nonrelativistic magnetohydrodynamic jets and winds. ApJ **596**, 1270–1294 (2003c)

Hillebrandt, W., Müller, E.: Matter in superstrong magnetic fields and the structure of a neutron star's surface. ApJ **207**, 589–591 (1976)

Hirotani, K., Iguchi, S., Kimura M., Wajima K.: Pair plasma dominance in the 3C 279 jet on parsec scales. Publ. Astron. Soc. Japan V. **51**, 263–267 (1999)

Hirotani, K., Okamoto, I.: Pair plasma production in a force-free magnetosphere around a supermassive black hole. ApJ **497**, 563–572 (1998)

Hirotani, K., Shibata, S.: One-dimensional electric field structure of an outer gap accelerator - III. Location of the gap and the gamma-ray spectrum. MNRAS **325**, 1228–1240 (2001)

Hirotani, K., Takahashi, M., Nitta, S., Tomimatsu, A.: Accretion in a Kerr black hole magnetosphere – Energy and angular momentum transport between the magnetic field and the matter. ApJ **386**, 455–463 (1992)

Hirotani, K., Tomimatsu, A.: Collimated propagation of magnetohydrodynamic waves near to a Schwarzschild black hole. Publ. Astron. Soc. Japan **46**, 643–651 (1994)

Holloway, N.J.: Angular momentum and energy loss from pulsars. MNRAS **181**, 9P–12P (1977)

Horiuchi, S., Mestel L., Okamoto I.: The axisymmetric non-degenerate black-hole magnetosphere. MNRAS **275**, 1160–1170 (1995)

Hunt, R.: Accretion of gas having specific heat ratio 4/3 by a moving gravitating body. MNRAS **198**, 83–91 (1979)

Hu, Y.Q., Esser, R., Habbal S.R.: A fast solar wind model with anisotropic proton temperature. J. Geophys. Res. **102**, 14661–14676 (1997)

Ichimaru, S.: Bimodal behavior of accretion disks – Theory and application to Cygnus X-1 transitions. ApJ **214**, 840–855 (1977)

Igumenshchev, I.V., Abramowicz, M.A., Narayan, R.: Numerical simulations of convective accretion flows in three dimensions. ApJ **537**, L27–L30 (2000)

Igumenshchev, I.V., Abramowicz, M.A., Novikov, I.D.: Slim accretion discs: a model for ADAF-SLE transitions. MNRAS **298**, 1069–1078 (1998)

Igumenshchev, I.V., Beloborodov, A.M.: Numerical simulation of thick disc accretion on to a rotating black hole. MNRAS **284**, 767–772 (1997)

Ilyasov, Yu.P., Kopeikin, S.M., Rodin, A.E.: The astronomical timescale based on the orbital motion of a pulsar in a binary system. Astron. Lett. **24**, 228–236 (1998)

Ingraham, R.: Algorithm for solving the nonlinear pulsar equation. ApJ **186**, 625–630 (1973)

Istomin, Ya.N., Pariev, V.I.: Stability of a relativistic rotating electron/positron jet. MNRAS **267**, 629–636 (1994)

Istomin, Ya.N., Pariev, V.I.: Stability of a relativistic rotating electron-positron jet: non-axisymmetric perturbations. MNRAS **281**, 1–26 (1996)

Istomin, Ya. N.; Sobyanin, D. N.: Electron-positron plasma generation in a magnetar magnetosphere. Astron. Lett. **33**, 660–672 (2007)

Iwasawa, K., Fabian, A.C., Brandt, W.N. et al.: Detection of an X-ray periodicity in the Seyfert galaxy IRAS 18325-5926. MNRAS **295**, L20-L24 (1998)

Johnston, S., Galloway, D.: Pulsar braking indices revisited. MNRAS **306**, L50–L54 (1999)

Johnston, S., Koribalski, B., Weisberg, J.M., Wilson, W.: Hi line measurements of pulsars towards the Galactic Centre and the electron density in the inner Galaxy. MNRAS **322**, 715–722 (1999)

Jones, P.B.: Pair formation and electric field boundary conditions at neutron star magnetic poles. MNRAS **192**, 847–860 (1980)

Junor, W., Biretta, J.A.: The radio jet in 3C274 at 0.01 PC resolution. AJ **109**, 500–506 (1995)

Junor, W., Biretta, J.A., Livio, M.: Formation of the radio jet in M87 at 100 Schwarzschild radii from the central black hole. Nature **401**, 891–892 (1999)

Kaburaki O.: Analytic model for advection-dominated accretion flows in a global magnetic field. ApJ **531**, 210–216 (2000)

Kato, S., Fukui, J.: Trapped radial oscillations of gaseous disks around a black hole. PASJ **32**, 377–388 (1980)

Kadomtsev, B.B.: Collective phenomena in plasma. Nauka, Moscow (1988)

Kadomtsev, B.B., Kudryavtsev, V.S.: Atoms in a superstrong magnetic field. JETP Lett. **13**, 42–44 (1971)

Kardashev, N.S.: Nature of the Radio Galaxy Cygnus A. Sov. Astron. **7**, 740–747 (1964)

Kardashev, N.S., Mitrofanov, I.G., Novikov, I.D.: Photon -e+/- interactions in neutron star magnetospheres. Sov. Astron. **28**, 651–657 (1984)

Kantor, E.M., Tsygan, A.I.: The electric fields of radio pulsars with asymmetric nondipolar magnetic fields. Astron. Rep. **47**, 613–620 (2003)

Katz J.I.: Yet another model of gamma-ray bursts. ApJ **490**, 633–641 (1997)

Kennel, C.F., Coroniti, F.V.: Confinement of the Crab pulsar's wind by its supernova remnant. ApJ **283**, 694–709 (1984a)

Kennel, C.F., Coroniti, F.V.: Magnetohydrodynamic model of Crab nebula radiation. ApJ **283**, 710–730 (1984b)

Kennel, C.F., Fujimura, F.S., Okamoto, I.: Relativistic magnetohydrodynamic winds of finite temperature. Geophys. Ap Fluid Dyn. **26**, 147–222 (1983)

Khanna, R.: On Cowling's theorem in the Kerr metric. In: Dremin, I.M., Semikhatov, A.M. (eds.) Proceedings of 2nd International Sakharov Conference on Physics, pp. 134–137. World Scientific, Moscow (1997)

Khanna, R.: On the magnetohydrodynamic description of a two-component plasma in the Kerr metric. MNRAS **294**, 673–681 (1998)

Khanna, R., Camenzind, M.: The gravitomagnetic dynamo effect in accretion disks of rotating black holes. ApJ **435**, L129–L132 (1994)

Khanna, R., Camenzind, M.: The ω-dynamo in accretion disks of rotating black holes. A&A **307**, 665–685 (1996a)

Khanna, R., Camenzind, M.: Erratum The ω-dynamo in accretion disks of rotating black holes. A&A **313**, 1028 (1996b)

Khokhlov, A., Novikov, I.D., Pethick, C.J.: Weak tidal encounters of a star with a massive black hole. ApJ **418**, 163–180 (1993)

Khlopov, M.Yu., Konoplich, R.V., Rubin, S.G., Sakharov, A.S.: Phase transitions as a source of black holes. Grav. Cosmol. **6**, 153–156 (2000)

Kim, H., Lee, H.M., Lee, C.H., Lee, H.K: Pulsar magnetosphere: a general relativistic treatment. MNRAS **358**, 998–1018 (2005)

Kirk, J.G., Skjaeraasen, O.: Dissipation in poynting-flux-dominated flows: The σ-problem of the Crab pulsar wind. ApJ **591**, 366–379 (2003)

Kirzhnits, D.A., Yudin, S.N.: Paradoxes of superfluid rotation. Phys. Uspekhi **38**, 1283–1288 (1995)

Kley, W., Papaloizou, J.C.B.: Causal viscosity in accretion disc boundary layers. MNRAS **285**, 239–252 (1997)

Koide, S.: Magnetic extraction of black hole rotational energy: Method and results of general relativistic magnetohydrodynamic simulations in Kerr space-time. Phys. Rev. D **67**, 104010 (2003)

Koide, S., Meier, D.L., Shibata, K., Kudo, T.: General relativistic simulations of early jet formation in a rapidly rotating black hole magnetosphere. ApJ **536**, 668–674 (2000)

Koide, S., Shibata, K., Kudo, T.: Relativistic jet formation from black hole magnetized accretion disks: method, tests, and applications of a general relativistic magnetohydrodynamic numerical code. ApJ **522**, 727–752 (1999)

Komissarov, S.S.: Direct numerical simulations of the Blandford-Znajek effect. MNRAS **326**, L41–L44 (2001)

Komissarov, S.S.: On the nature of the Blandford-Znajek mechanism. In: Semikhatov, A., Zaikin, V., Vasiliev, M. (eds.) Proceedings of 3rd International Sakharov Conference on Physics, pp. 392–400. World Scientific, Moscow (2003)

Komissarov, S.S.: Electrodynamics of black hole magnetospheres. MNRAS **350**, 427–448 (2004a)

Komissarov, S.S.: General relativistic magnetohydrodynamic simulations of monopole magnetospheres of black holes. MNRAS **326**, 1431–1936 (2004b)

Komissarov, S.S.: Observations of the Blandford-Znajek process and the magnetohydrodynamic Penrose process in computer simulations of black hole magnetospheres. MNRAS **359**, 801–808 (2005)

Komissarov, S.S.: Simulations of the axisymmetric magnetospheres of neutron stars. MNRAS **367**, 19–31 (2006)

Komissarov, S.S., Lyubarsky, Yu.E.: The origin of peculiar jet-torus structure in the Crab nebula. MNRAS **344**, L93–L96 (2003)

Komissarov, S.S., McKinney, J.C.: The 'Meissner effect' and the Blandford-Znajek mechanism in conducting black hole magnetosphere. MNRAS **377**, L49–L53 (2007)

Komissarov, S.S., Vlahakis, N., Königl, A., Barkov, M.: Magnetic acceleration of ultra-relativistic jets in gamma-bursts sources. MNRAS **394**, 1182–1212 (2009)

Königl, A.: Self-similar models of magnetized accretion disks. ApJ **342**, 208–223 (1989)

Königl, A., Kartje, J.F.: Disk-driven hydromagnetic winds as a key ingredient of active galactic nuclei unification schemes. ApJ **434**, 446–467 (1994)

Kormendy J., Richstone D.: Inward bound – The search for supermassive black holes in galactic nuclei. ARA&A **33**, 581–624 (1995)

Korn, G.A., Korn,T.M.: Mathematical handbook for scientists and engineers. McGraw-Hill, New York (1968)

Kouveliotou, C., Dieters, S., Strohmayer, T., et al.: An X-ray pulsar with a superstrong magnetic field in the soft γ-ray repeater SGR1806–20. Nature **393**, 235–237 (1998)

Kovalenko, I.G., Eremin, M.A.: Instability of spherical accretion – I. Shock-free Bondi accretion. MNRAS **298**, 861–870 (1998)

Kovalev, Yu. Yu: Intrinsic structure and kinematics of the sub-parsec scale jet of M87. In: Rector, T.A., De Young, D.S. (eds.) Extragalactic Jets: Theory and Observation from Radio to Gamma Ray, pp.155–160. ASP Conf. Ser. 386 (2008)

Kozlenkov, A. A., Mitrofanov, I.G.: Two-photon generation of electron-positron pairs in a strong magnetic field. Sov. Phys. JETP **64**, 1173–1179 (1986)

Kramer, M., Lyne, A.G., O'Brian, J.T., Jordan, C.A., Lorimer, D.R.: A periodically-active pulsar giving insight into magnetospheric physics. Science **312**, 549–551 (2006)

Krasheninnikov, S.I., Catto, P.J.: Effects of pressure anisotropy on plasma equilibrium in the magnetic field of a point dipole. Phys. Plasma **7**, 626–628 (2000)

Krause-Polstorff, J., Michel, F.C.: Electrosphere of an aligned magnetized neutron star. MNRAS **213**, 43P–49P (1995)

Krause-Polstorff, J., Michel, F.C.: Pulsar space charging. A&A **144**, 72–80 (1984)

Krolik, J.: Magnetized accretion inside the marginally stable orbit around a black hole. ApJ **515**, L73–L76 (1999a)

Krolik, J.: Active Galactic Nuclei: from the Central Black Hole to the Galactic Environment. Princeton University Press, Princeton (1999b)

Krolik, J.H., Hawley, J.F.: Where is the inner edge of an accretion disk around a black hole? ApJ **573**, 754–763 (2002)

Kudoh, T., Matsumoto, R., Shibata, K.: Magnetically driven jets from accretion disks. III. 2.5-dimensional nonsteady simulations for thick disk case. ApJ **508**, 186–199 (1998)

Kuncic, Z., Celotti, A., Rees, M.J.: Dense, thin clouds and reprocessed radiation in the central regions of active galactic nuclei. MNRAS **284**, 717–730 (1997)

Kuznetsov, O.A., Lovelace, R.V.E., Romanova, M.M., Chechetkin, V.M.: Hydrodynamic simulations of counterrotating accretion disks. ApJ **514**, 691–703 (1999)

Kuznetsova, I.V.: Relativistic Grad-Shafranov equation with anisotropic pressure. ApJ **618**, 432–437 (2005)
Lada, C.J.: Cold outflows, energetic winds, and enigmatic jets around young stellar objects. ARA&A **23**, 267–317 (1985)
Lammers, H.J.G.L.M., Cassinelli J.P.: Introduction to Stellar Wind. Cambridge University Press, Cambridge (1999)
Landau, L.D., Lifshits, E.M.: Mechanics. Pergamon Press, Oxford (1976)
Landau, L.D., Lifshits, E.M.: Fluid Mechanics. Pergamon Press, Oxford (1987)
Landau, L.D., Lifshits, E.M.: The Classical Theory of Fields. Pergamon Press, Oxford (1989)
Lasota, J.-P., Abramowicz, M.A., Chen, X., Krolik, J., Narayan, R., Yi, I.: Is the accretion flow in NGC 4258 advection dominated? ApJ **462**, 142–146 (1996)
Lacy, J.H., Townes, C.H., Hollenbach, D.J.: The nature of the central parsec of the Galaxy. ApJ **262**, L120–L123 (1982)
Landau, L.D.: On the theory of stars. Phys. Zeit. Sow. **1**, 271–273 (1932)
Lee, H.K., Lee, C.H., van Putten, M.H.P.M.: MNRAS **324**, 781–784 (2001)
Lee, H.K. Park, J.: Two-dimensional Poynting flux dominated flow onto a Schwarzschild black hole. Phys. Rev. D **70**, 063001 (2004)
Lee, H.K., Wijers, R.A.M.J., Brown G.E.: The Blandford-Znajek process as a central engine for a gamma-ray burst. Phys. Rep. **325**, 83–114 (2000)
Leer, L., Axford, W.I.: A two-fluid solar wind model with anisotropic proton temperature. Sol. Phys. **23**, 238–250 (1972)
Lery, T., Heyvaerts, J., Appl, S., Norman, C.A.: Outflows from magnetic rotators. I. Inner structure. A&A **337**, 603–624 (1998)
Lery, T., Heyvaerts, J., Appl, S., Norman, C.A.: Outflows from magnetic rotators. II. Asymptotic structure and collimation. A&A **347**, 1055–1068 (1999)
Li, L.-X.: Evolution of magnetic fields around a Kerr black hole. Phys. Rev. D **67**, 044007 (2003)
Li, Zh.-Yu.: Magnetohydrodynamic disk-wind connection: Self-similar solutions. ApJ **444**, 848–860 (1995)
Li, Zh.-Yu.: Magnetohydrodynamic disk-wind connection: magnetocentrifugal winds from ambipolar diffusion-dominated accretion disks. ApJ **465**, 855–868 (1996)
Li, Zh.-Yu., Begelman, M.C., Chiueh, T.: The effects of radiation drag on radial, relativistic hydromagnetic winds. ApJ **384**, 567–579 (1992)
Li, Zh.-Yu., Chiueh, T., Begelman, M.C.: Electromagnetically driven relativistic jets – A class of self-similar solutions. ApJ **394**, 459–471 (1992)
Liang, E.P.T.: Accretion disk corona and the nature of X-ray burster 3U 1820-30. ApJ **211**, L67–L70 (1977)
Liberman, M.A., Johansson, B.: Properties of matter in ultrahigh magnetic fields and the structure of the surface of neutron stars. Phys. Uspekhi **38**, 117–136 (1995)
Linet, B.: Electrostatics and magnetostatics in the Schwarzschild metric. J. Phys. A **9**, 1081–1087 (1976)
Lipunov, V.M.: Astrophysics of Neutron Stars. Springer, Heidelberg (1992)
Livio, M., Ogilvie, G.I., Pringle, J.E.: Extracting energy from black holes: the relative importance of the Blandford-Znajek mechanism. ApJ **512**, 100–104 (1992)
Lobanov, A., Hardee, P., Eilek, J.: Internal structure and dynamics of the kiloparsec-scale jet in M87. New Astron. Rev. **47**, 629–632 (2003)
Lovelace, R.V.E.: Dynamo model of double radio sources. Nature **262**, 649–652 (1976)
Lovelace, R.V.E., Mehanian, C., Mobarry, C.M., Sulkanen, M.E.: Theory of axisymmetric magnetohydrodynamic flows – Disks. ApJ Suppl. **62**, 1–37 (1986)
Lovelace, R.V.E., Wang, J.C.L., Sulkanen, M.E.: Self-collimated electromagnetic jets from magnetized accretion disks. ApJ **315**, 504–535 (1987)
Low, B.C., Tsinganos, K.: Steady hydromagnetic flows in open magnetic fields. I – A class of analytic solutions. ApJ **302**, 163–187 (1986)
Lubow, S.H., Papaloizou, J.C.B., Pringle, J.E. : Magnetic field dragging in accretion discs. MNRAS **267**, 235–240 (1994)

Lucek, S.G., Bell, A.R.: The stability, during formation, of magnetohydrodynamically collimated jets. MNRAS **290**, 327–333 (1997)
Lynden-Bell, D.: Galactic nuclei as collapsed old quasars. Nature **223**, 690–694 (1969)
Lynden-Bell, D.: Magnetic collimation by accretion discs of quasars and stars. MNRAS **279**, 389–401 (1996)
Lynden-Bell, D.: On why discs generate magnetic towers and collimate jets. MNRAS **341**, 1360–1372 (2003)
Lyne, A.G., Graham-Smith F.: Pulsar Astronomy. Cambridge University Press, Cambridge (1998)
Lyubarskii, Yu.E.: Equilibrium of the return current sheet and the structure of the pulsar magnetosphere. Sov. Astron. Lett. **16**, 16–20 (1990)
Lyubarskii, Yu.E.: Physics of Pulsars. Harwood Acad. Publ., Singapore (1995)
Lyubarskii, Yu.E.: On the electric current distribution in Poynting-dominated outflows. MNRAS **285**, 604–606 (1997)
Lyubarsky, Yu.E., Eichler, D.: The X-ray jet in the Crab Nebula: radical implications for pulsar theory? ApJ **562**, 494–498 (2001)
Lyubarsky, Yu.E., Kirk, J.G.: Reconnection in a striped pulsar wind. ApJ **547**, 437–448 (2001)
Lyutikov, M., Blandford, R.D., Machabeli, G.Z.: On the nature of pulsar radio emission. MNRAS **305**, 338–352 (2001)
Macdonald, D.: Numerical models of force-free black-hole magnetospheres. MNRAS **211**, 313–329 (1984)
Macdonald, D., Thorne K.S.: Black-hole electrodynamics – an absolute-space/universal-time formulation. MNRAS **198**, 345–382 (1982)
Mahadevan, R.: Scaling laws for advection-dominated flows: Applications to low-luminosity galactic nuclei. ApJ **477**, 585–601 (1997)
Mahadevan, R.: Reconciling the spectrum of Sagittarius A* with a two-temperature plasma model. Nature **394**, 651–653 (1998)
Malofeev,V.M., Malov, O.I., Teplykh,D.A.: Radio emission from AXP an XDINS. Ap&SS **308**, 211–216 (2007)
Marochnik, L.S., Suchkov, A.A.: The Milky Way Galaxy. Gordon and Breach, Amsterdam (1996)
Marsch, E., Schwenn, R., Rosenbauer, H., Muehlhaeuser, K.-H., Pilipp, W., Neubauer, F.M.: Solar wind protons – Three-dimensional velocity distributions and derived plasma parameters measured between 0.3 and 1 AU. J. Geophys. Res. **87**, 52–72 (1982)
Max, C., Perkins, F.: Strong electromagnetic waves in overdense plasmas. Phys. Rev. Lett. **27**, 1342–1345 (1971)
McKinney, J.C.: Relativistic force-free electrodynamic simulations of neutron star magnetospheres. MNRAS **368**, L30–L33 (2006a)
McKinney, J.C.: General relativistic magnetohydrodynamic simulations of the jet formation and large-scale propagation from black hole accretion systems. MNRAS **368**, 1561–1582 (2006b)
Medvedev, M.V., Narayan, R.: Axisymmetric self-similar equilibria of self-gravitating isothermal systems. ApJ **541**, 579–586 (2000)
Melatos, A.: Spin-down of an oblique rotator with a current-starved outer magnetosphere. MNRAS **288**, 1049–1059 (1997)
Melrose, D.: Amplified linear acceleration emission applied to pulsars. ApJ **225**, 557–573 (1978)
Mestel, L.: Magnetic braking by a stellar wind. MNRAS **138**, 359–391 (1968)
Mestel, L.: Pulsars – oblique rotator model with dense magnetosphere. Nature Phys. Sci. **233**, 149–152 (1971)
Mestel, L.: Force-free pulsar magnetospheres. Ap&SS **24**, 289–297 (1973)
Mestel, L.: Stellar Magnetism. Clarendon Press, Oxford (1999)
Mestel, L., Panagi, P., Shibata, S.: Model pulsar magnetospheres: the perpendicular rotator. MNRAS **309**, 388–394 (1999)
Mestel, L., Shibata, S.: The axisymmetric pulsar magnetosphere – a new model. MNRAS **271**, 621–638 (1994)

Mestel, L., Wang, Y.-M.: The axisymmetric pulsar magnetosphere. II MNRAS **188**, 799–812 (1979)
Mestel, L., Wang, Y.-M.: The non-aligned pulsar magnetosphere – an illustrative model for small obliquity. MNRAS **198**, 405–427 (1982)
Mészaros, P.: High-Energy Radiation from Magnetized Neutron Stars. University of Chicago Press, Chicago (1992)
Mészaros, P., Rees, M.J.: Poynting jets from black holes and cosmological gamma-ray bursts. ApJ **482**, L29–L32 (1997)
Michel, F.C.: Relativistic stellar-wind torques. ApJ **158**, 727–738 (1969)
Michel, F.C.: Accretion of matter by condensed objects. Ap&SS **15**, 153–160 (1972)
Michel, F.C.: Rotating magnetosphere: a simple relativistic model. ApJ **180**, 207–226 (1973a)
Michel, F.C.: Rotating magnetospheres: an exact 3-D solution. ApJ **180**, L133–L136 (1973b)
Michel, F.C.: Rotating magnetosphere: far-field solutions. ApJ **187**, 585–588 (1974)
Michel, F.C.: Theory of pulsar magnetospheres. Rev. Mod. Phys. **54**, 1–66 (1982)
Michel, F.C.: Theory of Neutron Star Magnetosphere. The University of Chicago Press, Chicago (1991)
Mihalas, D.: Stellar Atmospheres. Freeman & co., San Francisco (1978)
Mirabel, I.F., Rodriguez, L.F., Cordier, B. et al.: A double-sided radio jet from the compact Galactic Centre annihilator 1E140.7–2942. Nature **358**, 215–217 (1992)
Mirabel, I.F., Rodriguez, L.F.: A superluminal source in the Galaxy. Nature **371**, 46–48 (1994)
Mitrofanov, I.G., Pozanenko, A.S.: The generation of radiation in the quantum transitions of electrons in a strong magnetic field. Sov. Phys. JETP **66**, 1112–1120 (1987)
Moderski, R., Sikora, M.: On black hole evolution in active galactic nuclei. MNRAS **283**, 854–864 (1996)
Moderski, R., Sikora, M., Lasota, J.-P.: On the spin paradigm and the radio dichotomy of quasars. MNRAS **301**, 142–148 (1998)
Moss, I.G.: Singularity formation from colliding bubbles. Phys. Rev. D **50**, 676–681 (1994)
Müller, E.: Variational calculation of iron and helium atoms and molecular chains in superstrong magnetic fields. A&A **130**, 145–148 (1984)
Mushotzky, R.F., Done, C., Pounds, K. A.: X-ray spectra and time variability of active galactic nuclei. ARA&A **31**, 717–761 (1993)
Muslimov, A.G., Tsygan, A.I.: Influence of General Relativity effects on electrodynamics in the vicinity of a magnetic pole of a neutron star. Sov. Astron. **34**, 133–137 (1990)
Muslimov, A.G., Tsygan, A.I.: General relativistic electric potential drops above pulsar polar caps. MNRAS **255**, 61–70 (1992)
Narayan, R.: A flux-limited model of particle diffusion and viscosity. ApJ **394**, 261–267 (1992)
Narayan, R., Kato, S., Honma, F.: Global structure and dynamics of advection-dominated accretion flows around black holes. ApJ **479**, 49–60 (1997)
Narayan, R., Loeb, A., Kumar, P.: Casuality in strong shear flows. ApJ **431**, 359–279 (1994)
Narayan, R., Mahadevan, R., Grindlay, J.E., Popham, R.G., Gammie, C.: Advection-dominated accretion model of Sagittarius A*: evidence for a black hole at the Galactic center. ApJ **492**, 554–568 (1998)
Narayan, R., Mahadevan, R., Quataert, E.: Advection-dominated accretion around black holes. In: Abramowicz, M.A., Bjornsson, G., Pringle, J.E. (eds.) The Theory of Black Hole Accretion Disks, pp. 148–184. Cambridge Univtrsity Press, Cambridge (1998)
Narayan, R., McKinney, J., Farmer, A.F.: Self-similar force-free wind from an accretion disc. MNRAS **375**, 548–566 (2007)
Narayan, R., Yi, I.: Advection-dominated accretion: A self-similar solution. ApJ **428**, L13–L16 (1994)
Narayan, R., Yi, I.: Advection-dominated accretion: Self-similarity and bipolar outflows. ApJ **444**, 231–243 (1995a)

Narayan, R., Yi, I.: Advection-dominated accretion: underfed black holes and neutron stars. ApJ **452**, 710–735 (1995b)

Neuhauser, D., Langanke, K., Koonin, S.E.: Hartree-Fock calculations of atoms and molecular chains in strong magnetic fields. Phys. Rev. A **33**, 2084–2086 (1986)

Niemeyer, J.C., Jedamzik, K.: Dynamics of primordial black hole formation. Phys. Rev. D **59**, 124013 (1999)

Nishikawa, K.-I., Koide, S., Sakai, J.-I., Christodoulou, D.M., Sol, H., Mutel, R.L.: Three-dimensional magnetohydrodynamic simulations of relativistic jets injected into an oblique magnetic field. ApJ **498**, 166–169 (1998)

Nitta, S.: A non-force-free cylindrical jet embedded in a conical wind envelope. MNRAS **284**, 899–910 (1997)

Nitta, S., Takahashi, M., Tomimatsu, A.: Effects of magnetohydrodynamic accretion flows on global structure of a Kerr black-hole magnetosphere. Phys. Rev. D **44**, 2295–2305 (1991)

Nobili, L., Turolla, R., Zampieri, L.: Spherical accretion onto black holes – A complete analysis of stationary solutions. ApJ **383**, 250–262 (1991)

Nötzel, A., Schindler, K., Birn, J.:On the case of approximate pressure isotropy in the quiet near-earth plasma sheet. J. Geophys. Res. **90**, 8293–8300 (1985)

Novikov, I.D., Thorne, K.S.: Astrophysics of black holes. In: DeWitt, C., DeWitt, B. (eds.) Black Holes, pp. 343–450. Gordon and Breach, New York (1973)

Nowak, M.A., Wagoner, R.V.: Diskoseismology: Probing accretion disks. II – G-modes, gravitational radiation reaction, and viscosity. ApJ **393**, 697–707 (1992)

Nowak, M.A., Wagoner, R.V.: Turbulent generation of trapped oscillations in black hole accretion disks. ApJ **418**, 187–201 (1983)

Nowak, M.A., Wagoner, R.V., Begelman M.C., Lehr D.A.: The 67 HZ feature in the black hole candidate GRS 1915+105 as a possible "diskoseismic" mode. ApJ **477**, L91-L94 (1997)

O'Dell, S.L.: Radiation force on a relativistic plasma and the Eddington limit. ApJ **243**, L147–L149 (1981)

Ogilvie, G.I.: The equilibrium of a differentially rotating disc containing a poloidal magnetic field. MNRAS **288**, 63–77 (1997)

Ogilvie, G.I.: Time-dependent quasi-spherical accretion. MNRAS **306**, L9–L13 (1999)

Ogilvie, G.I., Livio, M.: On the difficulty of launching an outflow from an accretion disk. ApJ **499**, 329–339 (1998)

Ogura, J., Kojima, Y.: Some properties of an axisymmetric pulsar magnetosphere constructed by numerical calculation. Prog. Theor. Phys. **109**, 619–630 (2003)

Okamoto, I.: Force-free pulsar magnetosphere – I. The steady, axisymmetric theory for the charge-separated plasma. MNRAS **167**, 457–474 (1974)

Okamoto, I.: Magnetic braking by a stellar wind. V – Approximate determination of the poloidal field. MNRAS **173**, 357–379 (1975)

Okamoto, I.: Relativistic centrifugal winds. MNRAS **185**, 69–108 (1978)

Okamoto, I.: The evolution of a black hole's force-free magnetosphere. MNRAS **254**, 192–220 (1992)

Okamoto, I.: Do magnetized winds self-collimate? MNRAS **307**, 253–278 (1999)

Okamoto, I.: Acceleration and collimation of magnetized winds. MNRAS **318**, 250–262 (2001)

Ostriker, E.C.: Self-similar magnetocentrifugal disk winds with cylindrical asymptotics. ApJ **486**, 291–306 (1997)

Ouyed, R., Pudritz, R.: Numerical simulations of astrophysical jets from Keplerian disks. I. Stationary models. ApJ **482**, 712–732 (1997)

Owocki, S.P., Cranmer, S.R., Blondin, J.M.: Two-dimensional hydrodynamical simulations of wind-compressed disks around rapidly rotating B stars. ApJ **424**, 887–904 (1994)

Paatz, G., Camenzind, M.: Winds and accretion flows around T Tauri stars. A&A **308**, 77–90 (1996)

Pacini, F.: Energy emission from a neutron star. Nature **216**, 567–568 (1967)

Paczyński, B.: Cosmological gamma-ray bursts. Acta Astron. **41**, 257–267 (1991)

Paczyński, B.: Are Gamma-Ray Bursts in Star-Forming Regions? ApJ **499**, L45–L48 (1998)

Paczyński, B., Bisnovatyi-Kogan, G.S.: A model of a thin accretion disk around a black hole. Acta Astron. **31**, 283–291 (1981)
Paczyński, B., Wiita, P.J.: Thick accretion disks and supercritical luminosities. A&A **88**, 23–31 (1980)
Papadakis, I.E., Lawrence, A.: Quasi-periodic oscillations in X-ray emission from the Seyfert galaxy NGC55481993. Nature **361**, 233–236 (1993)
Papadakis, I.E., Lawrence, A.: A detailed X-ray variability study of the Seyfert galaxy NGC 4051. MNRAS **272**, 161–183 (1995)
Papaloizou, J., Szuszkiewicz, E.: A comparison of one-dimensional and two-dimensional models of transonic accretion discs around collapsed objects. MNRAS **268**, 29–39 (1994)
Pariev, V.I.: Hydrodynamic accretion on to a rapidly rotating Kerr black hole. MNRAS **283**, 1264–1280 (1996)
Park, S.J., Vishniac, E.T.: An axisymmetric nonstationary model of the central engine in an active galactic nucleus. I – Black hole electrodynamics. ApJ **337**, 78–83 (1989)
Park, S.J., Vishniac, E.T.: An axisymmetric, nonstationary model of the central engine in an active galactic nucleus. II – Accretion disk electrodynamics. ApJ **347**, 684–687 (1990)
Parker, E.N.: Dynamics of the interstellar gas and magnetic fields ApJ **128**, 664–676 (1958)
Peitz, J., Appl, S.: Viscous accretion discs around rotating black holes. MNRAS **286**, 681–695 (1997)
Peitz, J., Appl, S.: 3+1 formulation of non-ideal hydrodynamics. MNRAS **296**, 231–244 (1998)
Petri, J., Heyvaerts, J., Bonazzola, S.: Global static electrospheres of charged pulsars. A&A **384**, 414–432 (2002)
Phinney, E.S.: Black hole-driven hydromagnetic flows – flywheels vs. fuel. In: Ferrari, A., Pacholczyk, A.G. (eds.) Astrophysical Jets, pp. 201–212. D.Reidel, Dodrecht (1983)
Pelletier, G., Pudritz, R.E.: Hydromagnetic disk winds in young stellar objects and active galactic nuclei. ApJ **394**, 117–138 (1992)
Petrich, L.I., Shapiro, S., Stark, R.F., Teukolsky, S.: Accretion onto a moving black hole – A fully relativistic treatment. ApJ **336**, 313–349 (1989)
Petrich, L.I., Shapiro, S.L., Teukolsky, S.A.: Accretion onto a moving black hole - an exact solution. Phys. Rev. Lett. **60**, 1781–1784 (1988)
Pneuman, G.W., Kopp, R.A.: Gas-magnetic field interactions in the solar corona. Solar Phys. **18**, 258–270 (1971)
Polnarev, A.G., Khlopov, M.Yu.: Cosmology, primordial black holes, and supermassive particles. Sov. Phys. Uspekhi **28**, 213–232 (1985)
Polnarev, A.G., Zembowicz, R.: Formation of primordial black holes by cosmic strings. Phys. Rev. D **43**, 1106–1109 (1991)
Postnov, K.A.: Cosmic gamma-ray bursts. Phys. Usppkhi **42**, 469–480 (1999)
Proga, D., Stone, J.M., Kallman, T.R.: Dynamics of line-driven disk winds in active galactic nuclei. ApJ **543**, 686–696 (2000)
Pudritz, R.E., Norman, C.A.: Bipolar hydromagnetic winds from disks around protostellar objects. ApJ **301**, 571–586 (1986)
Punsly, B.: Black Hole Gravitohydromagnetics. Berlin: Springer, Berlin (2001)
Punsly, B., Coroniti, F.V.: Relativistic winds from pulsar and black hole magnetospheres. ApJ **350**, 518–535 (1990a)
Punsly, B., Coroniti, F.V.: Ergosphere-driven winds. ApJ **354**, 583–615 (1990b)
Quataert, E.: Particle heating by Alfvenic turbulence in hot accretion flows. ApJ **500**, 978–991 (1998)
Quinlan, G., Shapiro, S.: The dynamical evolution of dense star clusters in galactic nuclei. ApJ **356**, 483–500 (1990)
Radhakrishnan, V., Cooke, D.J.: Magnetic poles and the polarization structure of pulsar radiation. Astrophys. Lett. **3**, 225–229 (1969)
Rafikov, R.R., Gurevich, A.V., Zybin, K.P.: Inductive interaction of rapidly rotating conductive bodies with a magnetized plasma. JETP **115**, 297–308 (1999)

Rankin, J.: Toward an empirical theory of pulsar emission. I Morphological taxonomy. ApJ **274**, 333–368 (1983)
Rankin, J.: Toward an empirical theory of pulsar emission. IV Geometry of the core emission region. ApJ **352**, 247–257 (1990)
Rees, M.J.: Black hole models for Active Galactic Nuclei. ARA&A **22**, 471–506 (1984)
Rees, M.J.: Tidal disruption of stars by black holes of 10 to the 6th-10 to the 8th solar masses in nearby galaxies. Nature **333**, 523–528 (1988)
Rees, M.J.: Causes and effects of the first quasars. Proc. Nat. Acad. Sci. **90**, 4840–4847 (1993)
Rees, M.J.: Black holes in galactic nuclei. Rev. Mod. Astron. **10**, 179–190 (1997)
Rees, M.J., Gunn, J.E.: The origin of the magnetic field and relativistic particles in the Crab Nebula. MNRAS **167**, 1–12 (1974)
Reipurth, B., Bally, J.: Herbig-Haro flows: Probes of early stellar evolution. ARA&A. **39**, 403–455 (2001)
Renzini, A., Greggio, L., di Serego-Alighieri, S., Cappellari, M., Burstein, D., Bertola, F.: An ultraviolet flare at the centre of the elliptical galaxy NGC4552. Nature **378**, 39–40 (1995)
Reynolds, C.S., DiMatteo, T., Fabian, A.C., Hwang, U., Canizares, C.: The 'quiescent' black hole in M87. MNRAS, **283**, L111–L116 (1996a)
Reynolds, C.S., Fabian, A.C., Celotti, A., Rees, M.J.: The matter content of the jet in M87: evidence for an electron-positron jet. MNRAS **283**, 873–880 (1996b)
Riffert, H., Herold, H.: Relativistic accretion disk structure revisited. ApJ **450**, 508–511 (1995)
Romanova, M.M., Toropina, O.D., Toropin, Yu.M., Lovelace, R.V.E.: Magnetohydrodynamic simulations of accretion onto a star in the "propeller" regime. ApJ **588**, 400–407 (2003)
Romanova, M.M., Ustyugova, G.V., Koldoba A.V., Lovelace R.V.E.: Magnetohydrodynamic simulations of disk-magnetized star interactions in the quiescent regime: funnel flows and angular momentum transport. ApJ **578**, 420–438 (2002)
Ruderman, M.A.: Matter and magnetospheres in superstrong magnetic fields. Ann. NY Acad. Sci. **257**, 127–140 (1975)
Ruderman, M.A., Sutherland, P.G.: Theory of pulsars – Polar caps, sparks, and coherent microwave radiation. ApJ **196**, 51–72 (1975)
Ruffert, M., Arnett, D.: Three-dimensional hydrodynamic Bondi-Hoyle accretion. 2: Homogeneous medium at Mach 3 with gamma = 5/3. ApJ **427**, 351–376 (1994)
Ruffini, R., Salmonson, J.D., Wilson, J.R., Xue, S.-S.: On the pair electromagnetic pulse of a black hole with electromagnetic structure. A&A **350**, 334–343 (1999)
Ruzmaikin, A.A., Sokoloff, D.D., Shukurov, A.M.: Magnetic Fields of Galaxies. Kluwer, Dordrecht, (1988)
Sakurai, T.: Magnetic stellar winds – A 2-D generalization of the Weber-Davis model. A&A **152**, 121–129 (1985)
Sakurai, T.: Magnetohydrodynamic solar/stellar wind models. Computer Phys. Rep. **12**, 247–273 (1990)
Salpeter, E.E., Lai, D.: Hydrogen phases on the surface of a strongly magnetized neutron star. ApJ **491**, 270–285 (1997)
Sanders, R.: The effects of stellar collisions in dense stellar systems. ApJ **162**, 791–809 (1970)
Sauty, C., Tsinganos, K.: Nonradial and nonpolytropic astrophysical outflows III. A criterion for the transition from jets to winds. A&A. **287**, 893–926 (1994)
Sauty, C., Tsinganos, K., Trussoni, E.: Nonradial and nonpolytropic astrophysical outflows. IV. Magnetic or thermal collimation of winds into jets? A&A **348**, 327–349 (1999)
Sauty, C., Tsinganos, K., Trussoni, E.: Jet formation and collimation. In: Guthmann, A.W., Georganopoulos, M., Marcowith, A., Manolakou K. (eds.) Relativistic Flows in Astrophysics, pp. 41–65. Springer, Heidelberg (2002)
Sazhin, M.V.: Opportunities for detecting ultralong gravitational waves. Sov. Astron. **22**, 36–38 (1978)
Scharlemann, E.T., Fawley, W.M., Arons, J.: Potential drops above pulsar polar caps – Ultrarelativistic particle acceleration along the curved magnetic field. ApJ **222**, 297–316 (1978)

Scharlemann, E.T., Wagoner, R.V.: Aligned rotating magnetospheres. General analysis. ApJ **182**, 951–960 (1973)
Schmidt, W.K.H.: Distance limit for a class of model gamma-ray burst sources. Nature **271**, 525–527 (1978)
Schwartz, R.D.: Herbig-Haro objects. ARA&A **21**, 209–237 (1983)
Sedrakyan, D.M., Shakhabasyan, K.M.: Superfluidity and magnetic fields of pulsars. Sov. Phys. Uspekhi **34**, 555–571 (1991)
Seiradakis, J.H., Wielebinski, R.: Morphology and characteristics of radio pulsars. A&Ap Rev. **12**, 239–271, (2004)
Semenov, V.S., Dyadechkin, S.A., Ivanov, I.B., Biernat, H.K.: Energy confinement for a relativistic magnetic flux tube in the ergosphere of a Kerr black hole. Phys. Scripta **65**, 13–24 (2002)
Shabad, A.E., Usov, V.V.: Propagation of gamma-radiation in strong magnetic fields of pulsars. Ap&SS **102**, 327–358 (1984)
Shabad, A.E., Usov, V.V.: Gamma-quanta conversion into positronium atoms in a strong magnetic field. Ap&SS **117**, 309–325 (1985)
Shabad, A.E., Usov, V.V.: Photon dispersion in a strong magnetic field with positronium formation – Theory. Ap&SS **128**, 377–409 (1986)
Shafranov, V.D.: On the equilibrium MHD configurations. Sov. Phys. JETP **6** 642–654 (1958)
Shakura, N.I.: Disk model of gas accretion on a relativistic star in a close binary system. Sov. Astron. **16**, 756–762 (1973)
Shakura, N.I., Sunyaev, R.A.: Black holes in binary systems. Observational appearance. A&A **24**, 337–355 (1973)
Shalybkov, D., Rüdiger, G.: Magnetic field dragging and the vertical structure of thin accretion discs. MNRAS **315**, 762–766 (2000)
Shapiro, S.L., Teukolsky, S.A.: Black Holes, White Dwarfs, and Neutron Stars. A Wiley–Interscience Publication, New York (1983)
Shaviv, N.J., Heyl, J.S., Lithwick, Y.: Magnetic lensing near ultramagnetized neutron stars. MNRAS **306**, 333–347 (1999)
Shcherbakov, R.V.: Region of anomalous compression under Bondi-Hoyle accretion. Astron. Lett. **31**, 591–597 (2005)
Sherwin, B. D., Lynden-Bell, D.: Electromagnetic fields in jets. MNRAS **378**, 409–415 (2007)
Shibata, S.: The luminosity of an externally triggered discharge in the dead pulsar magnetosphere and the current closure problem. MNRAS **269**, 191–198 (1994)
Shibata, S.: The field-aligned accelerator in the pulsar magnetosphere. MNRAS **287**, 262–270 (1997)
Shibata, S., Miyazaki, J., Takahara, F.: On the electric field screening by electron-positron pairs in a pulsar magnetosphere. MNRAS **295**, L53–L58 (1998)
Shu, F.N., Najita, J., Ruden, S.P., Lizano, S.: Magnetocentrifugally driven flows from young stars and disks. 2: Formulation of the dynamical problem. ApJ **429**, 797–807 (1994)
Shukre, C.S., Radhakrishnan, V.: The diffuse gamma-ray background and the pulsar magnetic window. ApJ **258**, 121–130 (1982)
Shvartsman, V.F.: The influence of stellar wind on accretion. Sov. Astron. **14**, 527–528 (1970)
Sikora, M., Madejski, G.: On pair content and variability of subparsec jets in quasars. ApJ **534**, 109–113 (2000)
Smith, I.A., Michel, F.C., Thacker, P.D.: Numerical simulations of aligned neutron star magnetospheres. MNRAS **322**, 209–217 (2001)
Sol, H., Pelletier, G., Asséo, E.: Two-flow model for extragalactic radio jets. MNRAS **237**, 411–429 (1989)
Soldatkin, A.O., Chugunov, Yu.V.: Steady-state axisymmetric configurations of a weakly ionized plasma in the field of a rotating magnetized spherical body. Plasma Phys. Rep. **29**, 65–77 (2003)
Soloviev, L.S.: In: Leontovich, M.A. (ed.) Reviews of Plasma Physics, **3**, pp. 284–328. Consultants Bureau, New York (1965)
Spencer, R.E.: A radio jet in SS433. Nature **282**, 483–484 (1979)

Spitkovsky, A.: Time-dependent force-free pulsar magnetospheres: axisymmetric and oblique rotators. ApJ **648**, L51–L54 (2006)
Spitkovsky, A., Arons, J.: Simulations of pulsar wind formation. In: Slane, P.O., Gaensler, B.M. (eds.) Neutron Stars and Supernova Remnants, pp. 81–87. Astronomical Society of the Pacific, San Francisco (2003)
Spitzer, L.: Dynamical evolution of dense spherical star systems. In: O'Connel, D. (ed.) Galactic Nuclei, pp. 443–475. Elsevier, Dordrecht (1971)
Spitzer, L., Saslaw, W.C.: On the evolution of galactic nuclei. ApJ **143**, 400–419 (1966)
Spruit, H.C.: Magnetohydrodynamic jets and winds from accretion disks. In: Wijers, R.A.M.J., Davies, M.B., Tout C.A. (eds.) Evolutionary Processes in Binary Stars, pp. 249–286. Kluwer, Dordrecht (1996)
Stanyukovich, K. P.: Unsteady motion of continuous media. Gostekhizdat, Moscow (1955) (in Russian); In: Holt, M. (ed.), Unsteady Motion of Continuous Media. Pergamon Press, Oxford (1960)
Stone, J.M., Pringle, J.E.: Magnetohydrodynamical non-radiative accretion flows in two dimensions. MNRAS **322**, 461–472 (2001)
Stone, J.M., Hawley, J.F., Gammie, C.F., Balbus, S.A.: Three-dimensional magnetohydrodynamical simulations of vertically stratified accretion disks. ApJ **463**, 656–673 (1996)
Sturrock, P.A.: A model of pulsars. ApJ **164**, 529–556 (1971)
Sulkanen, M.E., Lovelace, R.V.E.: Pulsar magnetospheres with jets. ApJ **350**, 732–744 (1990)
Svensson, R.: Steady mildly relativistic thermal plasmas – Processes and properties. MNRAS **209**, 175–208 (1984)
Tagliaferri, G., Bao, G., Israel, G.L., Stella, L., Treves, A.: The soft and medium-energy X-ray variability of NGC 5548: a reanalysis of EXOSAT observations. ApJ **465**, 181–190 (1996)
Takahashi, M.: Transmagnetosonic accretion in a black hole magnetosphere. ApJ **570**, 264–276 (2002)
Takahashi, M., Nitta, S., Tatematsu, Ya., Tomimatsu, A.: Magnetohydrodynamic flows in Kerr geometry – Energy extraction from black holes. ApJ **363**, 206–217 (1990)
Takahashi, M., Rilett, D., Fukumura K., Tsuruta S.: Magnetohydrodynamic shock conditions for accreting plasma onto Kerr black holes. ApJ **572**, 950–961 (2002)
Tassoul, J.-L.: Theory of Rotating Stars. Princeton University Press, Princeton (1978)
Taylor, J.H., Stinebring, D.R.: Recent progress in the understanding of pulsars. ARA&A **24**, 285–327 (1986)
Taylor, J.H, Weisberg, J.M.: Further experimental tests of relativistic gravity using the binary pulsar PSR 1913 + 16. ApJ **345**, 434–450 (1989)
Thompson, C., Duncan, R.C.: Neutron star dynamos and the origins of pulsar magnetism. ApJ **408**, 194–217 (1993)
Thompson, C., Duncan, R.C.: The soft gamma repeaters as very strongly magnetized neutron stars – I. Radiative mechanism for outbursts. MNRAS **275**, 255–300 (1995)
Thorne, K.S., Flammang, R.A., Žytkov, A.N.: Stationary spherical accretion into black holes. I – Equations of structure. MNRAS **194**, 475–484 (1981)
Thorne, K.S., Macdonald, D.: Electrodynamics in curved spacetime – 3 + 1 formulation. MNRAS **198**, 339 (1982)
Thorne, K.S., Macdonald, D., Price, R.H., Black Holes: The Membrane Paradigm. Yale University Press, New Haven and London (1986)
Timokhin, A.N.: On the force-free magnetosphere of an aligned rotator. MNRAS **368**, 1055–1072 (2006)
Timokhin, A.N., Bisnovatyi-Kogan, G.S., Spruit, H.C.: The magnetosphere of an oscillating neutron star. Non-vacuum treatment. MNRAS **316**, 734–748 (1999)
Tomimatsu, A.: Asymptotic collimation of magnetized winds far outside the light cylinder. Proc. Astron. Soc. Japan **46**, 123–130 (1994)
Tomimatsu, A.: Relativistic dynamos in magnetospheres of rotating compact objects. ApJ **528**, 972–978 (2000)

Tomimatsu, A., Matsuoka, T., Takahashi, M: Screw instability in black hole magnetosphere and a stabilizing effect of field-line rotation. Phys. Rev. D **64**, 123003 (2001)

Tomimatsu, A., Takahashi, M.: Black hole magnetospheres around thin disks driving inward and outward winds. ApJ **552**, 710–717 (2001)

Toropin, Yu.M., Toropina, O.D., Saveliev, V.V., Romanova, M.M., Chechetkin, V.M., Lovelace, R.V.E.: Spherical Bondi accretion onto a magnetic dipole. ApJ **517**, 906–918 (1999)

Toropina, O.D., Romanova, M.M., Toropin Yu.M., Lovelace R.V.E.: Propagation of magnetized neutron stars through the interstellar medium. ApJ **561**, 964–979 (2001)

Toropina, O.D., Romanova, M.M., Toropin, Yu.M., Lovelace, R.V.E.: Magnetic inhibition of accretion and observability of isolated old neutron stars. ApJ **593**, 472–480 (2003)

Tsikarishvili, E.G., Rogava, A.D., Tsikauri, D.G.: Relativistic, hot stellar winds with anisotropic pressure. ApJ **439**, 822–827 (1995)

Tsinganos, K., Sauty, C.: Nonradial and nonpolytropic astrophysical outflows. I – Hydrodynamic solutions with flaring streamlines. A&A **255**, 405–419 (1992)

Tsinganos, K., Sauty, C., Surlantzis, G., Trussoni, E., Contopoulos, J.: On the relation of limiting characteristics to critical surfaces in magnetohydrodynamic flows. MNRAS **283**, 811–820 (1996)

Tsygan, A.I.: A radio pulsar in an external magnetic field. MNRAS **292**, 317–320 (1997)

Uchida, Y., Shibata, K.: Magnetically buffered accretion to a young star and the formation of bipolar flows. Pub. Astron. Soc. Japan **36**, 105–118 (1984)

Urry, C.M., Padovani, P.: Unified schemes for radio-loud active galactic nuclei. Pub. Astron. Soc. Pacific **107**, 803–845 (1995)

Usov, V.V.: Millisecond pulsars with extremely strong magnetic fields as a cosmological source of gamma-ray bursts. Nature **357**, 472–474 (1992)

Usov, V.V.: Photon splitting in the superstrong magnetic fields of pulsars. ApJ **572**, L87–L90 (2002)

Usov, V.V., Melrose, D.B.: Bound pair creation in polar gaps and gamma-ray emission from radio pulsars. ApJ **464**, 306–315 (1996)

Ustyugova, G.V., Koldoba, A.V., Romanova, M.M., Chechetkin, V.M., Lovelace, R.V.E.: Magnetohydrodynamic simulations of outflows from accretion disks. ApJ **439**, L39–L42 (1995)

Ustyugova, G.V., Lovelace, R.V.E., Romanova, M.M., Li, H., Colgate, S.A.: Poynting jets from accretion disks: Magnetohydrodynamic simulations. ApJ **541**, L21–L24 (2000)

Uzdensky, D.A.: On the axisymmetric force-free pulsar magnetosphere. ApJ **598**, 446–457 (2003)

Uzdensky, D.A.: Force-free magnetosphere of an accretion disk-black hole system. I. Schwarzschild geometry. ApJ **603**, 652–662 (2004)

Uzdensky, D.A.: Force-free magnetosphere of an accretion disk-black hole system. II. Kerr geometry. ApJ **620**, 889–904 (2005)

Vlahakis, N., Königl, A.: Relativistic magnetohydrodynamics with application to gamma-ray burst outflows. I. Theory and semianalytic trans-alfvenic solutions. ApJ **596**, 1080–1103 (2003)

Vlahakis, N., Tsinganos, K., Sauty, C., Trussoni, E.: A disc-wind model with correct crossing of all magnetohydrodynamic critical surfaces. MNRAS **318**, 417–428 (2000)

van Putten, M.H.P.M., Levinson A.: Theory and astrophysical consequences of a magnetized torus around a rapidly rotating black hole. ApJ **584**, 937–953 (2003)

Velikhov, E.P.: Stability of an ideally conducting liquid flowing between cylinders rotating in a magnetic field. Sov. Phys. JETP **36**, 995–998 (1959)

von Mises, R.: Mathematical Theory of Compressible Fluid Flow. Academic, New York (1958)

von Rekowski, B., Brandenburg, A., Dobler, W., Shukurov, A.: Structured outflow from a dynamo active accretion disc. A&A **398**, 825–844 (2003)

Wald, R.M.: Black hole in a uniform magnetic field. Phys. Rev. D **10**, 1680–1684 (1974)

Wampler, E.G., Scargle J.D., Miller J.S.: Optical observations of the Crab Nebula pulsar. ApJ **157**, L1–L4 (1969)

Wang, D.-X., Lei, W.-H., Xiao, K., Ma R.-Y.: An analytic model of black hole evolution and gamma-ray bursts. ApJ **580**, 358–367 (2002)

Wardle, M., Königl, A.: The structure of protostellar accretion disks and the origin of bipolar flows. ApJ **410**, 218–238 (1993)

Wardle, J.F.C., Homan, D.C., Ojha R., Roberts D.H.: Electron-positron jets associated with the quasar 3C279. Nature **395**, 457–461 (1998)

Weber, E.J., Davis, L.Jr.: The angular momentum of the solar wind. ApJ **148**, 217–227 (1967)

Weise, J.I., Melrose, D.B.: One-photon pair production in pulsars: non-relativistic and relativistic regimes. MNRAS **329**, 115–125 (2002)

Weltevrede, P., Stappers, B.W., Edwards, R.T.: The subpulse modulation properties of pulsars at 92 cm and the frequency dependence of subpulse modulation. A&A **469**, 607–631 (2007)

Weisskopf, M.C., Hester, J.J., Tennant, A.F. et al.: Discovery of spatial and spectral structure in the X-ray emission from the Crab Nebula. ApJ **536**, L81–L84 (2000)

Wilms, J., Reynolds, C.S., Begelman, M.C. et al.: XMM-EPIC observation of MCG-6-30-15: direct evidence for the extraction of energy from a spinning black hole? MNRAS **328**, L27–L31 (2001)

Xu, R.X., Qiao, G.J.: Pulsar braking index: A test of emission models? ApJ **561**, L85–L88 (2001)

Woosley, S.E.: Gamma-ray bursts from stellar mass accretion disks around black holes. ApJ **405**, 273–277 (1993)

Zavlin, V.E., Pavlov, G.G.: Modeling neutron star atmospheres. In: Becker, W., Lesch, H., Trümper, J. (eds.) Neutron Stars, Pulsars and Supernova Remnants, pp. 263–272. Proceedings of the 270 WE-Heraeus Seminar, Garching (2002)

Zel'dovich, Ya.B., Novikov, I.D.: Relativistic Astrophysics. University of Chicago Press, Chicago (1971)

Zel'dovich, Ya.B., Novikov, I.D.: The hypothesis of cores retarded during expansion and the hot cosmological model. Sov. Astron. **10**, 602–603 (1967a)

Zel'dovich, Ya.B., Novikov, I.D.: The uniqueness of the interpretation of isotropic cosmic radiation with T=3K. Sov. Astron. **11**, 526–527 (1967b)

Zhang, L., Cheng, K.S.: High-energy radiation from rapidly spinning pulsars with thick outer gaps. ApJ **487**, 370–379 (1997)

Zhang, H.Y., Gabuzda, D.C., Nan, R.D., Jin, C.J.: Parsec-scale rotation-measure distribution in the quasar 3C 147 at 8 GHz. A&A **415**, 477–481 (2004)

Ziegler, U., Rüdiger, G.: Angular momentum transport and dynamo-effect in stratified, weakly magnetic disks. A&A **356**, 1141–1148 (2000)

Znajek, R.L.: Black hole electrodynamics and the Carter tetrad. MNRAS **179**, 457–472 (1977)

Zheleznyakov, V.V.: Radiation in Astrophysical Plasmas. Kluwer, Boston (1996)

Index

ASTRONOMY AND ASTROPHYSICS LIBRARY

Series Editors: G. Börner · A. Burkert · W. B. Burton · M. A. Dopita
A. Eckart · E. K. Grebel · B. Leibundgut · A. Maeder
V. Trimble

The Stars By E. L. Schatzman and F. Praderie

Modern Astrometry 2nd Edition
By J. Kovalevsky

The Physics and Dynamics of Planetary Nebulae By G. A. Gurzadyan

Galaxies and Cosmology By F. Combes, P. Boissé, A. Mazure and A. Blanchard

Observational Astrophysics 2nd Edition
By P. Léna, F. Lebrun and F. Mignard

Physics of Planetary Rings Celestial Mechanics of Continuous Media
By A. M. Fridman and N. N. Gorkavyi

Tools of Radio Astronomy 4th Edition, Corr. 2nd printing
By K. Rohlfs and T. L. Wilson

Tools of Radio Astronomy Problems and Solutions 1st Edition, Corr. 2nd printing
By T. L. Wilson and S. Hüttemeister

Astrophysical Formulae 3rd Edition (2 volumes)
Volume I: Radiation, Gas Processes and High Energy Astrophysics
Volume II: Space, Time, Matter and Cosmology
By K. R. Lang

Galaxy Formation 2nd Edition
By M. S. Longair

Astrophysical Concepts 4th Edition
By M. Harwit

Astrometry of Fundamental Catalogues The Evolution from Optical to Radio Reference Frames
By H. G. Walter and O. J. Sovers

Compact Stars. Nuclear Physics, Particle Physics and General Relativity 2nd Edition
By N. K. Glendenning

The Sun from Space By K. R. Lang

Stellar Physics (2 volumes)
Volume 1: Fundamental Concepts and Stellar Equilibrium
By G. S. Bisnovatyi-Kogan

Stellar Physics (2 volumes)
Volume 2: Stellar Evolution and Stability
By G. S. Bisnovatyi-Kogan

Theory of Orbits (2 volumes)
Volume 1: Integrable Systems and Non-perturbative Methods
Volume 2: Perturbative and Geometrical Methods
By D. Boccaletti and G. Pucacco

Black Hole Gravitohydromagnetics
By B. Punsly

Stellar Structure and Evolution
By R. Kippenhahn and A. Weigert

Gravitational Lenses By P. Schneider, J. Ehlers and E. E. Falco

Reflecting Telescope Optics (2 volumes)
Volume I: Basic Design Theory and its Historical Development. 2nd Edition
Volume II: Manufacture, Testing, Alignment, Modern Techniques
By R. N. Wilson

Interplanetary Dust
By E. Grün, B. Å. S. Gustafson, S. Dermott and H. Fechtig (Eds.)

The Universe in Gamma Rays
By V. Schönfelder

Astrophysics. A New Approach 2nd Edition
By W. Kundt

Cosmic Ray Astrophysics
By R. Schlickeiser

Astrophysics of the Diffuse Universe
By M. A. Dopita and R. S. Sutherland

The Sun An Introduction. 2nd Edition
By M. Stix

Order and Chaos in Dynamical Astronomy
By G. J. Contopoulos

Astronomical Image and Data Analysis
2nd Edition By J.-L. Starck and F. Murtagh

The Early Universe Facts and Fiction
4th Edition By G. Börner

ASTRONOMY AND ASTROPHYSICS LIBRARY

A&A LIBRARY

Series Editors: G. Börner · A. Burkert · W. B. Burton · M. A. Dopita
A. Eckart · E. K. Grebel · B. Leibundgut · A. Maeder
V. Trimble

The Design and Construction of Large Optical Telescopes By P. Y. Bely

The Solar System 4th Edition
By T. Encrenaz, J.-P. Bibring, M. Blanc, M. A. Barucci, F. Roques, Ph. Zarka

General Relativity, Astrophysics, and Cosmology By A. K. Raychaudhuri, S. Banerji, and A. Banerjee

Stellar Interiors Physical Principles, Structure, and Evolution 2nd Edition
By C. J. Hansen, S. D. Kawaler, and V. Trimble

Asymptotic Giant Branch Stars
By H. J. Habing and H. Olofsson

The Interstellar Medium
By J. Lequeux

Methods of Celestial Mechanics (2 volumes)
Volume I: Physical, Mathematical, and Numerical Principles
Volume II: Application to Planetary System, Geodynamics and Satellite Geodesy
By G. Beutler

Solar-Type Activity in Main-Sequence Stars
By R. E. Gershberg

Relativistic Astrophysics and Cosmology
A Primer By P. Hoyng

Magneto-Fluid Dynamics
Fundamentals and Case Studies
By P. Lorrain

Compact Objects in Astrophysics
White Dwarfs, Neutron Stars and Black Holes
By Max Camenzind

Special and General Relativity
With Applications to White Dwarfs, Neutron Stars and Black Holes
By Norman K. Glendenning

Planetary Systems
Detection, Formation and Habitability of Extrasolar Planets
By M. Ollivier, T. Encrenaz, F. Roques F. Selsis and F. Casoli

The Sun from Space 2nd Edition
By Kenneth R. Lang

Tools of Radio Astronomy 5th Edition
By Thomas L. Wilson, Kristen Rohlfs and Susanne Hüttemeister

Astronomical Optics and Elasticity Theory
Active Optics Methods
By Gérard René Lemaitre

High-Redshift Galaxies
Light from the Early Universe
By I. Appenzeller

MHD Flows in Compact Astrophysical Objects Accretion, Winds and Jets
By Vasily S. Beskin

MIX
Papier aus verantwortungsvollen Quellen
Paper from responsible sources
FSC® C105338

If you have any concerns about our products,
you can contact us on
ProductSafety@springernature.com

In case Publisher is established outside the EU,
the EU authorized representative is:
Springer Nature Customer Service Center GmbH
Europaplatz 3, 69115 Heidelberg, Germany

Printed by Libri Plureos GmbH
in Hamburg, Germany